PREFACE TO THE SECOND (REVISED) EDITION OF VOLUME 2

Apart from a number of minor corrections and changes, a substantial reformulation and up-dating of Chapters 14 and 15 has taken place. This reformulation and up-dating is a major and very welcome contribution from my friend and colleague, Dr J.W. Sanders, to whom I express my sincere thanks. His efforts have produced a much better result than I could have achieved on my own. Warm thanks are also due to Dr Jo Ward, who checked some of the revised material.

New Sections 16.9 and 16.10 have also been added.

The bibliography has been expanded and brought up to date, though it is still not exhaustive.

In spite of these changes, the third paragraph in the Preface to the revised edition of Volume 1 is applicable here. What has been accomplished here is not a complete account of developments over the past 15 years; such an account would require many volumes. Even so, it may assist some readers who wish to appraise some of these developments. More ambitious readers should consult *Mathematical Reviews* from around Volume 50 onwards.

R. E. E.

CANBERRA, September 1981

Graduate Texts in Mathematics 85

PREFACE TO VOLUME 2

The substance of the first three paragraphs of the preface to Volume 1 of *Fourier Series: A Modern Introduction* applies equally well to this second volume. To what is said there, the following remarks should be added.

Volume 2 deals on the whole with the more modern aspects of Fourier theory, and with those facets of the classical theory that fit most naturally into a function-analytic garb. With their introduction to distributional concepts and techniques and to interpolation theorems, respectively, Chapters 12 and 13 are perhaps the most significant portions of Volume 2. From a pedagogical viewpoint, the carefully detailed discussion of Marcinkiewicz's interpolation theorem will, it is hoped, go some way toward making this topic more accessible to a beginner.

A major portion of Chapter 11 is devoted to the elements of Banach algebra theory and its applications in harmonic analysis. In Chapter 16 there appears what is believed to be the first reasonably connected introductory account of multiplier problems and related matters.

For the purposes of a short course, one might be content to cover Section 11.1, the beginning of Section 11.2, Section 11.4, Chapter 12 up to and including Section 12.10, Chapter 13 up to and including Section 13.6, Chapter 14, and Sections 15.1 to 15.3. Much of Chapters 13 to 15 is independent of Chapters 11 and 12, or is easily made so. While severe pruning might lead to a tolerable excision of Section 11.4, which is required but rarely in subsequent chapters, it would be a pity thus to omit all reference to Banach algebras.

I at one time cherished the hope of including in this volume a list of current research problems, but the available space will not accommodate such a list together with the necessary explanatory notes. The interested reader may go a long way toward repairing this defect by studying some of the articles appearing in [Bi] (see, most especially, pp. 351–354 thereof).

The cross-referencing system is as follows. With the exception of references to the appendixes, the numerical component of every reference to either volume appears in the form $a \cdot b \cdot c$, where a, b, and c are positive integers; the material referred to appears in Volume 1 if and only if $1 \leqslant a \leqslant 10$. In the case of references to the appendixes, all of which appear in Volume 1, a Roman numeral "I" has been prefixed as a reminder to the reader; thus, for example, "I,B.2.1" refers to Appendix B.2.1 in Volume 1.

An understanding of the main topics discussed in this book does not, I hope, hinge upon repeated consultation of the items listed in the bibliography. Readers with a limited aim should find strictly necessary only an occasional reference to a few of the book listed. The remaining items, and especially the numerous research papers mentioned, are listed as an aid to those readers who wish to pursue the subject beyond the limits reached in this book; such readers must be prepared to make the very considerable effort called for in making an acquaintance with current research literature. A few of the research papers listed cover developments that came to my notice too late for mention in the main text. For this reason, any attempted summary in the main text of the current standing of a research problem should be supplemented by an examination of the bibliography and by scrutiny of the usual review literature.

Finally, I take this opportunity to renew all the thanks expressed in the preface to Volume 1, placing special reemphasis on those due to Professor Edwin Hewitt for his sustained interest and help, to Dr. Garth Gaudry for his contributions to Chapter 13, and to my wife for her encouragement and help with the proofreading. My thanks for help in the latter connection are extended also to my son Christopher.

CANBERRA, 1967 *R. E. E.*

CONTENTS

R.E. Edwards

Fourier Series

A Modern Introduction

Volume 2

Second Edition

Springer-Verlag
New York Heidelberg Berlin

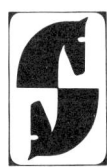

R.E. Edwards
The Australian National University
Department of Mathematics
(Institute for Advanced Studies)
P.O. Box 4
Canberra, A.C.T. 2600
Australia

AMS Subject Classification (1980): 42-01

Library of Congress Cataloging in Publication Data (Revised)
Edwards, Robert E
Fourier series, a modern introduction.
(Graduate texts in mathematics; 64, 85)
Bibliography: v. 1, p. 207–211; v. 2, p.
Includes indexes.
1. Fourier series. I. Title. II. Series.
QA404.E25 1979 515'.2433 79-11932
ISBN 0-387-90412-3 (v. 1) AACR2
ISBN 0-387-90651-7 (v. 2)

The first edition was published by Holt, Rinehart and Winston, Inc.

Printed in the United States of America.

9 8 7 6 5 4 3 2 1

ISBN 0-387-90651-7 Springer-Verlag New York Heidelberg Berlin
ISBN 3-540-90651-7 Springer-Verlag Berlin Heidelberg New York

CHAPTER 11

Spans of Translates. Closed Ideals. Closed Subalgebras. Banach Algebras

The first three sections of this chapter are devoted to some topics mentioned earlier, namely, the study of closed invariant subspaces and closed ideals [mentioned in 2.2.1 and 3.1.1(g)], and that of closed subalgebras [mentioned in 3.1.1(e) and (f)]. Throughout the discussion **E** will denote any one of the convolution algebras $\mathbf{L}^p$ $(1 \leqslant p < \infty)$ or **C** (see 3.1.1, 3.1.5, and 3.1.6) and we shall consider closed invariant subspaces, closed ideals, and closed subalgebras in **E**. The cases $\mathbf{E} = \mathbf{C}^k$ and $\mathbf{E} = \mathbf{L}^\infty$ could also be treated similarly, provided that in the last case one considered $\mathbf{L}^\infty$ with its so-called weak topology, in which a sequence or net (f_i) converges to f if and only if

$$\lim_i \frac{1}{2\pi}\int f_i g\,dx = \frac{1}{2\pi}\int fg\,dx$$

is true for each $g \in \mathbf{L}^1$. Compare I, B.1.7 and I, C.1.

For any compact group, Abelian or not, the structure theory for closed invariant subspaces and closed ideals is simple. For the group T the details are fully elucidated in 11.2.1. By contrast, except for the case $\mathbf{E} = \mathbf{L}^2$, the structure of closed subalgebras is not yet fully describable, even for the group T.

Subsections 11.2.3 and 11.2.4 are included on "cultural" grounds and are intended to show how the relatively simple problems treated in 11.2.1 and 11.2.2 lead to ones of considerable complexity and interest when the compact group T is replaced by a noncompact group such as R. (These subsections are not essential to an understanding of the rest of the book.) The relevant problems for the dual group Z are mentioned briefly in 11.2.5.

Section 11.3 is devoted to the problem of closed subalgebras in **E**.

The final section of this chapter (11.4) is devoted to a few of the fundamentals of commutative Banach algebra theory and some of its applications to harmonic analysis. When applications are made to the algebras **E** mentioned above, we find that the topics mentioned in Section 4.1 undergo natural development. Applications to other algebras will also be made and will provide proofs of results stated in Section 10.6.

Section 11.4 is in no sense a balanced introduction to the study of Banach algebras. References for further reading will be given in due course.

11.1 Closed Invariant Subspaces and Closed Ideals

By a *closed invariant subspace* of $\mathbf{E}$ is meant a linear subspace $\mathbf{V}$ of $\mathbf{E}$ which is (1) closed for the normal topology of $\mathbf{E}$ (see 2.2.4), and (2) invariant under translation, in the sense that $f \in \mathbf{V}$ entails $T_a f \in \mathbf{V}$ for all $a \in T$. (Compare the definition of invariant subspaces given in 2.2.1.)

Each $f \in \mathbf{E}$ is contained in a smallest closed invariant subspace $\bar{\mathbf{V}}_f$, which is none other than the closure in $\mathbf{E}$ of the invariant subspace $\mathbf{V}_f$ generated by f (as defined in 2.2.1). The reader will note that $\bar{\mathbf{V}}_f$ depends in general on the ambient space $\mathbf{E}$: for example, if f is continuous, the closure of $\mathbf{V}_f$ in $\mathbf{L}^p$ will in general be strictly larger than the closure of $\mathbf{V}_f$ in $\mathbf{C}$. Despite this, we do not think it necessary to complicate the notation accordingly.

In view of the fact that $\mathbf{E}$ is an algebra under convolution, we follow the usual algebraic terminology by describing as an *ideal* in $\mathbf{E}$, a linear subspace $\mathbf{I}$ of $\mathbf{E}$ with the property that $f * g \in \mathbf{I}$ whenever $f \in \mathbf{I}$ and $g \in \mathbf{E}$. A *closed ideal* in $\mathbf{E}$ is an ideal in $\mathbf{E}$ which is also a closed subset of $\mathbf{E}$.

As will be seen in 11.1.2, the closed invariant subspaces of $\mathbf{E}$ and the closed ideals in $\mathbf{E}$ are exactly the same things (although the invariant subspaces and the ideals are *not* the same things).

11.1.1. If $f \in \mathbf{E}$, then $\hat{f}(n)e_n \in \bar{\mathbf{V}}_f$ for all $n \in Z$.

Proof. Direct computation shows that

$$\hat{f}(n)e_n = e_n * f.$$

Since $e_n \in \mathbf{L}^1$, the assertion follows from 3.1.9. For an alternative proof, see Exercise 11.5. Yet another type of proof is described in 11.2.2.

11.1.2. A subset of $\mathbf{E}$ is a closed invariant subspace of $\mathbf{E}$ if and only if it is a closed ideal in $\mathbf{E}$. (Compare with 3.2.3.)

Proof. (1) Let $\mathbf{I}$ be a closed ideal in $\mathbf{E}$. We wish to show that $\mathbf{I}$ is translation-invariant. For this purpose, we utilize an argument appearing in 3.2.3. Choose an approximate identity $(k_n)_{n=1}^{\infty}$ comprised of elements of $\mathbf{E}$ (for example, the Fejér kernels introduced in Section 5.1). Since $\mathbf{I}$ is an ideal, $(T_a k_n) * f \in \mathbf{I}$ for all $n \geq 1$ and all $f \in \mathbf{I}$. But $(T_a k_n) * f = T_a(k_n * f)$ by 3.1.2, and $\lim_{n\to\infty} k_n * f = f$ in $\mathbf{E}$ by 3.2.2. Therefore $\lim_{n\to\infty} T_a(k_n * f) = T_a f$ in $\mathbf{E}$. $\mathbf{I}$ being closed, it follows that $T_a f \in \mathbf{I}$. This shows that $\mathbf{I}$ is translation-invariant and is therefore a closed invariant subspace of $\mathbf{E}$.

(2) Let $\mathbf{V}$ be a closed invariant subspace of $\mathbf{E}$. In order to prove that $\mathbf{V}$ is a closed ideal in $\mathbf{E}$, it suffices to show that $f * g \in \mathbf{V}$ whenever $f \in \mathbf{V}$ and $g \in \mathbf{E}$. In doing this we may, since $\mathbf{V}$ is closed in $\mathbf{E}$ and since the trigonometric

polynomials are everywhere dense in **E** (see 2.4.4), assume that g is a trigonometric polynomial; see 3.1.6. In that case, however, $f * g$ is a finite linear combination of terms $\hat{f}(n)e_n$, and 11.1.1 shows at once that $f * g \in \bar{\mathbf{V}}_f$. Finally, since $f \in \mathbf{V}$, $\bar{\mathbf{V}}_f \subset \mathbf{V}$, and therefore $f * g \in \mathbf{V}$. The proof is complete.

11.1.3. **Remarks.** (1) It has been noted in 3.2.3 that **E** is a module over $\mathbf{L}^1$; and in Section 12.7 it will appear that **E** is even a module over the superspace **M** of $\mathbf{L}^1$ composed of all Radon measures. It is quite simple to verify that the closed submodules of **E** (qua module over $\mathbf{L}^1$ or over **M**) are exactly the closed ideals in **E**.

(2) The reader will take care to remember that 11.1.2 is established only for the choices of **E** mentioned at the outset of this chapter; it is not true in all cases of interest. For example, if $\mathbf{L}^\infty$ is taken with its normed topology, there are closed ideals in the convolution algebra $\mathbf{L}^\infty$ that are not translation-invariant; see Exercises 11.22 and 11.23. Theorem 11.1.2 is also false for the measure algebra **M** introduced in Section 12.7; see Exercise 12.45.

11.2 The Structure of Closed Ideals and Related Topics

It can now be shown that a closed ideal **I** in **E** is characterized completely in terms of the common zeros of the Fourier transforms of elements of **I**.

For any $f \in \mathbf{E}$, we denote by Z_f the set of $n \in Z$ for which $\hat{f}(n) = 0$; and for any subset **S** of **E** we write

$$Z_{\mathbf{S}} = \bigcap \{Z_f : f \in \mathbf{S}\}.$$

11.2.1. Let **I** be any closed ideal in **E**, and let $f \in \mathbf{E}$. Then $f \in \mathbf{I}$ if and only if $Z_f \supset Z_{\mathbf{I}}$.

Proof. Obviously, $Z_f \supset Z_{\mathbf{I}}$ whenever $f \in \mathbf{I}$. Suppose conversely that $f \in \mathbf{E}$ and $Z_f \supset Z_{\mathbf{I}}$; we have to show that $f \in \mathbf{I}$. Let $n \notin Z_{\mathbf{I}}$ and choose $g \in \mathbf{I}$ such that $\hat{g}(n) \neq 0$. By 11.1.1, $e_n \in \bar{\mathbf{V}}_g$; and by 11.1.2, $\bar{\mathbf{V}}_g \subset \mathbf{I}$. Thus $e_n \in \mathbf{I}$, and this for any $n \notin Z_{\mathbf{I}}$. A fortiori, $e_n \in \mathbf{I}$ for any n for which $\hat{f}(n) \neq 0$. Now 6.1.1 shows that f is the limit in **E** of finite linear combinations of exponentials e_n with n restricted by the condition $\hat{f}(n) \neq 0$. Since **I** is a closed linear subspace of **E**, it appears that $f \in \mathbf{I}$, as was to be proved.

Remarks. (1) In view of 11.1.2, 11.2.1 may be reformulated in the following way. Let **V** be a closed invariant subspace of **E** and put $S = Z \backslash Z_{\mathbf{V}}$; then **V** is identical with the closed linear subspace of **E** generated by $\{e_n : n \in S\}$. In brief, **V** is generated (as a closed linear subspace, a closed invariant subspace, or a closed ideal) by the continuous characters it contains.

The equivalence of the two versions depends upon 6.1.1. As usual, the result remains true for $\mathbf{E} = \mathbf{L}^\infty$, provided the weak topology is used throughout; in this connection it is useful (although not essential) to note that $\lim_{N \to \infty} \sigma_N f = f$ weakly in $\mathbf{L}^\infty$ whenever $f \in \mathbf{L}^\infty$.

(2) In 11.2.1 it is essential that the ideal $\mathbf{I}$ be assumed to be closed. For example, if $\mathbf{I}$ is any everywhere dense and nonclosed ideal in $\mathbf{E}$, then $Z_{\mathbf{I}} = \varnothing = Z_{\mathbf{E}}$ but $\mathbf{I} \neq \mathbf{E}$. In such cases there is no known simple structure theorem.

(3) For a study of projections onto closed invariant subspaces of $\mathbf{L}^p(G)$, where G is a noncompact group, see Rosenthal [1].

11.2.2. **The Hahn-Banach Theorem Applied to 11.2.1.** A characteristically modern tool for the discovery and proof of theorems about linear approximation is the Hahn-Banach theorem, which is described briefly in I, B.5. We propose to indicate here how this theorem may be used to prove 11.2.1; it is equally useful in connection with the analogous problems mentioned in 11.2.3 and 12.11.4.

It must be admitted that its application to the proof of 11.2.1 does not appear to be particularly economical, and it must be stressed that the great merit of the theorem lies rather in the range of problems to which it provides a useful common approach (see [E], Chapter 2). No account of the methods of modern analysis can afford to ignore it.

The notation being as in 11.2.1, let us face anew the problem of showing that $f \in \mathbf{I}$ whenever $Z_f \supset Z_{\mathbf{I}}$. Since $\mathbf{I}$ is a closed linear subspace of $\mathbf{E}$, the Hahn-Banach theorem (specifically I, B.5.2) affirms that to do this it suffices (and is obviously necessary) to prove that, if F is any continuous linear functional on $\mathbf{E}$, and if

$$F(g) = 0 \qquad \text{for all } g \in \mathbf{I}, \tag{11.2.1}$$

and

$$Z_f \supset Z_{\mathbf{I}}, \tag{11.2.2}$$

then

$$F(f) = 0. \tag{11.2.3}$$

Now, since $\mathbf{I}$ is invariant, (11.2.1) entails that

$$F(T_a g) = 0 \qquad \text{for all } g \in \mathbf{I} \text{ and all } a. \tag{11.2.4}$$

This suggests that we look at the function ϕ_g defined by

$$\phi_g(a) = F(T_a g). \tag{11.2.5}$$

Since F is continuous on $\mathbf{E}$, while $a \to T_a g$ is continuous from R into $\mathbf{E}$ (see 2.2.4), ϕ_g is a continuous function. The reader will also observe for future use the fact that ϕ_g depends linearly and continuously on the variable $g \in \mathbf{E}$:

$$\|\phi_g\|_\infty \leqslant \|F\| \cdot \|g\|_{\mathbf{E}}.$$

The combination of these last remarks with a simple argument involving Riemann sums permits the computation of the Fourier coefficients of ϕ_g.

Thus (using an obvious notation),

$$\hat{\phi}_g(n) = \lim \frac{1}{2\pi} \sum \phi_g(a_k) e^{-ina_k} \Delta a_k$$

$$= \lim F\left(\frac{1}{2\pi} \sum T_{a_k} g \cdot e^{-ina_k} \Delta a_k\right),$$

by linearity of F, which in turn is equal to

$$F\left(\lim \frac{1}{2\pi} \sum T_{a_k} g \cdot e^{-ina_k} \Delta a_k\right)$$

on account of continuity of F. Now, if g is continuous, it is easy to check that the limit appearing in the last expression displayed is none other than the function

$$x \to \frac{1}{2\pi} \int g(x-a) e^{-ina}\, da,$$

which is, by virtue of the basic properties of the invariant integral recounted in 2.2.2, the function $\hat{g}(-n) \cdot e_{-n}$. Accordingly, the formula

$$\hat{\phi}_g(n) = \hat{g}(-n) \cdot F(e_{-n}) \tag{11.2.6}$$

is established for continuous $g \in \mathbf{E}$. However, for a fixed $n \in Z$, each side of (11.2.6) is a continuous linear functional of $g \in \mathbf{E}$; since the continuous functions are everywhere dense in $\mathbf{E}$ (a corollary of 2.4.4), (11.2.6) must hold for all $g \in \mathbf{E}$. The reader is urged to verify carefully all the steps in this computation of $\hat{\phi}_g$.

In view of (11.2.6), (11.2.4) entails that $F(e_n) = 0$ whenever $g \in \mathbf{I}$ and $\hat{g}(n) \neq 0$. Therefore

$$F(e_n) = 0 \qquad \text{for all } n \in Z \backslash Z_{\mathbf{I}}. \tag{11.2.7}$$

On the other hand, for any $f \in \mathbf{E}$ we have from 6.1.1

$$f = \lim_{N \to \infty} \sum_{|n| \leqslant N} \left(1 - \frac{|n|}{N+1}\right) \hat{f}(n) e_n.$$

So, by linearity and continuity of F,

$$F(f) = \lim_{N \to \infty} \sum_{|n| \leqslant N} \left(1 - \frac{|n|}{N+1}\right) \hat{f}(n) F(e_n). \tag{11.2.8}$$

Finally, by (11.2.2) and (11.2.7),

$$\hat{f}(n) F(e_n) = 0 \qquad \text{for all } n \in Z,$$

so that (11.2.3) follows from (11.2.8). This completes the proof.

Remarks. The computation of the Fourier coefficients of ϕ_g could be made to proceed more gracefully by appealing to the results of Chapter 12 and Appendix I, C.1 concerning the analytic representation of continuous linear functionals F on $\mathbf{E}$. However, we have preferred at this stage to sacrifice grace in favor of more elementary arguments.

11.2.3. **Closure of Translations Theorems.** The writer knows of no very significant applications of 11.2.1 to problems of concrete analysis, though it has its own interest as a structure theorem, albeit a simple one. However, it and certain corollaries one can deduce from it have analogues for other groups which are at once deeper and productive of genuinely significant results in concrete analysis. We propose to mention these analogues, devoting this subsection to so-called "closure of translations" theorems, and the next to some consequences of a Tauberian nature (see 5.3.5).

The position is that 11.2.1 and its derivatives, pertaining to the group T, are simple prototypes of bigger and better things which owe their significance to their applicability to noncompact groups.

When one contemplates replacing the compact group T by a noncompact (locally compact Abelian) group G, it is difficult to repress the hope that an analogue of the case $\mathbf{E} = \mathbf{L}^1$ of 11.2.1 lurks around the corner and awaits discovery. There is little difficulty in framing a plausible analogue, and this plausible analogue turns out to be "approximately true," or to be "true in spirit but false in detail." (Concerning $\mathbf{L}^1(G)$ for a general group, see, for example, [R], Chapter 1; [HR], Section 20; [E], Section 4.19; [Bo]; [Bo$_1$], Chapitre 2.)

The simplest choice for a noncompact group G would undoubtedly be the group Z. Despite this, the description immediately following is expressed in terms of the groups R^m (R the additive group of real numbers with its usual topology and m a natural number). One reason for this choice is that R^m is more typical of noncompact groups than is Z. Another reason is that the original "closure of translations" theorem of Wiener (see [Wi], pp. 99–100), which was the beginning of almost everything in this field, applies to the group R. The analogous problems for the technically somewhat simpler group Z will receive further attention in 11.2.5 and 12.11.4.

For $f \in \mathbf{L}^1(R^m)$, the Fourier transform of f is the function on R^m defined by

$$\hat{f}(\xi) = \int \cdots \int_{R^m} f(x_1, \cdots, x_m) e^{-2\pi i(\xi_1 x_1 + \cdots + \xi_m x_m)}\, dx_1 \cdots dx_m$$

for $\xi = (\xi_1, \cdots, \xi_m) \in R^m$; Z_f is defined to be the set of zeros of $\hat{f}$; and, for any ideal $\mathbf{I}$ in $\mathbf{L}^1(R^m)$, $Z_\mathbf{I}$ is defined to be the intersection of the sets Z_f when f ranges over $\mathbf{I}$. (A brief treatment of Fourier transforms of functions in $\mathbf{L}^1(R)$ and $\mathbf{L}^2(R)$ appears in Chapters 9 and 19 of [R$_1$]; see also [Wi] and [Ti], and the references cited therein.)

The Wiener *closure of translations theorem* for R^m asserts that an ideal $\mathbf{I}$ in $\mathbf{L}^1(R^m)$ is everywhere dense in $\mathbf{L}^1(R^m)$ if (and only if) $Z_\mathbf{I} = \varnothing$. This is a perfect analogue of the corresponding special case of 11.2.1, and is indeed encouraging.

For quite a while it remained tantalizingly in doubt whether a general closed ideal $\mathbf{I}$ in $\mathbf{L}^1(R^m)$ necessarily contains every $f \in \mathbf{L}^1(R^m)$ such that $Z_f \supset Z_\mathbf{I}$. The first example showing that this was *not* always the case was given by Laurent Schwartz [1] in 1948 and applied to R^m with $m \geqslant 3$; see also Reiter [1] and 12.11.5. Another decade was to elapse before similar examples pertaining first to R, and then to any noncompact G, were produced by Malliavin [1] in 1959.

Despite this disappointment, it turns out that if $Z_{\mathbf{I}}$ is topologically simple enough, then **I** does indeed contain every $f \in \mathbf{L}^1(R^m)$ for which $Z_f \supset Z_{\mathbf{I}}$; and that the conclusion stands, whatever $Z_{\mathbf{I}}$, if in addition $\hat{f}$ is subject to smoothness conditions. Results of this type permit the reader to judge for himself to what extent the analogue of 11.2.1 (for $\mathbf{E} = \mathbf{L}^1$) may be claimed to be "approximately true." See [HR], (39.24); [Re], p. 28; [Kz], p. 225; [R], 7.2.4; MR **37** # 6694; **40** # 6491; **46** # # 9650, 9652, **49** # 9542; **53** # 14025; **54** # # 10980, 13464.

A set $S \subset R^m$ having the property that

$$f \in \mathbf{L}^1(R^m), \quad Z_f \supset Z_{\mathbf{I}} \Rightarrow f \in \mathbf{I}$$

for every closed ideal **I** in $\mathbf{L}^1(R^m)$ for which $Z_{\mathbf{I}} = S$, is termed a *spectral* (or *harmonic*) *synthesis set* in R^m; Rudin ([R], p. 158) refers to them more briefly as *S-sets*. It is known that S is a spectral synthesis set in this sense if and only if there is but one closed ideal **I** in $\mathbf{L}^1(R^m)$ satisfying $Z_{\mathbf{I}} = S$.

Malliavin's result cited above asserts precisely that there exist closed subsets of R^m which are *not* spectral synthesis sets. On the other hand, the opening statement in the last paragraph but one amounts to saying that conditions of topological simplicity are known which ensure that a given closed set S is a spectral synthesis set; compare Exercise 12.52.

Malliavin's result cited above has given rise to many extensions, improvements and simplifications. For some (if not all) of the details, the reader should consult Malliavin [1], [2]; [R], Chapter 7; [KS], Chapitre IX; [Kz], pp. 229 ff; [HR], §42; de Leeuw and Herz [1]; MR **31** # 2567; **39** # 1977; Exercise 12.53 below. At this point we remark merely that Malliavin's original construction has been simplified by Kahane and Katznelson [2] and Richards [1]; and that Varopoulos [1], [2] introduced an entirely original (tensor product) approach to spectral synthesis problems in Banach algebras; see MR **41** # 830 and the remarks in 11.4.18(4) below.

As has been indicated, strictly analogous problems arise when attention is transferred from $\mathbf{L}^1(R^m)$ to $\ell^1(Z)$; concerning this particular extension we shall have a little more to say in Subsections 11.2.5, 12.11.4, 12.11.5, and 12.11.6.

Mention must also be made of analogues for noncompact groups G of the remaining cases covered by 11.2.1, namely, the closure of translations theorems in $\mathbf{E} = \mathbf{L}^p$ $(1 < p < \infty)$ and $\mathbf{E} = \mathbf{C}$. The results for $\mathbf{L}^\infty(G)$ with its weak topology (see the opening remarks to this chapter) go hand in hand with those for $\mathbf{L}^1(G)$ already discussed. For $\mathbf{L}^2(R)$ a complete solution was given by Wiener ([Wi], p. 100), and this extends without trouble to $\mathbf{L}^2(G)$. In all other cases, that is, for values of p different from 1, 2, and ∞, the known results are less complete. While conditions are known which are sufficient to ensure that the linear combinations of the translates of a given $f \in \mathbf{L}^p(G)$ are everywhere dense in that space, and yet others are known which are necessary for this to happen, there remains a gap between the two types of conditions. All attacks on this problem are bedevilled by the preliminary task of devising and handling a tractable definition of the Fourier transform of a function belonging to an arbitrary space $\mathbf{L}^p(G)$. This may be done in terms of pseudomeasures and similar objects (the periodic prototypes of which are mentioned in Section 12.11; see especially 12.11.4). There is, alas, no connected account in book form, but see Herz's survey article [2], Gaudry [1], [3], Edwards [4] and the references there cited, and Warner [1]. (The case of the group Z is discussed briefly in 11.2.5.). See also MR **38** # 4904.

One striking fact, applying when G is noncompact and $1 < p < 2$, is that there exists a closed invariant subspace $\mathbf{V} \neq \{0\}$ in $\mathbf{L}^2(G)$ which contains no nonzero element of $\mathbf{L}^p(G)$; see MR **52** # 14849. See also MR **48** # 11915.

Finally, an even wider diversity obtains when one turns to analogues of the case $\mathbf{E} = \mathbf{C}$ of 11.2.1. This is due to the fact that there are, in relation to a noncompact G, several natural spaces of continuous functions which coalesce for compact groups but which otherwise are widely different. The following four contenders have received attention:

(1) the space $\mathbf{C}(G)$ of all continuous functions on G, with the topology of locally uniform convergence;

(2) the space $\mathbf{BC}(G)$ of bounded, continuous functions on G, with the topology of uniform convergence;

(3) the space $\mathbf{BUC}(G)$ of bounded, uniformly continuous functions on G, with the topology of uniform convergence;

(4) the space $\mathbf{C}_0(G)$ of continuous functions which tend to zero at infinity, with the topology of uniform convergence.

For $\mathbf{C}_0(G)$ fairly complete results are known. For the remaining three, results are hard to come by; in the case of $\mathbf{BC}(G)$ and $\mathbf{BUC}(G)$, more progress has been made concerning approximation relative to a weaker (the so-called "strict") topology, originally suggested by ideas of Beurling; see Edwards [5] and Harasymiv [1]. In the case of $\mathbf{C}(G)$, most attention has been paid to functions, the linear combinations of translates of which are *not* everywhere dense in $\mathbf{C}(G)$: these were introduced and studied (for $G = R$) by Laurent Schwartz [2] in 1947, who christened them *mean periodic functions*; see also [Kah$_3$]. Some of Schwartz's results have since been extended to more general groups by Ehrenpreis [1], Elliott [1], Gilbert [1], and others.

11.2.4. About Tauberian Theorems.

We pass on to consider briefly some consequences of such closure of translations theorems as are typified by the case $Z_{\mathbf{I}} = \varnothing$ of 11.2.1 and the generalizations thereof mentioned in 11.2.3.

Let us begin with the group T. Suppose we take a subset A of T and a nonvoid collection Π of nonvoid subsets of A satisfying the following two conditions:

(1) the intersection of any two members of Π contains a member of Π (in M. Bourbaki's language, this signifies that Π is a *filter base* on A);

(2) if $a \in A$ and $P \in \Pi$, there exists $P' \in \Pi$ such that

$$P' \subset a + P \equiv \{a + x : x \in P\}.$$

If F is a real- or complex-valued function defined on A, we write

$$\lim_{\Pi} F = 0$$

if and only if to each $\varepsilon > 0$ corresponds a set $P_\varepsilon \in \Pi$ such that $|F(a)| \leqslant \varepsilon$ for $a \in P_\varepsilon$. Owing to condition (1), $\lim_\Pi (F_1 + F_2) = 0$ whenever $\lim_\Pi F_i = 0$ for $i = 1, 2$. Moreover, thanks to condition (2), $\lim_\Pi T_a F = 0$ whenever $a \in A$ and $\lim_\Pi F = 0$.

By way of example, one might take for A the set of all cosets $na + 2\pi Z$ obtained when a is a fixed real number and n ranges over Z, while for Π one might take the collection of sets P_k ($k = 1, 2, \cdots$), P_k being the set of cosets $na + 2\pi Z$ obtained for $|n| > k$. In this case, $\lim_\Pi F$ would signify what would normally be denoted by $\lim_{n\in Z, |n|\to\infty} F(na + 2\pi Z)$, or by $\lim_{n \in Z, |n| \to \infty} F(na)$, if F is first extended to T and then regarded as a periodic function on R.

Given $f_0 \in \mathbf{L}^1$, it may or may not be true to assert that the linear combinations of the translates $T_a f_0$ ($a \in A$) are everywhere dense in $\mathbf{L}^1$. As may be deduced from 11.2.1, this assertion will be true whenever $\hat{f}_0$ is nonvanishing on Z and A is everywhere dense in T. (This second condition is satisfied by the particular A mentioned in the last paragraph, if a/π is irrational; see Exercise 2.2.) The assertion may also be shown to be true for quite sparse subsets A of T, provided f_0 is, say, analytic and $\hat{f}_0$ is nonvanishing on Z (compare Exercise 11.9).

In any case, one may proceed without much difficulty to establish the following result.

(a) Suppose that A and Π are as above, that $f_0 \in \mathbf{L}^1$, that $\hat{f}_0$ is nonvanishing on Z, and that A is everywhere dense in T. If $g \in \mathbf{L}^\infty$, and if

$$\lim_{\Pi} f_0 * g = 0, \tag{11.2.9}$$

then also

$$\lim_{\Pi} f * g = 0 \qquad \text{for all } f \in \mathbf{L}^1. \tag{11.2.10}$$

It is true, but irrelevant at the moment, to say that (11.2.10) is equivalent to the assertion that $g = 0$ almost everywhere.

As far as the writer is aware, (a) has no especially significant consequences. However, (a) has a deeper analogue for the case in which T is replaced by any locally compact Abelian group G, and, when G is noncompact, very significant results are obtained in this way.

Let us state such an analogue for the typical case in which $G = R$, specializing on the way by taking $A = G = R$ and Π to be the set of complements in R of compact subsets of R. (This natural choice of Π is not permissible for compact groups, which is why (a) takes the rather complex form it does.) Then $\lim_\Pi F$ signifies what is normally written as $\lim_{x\in R, |x|\to\infty} F(x)$. The analogue runs as follows.

(b) If $f_0 \in \mathbf{L}^1(R)$ satisfies

$$\hat{f}_0(\xi) \equiv \int_R f(x) e^{-2\pi i \xi x}\, dx \neq 0 \qquad (\xi \in R),$$

and if $g \in \mathbf{L}^\infty(R)$ is such that

$$\lim_{|x| \to \infty} f_0 * g(x) = 0, \tag{11.2.11}$$

then also

$$\lim_{|x| \to \infty} f * g(x) = 0 \qquad \text{for all } f \in \mathbf{L}^1(R). \tag{11.2.12}$$

Statement (b), and its valid analogue for any locally compact Abelian group G in place of R, is an almost immediate corollary of the closure of translations theorem mentioned in 11.2.3; see [R], p. 163; [HR], (39.36); [Kz], p. 228; [Re], p. 10. The case $G = R$ is Wiener's famous *general Tauberian theorem*, so named because by design it includes as corollaries a number of results each of a Tauberian nature (see 5.3.5). The deductions of these special Tauberian theorems, which often require skillful choice of the "kernels" f_0 and f, followed by lengthy manipulations, would be out of place here. For an account the reader may be referred to Wiener's book [Wi] or, for more recent accounts, to [Ha] and/or [P]. (None of these references lays any stress on the relationship between the Wiener theorem and the ideal theory of $\mathbf{L}^1(R)$.) A brief proof of (b), using distributional techniques and the Hahn-Banach theorem, has been given by Korevaar [1]; the method is even more neatly expressible in terms of pseudomeasures (see Section 12.11 and compare the remarks in 11.2.5).

Weak versions of (b), in which more is assumed about the smallness of f_0 at infinity, can be proved by simpler arguments and retain some interest for applications; see, for example, Kac [1]. See also MR **50** # 7952; **52** # 1173.

Concerning abstract Tauberian theorems, see Subsection 11.4.18(3).

11.2.5. **The Case of the Dual Group Z.** The remarks in 11.2.3 that refer to analogues of 11.2.1, or of special cases of that theorem, may be further illustrated by looking at the situation in which the underlying group is Z. This we shall do very briefly.

The problem is that of classifying the closed invariant subspaces of $\ell^p = \ell^p(Z)$.

For $p = 1$, in which case we may speak equivalently of the closed ideals $\mathbf{I}$ in ℓ^1, some at least of the difficulties mentioned for general noncompact groups in 11.2.3 are already visible. However, it is true that a given $\phi \in \ell^1$ generates a dense ideal if and only if its Fourier transform $\hat{\phi}$ is nonvanishing; this, and more besides, is proved in Exercise 12.32.

The case $p = 2$ is completely solved, the solution being as follows. Given $\phi \in \ell^2$ and $\psi \in \ell^2$, ψ belongs to the closed invariant subspace of ℓ^2 generated by ϕ if and only if the set of zeros of $\hat{\psi}$ contains, modulo a null set, the set of zeros of $\hat{\phi}$; regarding the transforms $\hat{\phi}$ and $\hat{\psi}$, see 8.3.3.

For other finite values of p no complete solution is known, the remarks in 11.2.3 applying without modification. In Sections 12.11 and 12.12 we shall describe briefly the difficulties, unsolved problems, and partial successes in relation to this problem. It must here suffice to remark that the sharpest results available apply to a problem posed by Beurling, namely: Suppose that $\phi \in \ell^1(Z)$ ($\subset \ell^p(Z)$ for every $p \geqslant 1$); what can be said about the size of $E = \hat{\phi}^{-1}(\{0\})$ in relation to those values of p for which the closed invariant subspace of $\ell^p(Z)$ generated by ϕ is the whole of $\ell^p(Z)$? As we shall see, although the solution is not complete, quite precise results are expressible in terms of so-called "capacities" of E (or, what is essentially equivalent in the present context, the so-called "Hausdorff dimension" of E). Any reader who wishes to approach this problem independently will find the details in [KS], p. 112. See also Newman [3].

For $p = \infty$, it is the case that a given $\phi \in \ell^\infty$ has the linear combinations of its translates weakly dense in ℓ^∞ if and only if the pseudomeasure $\hat{\phi}$ has a full support (see 12.11.4). The same result holds for the uniformly closed invariant subspace of $\mathbf{c}_0$ generated by a given $\phi \in \mathbf{c}_0$. (A pseudomeasure σ on T has a full support if and only if the only continuous function u on T having an absolutely convergent Fourier series and satisfying $u \cdot \sigma = 0$ is the function $u = 0$.) See also [Kah], Chapter VIII.

11.3 Closed Subalgebras

We now turn to the consideration of structure theorems for closed subalgebras $\mathfrak{A}$ of $\mathbf{E}$, analogous to 11.2.1 for closed ideals.

The natural conjecture is that $\mathfrak{A}$ is fully determined by knowledge of those subsets of Z which are common sets of constancy of the Fourier transforms $\hat{f}$ of elements f of $\mathfrak{A}$; see 11.3.1 for details. When $\mathbf{E} = \mathbf{L}^2$, this conjecture is true, as will be shown in 11.3.6. When $\mathbf{E} = \mathbf{L}^1$, the conjecture is shown to be false by an example due to Kahane (1962); see 11.3.4. To the author's knowledge, its truth or falsity is undecided in all the remaining cases.

On the other hand, as will become apparent from subsequent results, the conjecture is "approximately true" (compare with 11.2.3).

For several other problems about subalgebras having close connections with harmonic analysis, see [R], Chapter 9.

11.3.1. Equivalence Relations and Idempotents in $\mathfrak{A}$. Let $\mathfrak{A}$ be a closed subalgebra of $\mathbf{E}$. We introduce the equivalence relation $r(\mathfrak{A})$ on Z which is defined by writing $nr(\mathfrak{A})n'$ if and only if $n, n' \in Z$ and $\hat{f}(n) = \hat{f}(n')$ for all $f \in \mathfrak{A}$. Accordingly, Z is partitioned into cosets S modulo $r(\mathfrak{A})$, of which there are at most countably many. We follow Kahane in terming each such coset S modulo $r(\mathfrak{A})$ a *Rudin class* of $\mathfrak{A}$. On each Rudin class of $\mathfrak{A}$, each function $\hat{f}$ ($f \in \mathfrak{A}$) assumes a constant value which we denote by $\hat{f}(S)$. Either $\hat{f}(S) = 0$ for all $f \in \mathfrak{A}$, or S is finite (see 2.3.8).

Given any equivalence relation r on Z, we shall write $\mathbf{L}^1_r$ for the set of all integrable functions f such that $\hat{f}(n) = \hat{f}(n')$ whenever $n, n' \in Z$ and nrn'. Then $\mathbf{E} \cap \mathbf{L}^1_r$ is a closed subalgebra of $\mathbf{E}$, which is plainly the largest closed subalgebra $\mathfrak{A}$ of $\mathbf{E}$ for which $r(\mathfrak{A}) = r$.

It follows that $\mathfrak{A} \subset \mathbf{E} \cap \mathbf{L}^1_{r(\mathfrak{A})}$ for any closed subalgebra $\mathfrak{A}$ of $\mathbf{E}$.

The natural conjecture is that $\mathfrak{A} = \mathbf{E} \cap \mathbf{L}^1_{r(\mathfrak{A})}$ for any closed subalgebra $\mathfrak{A}$ of $\mathbf{E}$.

Save for the cases mentioned in 11.3.4 and 11.3.6, the truth or falsity of this conjecture is an open problem. The basic positive result in favor of this conjecture asserts that in all cases $\mathfrak{A}$ contains each of the trigonometric polynomials

$$e_S = \sum_{n \in S} e_n,$$

where S is any coset modulo $r(\mathfrak{A})$ such that $\hat{f}(S) \neq 0$ for some $f \in \mathfrak{A}$. These elements e_S are easily seen to be precisely the minimal idempotents of the algebra $\mathfrak{A}$, that is, the idempotent elements e of $\mathfrak{A}$ which are not expressible as sums of two or more nonzero idempotents of $\mathfrak{A}$. Each idempotent element of $\mathfrak{A}$ is a finite sum of those minimal idempotents.

In order to prove that each e_S does indeed belong to $\mathfrak{A}$, we shall need to call upon the spectral radius formula

$$\lim_{k \to \infty} \|f^{*k}\|^{1/k} = \|\hat{f}\|_\infty \tag{11.3.1}$$

for elements f of $\mathbf{E}$, the norm appearing on the left-hand side being that on $\mathbf{E}$. For $\mathbf{E} = \mathbf{L}^2$ this result is established in Exercise 8.8. The formula for $\mathbf{E} = \mathbf{C}$ is easily deducible from the case $\mathbf{E} = \mathbf{L}^2$. A proof covering all the required cases will appear in 11.4.14. Meanwhile, (11.3.1) will be taken on trust.

11.3.2. Let $\mathfrak{A}$ be a closed subalgebra of $\mathbf{E}$. Then $e_S \in \mathfrak{A}$ for any coset S modulo $r(\mathfrak{A})$ for which $\hat{f}(S) \neq 0$ for some (possibly S-dependent) element f of $\mathfrak{A}$.

Proof. This is taken from [R], p. 232.

Enumerate the cosets S modulo $r(\mathfrak{A})$, for which $\hat{f}(S) \neq 0$ for some S-dependent $f \in \mathfrak{A}$, as S_α, where α runs over some set of positive integers.

Choose and fix any α and then any $f \in \mathfrak{A}$ such that $\hat{f}(S_\alpha) \neq 0$. There are at most a finite number of indices α' distinct from α for which $\hat{f}(S_{\alpha'}) = \hat{f}(S_\alpha)$. If no such indices α' exist, part (1) of the proof to follow simplifies and completes the proof. We shall proceed on the hypothesis that such indices α' exist, labeling them $\alpha_1, \cdots, \alpha_n$, and putting

$$t = e_{S_\alpha} + e_{S_{\alpha 1}} + \cdots + e_{S_{\alpha n}}.$$

(In case no indices α' exist, the appropriate definition would read: $t = e_{S_\alpha}$.)

(1) The first step is to show that $t \in \mathfrak{A}$. According to 2.3.8, $\hat{f}(S_\alpha)$ is an isolated point of $\hat{f}(Z)$, and $\hat{f}(Z)$ has no limit points other than zero. It follows that a polynomial F in one complex variable may be chosen so that

$$F(0) = 0, \qquad F(\hat{f}(S_\alpha)) = 1, \qquad |F(z)| < \tfrac{1}{2}$$

for $z \in \hat{f}(Z)\backslash\hat{f}(S_\alpha)$. Indeed, this may be reduced to the following constructional problem: given complex numbers z_m $(m = 0, 1, 2, \cdots)$ such that $z_0 \neq 0$, $z_m \neq z_0$ $(m > 0)$, $z_m \to 0$ as $m \to \infty$, it is required to find a polynomial F such that $F(0) = 0$, $F(z_0) = 1$, $|F(z_m)| < \frac{1}{2}$ for $m > 0$. To this end, choose an integer k so large that $|z_m| \leqslant \frac{1}{2}|z_0|$ for $m > k$, and put

$$F(z) = \left(\frac{z}{z_0}\right)^N \cdot \frac{(z - z_1)\cdots(z-z_k)}{(z_0 - z_1)\cdots(z_0 - z_k)},$$

where the positive integer N is to be chosen in a moment. Then, plainly, $F(0) = 0$, $F(z_0) = 1$, and $F(z_m) = 0$ for $0 < m \leqslant k$. Moreover, if $m > k$,

$$|F(z_m)| \leqslant \frac{2^{-N}(|z_1| + \frac{1}{2}|z_0|)\cdots(|z_k| + \frac{1}{2}|z_0|)}{|(z_0 - z_1)\cdots(z_0 - z_k)|},$$

which can be made less than $\frac{1}{2}$ if only N be chosen sufficiently large (depending upon k). If this be done, F is a polynomial satisfying all the requirements.

Suppose that

$$F(z) = c_1 z + \cdots + c_M z^M,$$

and consider the function

$$g = c_1 f + c_2 f^{*2} + \cdots + c_M f^{*M},$$

where, as usual, f^{*k} is the k-th convolution power of f. Since $f \in \mathfrak{A}$ and since $\mathfrak{A}$ is a subalgebra of E, $g \in \mathfrak{A}$. Plainly, $\hat{g} = F \circ \hat{f}$. Since both $\hat{g}$ and $\hat{t}$ take the value 1 on $S_\alpha \cup S_{\alpha_1} \cup \cdots \cup S_{\alpha_n}$, our choice of F ensures that $\|\hat{g} - \hat{t}\|_\infty < \frac{1}{2}$. The spectral radius formula for elements of $\mathbf{E}$ therefore entails that

$$\lim_{k \to \infty} \|(g - t)^{*k}\|^{1/k} < \tfrac{1}{2}. \tag{11.3.2}$$

On the other hand we have

$$\hat{t} = \hat{t}^k = \hat{g} \cdot \hat{t}^{k-1} = \hat{g}^2 \cdot \hat{t}^{k-2} = \cdots = \hat{g}^{k-1} \cdot \hat{t},$$

since $\hat{t}$ is the characteristic function of $S_\alpha \cup S_{\alpha_1} \cup \cdots \cup S_{\alpha_n}$, on which set $\hat{g}$ assumes the value 1. Therefore

$$(\hat{g} - \hat{t})^k = \hat{g}^k - \hat{t},$$

or, by the uniqueness theorem 2.4.1,

$$(g - t)^{*k} = g^{*k} - t.$$

Formula (11.3.2) now shows that

$$\lim_{k \to \infty} \|g^{*k} - t\|^{1/k} < \tfrac{1}{2},$$

so that

$$\|g^{*k} - t\| < 2^{-k}$$

for all sufficiently large k. Since $\mathfrak{A}$ is a closed subalgebra of $\mathbf{E}$, it appears thence that $t \in \mathfrak{A}$.

(2) The next step is to show that $\mathfrak{A}$ contains trigonometric polynomials t_i ($i = 1, 2, \cdots, n$) such that

$$\hat{t}_i(S_\alpha) = 1, \qquad \hat{t}_i(S_{\alpha_i}) = 0.$$

It will suffice to exhibit the construction of t_1.

Choose $h \in \mathfrak{A}$ such that $\hat{h}(S_\alpha) \neq \hat{h}(S_{\alpha_1})$ and put

$$h_1 = \frac{[h - \hat{h}(S_{\alpha_1})t]}{[\hat{h}(S_\alpha) - \hat{h}(S_{\alpha_1})]}.$$

by (1), $h_1 \in \mathfrak{A}$; and plainly

$$\hat{h}_1(S_\alpha) = 1, \qquad \hat{h}_1(S_{\alpha_1}) = 0.$$

If the construction described in (1) be applied on starting with h_1 in place of f, the result is easily seen to be a trigonometric polynomial $t_1 \in \mathfrak{A}$ satisfying the required conditions.

(3) By comparing Fourier transforms and using the uniqueness theorem 2.4.1, it is clear that $e_{S_\alpha} = t * t_1 * t_2 \cdots * t_n$, which makes it evident that $e_{S_\alpha} \in \mathfrak{A}$ and thus completes the proof.

11.3.3. The Natural Conjecture as an Approximation Problem. Given a closed subalgebra $\mathfrak{A}$ of $\mathbf{E}$, denote by $\mathfrak{A}_0$ the closed subalgebra of $\mathbf{E}$ generated by the idempotents e_S, S being any coset modulo $r(\mathfrak{A})$ such that $\hat{f}(S) \neq 0$ for some $f \in \mathfrak{A}$. By 11.3.2, $\mathfrak{A}_0 \subset \mathfrak{A}$. It is otherwise clear that $r(\mathfrak{A}_0) = r(\mathfrak{A})$.

The natural conjecture spoken of in 11.3.1 is thus to the effect that $\mathfrak{A}_0 = \mathfrak{A}$ for all closed subalgebras $\mathfrak{A}$ of $\mathbf{E}$.

Yet another way of expressing the conjecture is in the form of the following assertion about approximation.

Given disjoint nonvoid finite subsets S_α of Z, each $f \in \mathbf{E}$, such that $\hat{f}$ is constant on each S_α and vanishes on $Z \backslash \bigcup_\alpha S_\alpha$, is the limit in $\mathbf{E}$ of finite linear combinations of the idempotent trigonometric polynomials e_{S_α}.

Further discussion of the conjecture will often make use of this last version thereof.

11.3.4. Kahane's Results about Closed Subalgebras of $\mathbf{L}^1$. Let us agree to write $\#S = k$, or $\#S = \infty$, according as the set S is finite and has k elements, or is infinite, respectively; and to write $k < \infty$ for every real number k. Then the main results established by Kahane [1] may be stated as follows:

(a) There exist closed subalgebras $\mathfrak{A}$ of $\mathbf{L}^1$ which are not generated by their idempotent elements.

(b) If $\mathfrak{A}$ is a closed subalgebra of $\mathbf{L}^1$ whose Rudin classes are of bounded lengths (that is, are contained in intervals of bounded lengths), then $\mathfrak{A}$ is generated by its idempotents.

(c) There exists a closed subalgebra $\mathfrak{A}$ of $\mathbf{L}^1$, whose finite Rudin classes S satisfy $\sup \#S < \infty$, and which is nevertheless not generated by its idempotents.

Kahane also raises a number of problems demanding attention.

Problem 1. Are there any (infinite) subsets Q of Z with the property that, for any closed subalgebra $\mathfrak{A}$ of $\mathbf{L}^1$, each $f \in \mathfrak{A}$ whose Fourier transform vanishes on $Z \backslash Q$ is the limit in $\mathbf{L}^1$ of finite linear combinations of the idempotents in $\mathfrak{A}$?

Kahane remarks that his proof of (a) can be adapted to show that a set Q with this property cannot contain arbitrarily long arithmetic progressions. On the other hand, the results of Chapter 15 (see especially 15.3.1) show that Sidon sets Q have the desired property. A similar result is true for closed subalgebras of $\mathbf{E} = \mathbf{C}$ or $\mathbf{L}^p$, and may be obtained by combining 11.3.5 and the proof of 11.3.6 with 15.1.4 and 15.3.1, respectively. Nothing appears to be known concerning sets Q which fail to contain arbitrarily long arithmetic progressions and which are yet too "thick" to fall into the category of Sidon sets.

Problem 2. Kahane has shown that algebras $\mathfrak{A}$ exist with the property mentioned in (c) and whose finite Rudin classes contain at most four elements. Is it possible to reduce "four" to "two" in this assertion?

Problem 3. Which equivalence relations r on Z have the synthesis property, that is, are such that any closed subalgebra $\mathfrak{A}$ of $\mathbf{L}^1$, for which $r(\mathfrak{A}) = r$, is generated by its idempotents? Result (b) seems to be as much as is as yet known in this connection.

Although (a) decisively negatives the natural conjecture, the next four results use 11.3.2 in order to salvage something in the positive direction; they show that the natural conjecture is not absurdly wide of the mark.

11.3.5. Let $\mathfrak{A}$ be a closed subalgebra of $\mathbf{E}$. Then $\mathfrak{A}$ contains each $f \in \mathbf{L}^1_{r(\mathfrak{A})}$ for which $\sum_{n \in Z} |\hat{f}(n)| < \infty$.

Proof. Using the notation introduced at the beginning of the proof of 11.3.2, the absolutely convergent Fourier series of f may be regrouped to appear as

$$\sum_{\alpha} \hat{f}(S_\alpha) \sum_{n \in S_\alpha} e_n = \sum_{\alpha} \hat{f}(S_\alpha) e_{S_\alpha},$$

the regrouped series being again absolutely convergent. It follows that f is equal almost everywhere to the limit of a uniformly convergent sequence of finite linear combinations g_k $(k = 1, 2, \cdots)$ of the e_{S_α}; and if f is continuous it is equal everywhere to this limit. So in any case f is the limit in $\mathbf{E}$ of the g_k. By 11.3.2, each $g_k \in \mathfrak{A}$. Since $\mathfrak{A}$ is closed in $\mathbf{E}$, it follows that $f \in \mathfrak{A}$.

11.3.6. Let $\mathfrak{A}$ be a closed subalgebra of $\mathbf{L}^2$. Then $\mathfrak{A} = \mathbf{L}^2 \cap \mathbf{L}^1_{r(\mathfrak{A})}$.

Proof. It suffices to verify that the approximation assertion appearing in 11.3.3 is true. Suppose then that $f \in \mathbf{L}^2$, that $\hat{f}$ takes the constant value c_α on S_α, and that $\hat{f}$ vanishes outside $S = \bigcup_\alpha S_\alpha$. Writing e_α in place of e_{S_α}, the Parseval formula (8.2.2) gives for any finite set F of indices α

$$\|f - \sum_{\alpha \in F} c_\alpha e_\alpha\|_2^2 = \sum_{n \in Z} |\hat{f}(n) - \sum_{\alpha \in F} c_\alpha \hat{e}_\alpha(n)|^2$$

$$= \sum_\beta \sum_{n \in S_\beta} |c_\beta - \sum_{\alpha \in F} c_\alpha \hat{e}_\alpha(n)|^2.$$

Now

$$\sum_{\alpha \in F} c_\alpha \hat{e}_\alpha(n) = \begin{cases} c_\beta & \text{if } n \in S_\beta \text{ for some } \beta \in F, \\ 0 & \text{otherwise}, \end{cases}$$

so that

$$\|f - \sum_{\alpha \in F} c_\alpha e_\alpha\|_2^2 = \sum_{\beta \notin F} \sum_{n \in S_\beta} |c_\beta|^2$$

$$= \sum_{n \in S'_F} |\hat{f}(n)|^2 = \sum_{n \in S \setminus S_F} |\hat{f}(n)|^2,$$

where $S'_F = \bigcup_{\beta \notin F} S_\beta$ and $S_F = \bigcup_{\beta \in F} S_\beta$. The regrouping of terms of the series is justified, because $\sum_{n \in Z} |\hat{f}(n)|^2 < \infty$. This same condition ensures also that, given any $\varepsilon > 0$,

$$\sum_{n \in S \setminus S_F} |\hat{f}(n)|^2 < \varepsilon^2$$

for all sufficiently large finite sets F of indices. For such finite sets F, it is therefore the case that

$$\|f - \sum_{\alpha \in F} c_\alpha e_\alpha\|_2 < \varepsilon,$$

which is what we had to show.

11.3.7. The proof of 11.3.6 shows in fact that, if $1 \leqslant p \leqslant 2$, and if $\mathfrak{A}$ is a closed subalgebra of $\mathbf{L}^p$, then

$$\mathfrak{A} \supset \mathbf{L}^2 \cap \mathbf{L}^1_{r(\mathfrak{A})}.$$

(Recall that convergence in $\mathbf{L}^2$ implies convergence in $\mathbf{L}^p$ for any p satisfying $0 < p < 2$.) See Exercise 11.6.

11.3.8. If $\mathfrak{A}$ is a closed subalgebra of $\mathbf{E}$, then $f * g \in \mathfrak{A}$ whenever

$$f, g \in \mathbf{L}^2 \cap \mathbf{L}^1_{r(\mathfrak{A})}.$$

Proof. As the proof of 11.3.6 shows, each of f and g is the limit in $\mathbf{L}^2$ of finite linear combinations of the e_{S_α}. Since $e_{S_\alpha} * e_{S_\beta} = 0$ or e_{S_α} according as $\beta \neq \alpha$ or $\beta = \alpha$, it follows that $f * g$ is the uniform limit (a fortiori, the limit in $\mathbf{E}$) of finite linear combinations of the e_{S_α} and therefore, by 11.3.2, belongs to $\mathfrak{A}$.

See also Exercise 11.7, where a somewhat more general result is given.

11.3.9. **Existence of Zero Divisors.** For closed subalgebras $\mathfrak{A}$ of $\mathbf{E}$, an answer can now be given to the question [raised in 3.1.1(e)] about the existence in $\mathfrak{A}$ of zero divisors. The answer is as follows: a closed subalgebra $\mathfrak{A}$ of E has no zero divisors if and only if $\mathfrak{A}$ is either $\{0\}$ or consists of the scalar multiples of some nonzero idempotent trigonometric polynomial.

Proof. It is evident that if $\mathfrak{A}$ is of the stated type, then it possesses no zero divisors.

Suppose on the other hand that $\mathfrak{A} \neq \{0\}$. Then there exists $n_0 \in Z$ and $f_0 \in \mathfrak{A}$ such that $\hat{f}_0(n_0) \neq 0$. Let S_0 be the coset modulo $r(\mathfrak{A})$ containing n_0. By 11.3.2, $e_{S_0} \in \mathfrak{A}$. If $\mathfrak{A}$ were not exhausted by the scalar multiples of e_{S_0}, there would exist $n_1 \in Z \backslash S_0$ and $f_1 \in \mathfrak{A}$ such that $\hat{f}_1(n_1) \neq 0$. Then, by 11.3.2 again, if S_1 is the coset modulo $r(\mathfrak{A})$ containing n_1, we have $e_{S_1} \in \mathfrak{A}$. Since $S_1 \neq S_0$, $S_1 \cap S_0 = \varnothing$, and therefore $e_{S_1} * e_{S_0} = 0$. Each of e_{S_0} and e_{S_1} would thus be a zero divisor belonging to $\mathfrak{A}$. Consequently, $\mathfrak{A}$ must be exhausted by, and so be identical with, the set of scalar multiples of e_{S_0}.

11.3.10. **Maximal Subalgebras.** A closed subalgebra $\mathfrak{A}$ of $\mathbf{E}$ is termed a *maximal subalgebra* of $\mathbf{E}$ if $\mathfrak{A} \neq \mathbf{E}$ and if the only closed subalgebras of $\mathbf{E}$ which contain $\mathfrak{A}$ are $\mathfrak{A}$ and $\mathbf{E}$ itself.

These maximal subalgebras of $\mathbf{E}$ can be simply and fully characterized, as in the second half of the following statement.

11.3.11. Let $\mathfrak{A}$ be a closed subalgebra of $\mathbf{E}$, $\mathscr{F}\mathfrak{A}$ the set of Fourier transforms of elements of $\mathfrak{A}$, and $Z_{\mathfrak{A}}$ the set of common zeros of elements of $\mathfrak{A}$ (as in Section 11.2).

(i) If $\mathscr{F}\mathfrak{A}$ separates points of Z (that is, if, whenever $n_1 \neq n_2$ belong to Z, there exists $f \in \mathfrak{A}$ such that $\hat{f}(n_1) \neq \hat{f}(n_2)$), then *either* $Z_{\mathfrak{A}} = \varnothing$ and $\mathfrak{A} = \mathbf{E}$, *or* there exists $n_0 \in Z$ such that $Z_{\mathfrak{A}} = \{n_0\}$ and

$$\mathfrak{A} = \mathfrak{A}_{n_0} \equiv \{f \in \mathbf{E} : \hat{f}(n_0) = 0\}. \tag{11.3.3}$$

(ii) The maximal subalgebras of $\mathbf{E}$ are precisely the subalgebras $\mathfrak{A}_{n_0}$ $(n_0 \in Z)$ and the subalgebras

$$\mathfrak{A}_{n_1, n_2} \equiv \{f \in \mathbf{E} : \hat{f}(n_1) = \hat{f}(n_2)\}, \tag{11.3.4}$$

where $n_1, n_2 \in Z$ and $n_1 \neq n_2$. (Notice that each $\mathfrak{A}_{n_0}$ is actually an ideal in $\mathbf{E}$, and not merely a subalgebra.)

Proof. (i) To say that $\mathscr{F}\mathfrak{A}$ separates points of Z is equivalent to saying that each Rudin class of $\mathfrak{A}$ is a singleton. This being so, two cases arise according as $Z_{\mathfrak{A}}$ is or is not void.

If $Z_{\mathfrak{A}}$ is void, 11.3.2 shows that $e_n \in \mathfrak{A}$ for all $n \in Z$, and 6.1.1 then shows that $\mathfrak{A} = \mathbf{E}$.

Otherwise, $Z_{\mathfrak{A}}$ is a Rudin class of $\mathfrak{A}$ and hence takes the form $\{n_0\}$ for some $n_0 \in Z$. If $n \in Z$ is distinct from n_0, then $\{n\}$ is a Rudin class of $\mathfrak{A}$ and $\hat{f}(n) \neq 0$ for some $f \in \mathfrak{A}$. So 11.3.2 entails that $e_n \in \mathfrak{A}$ for all integers n different from n_0, and 6.1.1 shows that $\mathfrak{A} \supset \mathfrak{A}_{n_0}$ and so that in fact $\mathfrak{A} = \mathfrak{A}_{n_0}$.

(ii) The proof is broken into several steps.

(1) $\mathfrak{A}_{n_0}$ is maximal. For if $\mathfrak{B}$ is a closed subalgebra of $\mathbf{E}$ containing $\mathfrak{A}_{n_0}$, then $\mathscr{F}\mathfrak{B}$ plainly separates points of Z. So, by (i), $\mathfrak{B}$ is either $\mathbf{E}$ or is $\mathfrak{A}_n$ for some $n \in Z$. Since $\mathfrak{B} \supset \mathfrak{A}_{n_0}$, n must coincide with n_0 and $\mathfrak{B}$ with $\mathfrak{A}_{n_0}$. Whence results the maximality of $\mathfrak{A}_{n_0}$.

(2) $\mathfrak{A}_{n_1,n_2}$ is maximal. For if $\mathfrak{B}$ is a closed subalgebra of $\mathbf{E}$ containing $\mathfrak{A}_{n_1,n_2}$ properly, it is again clear that $\mathscr{F}\mathfrak{B}$ separates points of Z. So, by (i) again, $\mathfrak{B}$ is either $\mathbf{E}$ or $\mathfrak{A}_{n_0}$ for some $n_0 \in Z$. The latter alternative cannot arise since $\mathfrak{B} \supset \mathfrak{A}_{n_1,n_2}$. Hence $\mathfrak{B} = \mathbf{E}$.

(3) Suppose now that $\mathfrak{A}$ is a maximal subalgebra of $\mathbf{E}$, and suppose that *either* (a) there exists a Rudin class S of $\mathfrak{A}$ having at least three elements, n_1, n_2, n_3, *or* (b) there exist at least two Rudin classes $S = \{n_1, n_2, \cdots\}$ and S' of $\mathfrak{A}$ each having at least two elements.

Let $\mathfrak{B}$ be the closed subalgebra of $\mathbf{E}$ generated by $\mathfrak{A} \cup \{e_{n_1}\}$. Then $\mathfrak{B}$ contains $\mathfrak{A}$ and $\mathfrak{B} \neq \mathfrak{A}$ (because e_{n_1} belongs to $\mathfrak{B}$ but not to $\mathfrak{A}$, since $\hat{e}_{n_1}$ is not constant on $S \supset \{n_1, n_2\}$). Any element of $\mathfrak{B}$ is the limit in $\mathbf{E}$ of elements $\alpha e_{n_1} + f$, where α is a scalar and $f \in \mathfrak{A}$, and so each element of $\mathscr{F}\mathfrak{B}$ is constant on each of $S \backslash \{n_1\}$ and S'. By (a) or (b), at least one of these two sets has at least two elements, so that $\mathfrak{B} \neq \mathbf{E}$. This would contradict the maximality of $\mathfrak{A}$ and so negates both (a) and (b).

Thus at most one Rudin class of $\mathfrak{A}$ contains two elements, all others being singletons.

If all are singletons, (i) shows that $\mathfrak{A} = \mathfrak{A}_{n_0}$ for some integer n_0. Otherwise, there is just one Rudin class $S = \{n_1, n_2\}$ of $\mathfrak{A}$ having two elements, all others being singletons. But then $\mathfrak{A} \subset \mathfrak{A}_{n_1,n_2}$ and, since $\mathfrak{A}_{n_1,n_2}$ is a closed subalgebra of $\mathbf{E}$ different from $\mathbf{E}$, the assumed maximality of $\mathfrak{A}$ entails that $\mathfrak{A} = \mathfrak{A}_{n_1,n_2}$.

This completes the proof.

11.3.12. **Remarks.** Maximal subalgebras were first discussed in a different context by Wermer; see Exercise 11.25 and Wermer [1]. For a discussion of analogues of 11.3.11 for more general groups, see Chapter 9 of [R], the references cited there, and Liu [1].

Problems concerning generators of the Banach algebra $\mathbf{A}$ (see 10.6.1 and 11.4.17) have been studied in Newman, Schwartz and Shapiro [1].

Further reading: Greenleaf [2]; Reiter [2]; MR **38** # 486; 39 ## 4608, 6024; **40** # 7730; **41** ## 4138, 4139, 7730; **42** # 2254; **54** # 3298.

11.4 Banach Algebras and Their Applications

In this section we are going to deal with some of the rudiments of the Gelfand theory of commutative complex Banach algebras that possess identity elements, the principal aim being to apply this theory in such a way as to derive the spectral radius formula (of which some instances are established in other ways in Exercises 8.8 and 3.12 and which has been stated and used in Section 11.3) and the theorems of Wiener and Lévy mentioned in 10.6.3. A few other loose ends will also be tied up.

While our treatment is intentionally brief and highly selective, it may fairly claim to cover a few of the high spots of the Gelfand theory and its applications. Another introductory account will be found in Chapter 18 of [Ri]. The reader who wishes to pursue Banach algebra theory further may do so by consulting any desired selection of the following references: [B], [Bo_1], [N], [Lo], [Ri], [HS], [GRS], [Kz], [HR], [Mo]; see also the comments in 11.4.18.

Among the algebras $\mathbf{E}$ in which we are primarily interested are $\mathbf{E} = \mathbf{L}^p$ ($1 \leqslant p \leqslant \infty$) and $\mathbf{E} = \mathbf{C}^k$ (k a nonnegative integer). These do not possess identity elements, however, and it is technically advantageous to adjoin to $\mathbf{E}$ a formal identity element to obtain an enlarged algebra $\mathbf{B}_{\mathbf{E}}$ to which the general Gelfand theory is then applied.

Useful applications to other algebras will also be possible; see also Section 16.6.

11.4.1. Definitions and Examples. All the general developments and results we make and obtain will refer to a complex commutative Banach algebra $\mathbf{B}$ with an identity element. By this it is meant that

(a) $\mathbf{B}$ is an associative and commutative algebra over the complex field that possesses an identity (or unit) element e relative to multiplication;

(b) $\mathbf{B}$ is also a Banach space with a norm $\|\cdot\|$;

(c) one has $\|e\| = 1$ and

$$\|xy\| \leqslant \|x\| \cdot \|y\|$$

for any two elements x, y of $\mathbf{B}$.

From (c) it follows that

$$\|x^n\| \leqslant \|x\|^n \qquad (n = 1, 2, \cdots).$$

It will be convenient to define x^0 to be e.

We consider some examples.

(1) The examples of such Banach algebras to which our first applications of the forthcoming general theory will be made are the algebras $\mathbf{B}_{\mathbf{E}}$ obtained in the following way. If $\mathbf{E}$ denotes any one of $\mathbf{L}^p$ ($1 \leqslant p \leqslant \infty$) or $\mathbf{C}^k$ (k a nonnegative integer), then it is known already from Chapters 2 and 3 that $\mathbf{E}$

fulfils the conditions (a) to (c) above, save the parts referring to the existence of an identity element e. This defect is repaired by enlarging $\mathbf{E}$ into the algebra $\mathbf{B_E}$ whose elements are by definition ordered pairs (α, f), where α is a complex scalar and $f \in \mathbf{E}$. The algebraic operations and the norm in $\mathbf{B_E}$ are defined thus:

$$\begin{aligned}(\alpha, f) + (\beta, g) &= (\alpha + \beta, f + g),\\ \beta(\alpha, f) &= (\beta\alpha, \beta f),\\ (\alpha, f) \cdot (\beta, g) &= (\alpha\beta, \alpha g + \beta f + f * g),\\ \|(\alpha, f)\| &= |\alpha| + \|f\|_{\mathbf{E}},\end{aligned}$$

where α and β are complex scalars, f and g are elements of $\mathbf{E}$, and where $\|\cdot\|_{\mathbf{E}}$ denotes the appropriate norm in $\mathbf{E}$ (see 2.2.4). The mapping $f \to (0, f)$ imbeds $\mathbf{E}$ isometrically and isomorphically into $\mathbf{B_E}$. We leave to the reader the simple task of verifying that $\mathbf{B_E}$ does indeed satisfy conditions (a) to (c) above, the identity element e being $(1, 0)$. The passage from $\mathbf{E}$ to $\mathbf{B_E}$ is spoken of as that of *adjoining a formal identity element.*

We remark in passing that in the language to be introduced in Chapter 12, $\mathbf{B_E}$ can be identified algebraically with a set of Radon measures by means of the correspondence $(\alpha, f) \leftrightarrow \alpha\varepsilon + f$, where ε denotes the Dirac measure at the origin. However, unless $\mathbf{E} = \mathbf{L}^1$, this correspondence does not preserve norms. We shall nowhere in this chapter make use of this identification.

(2) Perhaps the simplest nontrivial type of Banach algebra which engages the interest of the functional analyst is the algebra $\mathbf{C}(S)$ of all continuous complex-valued functions on a compact Hausdorff topological space S, the algebraic operations being pointwise ($x + y$, αx and xy being the functions $s \to x(s) + y(s)$, $s \to \alpha x(s)$ and $s \to x(s)y(s)$, respectively) and the norm being the supremum (or maximum modulus) norm:

$$\|x\| = \sup\{|x(s)| : s \in S\}.$$

The identity element in $\mathbf{C}(S)$ is just the constant function 1.

Most of the problems we shall mention for Banach algebras in general admit rather transparent solutions for the algebras $\mathbf{C}(S)$. For this reason, the Gelfand theory is largely concerned with displaying to what extent a general algebra $\mathbf{B}$ is similar to an algebra $\mathbf{C}(S)$ for a suitably chosen S (which will depend upon $\mathbf{B}$). Further comment on this matter will be made in 11.4.18(1).

Although, as we have said, $\mathbf{C}(S)$ itself is rather simple, the same is far from true of various subalgebras of $\mathbf{C}(S)$ whose norms may or may not be obtained by restricting the above norm on $\mathbf{C}(S)$. One such subalgebra of $\mathbf{C}(R/2\pi Z)$ has been encountered in Section 10.6 and there christened $\mathbf{A}$; this algebra will be examined again in 11.4.17. For a survey of other closed subalgebras and the attendant problems, see the references cited in 11.4.18(4) and (5).

(3) Some further examples will appear in 11.4.17 and in Exercises 11.10, 11.12 and 11.15. Meanwhile we turn to some generalities.

11.4.2. **Inverses and Spectra.** An element x of a given algebra **B** is said to be *inversible* (in **B**) if and only if there exists an element y of **B** such that $xy = e$; there is then precisely one such element y of **B**, this element of **B** being termed the *inverse* of x and denoted by x^{-1}.

It is evident that e is inversible and coincides with its inverse. Moreover, if x and y are both inversible, then xy is inversible and $(xy)^{-1} = x^{-1}y^{-1}$ (recall that **B** is commutative by hypothesis). If x is inversible, we shall usually write x^{-n} for $(x^{-1})^n$, n being any positive integer.

The concept of inversibility is, as we shall see, central in all subsequent developments. Major steps in the theory amount simply to criteria for inversibility, fruitful instances of which appear in 11.4.6 and 11.4.10.

It is necessary to consider, along with a given x and its inversibility, the family of elements $x - \lambda e$ obtained when λ varies over all complex scalars. Given x, the scalars fall into two complementary sets: the set $\sigma(x)$ of scalars λ such that $x - \lambda e$ is not inversible, and the set $R(x)$ of scalars λ such that $x - \lambda e$ is inversible. The sets $\sigma(x)$ and $R(x)$ are termed, respectively, the *spectrum* and the *resolvent set* of x, the terminology being taken over from the so-called spectral theory of operators. The nonnegative real number [see (11.4.4)]

$$\rho(x) = \sup \{|\lambda| : \lambda \in \sigma(x)\} \tag{11.4.1}$$

is termed the *spectral radius* of x.

The next two results collect together a number of basic properties of the set of inversible elements, the spectrum, the resolvent set, and the spectral radius. They provide also a proof that $R(x)$ is nonvoid and open, that $\sigma(x)$ is nonvoid and compact, and that $(x - \lambda e)^{-1}$ depends analytically on $\lambda \in R(x)$. This last statement means that, for any continuous linear functional F on **B**, the complex-valued function ϕ defined by

$$\phi(\lambda) = F[(x - \lambda e)^{-1}]$$

is analytic in the ordinary sense on the open set $R(x)$. In brief, 11.4.3 and 11.4.4 contain the analytic heart of the Gelfand theory.

11.4.3. (1) If x is an inversible element of **B**, and if y is any element of **B** such that

$$\|y - x\| < \|x^{-1}\|^{-1},$$

then y is inversible and

$$y^{-1} = \sum_{n=0}^{\infty} x^{-n-1}(x - y)^n; \tag{11.4.2}$$

in particular, any $z \in \mathbf{B}$ satisfying $\|z - e\| < 1$ is inversible.

(2) The set **U** of inversible elements of **B** is a nonvoid open subset of **B** and the mapping $x \to x^{-1}$ is continuous from **U** into **B**; more precisely,

$$\|y^{-1} - x^{-1}\| \leqslant \|x^{-1}\| \cdot \{1 - \|x^{-1}\| \|y - x\|\}^{-1} \tag{11.4.3}$$

whenever $x \in \mathbf{U}$ and $\|y - x\| < \|x^{-1}\|^{-1}$.

Proof. (1) Define $s_n \in \mathbf{B}$ for $n = 1, 2, \cdots$ by

$$s_n = \sum_{k=0}^{n} x^{-k-1}(x - y)^k.$$

For $n > m$ one has

$$\|s_n - s_m\| \leqslant \sum_{m<k\leqslant n} \|x^{-1}\|^{k+1}\|y - x\|^k.$$

Since $\|y - x\| < \|x^{-1}\|^{-1}$, it appears that the sequence $(s_n)_{n=1}^{\infty}$ is a Cauchy sequence in **B**. By 11.4.1(b), therefore, $s = \lim_{n\to\infty} s_n$ exists in **B**. Now a direct calculation shows that

$$ys_n = e - x^{-n-1}(x - y)^{n+1},$$

whence it follows that $ys_n \to e$ as $n \to \infty$. On the other hand, 11.4.1(c) shows that multiplication is continuous in the pair of factors, so that from $s_n \to s$ follows $ys_n \to ys$. Thus ys must coincide with e, showing that y is inversible and that $y^{-1} = s$. This proves (11.4.2). The final statement in (1) ensues on taking $x = e$.

(2) That **U** is open and nonvoid follows at once from (1). Also, by (11.4.2),

$$y^{-1} - x^{-1} = \lim_{N\to\infty} \sum_{n=1}^{N} x^{-n-1}(x - y)^n.$$

Since

$$\Big\| \sum_{n=1}^{N} x^{-n-1}(x - y)^n \Big\| \leqslant \sum_{n=1}^{N} \|x^{-1}\|^{n+1}\|y - x\|^n,$$

(11.4.3) emerges on account of continuity of the norm.

11.4.4. (1) If $x \in \mathbf{B}$, then $R(x)$ is open and contains every complex number λ satisfying $|\lambda| > \|x\|$, $\sigma(x)$ is compact and nonvoid, and

$$\rho(x) \leqslant \|x\|. \tag{11.4.4}$$

(2) If $x \in \mathbf{B}$, then $(x - \lambda e)^{-1}$ depends analytically on $\lambda \in R(x)$ and tends to zero as $|\lambda| \to \infty$.

Proof. (1) Since $R(x)$ consists precisely of those scalars λ such that $x - \lambda e \in \mathbf{U}$, that $R(x)$ is open follows from 11.4.3(2) and the continuity of the mapping $\lambda \to x - \lambda e$. The complementary set $\sigma(x)$ is therefore closed. Also, $x - \lambda e = \lambda(\lambda^{-1}x - e)$ is, by 11.4.3(1), inversible whenever $\lambda \neq 0$ and $\|\lambda^{-1}x\| < 1$. Thus $\lambda \in R(x)$ whenever $|\lambda| > \|x\|$. This in turn entails that (11.4.4) holds. So $\sigma(x)$ is closed and bounded, and therefore compact. That $\sigma(x)$ is nonvoid will be established after the proof of (2) is finished.

(2) Take any continuous linear functional F on $\mathbf{B}$ and define the function ϕ on $R(x)$ by

$$\phi(\lambda) = F\{(x - \lambda e)^{-1}\}.$$

To see that ϕ is analytic on $R(x)$, we apply (11.4.2) on taking $\lambda_0 \in R(x)$ and replacing x by $x - \lambda_0 e$ and y by $x - \lambda e$: it then appears that

$$\phi(\lambda) = F\{\sum_{n=0}^{\infty} (x - \lambda_0 e)^{-n-1}(\lambda - \lambda_0)^n\},$$

for $|\lambda - \lambda_0| < \|(x - \lambda_0 e)^{-1}\|^{-1}$. Since F is continuous and linear, one has for such values of λ the power series expansion

$$\phi(\lambda) = \sum_{n=0}^{\infty} F\{(x - \lambda_0 e)^{-n-1}\}(\lambda - \lambda_0)^n,$$

which establishes the analytic character of ϕ. The relations

$$\lambda(x - \lambda e)^{-1} = (\lambda^{-1}x - e)^{-1}$$

and $\lambda^{-1}x - e \to -e$ as $|\lambda| \to \infty$ combine with 11.4.3(2) to show that

$$\|(x - \lambda e)^{-1}\| = O(|\lambda|^{-1}) \quad \text{as } |\lambda| \to \infty,$$

and the proof of (2) is finished.

Finally, let us return to complete the proof of (1) by showing that $\sigma(x)$ is never void. If $\sigma(x)$ were void, for any continuous linear functional F on $\mathbf{B}$, the function ϕ would be entire analytic. By Liouville's theorem, combined with the fact that $\phi(\lambda) \to 0$ as $|\lambda| \to \infty$, it would appear that ϕ is constantly zero. But then, by the Hahn-Banach theorem (I, B.5), $(x - \lambda e)^{-1}$ would be zero for all λ, which is evidently absurd. Thus $\sigma(x)$ must be nonvoid and the proof of 11.4.4 is complete.

11.4.5. **Ideals, Maximal and Otherwise.** We now turn to topics of a more algebraic nature.

By an *ideal* in $\mathbf{B}$ is meant a subset $\mathbf{I}$ of $\mathbf{B}$ which is a linear subspace and which is stable under multiplication, the latter clause meaning that $xy \in \mathbf{I}$ whenever $x \in \mathbf{I}$ and $y \in \mathbf{B}$. An ideal $\mathbf{I}$ in $\mathbf{B}$ is said to be *proper* if it does not exhaust $\mathbf{B}$; this is so if and only if $e \notin \mathbf{I}$, or again if and only if $\mathbf{I}$ contains no inversible element of $\mathbf{B}$.

For future use we observe that the closure in $\mathbf{B}$ of any ideal in $\mathbf{B}$ is again an ideal in $\mathbf{B}$.

Gelfand's theory lays special stress on the *maximal ideals* in $\mathbf{B}$, an ideal $\mathbf{m}$ being termed maximal if it is proper and if $\mathbf{m}$ and $\mathbf{B}$ are the only ideals in $\mathbf{B}$ which contain $\mathbf{m}$. The first step is to prove the existence of maximal ideals in $\mathbf{B}$.

To this end we consider any nonvoid set $\mathscr{F}$ of proper ideals in $\mathbf{B}$ having the property that any two members of $\mathscr{F}$ are contained in some one member of $\mathscr{F}$. Consider the union $\mathbf{J}$ of all members of $\mathscr{F}$. It is easy to check that $\mathbf{J}$

is an ideal in **B**, and (by showing that $e \notin \mathbf{J}$) that **J** is proper. Plainly, **J** contains each member of $\mathscr{F}$.

The preceding paragraph leads to this conclusion: if we partially order, by set-inclusion, the set of all proper ideals in **B**, then any linearly ordered subset thereof admits a supremum. As a consequence of this we may apply Zorn's lemma (see, for example [E], p. 6 or [HS], p. 14) to infer that:

(1) Any proper ideal in **B** is contained in some maximal ideal in **B**.

As we have seen, a proper ideal **I** in **B** can contain no inversible elements. By 11.4.3(1), therefore, **I** cannot be everywhere dense in **B**; the closure $\bar{\mathbf{I}}$ is thus a proper ideal in **B**. In particular, it follows that $\bar{\mathbf{m}}$ must coincide with **m** whenever **m** is a maximal ideal in **B**. Thus:

(2) Every maximal ideal in **B** is closed in **B**.

We can now state the first criterion of inversibility.

11.4.6. Let $x \in \mathbf{B}$. Then x is inversible in **B** if and only if x belongs to no maximal ideal in **B**.

Proof. If x is inversible, it can belong to no proper ideal (maximal or not) in **B**. Suppose on the other hand that x belongs to no maximal ideal. The set $\mathbf{I} = \{xy : y \in \mathbf{B}\}$, the *principal ideal* generated by x, cannot be proper: for otherwise 11.4.5(1) announces the existence of a maximal ideal **m** containing **I**, and **m** would then contain x. Thus $\mathbf{I} = \mathbf{B}$. In particular, $e \in \mathbf{I}$. This entails that x is inversible.

Subsequent applications of 11.4.6 depend upon setting up a close relationship between maximal ideals in **B** and complex homomorphisms of **B** (recall Exercise 4.1), and this is our next objective. The result we want appears as 11.4.10; the following intermediate results are directed to this end.

11.4.7. **Quotient Algebras.** Let **B** satisfy (as always) conditions (a) to (c) in 11.4.1, and let **I** be any proper closed ideal in **B**. The quotient set **B**/**I**, whose elements are the cosets $\dot{x} = x + \mathbf{I}$ modulo **I** of elements of **B**, can be formed into an algebra of the same type in a manner now to be described.

The algebraic operations and norm in **B**/**I** are defined thus (compare I, B.1.8):

$$\begin{aligned} \dot{x} + \dot{y} &= (x + y)^{\cdot}, \\ \alpha\dot{x} &= (\alpha x)^{\cdot}, \\ \dot{x}\dot{y} &= (xy)^{\cdot}, \\ \|\dot{x}\| &= \inf\{\|x + y\| : y \in \mathbf{I}\}. \end{aligned} \tag{11.4.5}$$

It is then very simple to verify that 11.4.1(a) is fulfilled, the identity element in **B**/**I** being $\dot{e}$. The only norm property which is not evident is the one asserting that $\|\dot{x}\| > 0$ whenever $\dot{x} \neq \dot{0}$. However, if $\|\dot{x}\| = 0$, there exist elements $y_n \in \mathbf{I}$ such that $\|x + y_n\| \to 0$ as $n \to \infty$. Since $-y_n \in \mathbf{I}$, and since $\mathbf{I}$ is closed, it follows that $x \in \mathbf{I}$ and therefore $\dot{x} = \dot{0}$.

Property 11.4.1(c) is plain to see, save perhaps the assertion concerning $\|\dot{e}\|$. As to that, (11.4.5) yields quite generally

$$\|\dot{x}\| \leqslant \|x\|. \tag{11.4.6}$$

Thus $\|\dot{e}\| \leqslant 1$. Were it the case that $\|\dot{e}\| < 1$, (11.4.5) would show that $\|e + y\| < 1$ for some $y \in \mathbf{I}$. But then $-y \in \mathbf{I}$ and, by 11.4.3(1), $-y$ is invertible. Since $\mathbf{I}$ is proper, this is a contradiction (see the outset of 11.4.5). Thus $\|\dot{e}\| = 1$.

Finally we verify 11.4.1(b) for $\mathbf{B}/\mathbf{I}$. Suppose that $(\dot{x}_n)$ is a sequence extracted from $\mathbf{B}/\mathbf{I}$ such that

$$\lim_{m,n \to \infty} \|\dot{x}_m - \dot{x}_n\| = 0; \tag{11.4.7}$$

we must show that this sequence is convergent in $\mathbf{B}/\mathbf{I}$ relative to the norm (11.4.5). Thanks to (11.4.7), a subsequence $(\dot{x}_{n_k})$ may be determined such that

$$\sum_{k=1}^{\infty} \|\dot{x}_{n_{k+1}} - \dot{x}_{n_k}\| < \infty. \tag{11.4.8}$$

Put $y_k = x_{n_{k+1}} - x_{n_k}$. By (11.4.5) and (11.4.8), elements z_k of $\mathbf{I}$ may be chosen so that

$$\sum_{k=1}^{\infty} \|y_k - z_k\| < \infty.$$

Then, by the assumed completeness of $\mathbf{B}$,

$$y = \sum_{k=1}^{\infty} (y_k - z_k)$$

exists as an element of $\mathbf{B}$, so that

$$\Big\|y - \sum_{k=1}^{r} y_k + \sum_{k=1}^{r} z_k\Big\| \to 0 \quad \text{as } r \to \infty. \tag{11.4.9}$$

From (11.4.9) and (11.4.5) it appears that

$$\Big\|\dot{y} - \sum_{k=1}^{r} \dot{y}_k\Big\| \to 0 \qquad \text{as } r \to \infty,$$

that is, that

$$\|\dot{y} - \dot{x}_{n_{r+1}} - \dot{x}_{n_1}\| \to 0 \qquad \text{as } r \to \infty. \tag{11.4.10}$$

This shows that the subsequence $(\dot{x}_{n_k})_{k=1}^{\infty}$ converges in $\mathbf{B}/\mathbf{I}$ to $\dot{x} = \dot{y} - \dot{x}_{n_1}$. But then (11.4.7) shows that the original sequence $(\dot{x}_n)$ is convergent in $\mathbf{B}/\mathbf{I}$ to the same limit.

This completes the verification that $\mathbf{B}/\mathbf{I}$ satisfies conditions (a) to (c) of 11.4.1 whenever $\mathbf{B}$ does so and $\mathbf{I}$ is a proper closed ideal in $\mathbf{B}$. The conclusion therefore applies when $\mathbf{I}$ is assumed to be a maximal ideal in $\mathbf{B}$ [see 11.4.5(2)]. However, as the next result shows, much more can be said in this case.

11.4.8. If $\mathbf{m}$ is a maximal ideal in $\mathbf{B}$, then to each $\dot{x} \in \mathbf{B}/\mathbf{m}$ corresponds a unique complex number ξ such that $\dot{x} = \xi\dot{e}$.

Proof. Granted the existence of ξ, its uniqueness is evident. On the other hand, if no such ξ existed, then $\dot{x} - \lambda\dot{e} \neq \dot{0}$ for all complex λ. For given λ, the set

$$\mathbf{I} = \{(x - \lambda e)y + m : y \in \mathbf{B}, m \in \mathbf{m}\}$$

is an ideal in $\mathbf{B}$ which contains $\mathbf{m}$ properly, $x - \lambda e$ belonging to $\mathbf{I}$ but not to $\mathbf{m}$. Maximality of $\mathbf{m}$ would imply that $\mathbf{I} = \mathbf{B}$, which in turn would imply that $\dot{x} - \lambda\dot{e}$ is inversible in $\mathbf{B}/\mathbf{m}$. This being true for any complex λ, it would appear that $\sigma(\dot{x})$ is void, which would contradict 11.4.4(1) applied to $\mathbf{B}/\mathbf{m}$. This proves 11.4.8.

Remark. The preceding argument really goes to show that if $\mathbf{B}$ is a *division algebra* (that is, is such that every nonzero element of $\mathbf{B}$ is inversible in $\mathbf{B}$), then $\mathbf{B}$ is isomorphic to the complex field (considered as an algebra over itself). This result, the *Gelfand-Mazur theorem*, admits various extensions and variants; see [R_1], pp. 354–355 and [Ri], pp. 37–40, 109–110.

11.4.9. **Maximal Ideals and Complex Homomorphisms.** Retaining the notations of 11.4.8, it is very simple to see that the mapping $\iota_{\mathbf{m}}: \dot{x} \to \xi$ is both an algebraic isomorphism and an isometry of $\mathbf{B}/\mathbf{m}$ onto the complex field. As a consequence the composite map $\gamma_{\mathbf{m}}$ of $\mathbf{B}$ defined by

$$\gamma_{\mathbf{m}}(x) = \iota_{\mathbf{m}}(\dot{x}) \tag{11.4.11}$$

proves to be a nontrivial continuous complex homomorphism of $\mathbf{B}$ whose kernel is exactly $\mathbf{m}$:

$$\left.\begin{aligned} \gamma_{\mathbf{m}}(x + y) &= \gamma_{\mathbf{m}}(x) + \gamma_{\mathbf{m}}(y), \\ \gamma_{\mathbf{m}}(\lambda x) &= \lambda \cdot \gamma_{\mathbf{m}}(x), \\ \gamma_{\mathbf{m}}(xy) &= \gamma_{\mathbf{m}}(x)\gamma_{\mathbf{m}}(y), \\ \gamma_{\mathbf{m}}(e) &= 1, \\ |\gamma_{\mathbf{m}}(x)| &\leqslant \|x\|, \end{aligned}\right\} \tag{11.4.12}$$

$$\mathbf{m} = \{x \in \mathbf{B} : \gamma_{\mathbf{m}}(x) = 0\}. \tag{11.4.13}$$

Conversely, if γ is a nontrivial complex homomorphism of $\mathbf{B}$, its kernel

$$\mathbf{m}_\gamma = \{x \in \mathbf{B} : \gamma(x) = 0\} \tag{11.4.14}$$

is easily verifiable to be a maximal ideal in $\mathbf{B}$. Moreover,

$$\gamma_{\mathbf{m}_\gamma} = \gamma, \qquad \mathbf{m}_{\gamma_{\mathbf{m}}} = \mathbf{m}; \tag{11.4.15}$$

the first relation combines with (11.4.12) to show that any complex homomorphism of $\mathbf{B}$ is necessarily continuous.

The relations (11.4.11) and (11.4.13) thus set up a one-to-one correspondence between the set of maximal ideals **m** in **B** and the set $\Gamma = \Gamma(\mathbf{B})$ of all nontrivial continuous complex homomorphisms of **B**. This correlation is one of the cornerstones of the Gelfand theory. The set $\Gamma(\mathbf{B})$ is termed the *Gelfand space* (or *representation space*) of **B**.

In viewing the symbol "$\gamma(x)$" we have so far thought of "x" as the variable. It is, however, also possible and useful to think of "γ" as the variable. In other words, with each $x \in \mathbf{B}$ one may associate the complex-valued function $\hat{x}$ on the set $\Gamma(\mathbf{B})$ defined by

$$\hat{x}(\gamma) = \gamma(x).$$

The function $\hat{x}$ is termed the *Gelfand transform* of x, and the mapping $x \to \hat{x}$ is referred to as the *Gelfand transformation*. (The notation is suggested by the circumstance that, when $\mathbf{B} = \mathbf{B}_{\mathrm{E}}$, the function $\hat{x}$ is very closely related to the Fourier transform; see 11.4.11.) The Gelfand transformation is an algebraic homomorphism of **B** into the algebra $\mathbf{B}(\Gamma)$ (with pointwise operations and supremum norm) of bounded complex-valued functions on Γ; compare 11.4.1(2). We shall return to this matter in 11.4.18(1).

Meanwhile we derive the promised crucial reformulation of 11.4.6 which is made possible by the substance of 11.4.8 and 11.4.9.

11.4.10. An element x of **B** is inversible if and only if $\gamma(x) \neq 0$ for all $\gamma \in \Gamma(\mathbf{B})$. More generally,

$$\sigma(x) = \{\gamma(x) : \gamma \in \Gamma(\mathbf{B})\}, \tag{11.4.16}$$

$$\rho(x) = \sup\{|\gamma(x)| : \gamma \in \Gamma(\mathbf{B})\}. \tag{11.4.17}$$

Proof. By 11.4.6, $\lambda \in \sigma(x)$ if and only if $x - \lambda e$ belongs to some maximal ideal **m** in **B**, that is, by (11.4.13), if and only if $\gamma_{\mathbf{m}}(x - \lambda e) = 0$ for some **m**. By (11.4.12), this is the case if and only if $\lambda = \gamma_{\mathbf{m}}(x)$ for some **m**. Since $\gamma_{\mathbf{m}}$ ranges over $\Gamma(\mathbf{B})$ when **m** ranges over all maximal ideals in **B**, (11.4.16) is established. The first assertion is a special case of (11.4.16), since x is inversible if and only if $0 \notin \sigma(x)$. Finally, (11.4.17) results on combining (11.4.16) and (11.4.1).

Remarks. From 11.4.10 it appears that if $x \in \mathbf{B}$ satisfies $\gamma(x) \neq 0$ for all $x \in \Gamma(\mathbf{B})$, then there exists $y \in \mathbf{B}$ satisfying $\gamma(y) = \gamma(x)^{-1}$ for all $\gamma \in \Gamma(\mathbf{B})$. An extension of this result is contained in Exercise 11.20.

Regarding (11.4.16), it may be noted incidentally that a continuous linear functional γ on **B** is multiplicative (that is, is a complex homomorphism of **B**) provided $\gamma(x) \in \sigma(x)$ for all $x \in \mathbf{B}$; see MR **37** # 4620.

11.4.11. **Example: The Algebra $\mathbf{C}(S)$.** The notation is as in 11.4.1(2). By adapting the hints attached to Exercise 11.16, the reader should experience no trouble in verifying that the maximal ideals in $\mathbf{C}(S)$ are precisely the sets

$$\mathbf{m}_s = \{x \in \mathbf{C}(S) : x(s) = 0\}$$

obtained when s varies over S; the corresponding homomorphism $\gamma_{\mathbf{m}_s}$ is defined by

$$\gamma_{\mathbf{m}_s}(x) = x(s).$$

The correspondence $s \leftrightarrow \mathbf{m}_s$ is one-to-one.

It is to be noted that this identification of the maximal ideals in $\mathbf{C}(S)$ depends on the elementary fact that the multiplicative inverse of a non-vanishing continuous function is continuous; and that this fact is seen a posteriori to be just what is asserted by the opening sentence of 11.4.10 for the algebra $\mathbf{B} = \mathbf{C}(S)$. Thus 11.4.10, when applied to $\mathbf{B} = \mathbf{C}(S)$, tells us nothing new.

Any disappointment the reader may feel because of this apparent anti-climax can be relieved by hurrying on to a more fruitful application of 11.4.10.

11.4.12. **Application to $\mathbf{B}_\mathbf{E}$.** Suppose that $\mathbf{E} = \mathbf{L}^p$ $(1 \leqslant p \leqslant \infty)$ or $\mathbf{C}^k$ (k a nonnegative integer). In order to apply 11.4.10 to $\mathbf{B}_\mathbf{E}$ (see 11.4.1(1)), we need to identify the elements of $\Gamma(\mathbf{B}_\mathbf{E})$. The essential step has already been carried out in Section 4.1.

Let $\gamma \in \Gamma(\mathbf{B}_\mathbf{E})$ and define γ' on $\mathbf{E}$ by

$$\gamma'(f) = \gamma((0, f)),$$

so that

$$\gamma((\alpha, f)) = \alpha + \gamma'(f) \tag{11.4.18}$$

for a general element (α, f) of $\mathbf{B}_\mathbf{E}$.

Two cases arise, namely:

(1) $\gamma' = 0$, in which case (11.4.18) reads

$$\gamma((\alpha, f)) = \alpha,$$

and we write $\gamma = \gamma_\infty$;

(2) $\gamma' \neq 0$, in which case 4.1.3 shows that there exists an integer $n \in Z$ such that

$$\gamma'(f) = \hat{f}(n)$$

for all $f \in \mathbf{E}$, and therefore [by (11.4.18) again]

$$\gamma((\alpha, f)) = \alpha + \hat{f}(n);$$

we denote this γ by γ_n.

Bearing in mind the Riemann-Lebesgue lemma 2.3.8, we see that $\Gamma(\mathbf{B}_\mathbf{E})$ may be identified with $Z \cup \{\infty\}$ in such a way that

$$\gamma_n((\alpha, f)) = \alpha + \hat{f}(n), \tag{11.4.19}$$

$\hat{f}(\infty)$ being interpreted as 0.

It is also worth observing at this point that, if γ' is any complex homomorphism of $\mathbf{E}$, the mapping $\gamma: (\alpha, f) \to \alpha + \gamma'(f)$ is a complex homomorphism of $\mathbf{B_E}$. By 11.4.9, γ is continuous. It follows at once that γ' is continuous and we have thus verified a remark made in 4.1.1, namely: any complex homomorphism of $\mathbf{E}$ is continuous.

A direct appeal to 11.4.10 now results in a conclusion which is the analogue of a theorem established by Wiener for $\mathbf{L}^1(R)$:

11.4.13. Suppose that $\mathbf{E}$ denotes $\mathbf{L}^p$ $(1 \leqslant p \leqslant \infty)$ or $\mathbf{C}^k$ (k a nonnegative integer). If $f \in \mathbf{E}$ and α is a complex number, and if

$$\alpha \neq 0, \qquad \alpha + \hat{f}(n) \neq 0 \qquad \text{for all } n \in Z, \tag{11.4.20}$$

then there exists $g \in \mathbf{E}$ such that

$$\hat{g}(n) = \frac{\hat{f}(n)}{\{\alpha + \hat{f}(n)\}} \qquad \text{for all } n \in Z, \tag{11.4.21}$$

Proof. Consider $x = (\alpha, f) \in \mathbf{B_E}$. In view of 11.4.12, the conditions (11.4.20) express precisely that $\gamma(x) \neq 0$ for each $\gamma \in \Gamma(\mathbf{B_E})$. According to 11.4.10, therefore, x is inversible in $\mathbf{B_E}$. Let $x^{-1} = (\beta, h)$, where β is a complex number and $h \in \mathbf{E}$. The properties (11.4.12) show that one has for all $\gamma \in \Gamma(\mathbf{B_E})$ the relation

$$\gamma((\beta, h)) = \frac{1}{\gamma((\alpha, f))}.$$

Using (11.4.19) and taking $\gamma = \gamma_\infty$, this relation gives $\beta = 1/\alpha$; taking $\gamma = \gamma_n$ $(n \in Z)$, it gives $\beta + \hat{h}(n) = 1/\{\alpha + \hat{f}(n)\}$ for all $n \in Z$. It appears therefore that

$$\begin{aligned} \hat{h}(n) &= \frac{1}{\{\alpha + \hat{f}(n)\}} - \frac{1}{\alpha} \\ &= \frac{-\alpha^{-1}\hat{f}(n)}{\{\alpha + \hat{f}(n)\}} \end{aligned}$$

for all $n \in Z$, and it suffices to take $g = -\alpha h$.

Remarks. (1) The results of 11.4.12 and 11.4.13 apply with other choices of $\mathbf{E}$: see 4.1.3(1) and Exercise 11.10.

(2) The case $\mathbf{E} = \mathbf{L}^1$ of 11.4.13 expresses the fact that $\phi/(\alpha + \phi) \in \mathbf{A}(Z)$ whenever $\phi \in \mathbf{A}(Z)$, $\alpha \neq 0$ is a complex number, and $\alpha + \phi(n) \neq 0$ for all $n \in Z$.

(3) Generalizations of 11.4.10 and 11.4.13 appear in 11.4.15 and 11.4.16, respectively. Our approach to these extensions is based upon a study of the spectral radius formula, which constitutes our next objective.

11.4.14. **The Spectral Radius Formula.** For an algebra **B** of the type specified in 11.4.1, this formula reads

$$\rho(x) = \sup_{\gamma \in \Gamma(\mathbf{B})} |\gamma(x)| = \lim_{n \to \infty} \|x^n\|^{1/n}. \tag{11.4.22}$$

When this is applied to the special case of an element $x = (0, f)$ of $\mathbf{B}_{\mathbf{E}}$, the formula takes the form

$$\|\hat{f}\|_{\infty} = \lim_{n \to \infty} \|f^{*n}\|_{\mathbf{E}}^{1/n},$$

which is the version encountered elsewhere in this book [see Exercises 8.8 and 3.12 and equation (11.3.1)].

Proof of (11.4.22). The first equality in (11.4.22) is just (11.4.17). Since $\gamma(x^n) = \gamma(x)^n$ for each x and each γ, this first equality entails that

$$\rho(x) = \rho(x^n)^{1/n}.$$

Accordingly, (11.4.4) shows that

$$\rho(x) \leqslant \liminf_{n \to \infty} \|x^n\|^{1/n}.$$

To establish (11.4.22) it will therefore suffice to show that

$$\limsup_{n \to \infty} \|x^n\|^{1/n} \leqslant \rho(x). \tag{11.4.23}$$

To this end, write $\rho = \rho(x)$ and fix any $\rho' > \rho$. By definition of ρ, $(e - \lambda x)^{-1}$ exists whenever $|\lambda| \leqslant 1/\rho'$. Choose any continuous linear functional F on **B**. Reference to 11.4.4(2) shows that

$$\phi(\lambda) = F\{(e - \lambda x)^{-1}\}$$

is analytic on some open set containing the disk $|\lambda| \leqslant 1/\rho'$.

On the other hand, a special case of (11.4.2) shows that

$$(e - \lambda x)^{-1} = \sum_{n=0}^{\infty} \lambda^n x^n$$

for $|\lambda| < \|x\|^{-1}$. Since F is linear and continuous, we have correspondingly the Taylor expansion

$$\phi(\lambda) = \sum_{n=0}^{\infty} F(x^n)\lambda^n \tag{11.4.24}$$

holding for $|\lambda| < \|x\|^{-1}$. On account of (11.4.24), the Cauchy integral formulae for ϕ, and the analyticity of ϕ on a neighborhood of the disk $|\lambda| \leqslant 1/\rho'$, it results that

$$F(x^n) = \frac{\{(d/d\lambda)^n \phi(\lambda)\}_{\lambda=0}}{n!} = O(\rho'^n) \tag{11.4.25}$$

as $n \to \infty$. The reader will notice that the derivation of (11.4.25) does not depend on knowing that (11.4.24) is valid for all values of λ satisfying

$|\lambda| \leqslant 1/\rho'$: this last statement is in fact true (see Exercise 11.14), but we are neither asserting nor using it here.

Armed with (11.4.25) we introduce the function p on $\mathbf{B}'$, the dual of the Banach space $\mathbf{B}$ (see I, B.1.7), defined by

$$p(F) = \sup_n \rho'^{-n} |F(x^n)|.$$

It is simple to verify that p is a seminorm on $\mathbf{B}'$ (see I, B.1.2), and that p is lower semicontinuous relative to the dual norm on $\mathbf{B}'$, with respect to which norm $\mathbf{B}'$ is a Banach space (I, B.1.7 again). The boundedness principle appearing in I, B.2.1 affirms that p is continuous on $\mathbf{B}'$, that is, that there exists a number $c = c(x)$ such that $p(F) \leqslant c \cdot \|F\|$ for all $F \in \mathbf{B}'$. Appeal to a corollary of the Hahn-Banach theorem (I, B.5.3) leads from this to the inequality

$$\rho'^{-n} \|x^n\| \leqslant c \qquad (n = 1, 2, \cdots),$$

which in turn shows at once that

$$\|x^n\|^{1/n} \leqslant \rho' c^{1/n} \qquad (n = 1, 2, \cdots).$$

From this it follows immediately that

$$\limsup_{n \to \infty} \|x^n\|^{1/n} \leqslant \rho'.$$

Since ρ' is freely chosen in excess of ρ, (11.4.23) follows and the proof of (11.4.22) is complete.

Remark. The spectral radius formula was discovered first by Beurling [1] for Fourier transforms of functions integrable over the additive group R of real numbers. The extension to Banach algebras is due to Gelfand [1]. Beurling's paper has had a profound influence on many subsequent developments in harmonic analysis.

11.4.15. **An Extension of 11.4.10.** We are now going to make fuller use of the analytic nature of inversion in order to derive the following extension of 11.4.10.

Suppose that $x \in \mathbf{B}$ and that Φ is a complex-valued function defined and analytic on some open set Ω containing $\sigma(x)$. Then there exists an element y of $\mathbf{B}$ such that

$$\gamma(y) = \Phi(\gamma(x)) \qquad (\gamma \in \Gamma(\mathbf{B})). \tag{11.4.26}$$

Proof. Let $\varepsilon > 0$ be the distance between $\sigma(x)$ and the frontier of Ω. By covering the compact set $\sigma(x)$ by a finite number of disks of radius $\varepsilon/4$ and then taking the frontier of the union of the concentric disks of radius

$\varepsilon/2$, one can obtain an oriented curve L (composed of a finite number of circular arcs) lying in $\Omega\backslash\sigma(x)$ and such that the Cauchy formula

$$\Phi(\lambda) = \frac{1}{2\pi i}\int_L \frac{\Phi(\zeta)\,d\zeta}{\zeta - \lambda} \tag{11.4.27}$$

holds for any $\lambda \in \sigma(x)$.

The crucial step now is to assign a meaning to the integral

$$y = \frac{1}{2\pi i}\int_L \Phi(\zeta)(\zeta e - x)^{-1}\,d\zeta, \tag{11.4.28}$$

which will make it an element of **B** having the property that

$$\gamma(y) = \frac{1}{2\pi i}\int_L \gamma\{\Phi(\zeta)(\zeta e - x)^{-1}\}\,d\zeta \tag{11.4.29}$$

for each $\gamma \in \Gamma(\mathbf{B})$. If this can be done, the properties (11.4.12) lead from (11.4.29) to

$$\gamma(y) = \frac{1}{2\pi i}\int_L \frac{\Phi(\zeta)\,d\zeta}{\{\zeta - \gamma(x)\}},$$

and then (11.4.27) and (11.4.16) will carry us straight to (11.4.26).

To define the integral (11.4.28) with the desired properties, we first partition L into its component arcs L_k $(k = 1, 2, \cdots)$, which are evidently finite in number. A little thought will show that it will suffice to define the integrals $\int_{L_k}$ as elements of **B** in such a way that they each possess the property corresponding to (11.4.29), and then set $\int_L = \sum_k \int_{L_k}$.

Now L_k can be parametrized by means of a continuously differentiable complex-valued function $t \to \zeta_k(t)$ defined on the interval $0 \leqslant t \leqslant 1$. The natural definition of $z_k = \int_{L_k}$ is then

$$\begin{aligned} z_k &= \int_0^1 \Phi(\zeta_k(t))\{\zeta_k(t)e - x\}^{-1}\zeta_k'(t)\,dt \\ &= \int_0^1 \Phi_k(t)\,dt, \end{aligned}$$

say, where

$$\Phi_k(t) = \Phi(\zeta_k(t))\{\zeta_k(t)e - x\}^{-1}\zeta_k'(t)$$

and the prime denotes differentiation, provided the existence of this integral can be ensured as an element of **B** and that the result of applying any $\gamma \in \Gamma(\mathbf{B})$ to z_k is obtainable by applying γ to the integrand $\Phi_k(t)$ followed by integration. However, by 11.4.3(2) and the choice of L, the integrand Φ_k is continuous from $[0, 1]$ into **B**. Accordingly, the obvious procedure to be used for defining z_k is the use of approximative Riemann sums, especially so since it is almost evident that this process, if effective in defining $z_k \in \mathbf{B}$, will certainly arrange that the desired property

$$\gamma(z_k) = \int_0^1 \gamma\{\Phi_k(t)\}\,dt$$

will in fact hold for any continuous linear functional γ on $\mathbf{B}$. That this procedure does indeed work satisfactorily, the reader is asked to verify in the manner proposed in Exercise 11.11. When this is done, the proof of 11.4.14 will be complete.

Remark. There is in existence a general theory of the integration of $\mathbf{B}$-valued functions of a much more general character than the ad hoc procedure suggested above; see, for example [E], Sections 8.14 to 8.20.

On applying this last result to the case in which $\mathbf{B} = \mathbf{B}_\mathbf{E}$ and referring to 11.4.12, we shall obtain the following extension of 11.4.13.

11.4.16. Suppose that $\mathbf{E}$ denotes $\mathbf{L}^p$ $(1 \leqslant p \leqslant \infty)$ or $\mathbf{C}^k$ (k a nonnegative integer), that $f \in \mathbf{E}$ and that Φ is a complex-valued function defined and analytic on some open set containing $\hat{f}(Z) \cup \{0\}$ and satisfying $\Phi(0) = 0$. Then there exists an element g of $\mathbf{E}$ such that

$$\hat{g}(n) = \Phi(\hat{f}(n)) \qquad (n \in Z). \tag{11.4.30}$$

Proof. Take $x = (0, f) \in \mathbf{B}_\mathbf{E}$. From 11.4.10 and 11.4.12 it appears without difficulty that $\sigma(x) = \hat{f}(Z) \cup \{0\}$. Applying 11.4.15, we obtain the existence of $y = (\beta, g)$ in $\mathbf{B}_\mathbf{E}$ such that (11.4.26) holds for all $\gamma \in \Gamma(\mathbf{B}_\mathbf{E})$. Referring again to 11.4.12 and taking $\gamma = \gamma_\infty$, we see that $\beta = \Phi(0) = 0$. On taking $\gamma = \gamma_n$, it then appears from (11.4.26) that (11.4.30) holds for each $n \in Z$.

Remark. On taking $\mathbf{E} = \mathbf{L}^1$, 11.4.16 entails that $\Phi \circ \phi \in \mathbf{A}(Z)$ whenever $\phi \in \mathbf{A}(Z)$, Φ is analytic on some neighborhood of $\phi(Z) \cup \{0\}$ and $\Phi(0) = 0$. The result dual to this (with the group Z replacing T) has been mentioned in 10.6.3 and will now be proved by applying 11.4.15 to the algebra $\mathbf{A} = \mathbf{A}(T)$.

11.4.17. **A as a Banach Algebra: the Theorems of Wiener and Lévy.** We recall from 2.5.3 and Section 10.6 that $\mathbf{A} = \mathbf{A}(T)$ denotes the set of continuous complex-valued functions f on T which have absolutely convergent Fourier series. As we then saw, $\mathbf{A}$ is a Banach algebra under pointwise operations, the norm being

$$\|f\|_\mathbf{A} = \sum_{n \in Z} |\hat{f}(n)|.$$

The conditions (a) to (c) of 11.4.1 are satisfied by $\mathbf{A}$, the identity element being the constant function 1.

Let us identify the elements of $\Gamma(\mathbf{A})$, that is, let us determine an expression for a general nonzero continuous complex homomorphism γ of $\mathbf{A}$ (compare

Exercise 4.7). To this end, write u for the element e_1 of **A**. Then for any element f of **A** one may write

$$f = \sum_{n \in Z} \hat{f}(n) u^n, \tag{11.4.31}$$

the series being convergent in **A** because $\|f\|_{\mathbf{A}} < \infty$. Evidently, u is an inversible element of A and

$$\|u\|_{\mathbf{A}} = \|u^{-1}\|_{\mathbf{A}} = 1.$$

The properties (11.4.12) show that $|\gamma(u)| \leqslant 1$ and $|\gamma(u^{-1})| = |\gamma(u)|^{-1} \leqslant 1$, so that $|\gamma(u)| = 1$. Accordingly, $\gamma(u) = e^{ix}$ holds for precisely one $x \in R/2\pi Z$. From (11.4.31) we then obtain via (11.4.12)

$$\begin{aligned} \gamma(f) &= \sum_{n \in Z} \hat{f}(n)\gamma(u^n) = \sum_{n \in Z} \hat{f}(n)\gamma(u)^n \\ &= \sum_{n \in Z} \hat{f}(n) e^{inx} = f(x). \end{aligned}$$

Conversely and trivially, given $x \in T$, the mapping γ_x of **A** defined by

$$\gamma_x(f) = f(x) \tag{11.4.32}$$

is evidently a nonzero continuous complex homomorphism of **A**. Thus $\Gamma(\mathbf{A})$ consists precisely of the maps γ_x obtained when x varies over T (or, if we wish, over all real numbers); the Gelfand space $\Gamma(\mathbf{A})$ can be identified with T.

From 11.4.10 we may now read off the theorem of Wiener: if $f \in \mathbf{A}$ satisfies $f(x) \neq 0$ for all real x, then $1/f \in \mathbf{A}$. ([Wi], p. 91.)

Likewise, from 11.4.15 we may read off Lévy's extension of Wiener's theorem, namely: if $f \in \mathbf{A}$, and if Φ is defined and analytic on some open set containing $f(T)$, then $\Phi \circ f \in \mathbf{A}$.

These results were stated without proof in 10.6.3. There are extensions to other interesting algebras; see Exercises 11.15 and 11.20 below and also MR **38** # 485; **41** # 5864; **51** # # 1255, 8728. In 10.6.3 we also remarked on partial converses of such theorems of Wiener-Lévy type; for instances see MR **53** # # 14017, 14018; **54** # # 858, 5747.

11.4.18. **Pointers to Further Developments.** We have carried our excursion into the theory of Banach algebras as far as is needed for our primary applications; it remains only to indicate a few of the further possible developments.

(1) It has been seen in 11.4.9 that the Gelfand transformation $x \to \hat{x}$ is an algebraic homomorphism of **B** into the algebra $\mathbf{B}(\Gamma)$ (with pointwise operations and supremum norm) of bounded complex-valued functions on $\Gamma = \Gamma(\mathbf{B})$. From the spectral radius formula (11.4.22), the Gelfand transformation is seen to be one-to-one if and only if $x = 0$ is the only element of **B** for which

$$\lim_{n \to \infty} \|x^n\|^{1/n} = 0. \tag{11.4.33}$$

Elements x of **B** which satisfy (11.4.33) are termed *generalized* (or *topological*) *nilpotents* (compare Exercise 3.12). The spectral radius formula shows that x is a generalized nilpotent if and only if it belongs to every maximal ideal. If in **B** there exist no generalized nilpotents other than 0, **B** is sometimes said to be *semisimple*; in this case the Gelfand transformation provides an algebraically faithful representation of **B**.

Unhappily, there are quite reasonable algebras **B** that are far from being semisimple; for an example, see Exercise 11.12. As Exercise 11.13 shows, such algebras may still be of interest to the analyst.

Even if **B** is semisimple, the Gelfand transformation is, in general, not an isometry; in general, $\rho(x) = \sup |\hat{x}|$ is a strictly weaker norm than $\|x\|$, that is, there exist in general sequences (x_i) of elements of **B** such that $\rho(x_i) \to 0$ but $\|x_i\| \nrightarrow 0$. Much effort has been expended (with considerable success) in determining categories of algebras **B** for which these norms are identical (or at least equivalent); and applications provide ample reward for these labors.

So far, although we have spoken of $\Gamma = \Gamma(\mathbf{B})$ as a "space," it has not been endowed with any topology. The standard way of topologizing Γ makes it into a compact Hausdorff space on which each of the Gelfand transforms $\hat{x}$ is continuous. In this topology a base of neighborhoods of a point γ_0 of Γ is formed of the sets

$$W(F, \varepsilon) = \{\gamma \in \Gamma : |\gamma(x) - \gamma_0(x)| < \varepsilon \quad \text{for } x \in F\},$$

where F ranges over finite subsets of **B** and ε over positive numbers. Then the Gelfand transformation is an algebraic homomorphism of **B** into the algebra (with pointwise operations) $\mathbf{C}(\Gamma)$ of continuous complex-valued functions on Γ. To determine conditions under which the image of **B** covers the whole of $\mathbf{C}(\Gamma)$ is again a fruitful problem which has been attacked with success (see Exercise 11.24; [N], p. 230; [Ri], p. 190; [Lo], pp. 78, 90–91) and which has many rewarding applications, two of which are mentioned in (2) and (3). There is a sense in which in any case the image of **B** contains all analytic functions on Γ.

Concerning the method just described for topologizing $\Gamma = \Gamma(\mathbf{B})$, one question will (or should) arise in the reader's mind, namely: what happens if $\mathbf{B} = \mathbf{C}(S)$ and S is a compact Hausdorff space (see 11.4.1(2))? He will doubtless expect that there should be a close relationship between $\Gamma = \Gamma(\mathbf{B})$ and S. Well, as has appeared in 11.4.11, $\Gamma = \Gamma(\mathbf{B}) = \Gamma(\mathbf{C}(S))$ can be identified set-theoretically with S via the correspondence $s \leftrightarrow \gamma_s \equiv \gamma_{\mathbf{m}_s}$, and this in such a way that

$$\hat{x}(\gamma_s) = x(s).$$

It is not at all difficult to show that, once this identification is made, the topology on Γ described above is identical with the initial topology on S. For certain noncompact Hausdorff spaces S, $\Gamma(\mathbf{C}(S))$ proves to be homeomorphic with the Stone-Čech compactification of S.

It is within the framework of the Gelfand transform that one finds the Banach algebra-based approach to extensions of the Bochner representation theorem mentioned in Section 9.4. This approach is very largely due to the Russian mathematicians Gelfand and Raikov, who employed it as a basis for harmonic analysis. For details, see [N], pp. 404–425 or [Lo], Chapter VII.

The study of versions of the Weierstrass-Stone theorem (mentioned in Subsection 6.2.3) applicable to compact differentiable manifolds is influenced markedly by the Gelfand theory; see Freeman [1] and the references cited there. It should perhaps be remarked that parts of the theory of Banach algebras have been extended to more general topological algebras; see [Mi], [C], Neubauer [1], [2], Allen [1], [2], Waelbroeck [1], [2], Benedetto [1].

(2) Another of the applications of the developments mentioned under (1) arises when **B** is a certain type of commutative algebra of continuous normal endomorphisms of a Hilbert space, the result being in this case a novel approach to the simultaneous spectral resolution theorem for commutative sets of such endomorphisms. See [N], p. 248; [Lo], pp. 92–95.

(3) A second application is to commutative harmonic analysis, **B** being taken to be $\mathbf{L}^1(G)$ with a formal identity adjoined (as in 11.4.1(1)). This has resulted in an almost autonomous approach to the Fourier inversion theorem and Bochner's theorem (see Chapter 9). For accounts of this application, see [N], Chapter VI (especially pp. 404 ff.); [Ri], pp. 325 ff.; and [Lo], Chapter VII. See also Helson [6].

An effective treatment of other problems in harmonic analysis necessitates the treatment of algebras **B** which, like $\mathbf{L}^1(G)$ in the general case, are deprived of an identity element, and this in cases where the adjunction of a formal identity serves no useful purpose. The treatment of such problems has led to abstract versions of the Tauberian theorem mentioned in Subsection 11.2.4.

In such an algebra **B**, the nontrivial complex homomorphisms γ of **B** retain their fundamental significance, but now they correspond with those maximal ideals **m** in **B** which are *regular* or *modular* (that is, which possess "relative identities" $e \in \mathbf{B}$ having the property that $ex - x \in \mathbf{m}$ for all $x \in \mathbf{B}$; compare Exercise 4.1). The correspondence $\gamma \leftrightarrow \mathbf{m}$ remains exactly as described in Subsection 11.4.9. The Gelfand space $\Gamma(\mathbf{B})$ and the Gelfand transformation $x \to \hat{x}$ are also defined as described in Subsection 11.4.9, but now $\Gamma(\mathbf{B})$, when topologized in the fashion described in (1) immediately above, proves to be a locally compact Hausdorff space which is in general noncompact. However, there may be maximal ideals in **B** which are not modular; see Exercises 11.26 and 11.27.

Certain algebras **B** of this type appear to form a natural setting for abstract Tauberian theorems which include Wiener's theorem applying when $\mathbf{B} = \mathbf{L}^1(G)$ (see Subsection 11.2.4). Thus it is known (see, for example, [Lo], p. 85 and [HR], (39.27)) that, under certain conditions which we do not specify here, any proper closed ideal **I** in **B** is contained in some modular maximal ideal in **B**; in other words, a closed ideal **I** in **B** exhausts **B** if and only if it is annulled by no $\gamma \in \Gamma(\mathbf{B})$. This statement is the abstract Tauberian theorem: it includes the special case $Z_{\mathbf{I}} = \varnothing$ of 11.2.1, as well as the extensions thereof mentioned in Subsections 11.2.3 and 11.2.4 and associated with Wiener's name. (As Loomis has wryly remarked, this abstract version is a Tauberian theorem in almost perfect disguise.) Tauberian theorems have been considered in still more general contexts; see, for example, Benedetto [1] (where, however, the exposition is such as to demand a good deal of care on the reader's part). See also MR **39** # 779.

(4) The problems spoken of in 11.2.3 for the case in which $\mathbf{E} = \mathbf{L}^1(G)$ with G noncompact can and have been viewed from the point of view of Banach algebra theory. In this context the question might be formulated thus: given a Banach algebra (as in 11.4.1) and a closed ideal **I** in **B**, is **I** expressible as the intersection of a suitable set of maximal ideals in **B**?

There are nontrivial examples of algebras **B** for which the answer is "Yes": in fact, every algebra of the type $\mathbf{C}(S)$ (see 11.4.1(2)) has this property. (One proof of this is suggested in Exercise 11.17; a different proof appears in 10.4.6 of [E].) The cases mentioned in 11.2.3 are ones in which the answer has turned out to be "No," but only after doubts persisted for a considerable time despite very close attention.

A more fruitful guide is provided by other examples in which the answer is transparently "No" and where the cause of failure is easier to detect. A typical such example is provided by the algebra $\mathbf{B} = \mathbf{C}^1(K)$ formed of those complex-valued functions defined and continuously differentiable on the compact interval $K = [0, 1]$, the algebraic operations being pointwise and the norm being defined by

$$\|f\| = \sup_{x \in K} |f(x)| + \sup_{x \in K} |Df(x)|.$$

The maximal ideals in this algebra are just the sets

$$\mathbf{m}_x = \{f \in \mathbf{C}^1(K) : f(x) = 0\}$$

obtained when x varies over K; see Exercise 11.16. On the other hand, for each $x \in K$ the set

$$\mathbf{I}_x = \{f \in \mathbf{C}^1(K) : f(x) = Df(x) = 0\}$$

is a closed ideal in $\mathbf{C}^1(K)$, and it is evident that $\mathbf{I}_x$ is not expressible as the intersection of any set of maximal ideals $\mathbf{m}_y$.

In the above example the ideal $\mathbf{I}_x$ is *primary*, in the sense that it is contained in just one maximal ideal (to wit, $\mathbf{m}_x$), and the example suggests that it may be fruitful to reformulate the general question by asking whether ideals can always be expressed as intersections of primary ideals (which may or may not themselves be required to be closed).

To this question the answer is known to be "Yes" for a number of algebras of differentiable functions in one or several variables; see [Ri], pp. 300–302 and the references there cited; see also [Ho], Chapters 6 and 10; Srinivasan and Wang [2]; MR **37** # 1997. In $\mathbf{L}^1(G)$, where the group G is noncompact, the answer is still "No," if only *closed* primary ideals are to be admitted. [This is a consequence of Malliavin's work, taken in conjunction with Wiener's Tauberian theorem (see 11.2.3) and a theorem of Kaplansky asserting that any closed primary ideal in $\mathbf{L}^1(G)$ is necessarily of the form $\{f \in \mathbf{L}^1(G) : \hat{f}(\xi) = 0\}$ for some ξ in the character group of G; see, for example, [Bo_1], p. 144, Corollaire. Kaplansky's theorem is itself a special case of known properties of spectral synthesis sets; see 11.2.3 again.]

Most of the general work concerning primary decompositions is due to G. E. Silov; for an account of this, see Mirkil [1]. See also Glaeser [1]; MR **33** # 3053; **37** # # 1897, 3361; **50** # 10689.

Beurling [2] has introduced a class of convolution algebras and has analyzed certain of these algebras in case the underlying group is R. His work shows, in particular, that the closed ideals in these algebras are characterized entirely by the set of common zeros of the Fourier transforms of their elements (compare 11.2.3). The study of these algebras has been continued by Igari [1].

The reader versed in abstract algebra will recognize that the questions under discussion are suggested quite naturally by the relatively elementary primary decomposition theory for ideals in Noetherian rings (although there the term "primary" is usually defined in a different way). Of course, one cannot expect to take this purely algebraic theory over to Banach algebras without important changes; almost all Banach algebras of lasting interest to the analyst contain a multitude of ideals that are not finitely generated.

(5) In 11.4.1(2) we made passing mention of various subalgebras of $\mathbf{C}(S)$. Among these appears the algebra $\mathbf{A}$ of 11.4.17 and its relatives (the Beurling algebras mentioned in Exercise 11.15) and the algebras of differentiable functions mentioned in (4) immediately above (for details of which the reader has been referred to Mirkil [1]; see also [Ma]).

In addition to these "real variables" examples, there are similar algebras having their roots in complex variables theory which present many fascinating problems. For a brief introduction to these the reader is referred to the survey article of Wermer [1].

(6) The reader is recommended to examine P. J. Cohen's Banach algebra-based approach to factorization theorems (which have been otherwise treated in Section 7.5) and its subsequent developments (Cohen [4], Hewitt [1], Curtis and Figà-Talamanca [1], and Bryant [1]).

In Hewitt's formulation (which is closely akin to [HR], (32.22)) one is concerned with the situation in which $\mathbf{B} = \{x, y, \cdots\}$ is a real (respectively, complex) Banach algebra, which need not be commutative and which may fail to possess an identity element, together with a real (respectively, complex) Banach space $\mathbf{L} = \{f, g, \cdots\}$. It is assumed that there is given a mapping $(x, f) \to x \cdot f$ of $\mathbf{B} \times \mathbf{L}$ into $\mathbf{L}$ such that the following conditions (a) to (d) are fulfilled:

(a) $(x, f) \to x \cdot f$ is linear in $x \in \mathbf{B}$ for any fixed $f \in \mathbf{L}$;

(b) $(xy) \cdot f = x \cdot (y \cdot f)$ for $x, y \in \mathbf{B}$ and $f \in \mathbf{L}$;

(c) $\|x \cdot f\|_{\mathbf{L}} \leqslant c \cdot \|x\|_{\mathbf{B}} \cdot \|f\|_{\mathbf{L}}$ for $x \in \mathbf{B}$ and $f \in \mathbf{L}$, c being independent of x and f;

(d) there exists a number $d > 0$ such that for any finite subset $\{x_1, \cdots, x_m\}$ of $\mathbf{B}$, any $f \in \mathbf{L}$ and any $\varepsilon > 0$, there exists $y \in \mathbf{B}$ such that $\|y\|_{\mathbf{B}} \leqslant d$, $\|yx_j - x_j\|_{\mathbf{B}} < \varepsilon$ $(j = 1, 2, \cdots, m)$ and $\|y \cdot f - f\|_{\mathbf{L}} < \varepsilon$.

Hewitt's conclusion is that to each $f \in \mathbf{L}$ and each $\varepsilon > 0$ correspond elements $x \in \mathbf{B}$ and $g \in \mathbf{L}$ satisfying the following four conditions:

(e) $f = x \cdot g$;

(f) g belongs to the closure in $\mathbf{L}$ of the set $\{y \cdot f : y \in \mathbf{B}\}$;

(g) $\|g - f\|_{\mathbf{L}} \leqslant \varepsilon$;

(h) $\|x\|_{\mathbf{B}} \leqslant d$.

Further reading on factorization: [HR], §32; [DW]; MR **36** # 5714; **37** # 2003; **39** # # 1982, 3311; **40** # # 703, 4779; **46** # 16032; **49** # 9537; **53** # 1171.

(7) Various group algebras of vector-valued functions have been considered; see MR **37** # # 2001, 2002.

(8) A good deal of attention has been directed towards homogeneous Banach spaces and algebras of functions (the term "homogeneous" being either as in 6.1.2 or in a slightly generalised sense) and the closely-related Segal algebras. For accounts of some of this work, see [Re], [Re_1], [Wa], [War], [Go_1], Burnham [1], [2], [3], Burnham and Goldberg [1], de Leeuw [2].

The reader should derive pleasure from verifying that this factorization theorem leads to those in Section 7.5, if **B** and **L** are suitably chosen sets of functions and "·" is interpreted as convolution.

EXERCISES

11.1. Define **V** to be the set of $f \in \mathbf{L}^1$ such that

$$\|s_N f - f\|_1 \to 0 \qquad \text{as } N \to \infty.$$

Is **V** an invariant subspace of $\mathbf{L}^1$? Is it an ideal in $\mathbf{L}^1$? Is it closed in $\mathbf{L}^1$? Justify your answers.

11.2. Suppose that $1 \leqslant p < \infty$, and that $(\lambda_N)_{N=1}^{\infty}$ is a given sequence of positive numbers converging to infinity with N. Define **V** to be the set of $f \in \mathbf{L}^p$ such that

$$\lambda_N \|s_N f - f\|_p \to 0 \qquad \text{as } N \to \infty.$$

Give (justified) answers to the questions posed in the preceding exercise, with $\mathbf{L}^p$ in place of $\mathbf{L}^1$ throughout.

Do likewise for the case in which $s_N f$ is replaced by $\sigma_N f$ in the definition of **V**.

Note: It follows from Exercise 10.2 that the relation $\|f - s_N f\|_1 = O(1)$ is false for a general $f \in \mathbf{L}^1$. It is true (but the proof does not appear until 12.10.1) that $\|f - s_N f\|_p = o(1)$ when $1 < p < \infty$ and $f \in \mathbf{L}^p$.

11.3. Suppose that **E** is **C** or $\mathbf{L}^p$ $(1 \leqslant p \leqslant \infty)$, and that $(k_i)_{i \in I}$ is an arbitrary family of functions in $\mathbf{L}^1$. Show that the set of solutions $f \in \mathbf{E}$ of the equations $k_i * f = 0$ $(i \in I)$ is a closed ideal in **E**.

11.4. Suppose that **E** is **C** or $\mathbf{L}^p$ $(1 \leqslant p < \infty)$, and that $f \in \mathbf{E}$. Let **V** be the linear subspace of **E** generated by the translates of f. Show that **V** is closed in **E** if and only if f is a trigonometric polynomial.

Consider also the analogous problem arising when **V** is replaced by the principal ideal **I** in **E** generated by f (that is, **I** is the set of functions $f * g$ obtained when g ranges over **E**).

Note: Assume the known result (due to F. Riesz; see [E], p. 65) that any normed linear space having a compact neighborhood of zero is finite-dimensional.

Hints: Supposing that $\mathbf{V}$ is closed in $\mathbf{E}$, we consider the set $\mathbf{V}_n$ formed of all $g \in \mathbf{V}$ expressible in the form $g = \sum_{k=1}^{n} \alpha_k \cdot T_{a_k} f$, where $\sum_{k=1}^{n} |\alpha_k| \leqslant n$ and $a_1, \cdots, a_n \in T$. By using the category theorem (I, A.3) show that some $\mathbf{V}_n$ is a relative neighborhood of zero in $\mathbf{V}$, and hence that $\mathbf{V}$ has a compact neighborhood of zero. Apply Riesz's theorem cited above and 11.1.1.

11.5. Let $f \in \mathbf{E}$ and $\varepsilon > 0$ be given. Show that it is possible to choose a trigonometric polynomial P such that $\hat{P}(0) = 1$ and

$$\|f - P * f\| < \varepsilon.$$

Putting d for the degree of P, and supposing that N is an integer exceeding d, we deduce that

$$\left\|N^{-1} \sum_{k=0}^{N-1} T_{(2\pi k/N)} f - \hat{f}(0)1\right\| < \varepsilon.$$

Remark. This leads to another proof of 11.1.1; the argument is due to Salem (see $[Z_1]$, pp. 180–181).

11.6. Supply the details of the proof of the statement in 11.3.7.

11.7. Let $\mathfrak{A}$ be a closed subalgebra of $\mathbf{L}^p$, where $1 \leqslant p \leqslant 2$. Suppose that $f \in \mathbf{L}^q \cap \mathbf{L}^1_{r(\mathfrak{A})}$ for some $q > 1$. Assuming 13.5.1(1), prove that

$$f^{*N} \in \mathfrak{A}$$

for any integer N not less than $q/(2q - 2)$.

11.8. Suppose that $\mathbf{E}$ denotes $\mathbf{L}^p$ $(1 \leqslant p < \infty)$ or $\mathbf{C}$, that $f, g \in \mathbf{E}$, and that $\hat{f}(n) = 0$ for $n \in Z$ and $n < 0$. Let A be any set of real numbers having strictly positive Lebesgue measure. Show that if g belongs to the smallest closed ideal in $\mathbf{E}$ containing f, then g is the limit in $\mathbf{E}$ of finite linear combinations of those translates $T_a f$ of f corresponding to points $a \in A$.

Hints: Use the Hahn-Banach theorem (I, B.5) in conjunction with Exercise 8.15.

11.9. State and prove an analogue of the preceding exercise applying to the case in which the hypothesis

$$\hat{f}(n) = 0 \qquad \text{for } n \in Z \text{ and } n < 0$$

is replaced by the assumption that

$$\hat{f}(n) = O(e^{-\varepsilon|n|}) \qquad \text{for } n \in Z, \quad |n| \to \infty$$

for some $\varepsilon > 0$.

11.10. State and prove the analogue of 11.4.13 for the case in which $\mathbf{E} = \mathbf{H}^p$ $(1 \leqslant p \leqslant \infty)$, defined as in Exercise 3.9. (For more about $\mathbf{H}^p$ spaces, see MR **56** # # 6263, 6264.)

11.11. Suppose that **B** is a Banach space and that Φ is a continuous **B**-valued function on $[0, 1]$. By following the procedure suggested immediately below, show how to define the integral

$$z = \int_0^1 \Phi(t)\,dt$$

as an element of **B** with the property that

$$\gamma(z) = \int_0^1 \gamma(\Phi(t))\,dt$$

for each continuous linear functional γ on **B**.

Suggestions: Consider partitions $\Delta : 0 = t_0 < t_1 \cdots < t_n = 1$ of $[0, 1]$ with "division points" t_i, and associate with each Δ a Riemann sum

$$z_\Delta = \sum_{i=1}^{n} \Phi(t_i)(t_i - t_{i-1}).$$

A partition Δ' is a basic refinement of Δ if Δ' is obtained from Δ by inserting just one new division point; a (general) refinement Δ' of Δ is obtained as the result of a finite sequence of basic refinements, starting from Δ. Put $|\Delta| = \max\,(t_i - t_{i-1})$ and

$$\omega(\delta) = \sup\{\|\Phi(t) - \Phi(t')\| : 0 \leqslant t, t' \leqslant 1, |t - t'| \leqslant \delta\},$$

so that $\omega(\delta) \to 0$ with δ. Verify that

$$\|z_\Delta - z_{\Delta'}\| \leqslant \omega(|\Delta|)(t' - t_{i-1})$$

if Δ' is a basic refinement of Δ obtained by inserting a division point t' in the interval (t_{i-1}, t_i). Deduce that

$$\|z_\Delta - z_{\Delta'}\| \leqslant 2\omega(\delta)$$

for any two partitions Δ and Δ' satisfying $|\Delta| \leqslant \delta$ and $|\Delta'| \leqslant \delta$. Show finally that there exists $z \in \mathbf{B}$ with the property: given $\varepsilon > 0$, there exists $\delta = \delta(\varepsilon) > 0$ such that $\|z - z_\Delta\| < \varepsilon$ for all partitions Δ satisfying $|\Delta| \leqslant \delta$, and check that this z satisfies all requirements.

11.12. Let a be a positive real number and let **B** denote $\mathbf{L}^1(0, a)$ with its usual norm and linear space structure. As the product in **B** take the *truncated convolution*

$$f * g(t) = \int_0^t f(t - s)g(s)\,ds.$$

Verify that **B** is thus made into a Banach algebra satisfying conditions (a) to (c) of 11.4.1, save the parts referring to the existence and properties of an identity element.

Show that there exist no nontrivial continuous complex homomorphisms of **B**.

Hints: Let γ be a continuous complex homomorphism of **B** and put $c_n = \gamma(u_n)$ for $n = 1, 2, \cdots$, where u_n is that element of B defined by $u_n(t) = t^{n-1}$. Verify that

$$c_m c_n = \frac{\Gamma(m)\Gamma(n)c_{m+n}}{\Gamma(m+n)}.$$

Using continuity of γ, conclude that $c_n = 0$ for all n, and thence that $\gamma = 0$.

11.13. Prove that if $f \in \mathbf{L}^1(0, a)$ and $*$ denotes truncated convolution, as in the preceding exercise, then

$$\lim_{n \to \infty} \|f^{*n}\|_1 = 0,$$

where

$$f^{*1} = f \quad \text{and} \quad f^{*(n+1)} = f * f^{*n} \qquad \text{for } n = 1, 2, \cdots.$$

Remark. This is a basic result in the Mikusiński operational calculus; see [Er], p. 46.

Hint: Apply the spectral radius formula to the algebra obtained by adjoining a formal identity element to the algebra **B** described in the preceding exercise.

11.14. Prove that the formula

$$(e - \lambda x)^{-1} = \sum_{n=0}^{\infty} \lambda^n x^n$$

makes sense and is valid for $|\lambda| < 1/\rho(x)$, the notations and hypotheses being as in 11.4.14.

Hint: Use the spectral radius formula to show that both sides depend analytically on λ for $|\lambda| < 1/\rho(x)$.

11.15. Explore as far as you are able the possibility of analogues of 11.4.17 for the case in which **A** is replaced by the set $\mathbf{A}_W$ of continuous functions f for which

$$\sum_{n \in Z} W(n)|\hat{f}(n)| < \infty,$$

where W is a positive "weight function" defined on Z.

Remark. As with many other things in harmonic analysis, the algebras $\mathbf{A}_W$ originated in the work (much of it unpublished) of Beurling. An account of these so-called *Beurling algebras*, and of their extensions relating to more general groups is to be found in Domar [1]. See also [War], Chapter 2.

The algebra

$$\mathbf{A}^+ = \{f \in \mathbf{A} : \hat{f}(n) = 0 \quad \text{for all } n \in Z \text{ such that } n < 0\}$$

is also of interest. See [Kah_2], Chapter XI.

11.16. Suppose that $\mathbf{C}^1(K)$ is defined as in 11.4.18(4). Prove that if **m** is a maximal ideal in $\mathbf{C}^1(K)$, then

$$\mathbf{m} = \mathbf{m}_x \equiv \{f \in \mathbf{C}^1(K) : f(x) = 0\}$$

for some $x \in K$.

Hints: If $\mathbf{m}$ were distinct from every $\mathbf{m}_x$, to each $x \in K$ would correspond $f_x \in \mathbf{m}$ satisfying $f_x \notin \mathbf{m}_x$. Show that then there would exist points $x_1, \cdots, x_n$ of K such that

$$f = \sum_{k=1}^{n} f_{x_k} \cdot \bar{f}_{x_k}$$

is inversible in $\mathbf{C}^1(K)$ and so derive a contradiction.

11.17. Let $\mathbf{C}(S)$ be as in 11.4.1(2) and 11.4.11, and let $\mathbf{I}$ be a closed ideal in $\mathbf{C}(S)$. Define

$$F = \{s \in S : x(s) = 0 \quad \text{for all } x \in \mathbf{I}\}.$$

Prove that

$$\mathbf{I} = \mathbf{I}_F \equiv \{y \in \mathbf{C}(S) : y(s) = 0 \quad \text{for all } s \in F\}.$$

Hints: It is trivial that $\mathbf{I} \subset \mathbf{I}_F$. To prove the reverse inclusion, show first that $\mathbf{I}$ contains every $y \in \mathbf{C}(S)$ which vanishes on some open subset U of S which contains F. Do this by using compactness of S and the fact that to each $s \in S \backslash U$ corresponds $x_s \in \mathbf{I}$ which is nonvanishing on some neighborhood N_s of s. Cover $S \backslash U$ by suitably chosen neighborhoods $N_{s_1}, \cdots, N_{s_n}$ and consider the function equal to $x^{-1}y$ on $\bigcup_{k=1}^{n} N_{s_k}$ and to zero elsewhere, where $x = \sum_{k=1}^{n} x_{s_k} \bar{x}_{s_k}$.

Complete the proof by using Urysohn's lemma (see [E], 0.2.12 and 0.2.17(4); [HS], p. 75; [R$_1$], p. 39.) to show that any element of $\mathbf{C}(S)$ which vanishes on F is the uniform limit of functions in $\mathbf{C}(S)$, each of which vanishes on some open subset of S which contains F.

11.18. Let $\mathbf{A}$ be as in 11.4.17. Suppose that (U_α) is a family of open subsets of T forming a covering of T. Show that there exists a finite sequence $(u_k)_{k=1}^{r}$ of elements of $\mathbf{A}$ with the following properties:

(1) to each k corresponds an index $\alpha = \alpha_k$ such that $\operatorname{supp} u_k \subset U_\alpha$;

(2) $\sum_{k=1}^{r} u_k = 1$;

(3) $u_k \geqslant 0$.

In (1), $\operatorname{supp} u_k$ denotes the *support* of u_k, that is, the closure of the set of points of T at which $u_k \neq 0$.

Remark. A sequence (u_k) with the first two properties above is said to form a *partition of unity in* $\mathbf{A}$ *subordinate to the covering* (U_α). As the next exercise illustrates, the existence of such partitions of unity is a useful tool in passing from local to global assertions. It is quite simple to give explicit examples of partitions of unity (u_k) in $\mathbf{A}$ (see, for example, [Ba$_2$], p. 188), but the proof hinted at here has a much wider range of applicability. Compare Exercise 12.28.

Hints: For each $x \in T$ choose a neighborhood V_x of x and an index α_x such that $\bar{V}_x \subset U_{\alpha_x}$ and then a nonnegative $g_x \in \mathbf{A}$ such that $g_x(x) \neq 0$ and $g_x(y) = 0$ for all $y \in T \backslash V_x$. Show that $x_k \in T$ may be chosen so that $h = \sum_{k=1}^{r} g_{x_k}$ is nonvanishing and use 11.4.17.

11.19. Let **A** be as in 11.4.17 and let **I** be an ideal in **A**. Suppose that f is a function on T which *belongs locally to* **I** in the sense that to each $x \in T$ correspond an open neighborhood U_x of x and a function $f_x \in \mathbf{I}$ such that $f = f_x$ on U_x. Prove that $f \in \mathbf{I}$.

Remark. This result is due to Wiener, at least for the case in which $\mathbf{I} = \mathbf{A}$; see 10.6.2(6).

Hint: Use a partition of unity in **A** subordinate to the covering (U_x).

11.20. Suppose that **B** is as in 11.4.1 and that S is a subset of $\Gamma = \Gamma(\mathbf{B})$ with the property that to each $\gamma_0 \in \Gamma \backslash S$ corresponds a (possibly γ_0-dependent) $z \in \mathbf{B}$ such that $\gamma_0(z) \neq 0$ and $\gamma(z) = 0$ for all $\gamma \in S$.

Prove that, if $x \in \mathbf{B}$ and $\gamma(x) \neq 0$ for all $\gamma \in S$, then there exists $y \in \mathbf{B}$, satisfying $\gamma(y) = \gamma(x)^{-1}$ for all $\gamma \in S$.

Interpret this result when (1) $\mathbf{B} = \mathbf{B}_E$, as in 11.4.1 and (2) $\mathbf{B} = \mathbf{A}$, as in 11.4.17.

Hints: Introduce the ideal

$$\mathbf{I} = \{x \in \mathbf{B} : \gamma(x) = 0 \quad \text{for all } \gamma \in S\},$$

form the quotient algebra $\mathbf{B}/\mathbf{I}$ and determine all the complex homomorphisms of $\mathbf{B}/\mathbf{I}$.

11.21. Show that if $f \in \mathbf{L}^\infty$ has the property that the set of translates $\{T_a f : a \in R\}$ is a separable subset of $\mathbf{L}^\infty$, then f is equal almost everywhere to a continuous function. (Compare with Exercise 3.5.)

Hints: Use Exercises 3.16 and 3.5.

11.22. Let $f \in \mathbf{L}^\infty$. Show that the smallest closed ideal $\mathbf{I}_f$ in the convolution algebra $\mathbf{L}^\infty$ (taken with its usual norm) which contains f is identical with the set of elements of the form $\lambda f + g$, where λ is a complex number and g belongs to the closure in $\mathbf{L}^\infty$ of $f * \mathbf{L}^\infty$.

Deduce that $\mathbf{I}_f$ is translation-invariant if and only if f is equal almost everywhere to a continuous function.

Hints: For the first part, observe that $\mathbf{I}_f$ is the closure in $\mathbf{L}^\infty$ of the set of elements $\lambda f + h$, where λ is a complex number and $h \in f * \mathbf{L}^\infty$. Then consider separately two cases according to whether f is or is not equal almost everywhere to a continuous function.

For the second part, use the preceding exercise.

11.23. Let $(K_N)_{N=1}^\infty$ be an approximate identity in $\mathbf{L}^1$ (see 3.2.1), and let **S** denote the set of $f \in \mathbf{L}^\infty$ such that

$$\lim_{N \to \infty} \frac{1}{2\pi} \int K_N f \, dx$$

exists. Show that **S** is a closed ideal in the convolution algebra $\mathbf{L}^\infty$ and that **S** is not translation-invariant.

11.24. Let **B** be as in 11.4.1. Assume further that there exists a mapping $x \to x^*$ of **B** into itself such that $(x^*)^\wedge = (\hat{x})^-$, $(xy)^* = x^*y^*$, and $x^{**} = x$

for all $x, y \in \mathbf{B}$. Suppose finally that *either*

(1) there exists a number $c \geqslant 0$ such that

$$\|x\|^2 \leqslant c\|x^2\| \qquad \text{for each } x \in \mathbf{B},$$

or (2) there exists a number $c \geqslant 0$ such that

$$\|x\|^2 \leqslant c\|xx^*\| \qquad \text{for each } x \in \mathbf{B}.$$

Prove that the Gelfand transformation $x \to \hat{x}$ maps $\mathbf{B}$ one-to-one and bicontinuously onto $\mathfrak{C}(\Gamma)$, where $\Gamma = \Gamma(\mathbf{B})$. (If $c = 1$, the Gelfand transformation is also an isometry of $\mathbf{B}$ onto $\mathfrak{C}(\Gamma)$.)

Hints: First make acquaintance with the Weierstrass-Stone theorem, for which see [HS], pp. 94–98, or [E], Section 4.10.

Assuming condition (1), use 11.4.14 to show that

$$\|x\| \leqslant c\|\hat{x}\|_\infty \tag{3}$$

for each $x \in \mathbf{B}$. Assuming (2), establish equation (3) first for those $x \in \mathbf{B}$ which are self-adjoint in the sense that $x = x^*$. Then, observing that xx^* is always self-adjoint, derive equation (3) again for each $x \in \mathbf{B}$.

Topologize Γ as described in 11.4.18(1) and apply the Weierstrass-Stone theorem to show that the Gelfand transform maps $\mathbf{B}$ onto an everywhere dense subset of $\mathfrak{C}(\Gamma)$. Combine this with equation (3) to achieve the desired aim.

11.25. Let $\mathfrak{A}$ be (algebraically) a subalgebra of $\mathfrak{C}(T)$ containing $\mathbf{T}$, all algebraic operations being pointwise. Suppose that $\mathfrak{A}$ is a Banach algebra with respect to a norm (not necessarily that induced by the usual norm on $\mathfrak{C}(T)$), that $\mathbf{T}$ is everywhere dense in $\mathfrak{A}$, and that each element of $\Gamma(\mathfrak{A})$ is an evaluation map $f \to f(x)$ $(x \in T)$. Prove that

$$\mathfrak{A}^+ \equiv \{f \in \mathfrak{A} : \hat{f}(n) = 0 \quad \text{for } n \in Z, n < 0\}$$

is a maximal subalgebra of $\mathfrak{A}$.

Remarks. The case in which $\mathfrak{A}$ is identical with $\mathfrak{C}(T)$ (algebraically and topologically) is a prototype result due to Wermer; see 11.3.12. It is easy to reformulate the result in terms of convolution algebras over Z; in this form the result bears upon the problems dual to those handled by 11.3.11.

Hints: First show that $\mathfrak{A}^+$ is a closed subalgebra of $\mathfrak{A}$. Suppose that $\mathfrak{A}^+ \subset \mathfrak{B} \subset \mathfrak{A}$, $\mathfrak{B} \neq \mathfrak{A}$, where $\mathfrak{B}$ is a closed subalgebra of $\mathfrak{A}$. Show that there exists an integer $s > 0$ such that $\gamma(e_s) = 0$ for some $\gamma \in \Gamma(\mathfrak{B})$. Noting that, if $n > 0$ is an integer, one has $mn > s$ for some integer $m > 0$, deduce that

$$\gamma(e_n) = 0 \qquad (n \in Z, n > 0). \tag{1}$$

Show (compare the hints to Exercise 12.26 and recall (11.4.22)) that there exists a positive measure μ such that $\gamma(f) = \mu(f)$ for $f \in \mathfrak{B}$. Using (1), deduce that $\mu = 1$ (see the closing remarks in 12.2.3) and that

$$\gamma(f) = \hat{f}(0) \qquad (f \in \mathfrak{B}). \tag{2}$$

Observing that $fe_n \in \mathfrak{B}$ whenever $f \in \mathfrak{B}$ and $n \in Z$ and $n > 0$, deduce from (1) and (2) that $\mathfrak{B} \subset \mathfrak{A}^+$.

11.26. Suppose that $\mathbf{B}$ is as in 11.4.1, save that the existence of an identity element of $\mathbf{B}$ is *not* stipulated. Let $\mathbf{m}$ denote a maximal ideal in $\mathbf{B}$, $\mathbf{Q}$ the quotient algebra $\mathbf{B}/\mathbf{m}$, and $\mathbf{K}$ the set of $\dot{x} \in \mathbf{Q}$ such that $\dot{x}\dot{y} = \dot{0}$ for all $\dot{y} \in \mathbf{Q}$, where, for any $z \in \mathbf{B}$, $\dot{z}$ denotes the coset modulo $\mathbf{m}$ containing z.

Show that $\mathbf{Q}$ has no ideals other than $\{\dot{0}\}$ and $\mathbf{Q}$, so that, in particular, $\mathbf{K}$ is either $\{\dot{0}\}$ or $\mathbf{Q}$.

Prove that the following four statements are equivalent:

(a_1) $\mathbf{m}$ is modular;

(a_2) $\mathbf{Q}$ has an identity element;

(a_3) $\mathbf{K} = \{\dot{0}\}$;

(a_4) $\mathbf{m}$ is the kernel of some (nonzero) continuous complex homomorphism γ of $\mathbf{B}$.

Prove also the equivalence of the following four statements:

(b_1) $\mathbf{m}$ is nonmodular;

(b_2) $\mathbf{K} = \mathbf{Q}$;

(b_3) $\mathbf{m} \supset \mathbf{B}^2 = \{xy : x \in \mathbf{B}, y \in \mathbf{B}\}$;

(b_4) $\mathbf{m}$ is the kernel of some (nonzero) linear functional λ on $\mathbf{B}$ satisfying $\lambda(\mathbf{B}^2) = \{0\}$.

Conclude that every maximal ideal in $\mathbf{B}$ is modular if and only if the linear subspace of $\mathbf{B}$ generated by $\mathbf{B}^2$ is the whole of $\mathbf{B}$.

Hints: The equivalence of statements (a_1) and (a_2) is trivial, as also are the implications (a_2) $\Rightarrow$ (a_3) and (a_4) $\Rightarrow$ (a_1). To prove that (a_3) implies (a_2), show first that $\dot{x}\mathbf{Q} = \mathbf{Q}$ for any $\dot{x} \neq \dot{0}$ in $\mathbf{Q}$, and deduce that $\mathbf{Q}$ has no zero divisors. Choose any $\dot{a} \neq \dot{0}$ in $\mathbf{Q}$ and $\dot{e} \in \mathbf{Q}$ such that $\dot{a}\dot{e} = \dot{a}$; prove that $\dot{e}$ is an identity element in $\mathbf{Q}$.

It remains to prove that (a_2) implies (a_4). Since $\mathbf{Q}$ has no ideals other than $\{\dot{0}\}$ and $\mathbf{Q}$, (a_2) shows that every nonzero element of $\mathbf{Q}$ is inversible. Also, (a_1) implies that $\mathbf{m}$ is closed in $\mathbf{B}$: to see this, let e be an identity modulo $\mathbf{m}$ and show, by consideration of the element $y = -\sum_{n=1}^{\infty}(e - x)^n$, which satisfies $e \equiv x - xy \bmod \mathbf{m}$, that $\mathbf{m}$ contains no $x \in \mathbf{B}$ satisfying $\|e - x\| < 1$. Thus $\mathbf{Q}$ is complete and therefore satisfies all the conditions of 11.4.1, and 11.4.4(1) entails that each $\dot{x} \in \mathbf{Q}$ is uniquely expressible in the form $\dot{x} = \xi\dot{e}$, where ξ is a suitable complex number. Consider the mapping $\gamma : x \to \dot{x} \to \xi$.

The equivalence of (b_1) and (b_2) comes from that of (a_1) and (a_3); and it is trivial that (b_2) is equivalent to (b_3). If (b_4) holds, it is easy to verify that $\mathbf{m}$ is a maximal ideal in $\mathbf{B}$ satisfying (b_3). To show that (b_3) implies (b_4), choose any $x_0 \in \mathbf{B}\backslash\mathbf{m}$ and consider the set $\mathbf{I} = \{\alpha x_0 + m : \alpha \text{ a complex number}, m \in \mathbf{m}\}$.

11.27. Show that every maximal ideal in $\mathbf{L}^1$ is modular, but that there exist nonmodular maximal ideals in $\mathbf{L}^2$.

Hints: Use the preceding exercise, together with 7.5.1, 8.2.1, and 8.3.1.

Remark. That there exist nonmodular maximal ideals in $\mathbf{L}^p$ whenever $p > 1$ follows likewise from Exercise 13.20 below. On the other hand, it follows from (b_3) that every *closed* maximal ideal in $\mathbf{L}^p$ is modular. See also [HR], (38.23) and (39.41) and MR **40** # 4779.

CHAPTER 12

Distributions and Measures

In this lengthy chapter we are going to initiate the investigation of one way of handling and accounting for trigonometric series

$$\sum_{n \in Z} c_n e^{inx} \tag{12.1}$$

in which the coefficients c_n are subject merely to a relatively mild restriction on their rate of growth, namely,

$$c_n = O(|n|^k) \qquad (|n| \to \infty) \tag{12.2}$$

for some k which may vary from one series to the next. Such sequences $(c_n)_{n \in Z}$, and the corresponding series (12.1), will be said to be *tempered* or *temperate*.

Virtually all the trigonometric series considered in the classical theory referred to in Chapter 1 are tempered, the said classical theory being concerned mainly with the pointwise convergence or summability (everywhere or almost everywhere) of such series, and with the relationship between the given series and the Fourier-Lebesgue series of the sum-function whenever the latter is integrable.

The approach adopted in this chapter initially throws overboard all questions of pointwise convergence or summability in favor of a concept of convergence suggested by formula (D) in 1.3.2. The sum of the series will, as a result, no longer be an ordinary function at all, but rather an entity of the type now termed a *distribution* or a *generalized function* as introduced by Laurent Schwartz.

It will appear in 12.5.3 that any tempered trigonometric series converges in this new sense to such a distribution, in terms of which the coefficients c_n are expressible in a fashion that is an exact extension of the Fourier formulae (1.1.2*) for the case in which the sum is an ordinary integrable function. It will thus be natural to speak of the corresponding series as the *Fourier-Schwartz series* of its sum-distribution. To express the situation slightly differently, one may say that the theory of distributions provides one way of defining the Fourier transform $\hat{\phi}$ for any tempered function ϕ on the group Z, a problem that was already raised in Section 2.5. The related matters mentioned in Sections 3.4 and 6.7 will accordingly be seen in sharper focus.

Some of the operations performable on functions can be extended to distributions: this is notably the case with differentiation and convolution (see Sections 12.4 and 12.6). For what cannot be done, see 12.3.5. Indeed, distributions theory is usually approached for not-necessarily-periodic functions of one or more real variables, and the gay abandon enjoyed in differentiating distributions might almost be said to be their raison d'être. We are here concerned solely with what may be termed periodic distributions in one variable (distributions on T) and lay special stress on their connections with trigonometric series. A more balanced approach is to be found in the books by Schwartz [S] and those by Gelfand and Šilov [GS] and Gelfand and Vilenkin [GV], where distributions on the line R or on the product groups R^m, or on suitable subsets thereof, are the primary objects of study. Somewhat more leisurely accounts appear in [Ga], [MT], [Er], [Hal], [Lig], [Br], [Tr], [J]. See also [D].

The use of distributions will prove to be helpful in the discussion of certain questions existing in the classical theory of Fourier series—for example, in the study of so-called *conjugate series* (see Section 12.8).

In framing the definition of distributions, it is helpful to bear in mind two pointers:

(1) In view of 2.3.4 and (12.2) we may expect that a tempered trigonometric series should be correlated with the result of repeated differentiation, in some generalized sense, of a suitable function.

(2) If u is a sufficiently smooth function, and if (12.1) is the Fourier series of a function $f \in \mathbf{L}^1$, then (see 6.2.5 and 6.2.6)

$$\frac{1}{2\pi}\int f(x)u(x)\,dx = \sum_{n\in Z} c_n \cdot \hat{u}(-n); \tag{12.3}$$

in particular, this formula certainly holds for each $u \in \mathbf{C}^\infty$. It is vital to observe that the left-hand side of (12.3) defines a continuous linear functional on $\mathbf{C}^\infty$; and that, as follows from 2.4.1, knowledge of this linear functional determines the function f almost everywhere.

From this last remark we shall take our cue, a return to (1) being made via Section 12.4 and 12.5.7. Distributions will be introduced as continuous linear functionals on $\mathbf{C}^\infty$, but it is first of all necessary to consider the function-space $\mathbf{C}^\infty$ more closely.

The measures referred to in the title of this chapter constitute an especially important class of distributions; they are defined in 12.2.3 and studied in more detail in 12.5.10 and Section 12.7. The Fourier-Schwartz series of a measure is often termed a *Fourier-Stieltjes* series, the reasons for the name being discussed in 12.5.10.

As has been indicated, our approach to distributions is the analogue (for the periodic case) of that originally set forth by L. Schwartz (see [S]) for distributions on R^n; the reader is recommended to consult this reference, and/or [GS], frequently. A theory of distributions over an arbitrary locally compact Abelian group has been expounded by J. Riss [1]; as one might expect, this theory exhibits some rather weird features when the underlying group is locally very non-Euclidean. See also Reid [1]; MR **25** # 4354; **49** # # 11145, 11243; **51** # 11022. For references to even broader extensions of the theory, see [E], 5.11.5.

12.1 Concerning $\mathbf{C}^\infty$

In 2.2.4 we have defined the space $\mathbf{C}^\infty$ and its topology. Thus, if $u \in \mathbf{C}^\infty$ and if $(u_k)_{k=1}^\infty$ is a sequence extracted from $\mathbf{C}^\infty$, we shall write

$$\mathbf{C}^\infty - \lim_{k\to\infty} u_k = u \qquad \text{or} \qquad u_k \to u \text{ in } \mathbf{C}^\infty$$

if and only if any one (hence all) of the following three equivalent conditions is(are) fulfilled:

$$\lim_{k\to\infty} \|D^p u_k - D^p u\|_\infty = 0 \qquad (p = 0, 1, 2, \cdots); \tag{12.1.1}$$

$$\lim_{k\to\infty} \|u_k - u\|_{(p)} = 0 \qquad (p = 0, 1, 2, \cdots); \tag{12.1.2}$$

$$\lim_{k\to\infty} \|u_k - u\|_{(\infty)} = 0. \tag{12.1.3}$$

The equivalence of (12.1.1) and (12.1.2) is visible after reference to the defining formula (2.2.16); that of (12.1.2) and (12.1.3) depends on the defining formula (2.2.17) and a simple argument, which the reader is urged to supply.

A most important instance of this mode of convergence figures in the following result.

12.1.1. A continuous function u belongs to $\mathbf{C}^\infty$ if and only if

$$\lim_{|n|\to\infty} n^k \cdot \hat{u}(n) = 0 \qquad (k = 1, 2, \cdots), \tag{12.1.4}$$

in which case

$$u = \mathbf{C}^\infty - \lim_{N\to\infty} \sum_{|n|\leqslant N} \hat{u}(n)e^{inx}. \tag{12.1.5}$$

Proof. If $u \in \mathbf{C}^\infty$, the relation (12.1.4) follows from repeated use of 2.3.4, coupled with 2.3.2 (or with 2.3.8). Conversely, if u is continuous and (12.1.4) holds, then 2.4.3 shows that

$$u(x) = \sum_{n\in Z} \hat{u}(n)e^{inx},$$

and that this series, as well as those obtained from it by repeated termwise differentiations, are uniformly convergent. A well-known theorem in real analysis implies that the sum-function u therefore belongs to $\mathbf{C}^\infty$ and that

$$D^p u(x) = \sum_{n \in Z} (in)^p \hat{u}(n) e^{inx}$$

for $p = 0, 1, 2, \cdots$; see the proof of 12.1.3 below.

The final assertion appears from what has just been said concerning uniform convergence of the above series, coupled with the criteria for convergence in $\mathbf{C}^\infty$ expressed in (12.1.1).

Remark. If u is an integrable function satisfying (12.1.4), then u is equal almost everywhere to a function in $\mathbf{C}^\infty$ (see 2.4.2); and conversely.

12.1.2. It will cause the reader no pain to verify that if $u_k \to u$ and $v_k \to v$ in $\mathbf{C}^\infty$, and if (λ_k) is any sequence of scalars converging to λ, then $\lambda_k u_k \to \lambda u$ and $u_k + v_k \to u + v$ in $\mathbf{C}^\infty$.

12.1.3. That $\mathbf{C}^\infty$ is complete for the metric appearing in (12.1.3) will be vital in some of our subsequent arguments.

To establish this, suppose that (u_k) is a Cauchy sequence of elements of $\mathbf{C}^\infty$. Reference to (12.1.1), combined with Cauchy's general principle of convergence, shows that then $v_p = \lim_{k \to \infty} D^p u_k$ exists uniformly for each $p = 0, 1, 2, \cdots$. The limit function v_p is continuous. Now

$$D^p u_k(x) - D^p u_k(x') = \int_{x'}^{x} D^{p+1} u_k(y)\, dy$$

and uniform convergence yields in the limit

$$v_p(x) - v_p(x') = \int_{x'}^{x} v_{p+1}(y)\, dy,$$

which shows that $Dv_p = v_{p+1}$. Putting $v = v_0$, it appears thence that $v \in \mathbf{C}^\infty$ and that $v_p = D^p v$ $(p = 0, 1, 2, \cdots)$. Accordingly, $\lim_{k \to \infty} D^p u_k = D^p v$ uniformly for each p. In other words, $u_k \to v$ in $\mathbf{C}^\infty$. This shows that $\mathbf{C}^\infty$ is indeed complete.

12.1.4. From 12.1.2 and 12.1.3 we see that $\mathbf{C}^\infty$ is at once a linear space and a complete metric space, and that the linear operations are continuous.

Moreover, the topology of $\mathbf{C}^\infty$ is definable in terms of the seminorms $\|\cdot\|_{(p)}$ $(p = 0, 1, 2, \cdots)$, a base of neighborhoods of 0 in $\mathbf{C}^\infty$ consisting of the sets

$$\{u \in \mathbf{C}^\infty : \|u\|_{(p)} < \varepsilon\}$$

obtained when p ranges over the nonnegative integers and ε over the positive members. In other words (see I, B.1.3), C^∞ is a topological linear space of the type now customarily known as a *Fréchet space.*

12.1.5. Not only are the linear space operations continuous on $\mathbf{C}^\infty$, so too is the operation of pointwise multiplication. That is, the mapping $(u, v) \to uv$ is continuous from $\mathbf{C}^\infty \times \mathbf{C}^\infty$ into $\mathbf{C}^\infty$.

In addition, D (the differentiation operator) is a continuous endomorphism of $\mathbf{C}^\infty$.

Likewise, each translation operator T_a (see 2.2.1) is a continuous endomorphism of $\mathbf{C}^\infty$.

12.2 Definition and Examples of Distributions and Measures

12.2.1. **Definition of Distributions; the Space D.** By a *distribution* is meant a continuous linear functional on $\mathbf{C}^\infty$.

Henceforth we shall always denote by $\mathbf{D}$ the set of distributions. Since $\mathbf{D}$ is the set of continuous linear functionals on a topological linear space; it carries a natural linear space structure: if $F_1, F_2 \in \mathbf{D}$ and λ is a scalar, $F_1 + F_2$ and λF are the functionals defined by

$$(F_1 + F_2)(u) = F_1(u) + F_2(u), \qquad (\lambda F)(u) = \lambda \cdot F(u)$$

for $u \in \mathbf{C}^\infty$; compare I, B.1.7.

12.2.2. **Functions as Distributions.** The formula

$$F(u) = \frac{1}{2\pi} \int f(x)u(x)\,dx \tag{12.2.1}$$

associates with any integrable function f a linear functional F on $\mathbf{C}^\infty$. Inasmuch as

$$\left|\frac{1}{2\pi} \int f(x)u(x)\,dx\right| \leqslant \|f\|_1 \cdot \|u\|_\infty,$$

it is plain that this functional is continuous on $\mathbf{C}^\infty$. In this way we have associated with each $f \in \mathbf{L}^1$ a distribution. Knowledge of this distribution determines the function f a.e. and we shall identify the function (or, more accurately, the equivalence class, modulo null functions, determined by that function) and the distribution it generates. $\mathbf{L}^1$ thus appears as a linear subspace of $\mathbf{D}$.

12.2.3. **Definition of Measures; the Space M.** The distributions generated by integrable functions are not the only distributions F satisfying an inequality of the form

$$|F(u)| \leqslant \text{const}\, \|u\|_\infty. \tag{12.2.2}$$

Distributions of this type will be termed (Radon) *measures*.

The reason for the term "measure" is the fact that any functional F which is a measure in this sense can be expressed as an integral with respect to some uniquely determined regular Borel measure m on the underlying group $G = T$:

$$F(u) = \int_T u(x)\, dm(x).$$

This assertion, usually known as the Riesz-Markov-Kakutani theorem (for the compact space T), is a mild extension of Theorem 6.4d of [W]; a more detailed treatment in a more general setting appears in Chapter 4 of [E], especially Section 4.10 and Exercise 4.45; see also [E_1], Part 1; [HS], Chapter III, especially p. 177, and p. 364; [AB], Chapter 8; [R_1], pp. 40-47. This representation theorem confers much greater flexibility in the manipulation of Radon measures, largely because the expression of F as an integral combines with the appropriate integration theory to provide at once a good definition of $F(u)$ for each bounded Borel measurable function u on T, instead of merely for functions u in $\mathbf{C}^\infty$. (The possibility of extending F from $\mathbf{C}^\infty$ to $\mathbf{C}$ is established in a more elementary way in 12.2.8 and 12.2.9.) This added flexibility is almost essential for the discussion of some of the subtler properties of Radon measures that feature in a number of recent researches (such as those referred to in 12.7.4 and those appearing in Chapter 5 of [R]). For the principal results in this book, however, we shall not need to make any essential use of the representation theorem and its consequences.

An earlier and more concrete representation of a Radon measure was given by F. Riesz, who showed that such a functional F can be expressed as a Riemann-Stieltjes integral

$$F(u) = \frac{1}{2\pi} \int_{-\pi}^{\pi} u(x)\, d\phi(x),$$

where ϕ is a function of bounded variation[1] determined by F. A proof of this is to be found in the Appendix to [He]; see also [HS], (8.16), and [AB], p. 372. The sole explicit use to be made of this representation theorem appears in 12.5.10, where the use of the term "Fourier-Stieltjes series" receives some explanation.

Henceforth we shall denote by $\mathbf{M}$ the set of measures. Evidently, $\mathbf{M}$ is a linear subspace of $\mathbf{D}$.

In view of 12.2.2, $\mathbf{L}^1$ may be regarded as a linear subspace of $\mathbf{M}$.

Perhaps the simplest example of a measure that is not a function (that is, of an element of $\mathbf{M} \setminus \mathbf{L}^1$) is the so-called *Dirac measure* at the point x: this is the functional ε_x defined by

$$\varepsilon_x(u) = u(x).$$

The reader is urged to supply a proof that the measure ε_x is indeed not (generated by) a function in $\mathbf{L}^1$. In spite of this, the measure ε_x is often

[1] The function ϕ is in general not periodic.

improperly spoken of as "the Dirac δ-function placed at x"; see 3.2.2, 3.2.4, and 12.3.2(3). (Incidentally, we use the symbol ε in place of δ, partly to keep clearly in mind the correct terminology, and partly because the former symbol seems more appropriate for what is the identity element relative to convolution; see 12.6.7(1).)

Concerning the definition of Radon measures. Some readers, especially those familiar with the use of the term "measure" to describe a species of set-function (as described in Section 13.1), will surely feel upset by the choice of the term "Radon measure" to describe something that is evidently closer in nature to an integral (compare the discussion in 2.2.2, that in Sections 2.5 and 3.4 of [AB], the study by Schaefer in [Hi]); and see again [E_1], Part 1. The terminology, which is due to M. Bourbaki, is by now fairly well fixed and one must presumably make the best of it.

The fact is that each Radon measure can be extended into a Lebesgue-like integral (in much the same way as the Riemann integral can be extended into the Lebesgue integral and a Riemann-Stieltjes integral into a Lebesgue-Stieltjes integral), and that there is a one-to-one correspondence between these extended Radon measures μ and a species of set-function measures m_μ. Although we have no space to go into all the details, more comment will be made and references given in 12.5.10.

The execution of the details of the developments mentioned in the preceding paragraph are due to M. Bourbaki. The result is a complete theory of integration which, within its range of application, is at least as good as those based on a set-function-measure approach. In relation to the latter, Bourbaki's point of view amounts to a mental somersault: one takes a theorem (in this case, the Riesz-Markov-Kakutani theorem), hitherto well-hidden in the heart of a subject, and sets it up as a basic definition in a reformulated theory.

Notation for functions as measures. The Radon measure generated by a function $f \in \mathbf{L}^1$ would, in more traditional notations, bear a symbol different from f. The invariant integral I is, of course a special Radon measure, the associated set-function measure m_I being $(2\pi)^{-1}$ times Lebesgue measure on $[0, 2\pi)$ (where T is identified with $[0, 2\pi)$). To say that a Radon measure μ is generated by a function f is to say that m_μ is absolutely continuous with respect to m_I and that the Lebesgue-Radon-Nikodým derivative dm_μ/dm_I is f (see [HS], p. 328).

Again, one could as well write $\mu = f \cdot I$ in place of $\mu = f$; compare 12.3.4, 12.11.3, and [E], p. 235.

Yet another way of symbolizing the same relationship would amount to writing $d\mu = (2\pi)^{-1} f\,dx$.

We shall not adopt any of these notations, partly because they involve essentially the set-function measure approach, and partly because they tend to obstruct the view we wish to foster, namely, the view that measures and distributions are generalized functions.

12.2.4. **Distributions That Are Not Measures.** It is simple to give examples of distributions F that are neither functions nor measures. Consider, for instance, the distribution F defined by

$$F(u) = D^p u(0),$$

where p is a positive integer. It is apparent, on considering the functions $u = e_n$, that this F does not satisfy any inequality of the form (12.2.2) and therefore fails to be a measure.

12.2.5. **Continuity Expressed by Inequalities.** It is possible to classify distributions by means of inequalities of which (12.2.2) is a special instance, but in which higher derivatives of u appear.

The classification is based on the statement that a linear functional F on $\mathbf{C}^\infty$ is a distribution (that is, is continuous) if and only if there exists an integer $m \geqslant 0$ and a number c (both F-dependent) such that for all $u \in \mathbf{C}^\infty$ one has

$$|F(u)| \leqslant c \cdot \sup_{0 \leqslant p \leqslant m} \|D^p u\|_\infty . \tag{12.2.3}$$

Proof. It is evident that the inequality (12.2.3) ensures the continuity of F. Conversely, suppose that F is continuous. If no inequality (12.2.3) were valid, functions $u_k \in \mathbf{C}^\infty$ $(k = 1, 2, \cdots)$ would exist such that

$$|F(u_k)| > k \cdot \sup_{0 \leqslant p \leqslant k} \|D^p u_k\|_\infty . \tag{12.2.4}$$

This implies that $u_k \neq 0$, so that

$$\alpha_k = \sup_{0 \leqslant p \leqslant k} \|D^p u_k\|_\infty > 0 .$$

Define $v_k = (k\alpha_k)^{-1} u_k$. Then $v_k \in \mathbf{C}^\infty$ and

$$\sup_{0 \leqslant p \leqslant k} \|D^p v_k\|_\infty = k^{-1},$$

which entails that $v_k \to 0$ in $\mathbf{C}^\infty$ [compare (12.1.1)]. On the other hand, by (12.2.4) and the linearity of F,

$$|F(v_k)| = (k\alpha_k)^{-1} |F(u_k)| > 1 .$$

This would contradict the assumed continuity of F, which must therefore satisfy an inequality of the type (12.2.3).

12.2.6. **Order of a Distribution.** For a given distribution F, there exists therefore a least integer $m \geqslant 0$ such that (12.2.3) holds for a suitable (F-dependent) number c. We then say that F is a distribution of *order* m.

Reference to 12.2.3 shows that the measures are exactly the distributions of zero order.

12.2.7. **The Space $\mathbf{D}^m$.** We shall henceforth denote by $\mathbf{D}^m$ $(m = 0, 1, 2, \cdots)$ the set of distributions of order at most m. $\mathbf{D}^m$ is a linear subspace of $\mathbf{D}$ and we have the relations

$$\mathbf{L}^1 \subset \mathbf{M} = \mathbf{D}^0 \subset \mathbf{D}^1 \subset \cdots \subset \mathbf{D}^m \subset \mathbf{D}^{m+1} \subset \cdots, \qquad \mathbf{D} = \bigcup_{m=0}^{\infty} \mathbf{D}^m .$$

(One has here an extension of the chain of inclusions (2.2.18).) For each m, the inclusion $\mathbf{D}^m \subset \mathbf{D}^{m+1}$ is easily seen to be proper.

12.2.8. **$\mathbf{D}^m$ and Continuous Linear Functionals on $\mathbf{C}^m$.** There is another way of visualizing $\mathbf{D}^m$ which must be observed here. We shall verify that $\mathbf{D}^m$ can be thought of as the set of continuous linear functionals on $\mathbf{C}^m$, the latter being considered as a Banach space with the norm $\| \cdot \|_{(m)}$.

Indeed, on the one hand it is evident that the restriction to $\mathbf{C}^\infty \subset \mathbf{C}^m$ of a continuous linear functional L on $\mathbf{C}^m$ is a distribution of order at most m. Moreover, since $\mathbf{C}^\infty$ is dense in $\mathbf{C}^m$ (a corollary of 6.1.1), a continuous linear functional L on $\mathbf{C}^m$ is uniquely determined by its restriction to $\mathbf{C}^\infty$.

It thus remains only to verify that each $F \in \mathbf{D}^m$ can be extended into a continuous linear functional L on $\mathbf{C}^m$ (and of which it is the restriction to $\mathbf{C}^\infty$). But suppose $u \in \mathbf{C}^m$. Choose any sequence (u_k) from $\mathbf{C}^\infty$ such that

$$\lim_{k \to \infty} \|u - u_k\|_{(m)} = 0 .$$

Then (12.2.3) shows that $\lim_{k \to \infty} F(u_k)$ exists finitely. By the same token, the value of this limit is the same for any other sequence (u_k') extracted from $\mathbf{C}^\infty$ and such that $\|u - u_k'\|_{(m)} \to 0$ as $k \to \infty$. The required extension L of F is obtained by setting $L(u) = \lim_{k \to \infty} F(u_k)$. It is clear that L is thereby defined as a linear functional on $\mathbf{C}^m$ whose restriction to $\mathbf{C}^\infty$ is F. Beside this, (12.2.3) shows that

$$\begin{aligned} |L(u)| &= \lim |F(u_k)| \leqslant c \cdot \lim_k \|u_k\|_{(m)} \\ &= c \cdot \|u\|_{(m)}, \end{aligned}$$

so that L is indeed continuous on $\mathbf{C}^m$.

12.2.9. **Measures as Functionals on $\mathbf{C}$.** The case $m = 0$ of 12.2.8 is especially important. It asserts that each measure can be extended into a continuous linear functional on $\mathbf{C} = \mathbf{C}^0$, and that conversely each continuous linear functional on $\mathbf{C}$ is obtained by thus extending precisely one measure.

This marks one more step in bringing our definition of measures into line with the Riesz-Markov-Kakutani representation theorem referred to in 12.2.3.

More about measures will appear in 12.3.8, 12.3.9, 12.5.10, and Section 12.7.

12.3 Convergence of Distributions

12.3.1. **Definition of Distributional Convergence.** A sequence $(F_n)_{n=1}^{\infty}$ of distributions will be said to *converge in* **D** to a distribution F if and only if

$$\lim_{n \to \infty} F_n(u) = F(u)$$

for each $u \in \mathbf{C}^{\infty}$. A similar definition applies to the relationship

$$\lim_{t \to t_0} F_t = F \text{ in } \mathbf{D},$$

where $t \to F_t$ is a **D**-valued function of a real or complex variable t defined on a punctured neighborhood of t_0; t_0 may here be $-\infty$ or ∞.

This type of convergence of distributions is sometimes spoken of as *distributional convergence.*

Remark. We note, but will never use, the fact that the general theory of duality for topological linear spaces leads to several topologies on **D** with respect to any one of which the notion of sequential convergence (or the convergence of **D**-valued functions F_t as specified above) accords exactly with that prescribed in 12.3.1.

12.3.2. **Examples.** (1) If $F \in \mathbf{D}$ and $a \in T$ the translate $T_a F$ is the distribution defined by the formula

$$(T_a F)(u) = F(T_{-a} u)$$

for $u \in \mathbf{C}^{\infty}$, this definition being so chosen that if F is (generated by) a function $f \in \mathbf{L}^1$ (see 12.2.2), then $T_a F$ is the distribution (generated by) the function $T_a f$. That $T_a F$ so defined is indeed a distribution follows from 12.1.5. Notice that $T_{a+2\pi} F = T_a F$, so that the distributions we are speaking of may be said to have period 2π.

It is very simple to verify that $T_a F \to T_{a_0} F$ in **D** as $a \to a_0$.

(2) If the functions f_n $(n = 1, 2, \cdots)$ in $\mathbf{L}^1$ converge in mean in that space to f, then $f_n \to f$ in **D**. This follows immediately from the substance of 12.2.2.

As the next example shows, a sequence of functions in $\mathbf{L}^1$ may well be distributionally convergent without being convergent in $\mathbf{L}^1$.

(3) If $(f_n)_{n=1}^{\infty}$ is any approximate identity in $\mathbf{L}^1$ (see 3.2.1), then $f_n \to \varepsilon$ distributionally, where $\varepsilon = \varepsilon_0$ is the Dirac measure at the origin.

This is a reformulation of a special case of 3.2.2; see also the remarks in 3.2.4.

(4) The relation

$$n^{-1} \sum_{k=1}^{n} \varepsilon_{2\pi k/n} \to 1$$

holds distributionally. For the result of applying the measure appearing on the left-hand side to a function $u \in \mathbf{C}^\infty$ is

$$n^{-1} \sum_{k=1}^{n} u\left(\frac{2\pi k}{n}\right) = \frac{1}{2\pi} \sum_{k=1}^{n} u\left(\frac{2\pi k}{n}\right)\left[\frac{2\pi k}{n} - \frac{2\pi(k-1)}{n}\right],$$

which (even for any continuous u) tends to

$$\frac{1}{2\pi} \int_0^{2\pi} u(x)\, dx = 1(u).$$

(5) The following example is less transparent. In 12.2.2 we have seen how to associate a distribution with each integrable function. Now we shall illustrate a method of associating a distribution with certain nonintegrable functions. Viewed otherwise, it provides an instance of a sequence of integrable functions that converges to a distribution of order one.

The nonintegrable function to be treated is ω, defined almost everywhere on $[-\pi, \pi)$ as cosec $\frac{1}{2}x$, and then extended by periodicity. The corresponding distribution is obtained as the limit, as $\varepsilon \downarrow 0$, of the integrable functions ω_ε defined almost everywhere on $(-\pi, \pi)$ by

$$\omega_\varepsilon(x) = \begin{cases} \operatorname{cosec} \frac{1}{2}x & \text{if } \varepsilon < |x| < \pi, \\ 0 & \text{if } |x| \leqslant \varepsilon, \end{cases}$$

ω_ε being defined elsewhere by periodicity. The distributional limit does indeed exist, since we may write

$$\begin{aligned} \omega_\varepsilon(u) &= \frac{1}{2\pi} \int_{\varepsilon < |x| \leqslant \pi} u(x) \operatorname{cosec} \tfrac{1}{2}x\, dx \\ &= \frac{1}{2\pi} \int_{\varepsilon < |x| \leqslant \pi} [u(x) - u(0)] \operatorname{cosec} \tfrac{1}{2}x\, dx, \end{aligned}$$

because

$$\int_{\varepsilon < |x| \leqslant \pi} \operatorname{cosec} \tfrac{1}{2}x\, dx = 0$$

owing to the integrand being an odd function of x. Furthermore,

$$u(x) - u(0) = O(|x|)$$

for small x, so that the integrand remaining is integrable. Thus the ω_ε converge distributionally to the distribution Ω defined by

$$\Omega(u) = \frac{1}{2\pi} \int [u(x) - u(0)] \operatorname{cosec} \tfrac{1}{2}x\, dx \tag{12.3.1}$$

Since the first mean value theorem shows that

$$|u(x) - u(0)| \leqslant |x| \cdot \|u\|_{(1)},$$

we see that

$$|\Omega(u)| \leqslant \frac{1}{2\pi} \int |x \cdot \operatorname{cosec} \tfrac{1}{2}x|\, dx \cdot \|u\|_{(1)} \tag{12.3.2}$$

which confirms that Ω is of order at most one: $\Omega \in \mathbf{D}^1$.

This mode of defining Ω suggests naming it the *principal value* of ω and denoting it by P. V. ω.

The distribution Ω is genuinely of order one, that is (in view of what we already know), it is not a measure. Indeed, were Ω to be a measure, (12.2.2) and (12.3.1) would combine with Exercise 12.5(2) to show that

$$\frac{1}{2\pi} \int_a^b \operatorname{cosec} \tfrac{1}{2} x \, dx \leqslant \text{const}$$

for $0 < a < b < \pi$. But it is easily seen that

$$\frac{1}{2\pi} \int_a^b \operatorname{cosec} \tfrac{1}{2} x \, dx = \frac{1}{\pi} \log \frac{b}{a} + O(1),$$

and is therefore unbounded for $0 < a < b < \pi$.

In Example 12.4.3(3) we shall see how to represent $\Omega = \text{P. V. } \omega$ as the distributional derivative of an integrable function. Meanwhile, we return to some generalities.

12.3.3. It is very simple to verify that if $F_n \to F$ and $G_n \to G$ distributionally, then $F_n + G_n \to F + G$ and $cF_n \to cF$ distributionally, c denoting any constant.

12.3.4. **Product of a Distribution and a $\mathbf{C}^\infty$ Function.** The second assertion in 12.3.3 may be extended.

In the first place, we can define the product uF of any function $u \in \mathbf{C}^\infty$ and any distribution $F \in \mathbf{D}$ by writing

$$(uF)(v) = F(uv) \qquad \text{for } v \in \mathbf{C}^\infty.$$

The justification for this definition is contained in 12.1.5, the substance of which leads also to the conclusion that

$$u_n F \to uF \quad \text{distributionally}$$

if $u_n \to u$ in $\mathbf{C}^\infty$, and that

$$uF_n \to uF \quad \text{distributionally}$$

if $u \in \mathbf{C}^\infty$ and $F_n \to F$ distributionally.

It is also true that $u_n F_n \to uF$ distributionally whenever $u_n \to u$ in $\mathbf{C}^\infty$ and $F_n \to F$ distributionally, but this is less obvious. No use will be made of this fact and its proof is omitted.

12.3.5. **The Product in Other Cases.** In connection with 12.3.4 we may observe that uF can be defined for $u \in \mathbf{C}^m$ and $F \in \mathbf{D}^m$, the result being an element of $\mathbf{D}^m$. The basis for this statement lies in 12.2.8.

Another case in which the product is satisfactorily definable will appear in 12.11.3.

On the other hand there is no hope of giving a "reasonable" definition of the product of two arbitrary distributions—nor even of the product of two arbitrary measures. The qualification "reasonable" is here intended to cover a tacit assumption that the required definition shall retain one or more properties of the product as applied to smooth functions.

For example, one can show (see Exercise 12.9) that it is impossible to define a product $\alpha\beta$ of two arbitrary measures in such a way that $(\alpha, \beta) \to \alpha\beta$ is a bilinear mapping of $\mathbf{M} \times \mathbf{M}$ into $\mathbf{D}$ having the properties (1) if $\alpha(u) \geqslant 0$ and $\beta(u) \geqslant 0$ for nonnegative functions $u \in \mathbf{C}^\infty$ (in which case the measures α and β are said to be *positive*), then $\alpha\beta(u) \geqslant 0$ for such functions; and (2) if α and β are (generated by) functions f and g in $\mathbf{C}^\infty$, respectively, then $\alpha\beta$ is (generated by) the function fg (ordinary pointwise product).

12.3.6. **Compacity Principles.** Each of the next four results states in sequential form a compactness property of certain sets of distributions, of measures, or of functions. Each is a very close analogue of the Weierstrass-Bolzano theorem, which asserts that from any bounded sequence of real or complex numbers may be extracted a convergent subsequence. The common source of these four results is an abstract compacity principle which is discussed in I, B.4; the fourth result uses also a characterization of the continuous linear functionals on the space $\mathbf{L}^p$ $(1 \leqslant p < \infty)$ which is discussed in I, C.1.

In order to heighten the analogy with the Weierstrass-Bolzano theorem, we first introduce the appropriate concepts of boundedness. These are as follows:

(1) A set $\mathbf{S}$ of distributions is said to be *bounded in* $\mathbf{D}$, or to be *distributionally bounded*, if and only if

$$\sup\{|F(u)| : F \in \mathbf{S}\} < \infty \tag{12.3.3}$$

for each $u \in \mathbf{C}^\infty$.

(2) A set $\mathbf{S}$ of measures is said to be *bounded in* $\mathbf{M}$, if and only if

$$\sup\{|\mu(u)| : \mu \in \mathbf{S}\} < \infty \tag{12.3.4}$$

for each $u \in \mathbf{C}$.

(3) A set $\mathbf{S}$ of functions in $\mathbf{L}^p$ $(1 \leqslant p \leqslant \infty)$ is said to be *bounded in* $\mathbf{L}^p$, if and only if

$$\sup\{|\frac{1}{2\pi}\int fu\,dx| : f \in \mathbf{S}\} < \infty$$

for each $u \in L^{p'}$, where $1/p + 1/p' = 1$.

In each case the value of the supremum will in general depend upon u.

The concept of boundedness expressed in (2) [respectively (3)] is sometimes spoken of as *weak boundedness in* $\mathbf{M}$ [respectively in $\mathbf{L}^p$] (compare I, B.1.7), in contradistinction from *norm boundedness*. As we shall soon see, however, the two concepts are equivalent.

It is easily seen that, if $\mathbf{S} \subset \mathbf{L}^p$ and if we inject $\mathbf{L}^p \subset \mathbf{L}^1$ into $\mathbf{M}$, which is in turn injected into $\mathbf{D}$, then (3) implies (2), and (2) implies (1). Also, if $\mathbf{S} \subset \mathbf{L}^q$ and is bounded in $\mathbf{L}^q$, and if $1 \leqslant p < q$, then $\mathbf{S}$ is bounded in $\mathbf{L}^p$. The converse statements are false. For example, the sequence of measures $n[\varepsilon_{(1/n)} - \varepsilon]$ $(n = 1, 2, \cdots)$ is bounded in $\mathbf{D}$ but not in $\mathbf{M}$; again, the Fejér kernels F_n $(n = 1, 2, \cdots)$ form a bounded subset of $\mathbf{L}^1$ and of $\mathbf{M}$, but they are not bounded in $\mathbf{L}^p$ for any $p > 1$.

One might also define a subset $\mathbf{S}$ of $\mathbf{D}^m$ to be *bounded in* $\mathbf{D}^m$, if and only if

$$\sup\{|F(u)| : F \in \mathbf{S}\} < \infty \tag{12.3.5}$$

for each $u \in \mathbf{C}^m$. We shall have no special use for this concept of boundedness, however.

12.3.7. Let $(F_n)_{n=1}^{\infty}$ be a sequence of distributions forming a bounded subset of $\mathbf{D}$. Then there exists a subsequence $(F_{n_k})_{k=1}^{\infty}$ and a distribution F such that

$$\lim_{k \to \infty} F_{n_k} = F$$

in $\mathbf{D}$.

Proof. Since $\mathbf{C}^\infty$ is complete (see 12.1.3), and since $\mathbf{D}$ is defined to be the set of all continuous linear functionals on $\mathbf{C}^\infty$, the assertion is a special case of 1, B.4.1; separability of $\mathbf{C}^\infty$ follows from 12.1.1.

12.3.8. **M as a Normed Linear Space.** There is an analogue of 12.3.7 for bounded sequences in $\mathbf{D}^m$. Especially significant for future developments is the case $m = 0$, to which the next result applies.

Let us first define the *norm* of a measure $\mu \in \mathbf{M}$ by the equation

$$\|\mu\|_1 = \sup\{|\mu(u)| : u \in \mathbf{C}, \|u\|_\infty \leqslant 1\}. \tag{12.3.6}$$

In other words, $\|\mu\|_1$ is the smallest number $c \geqslant 0$ for which it is true that

$$|\mu(u)| \leqslant c \cdot \|u\|_\infty$$

for each $u \in \mathbf{C}$ (or, what is equivalent, for each $u \in \mathbf{C}^\infty$); compare equation (12.2.2). The reader will be able to verify without trouble that $\|\cdot\|_1$ is indeed a norm on $\mathbf{M}$. The notation is suggested by the fact that, if μ is (generated by) a function $f \in \mathbf{L}^1$, then $\|\mu\|_1$ turns out to be none other than $\|f\|_1$ as defined in (2.2.13); see Exercise 12.10.

12.3.9. A sequence $(\mu_n)_{n=1}^{\infty}$ of measures is bounded in $\mathbf{M}$, if and only if

$$\sup_n \|\mu_n\|_1 < \infty,$$

in which case a subsequence $(\mu_{n_k})_{k=1}^{\infty}$ and a measure μ exist such that $\lim_{k \to \infty} \mu_{n_k} = \mu$ *weakly in* $\mathbf{M}$, by which it is meant (see I, B.1.7) that

$$\lim_{k \to \infty} \mu_{n_k}(u) = \mu(u)$$

for each $u \in \mathbf{C}$. (See also Exercises 12.13 and 12.43.)

Proof. Since $\mathbf{C}$ is complete, the first statement is a special case of the uniform boundedness principle (I, B.2.1). The second statement follows from I, B.4.1 because $\mathbf{C}$ is separable (see 6.1.1) and because of the identification of $\mathbf{M}$ with the set of continuous linear functionals on $\mathbf{C}$ established in 12.2.8.

12.3.10. (1) If $1 \leqslant p \leqslant \infty$, a sequence $(f_n)_{n=1}^{\infty}$ of functions in $\mathbf{L}^p$ is bounded in $\mathbf{L}^p$, if and only if

$$\sup_n \|f_n\|_p < \infty.$$

(2) If $1 < p \leqslant \infty$, and if $(f_n)_{n=1}^{\infty}$ is a bounded sequence in $\mathbf{L}^p$, there exists a subsequence $(f_{n_k})_{k=1}^{\infty}$ and a function $f \in \mathbf{L}^p$ such that $\lim_{k \to \infty} f_{n_k}$ *weakly in* $\mathbf{L}^p$, by which it is meant (see I, B.1.7) that

$$\lim_{k \to \infty} \frac{1}{2\pi} \int f_{n_k} u \, dx = \frac{1}{2\pi} \int fu \, dx$$

for each $u \in \mathbf{L}^{p'}$ (where $1/p + 1/p' = 1$). (See also Exercise 12.14.)

Proof. Statement (1) follows from the uniform boundedness principle exactly as does the corresponding assertion in 12.3.9, provided one observes that, if $f \in \mathbf{L}^p$,

$$\|f\|_p = \sup \{|\frac{1}{2\pi} \int fu \, dx| : u \in \mathbf{L}^{p'}, \|u\|_{p'} \leqslant 1\},$$

which is the converse of Hölder's inequality (see Exercise 3.6).

Statement (2) again follows from the compacity principle in I, B.4.1, coupled with the identification of the set of continuous linear functionals on $\mathbf{L}^q$ $(1 \leqslant q < \infty)$ with $\mathbf{L}^{q'}$ established in I, C.1, q here being taken to be p'.[1]

12.3.11. **Remarks.** Part (2) of 12.3.10 is false for $p = 1$. For example, the sequence of Fejér kernels $(F_N)_{N=1}^{\infty}$ is bounded in $\mathbf{L}^1$, but no subsequence $(F_{N_k})_{k=1}^{\infty}$ converges weakly in $\mathbf{L}^1$. (The reader should prove this, bearing in mind the substance of 3.2.4.) Indeed, by appeal to I, C.2, it can be seen that no such subsequence is a weak Cauchy sequence in $\mathbf{L}^1$, that is, that no such subsequence has the property that

$$\lim_{k \to \infty} \frac{1}{2\pi} \int F_{n_k} u \, dx$$

exists finitely for each $u \in \mathbf{L}^{\infty}$.

[1] That $\mathbf{L}^q$ $(1 \leqslant q < \infty)$ is separable, is shown by 6.1.1.

The same remarks apply when $(F_N)_{N=1}^{\infty}$ is replaced by any approximate identity in $\mathbf{L}^1$.

12.4 Differentiation of Distributions

If f is an absolutely continuous function, the derived function Df is defined almost everywhere and is integrable. Furthermore, partial integration (see [W], Theorem 5.4a, Exercise 16 on p. 111, Theorem 6.3d; [HS], p. 287) shows that

$$\frac{1}{2\pi}\int Df \cdot u \, dx = -\frac{1}{2\pi}\int f \cdot Du \, dx$$

for each $u \in \mathbf{C}^{\infty}$. This circumstance prompts the following definition.

12.4.1. **Definition of Derivative.** If F is a distribution, its (*distributional*) *derivative* DF is the distribution defined by

$$DF(u) = -F(Du) \tag{12.4.1}$$

for $u \in \mathbf{C}^{\infty}$.

12.4.2. **Remarks on the Definition of Derivative.** We have taken care that this notion of derivative, the *distributional derivative*, coincides with the ordinary one when applied to distributions generated by absolutely continuous functions. It must therefore be made quite clear that a divergence appears when nonabsolutely continuous functions f are involved: in such cases it may well happen that the pointwise derivative f' will exist almost everywhere and be integrable, and yet the distributional derivative Df will be quite different from the distribution generated by the integrable function f'.

As an illustration, consider the function f defined to be 0 on $[-\pi, 0]$, to be 1 on $(0, \pi)$, and elsewhere so as to have period 2π. To compute the distributional derivative Df we have

$$\begin{aligned} Df(u) &= -\frac{1}{2\pi}\int_{-\pi}^{\pi} f \cdot Du \, dx \\ &= -\frac{1}{2\pi}\int_{-\pi}^{0} 0 \cdot Du \, dx - \frac{1}{2\pi}\int_{0}^{\pi} 1 \cdot Du \, dx \\ &= 0 - \frac{1}{2\pi}[u(\pi) - u(0)] = \frac{1}{2\pi}[u(0) - u(\pi)], \end{aligned}$$

which shows that $Df = (1/2\pi)[\varepsilon - \varepsilon_{\pi}]$. On the other hand, $f' = 0$ a.e. and so generates the zero distribution.

Support for regarding the distributional derivative as an appropriate concept lies in the fact that if we wished to evaluate

$$-\frac{1}{2\pi}\int_{-\pi}^{\pi} f \cdot Du\, dx$$

by using partial integration, it would be necessary to decompose the range of integration into subintervals $[-\pi, -\varepsilon]$, $[-\varepsilon, \varepsilon]$, $[\varepsilon, \pi - \varepsilon]$, $[\pi - \varepsilon, \pi]$, on the first and third of which f is absolutely continuous, apply partial integration to the first and third subintervals, and notice that the remaining integrals are in any case $o(1)$ as $\varepsilon \to 0$. The result would be $(1/2\pi)[u(0) - u(\pi)]$. In other words, a *correct* application of partial integration leads to a demand for the distributional derivative rather than the pointwise one.

It is to be noticed that the jump discontinuities of f at the origin and at π introduce into the distributional derivative terms involving Dirac measures at these points. This feature is quite typical. Further differentiations in the distributional sense will introduce distributions of higher and higher order.

In the sequel, failing any indication to the contrary, differentiation will always be performed in the distributional sense. With this convention one may (truthfully) say that a function $f \in \mathbf{L}^1$ is equal almost everywhere to an absolutely continuous function, if and only if $Df \in \mathbf{L}^1$. (At the risk of over-repetition, we reaffirm that this statement is *not* true, if the derivative is interpreted in the pointwise sense.) Compare Exercise 12.12.

It is similarly true that a function $f \in \mathbf{L}^1$ is equal amost everywhere to a function of bounded variation, if and only if $Df \in \mathbf{M}$. (And again Df *must* here be interpreted distributionally.) A proof appears in 12.5.10.

12.4.3. **Examples.** (1) The translates $T_a F$ of a distribution F have been defined in 12.3.2(1). It is simple to verify that

$$DF = \lim_{a \to 0} a^{-1}(T_{-a}F - F)$$

distributionally. For, if $u \in \mathbf{C}^\infty$,

$$[a^{-1}(T_{-a}F - F)](u) = a^{-1}[F(T_a u) - F(u)] = F[a^{-1}(T_a u - u)]$$

by the definition of $T_{-a}F$ and by linearity of F. It will therefore suffice to verify that

$$\lim_{a \to 0} a^{-1}(T_a u - u) = -Du$$

in $\mathbf{C}^\infty$, which follows by application of the first mean value theorem.

(2) Let A be a given distribution and let us examine the possibility of solving the equation

$$DF = A \tag{12.4.2}$$

for the unknown distribution F.

This equation is not always soluble. Indeed, since $DF(1) = -F(D1) = -F(0) = 0$, a necessary condition for solubility is that $A(1) = 0$. We shall show how to solve (12.4.2) whenever $A(1) = 0$.

Assuming that F is a solution of (12.4.2) we have, for each $u \in \mathbf{C}^\infty$, $F(Du) = -A(u)$. Putting $u = Jv$, where J is the endomorphism of $\mathbf{C}^\infty$ defined by

$$Jv(x) = \int_0^x [v(y) - \hat{v}(0)]\, dy,$$

we have $Du = v - \hat{v}(0)1$ and so

$$F[v - \hat{v}(0)1] = -A(Jv).$$

This may also be written as

$$F(v) = F(1) \cdot 1(v) - A \circ J(v),$$

or

$$F = F(1) \cdot 1 - A \circ J. \tag{12.4.3}$$

The reader will observe that, since J is a continuous endomorphism of $\mathbf{C}^\infty$, $A \circ J$ is indeed a distribution. Formula (12.4.3) gives the solution of (12.4.2), assumed to exist. It remains to verify that (12.4.3) really *is* a solution, provided $A(1) = 0$.

But, if F is given by (12.4.3), we have for any $u \in \mathbf{C}^\infty$

$$DF(u) = -F(Du) = -F(1) \cdot 1(Du) + A(JDu).$$

Herein, $1(Du) = 0$ and $JDu = u - u(0)1$. So, since $A(1) = 0$, we obtain

$$DF(u) = -F(1) \cdot 0 + A[u - u(0)1] = A(u),$$

which shows that $DF = A$.

An alternative discussion of the equation (12.4.2) can be based on the use of Fourier series; see Example 12.5.9 and Exercise 12.15.

(3) We revert temporarily to the distribution

$$\Omega = \text{P.V.}\, \omega$$

defined in Example 12.3.2(5). We have seen that, for $u \in \mathbf{C}^\infty$,

$$\Omega(u) = \lim_{\varepsilon \to 0} \frac{1}{2\pi} \int_{\varepsilon < |x| \leq \pi} u(x) \operatorname{cosec} \tfrac{1}{2} x\, dx = \lim_{\varepsilon \to 0} \omega_\varepsilon(u).$$

Now one may write

$$\omega_\varepsilon(u) = \frac{1}{2\pi} \int_\varepsilon^\pi u(x) \operatorname{cosec} \tfrac{1}{2} x\, dx - \frac{1}{2\pi} \int_\varepsilon^\pi u(-x) \operatorname{cosec} \tfrac{1}{2} x\, dx. \tag{12.4.4}$$

Introduce the function φ on $(0, \pi]$ defined by

$$\varphi(x) = -\int_x^\pi \operatorname{cosec} \tfrac{1}{2}y \, dy .$$

Then φ is integrable over $(0, \pi)$, is $O(\log x^{-1})$ as $x \downarrow 0$, and is absolutely continuous on $[\varepsilon, \pi]$ for any $\varepsilon > 0$. Applying partial integration to each of the integrals appearing in (12.4.4), it is found that

$$2\pi\omega_\varepsilon(u) = -u(\varepsilon)\varphi(\varepsilon) + u(-\varepsilon)\varphi(\varepsilon) - \int_\varepsilon^\pi \varphi(x)Du(x)\,dx - \int_\varepsilon^\pi \varphi(x)Du(-x)\,dx . \tag{12.4.5}$$

The integrated terms are together $o(1)$ as $\varepsilon \to 0$ (since u is differentiable at the origin). So, since φ is integrable over $(0, \pi)$, (12.4.5) leads to the relation

$$\begin{aligned} 2\pi\Omega(u) &= -\int_0^\pi \varphi(x)Du(x)\,dx - \int_0^\pi \varphi(x)Du(-x)\,dx \\ &= -\int_{-\pi}^\pi \psi(x)Du(x)\,dx , \end{aligned} \tag{12.4.6}$$

where we have defined

$$\psi(x) = \varphi(|x|),$$

first for $0 < |x| \leqslant \pi$; $\psi(0)$ may be defined arbitrarily, and the definition completed by requiring ψ to be periodic. The resulting function ψ is integrable, and (12.4.6) signifies precisely that

$$\Omega \equiv \text{P.V.}\,\omega = D\psi$$

distributionally.

12.4.4. **Properties of Differentiation.** There are a number of simple properties of the differentiation operator acting on distributions, each of which is a direct and simple consequence of (12.4.1), together with the contents of Section 12.1.

First, D is linear:

$$D(F + G) = DF + DG, \qquad D(cF) = c \cdot DF,$$

c being any constant.

Differentiation is also continuous: if $F_n \to F$ in $\mathbf{D}$ then $DF_n \to DF$ in $\mathbf{D}$. This property appears in marked contrast to the situation prevailing in relation to (say) uniform convergence or mean convergence in $\mathbf{L}^p$.

Combining the preceding properties, we infer that

$$D(\sum_{n=1}^\infty F_n) = \sum_{n=1}^\infty DF_n$$

whenever the series $\sum_{n=1}^{\infty} F_n$ converges distributionally, in which case the series $\sum_{n=1}^{\infty} DF_n$ is likewise convergent.

12.4.5. **Leibnitz's formula** for differentiating a product function has a perfect analogue. Thus, if $F \in \mathbf{D}$ and $u \in \mathbf{C}^{\infty}$, then

$$D(uF) = (Du)F + uDF;$$

see 12.3.4.

12.4.6. Reference to 12.2.6 will make it plain that $DF \in \mathbf{D}^{m+1}$ whenever $F \in \mathbf{D}^m$. More precisely, if $m > 0$ and $F \in \mathbf{D}^m \backslash \mathbf{D}^{m-1}$, then $DF \in \mathbf{D}^{m+1} \backslash \mathbf{D}^m$; see Exercise 12.54.

12.5 Fourier Coefficients and Fourier Series of Distributions

Reference to 12.2.2 and formula (2.3.1) should render the following definition of the Fourier coefficients and series of a distribution seem entirely natural (and indeed obligatory, if a consistent extension is to be achieved).

12.5.1. **Definition.** If $F \in \mathbf{D}$, we define its *Fourier coefficients* by the formula

$$\hat{F}(n) = F(e_{-n}) \qquad (n \in Z),$$

where, as usual, $e_m \in \mathbf{C}^{\infty}$ is the function $x \to e^{imx}$. The series

$$\sum_{n \in Z} \hat{F}(n) e^{inx}$$

is called the *Fourier series* of F.

12.5.2. **Fourier-Lebesgue, Fourier-Stieltjes, and Fourier-Schwartz Series.** In order to avoid possible confusion in certain statements, a series $\sum_{n \in Z} c_n e^{inx}$ is spoken of as a *Fourier-Lebesgue* series if and only if there is some (perhaps unspecified) integrable function f such that $c_n = \hat{f}(n)$ $(n \in Z)$; in a similar vein, the series will be described as a *Fourier-Schwartz* series if and only if there exists a (perhaps unspecified) distribution F such that $c_n = \hat{F}(n)$ $(n \in Z)$. In addition, if there is a measure μ such that $c_n = \hat{\mu}(n)$ $(n \in Z)$, it is customary to speak of the series as a *Fourier-Stieltjes series.* Likewise, a function on Z of the form $\hat{\mu}$, where $\mu \in \mathbf{M}$, is often spoken of as a *Fourier-Stieltjes transform.* A more detailed explanation of the use of this term appears in 12.5.10.

The problem of deciding whether a given series is a Fourier-Lebesgue or a Fourier-Stieltjes series, is often extremely difficult (see the remarks in

2.3.9). It may come as a surprise, therefore, to discover that the corresponding decision problem for Fourier-Schwartz series is comparatively trivial, as the next result shows. This result also makes it plain that most trigonometric series which will arise in practice, and certainly those normally considered in the classical Riemann theory, are Fourier-Schwartz series.

12.5.3. (1) Suppose that $F \in \mathbf{D}$ is of order at most m. Then

$$\hat{F}(n) = O(|n|^m) \qquad \text{as } |n| \to \infty, \tag{12.5.1}$$

so that the Fourier series of F is tempered. Moreover,

$$s_N F \equiv \sum_{|n| \leqslant N} \hat{F}(n) e_n \to F \qquad \text{as } N \to \infty$$

in $\mathbf{D}$ [see relation (D) in Subsection 1.3.2].

(2) Given any tempered sequence $(c_n)_{n \in Z}$, the distributions

$$s_N = \sum_{|n| \leqslant N} c_n e_n$$

converge in $\mathbf{D}$ as $N \to \infty$ to a distribution F such that $\hat{F}(n) = c_n$ $(n \in Z)$, so that the given series is the Fourier series of F.

Proof. (1) The statement (12.5.1) follows immediately from the inequality (12.2.3) if we take therein $u = e_{-n}$. Next we have for $u \in \mathbf{C}^\infty$

$$s_N F(u) = \sum_{|n| \leqslant N} \hat{F}(n) \hat{u}(-n) = F\Big[\sum_{|n| \leqslant N} \hat{u}(-n) e_{-n}\Big],$$

the last step by definition of the Fourier coefficients of F and by linearity of F. Thus

$$s_N F(u) = F\Big[\sum_{|n| \leqslant N} \hat{u}(n) e_n\Big].$$

This combines with (12.1.5) and the continuity of F to show that

$$\lim_{n \to \infty} s_N F(u) = F(u),$$

which says precisely that $s_N F \to F$ in $\mathbf{D}$.

(2) To say that (c_n) is tempered signifies that

$$|c_n| \leqslant A |n|^k \qquad (n \neq 0)$$

for a suitable number $A \geqslant 0$ and a suitable integer $k \geqslant 0$. For $u \in \mathbf{C}^\infty$, we have

$$s_N(u) = \sum_{|n| \leqslant N} c_n \hat{u}(-n).$$

The series $\sum_{n \in Z} c_n \hat{u}(-n)$ is absolutely convergent, as follows from (12.1.4). Thus the formula

$$F(u) \equiv \lim_{N \to \infty} s_N(u) = \sum_{n \in Z} c_n \hat{u}(-n) \tag{12.5.2}$$

defines F as a functional on $\mathbf{C}^\infty$ which is evidently linear. That F is continuous could be deduced from 12.3.7. A more direct argument uses 2.3.4 thus:

$$\begin{aligned} \left|\sum_{n \in Z} c_n \hat{u}(-n)\right| &\leqslant |c_0 \hat{u}(0)| + A \sum_{n \neq 0} |n|^k |\hat{u}(-n)| \\ &\leqslant |c_0 \hat{u}(0)| + A \sum_{n \neq 0} |n|^k \cdot \frac{\|D^{k+2}u\|_\infty}{|n|^{k+2}} \\ &= |c_0 \hat{u}(0)| + [A \sum_{n \neq 0} |n|^{-2}] \|D^{k+2}u\|_\infty, \end{aligned}$$

which clearly shows that F is a distribution of order at most $k + 2$ (see 12.2.5 and 12.12.6).

The reader will observe that the preceding argument shows even that the Fourier series of F is *unconditionally* convergent in $\mathbf{D}$.

Finally, (12.5.2) applied with $u = e_{-m}$ shows that $\hat{F}(m) = c_m$ for all $m \in Z$.

12.5.4. **Remarks.** (1) The reader will notice that we have established en route the Parseval formula

$$F(u) = \sum_{n \in Z} \hat{F}(n) \hat{u}(-n) \tag{12.5.3}$$

for $F \in \mathbf{D}$ and $u \in \mathbf{C}^\infty$, the series being absolutely convergent. Compare also with equation (12.3). From this, or from 12.5.3(1), there follows an extension of the uniqueness theorem 2.4.1 from functions to distributions.

(2) Armed with 12.5.3(2), the reader may with profit glance again at Sections 2.5 and 6.7. It can now be said that $\hat{\phi}$ is defined as a distribution for any tempered function ϕ on Z, and that the inversion formula $(\hat{\phi})^\wedge = \phi$ is valid. In particular, $\hat{\phi}$ is defined whenever ϕ belongs to ℓ^p for some p satisfying $0 < p \leqslant \infty$.

(3) Nowhere in the sequel shall we turn aside to discuss conditions on a distribution that will ensure the pointwise convergence or summability of its Fourier series; concerning this and related questions, see Walter [1]. Related to this is the problem of assigning a numerical value to certain distributions at certain points; for this, see Łojasiewicz [1], [2].

12.5.5. From the definition of $T_a F$ set out in 12.3.2(1) it follows immediately that (compare 2.3.3)

$$(T_a F)^\wedge(n) = e^{-ina} \hat{F}(n) \qquad (n \in Z).$$

12.5.6. From 12.4.1 it is immediate that (compare 2.3.4)

$$(DF)^{\wedge}(n) = in \cdot \hat{F}(n) \qquad (n \in Z). \tag{12.5.4}$$

Moreover, 12.4.4 affirms that a Fourier-Schwartz series may be differentiated termwise—provided, of course, that differentiation and convergence are each interpreted in the distributional sense. This conclusion appears also from (12.5.4) and 12.5.3(2).

12.5.7. **Distributions as Derivatives of Functions.** If F is a distribution of order at most m, (12.5.1) shows that

$$f(x) = \hat{F}(0) + \sum_{n \neq 0} \frac{\hat{F}(n)e^{inx}}{(in)^{m+2}}$$

is a continuous function, the series being absolutely and uniformly convergent. Repeated application of (12.5.4) leads to the conclusion that

$$D^{m+2}f = F - \hat{F}(0).$$

Thus F is, apart from the additive constant $\hat{F}(0) = F(1)$, the $(m + 2)$-nd distributional derivative of a continuous function. Compare with (1) in the introduction to this chapter.

12.5.8. **Example.** We return momentarily to consider the distribution $\Omega = \text{P.V.}\ \omega$ discussed in Examples 12.3.2(5) and 12.4.3(3). From 12.5.1 and (12.3.1) we have

$$\begin{aligned}\hat{\Omega}(n) &= \frac{1}{2\pi}\int (e^{-inx} - 1) \operatorname{cosec} \tfrac{1}{2} x \, dx \\ &= \frac{1}{2\pi}\int e^{-\frac{1}{2}inx} \cdot -2i \sin \tfrac{1}{2} nx \cdot \operatorname{cosec} \tfrac{1}{2} x \, dx \\ &= -\frac{i}{\pi}\int \cos \tfrac{1}{2} nx \cdot \sin \tfrac{1}{2} nx \cdot \operatorname{cosec} \tfrac{1}{2} x \, dx \\ &\quad - \frac{1}{\pi}\int \sin^2 \tfrac{1}{2} nx \cdot \operatorname{cosec} \tfrac{1}{2} x \, dx.\end{aligned}$$

The last-written integral vanishes, since the integrand is an odd function, and so

$$\hat{\Omega}(n) = -\frac{i}{2\pi}\int_{-\pi}^{\pi} \sin nx \cdot \operatorname{cosec} \tfrac{1}{2} x \, dx. \tag{12.5.5}$$

From this formula the behavior of $\hat{\Omega}(n)$ for large values of $|n|$ is readily inferred. Since $(\frac{1}{2}x)^{-1} - \operatorname{cosec} \frac{1}{2}x$ is integrable over $(-\pi, \pi)$, the Riemann-

Lebesgue lemma 2.3.8 shows that

$$\begin{aligned}\hat{\Omega}(n) &= -\frac{i}{\pi}\int_{-\pi}^{\pi}\frac{\sin nx}{x}\,dx + o(1)\\ &= -\frac{i}{\pi}\int_{-n\pi}^{n\pi}\frac{\sin y}{y}\,dy + o(1).\end{aligned}$$

Since

$$\int_{-\infty}^{\infty}\frac{\sin y\,dy}{y} = \pi,$$

it follows that

$$\hat{\Omega}(n) = -i\cdot\operatorname{sgn} n + o(1). \tag{12.5.6}$$

The result (12.5.6) should be compared with that expressed by (12.8.1) for the closely related Hilbert distribution H = P.V. cot $\frac{1}{2}x$ to be introduced in Section 12.8.

12.5.9. **Example.** We reconsider the equation

$$DF = A \tag{12.5.7}$$

examined in Example 12.4.3(2), where it was seen that a solution F exists if and only if $A(1) = 0$. The same conclusion may be reached by looking at the transformed equation, namely,

$$in\cdot\hat{F}(n) = \hat{A}(n) \qquad (n \in Z), \tag{12.5.8}$$

which is attained by application of 12.5.6. This shows that indeed a solution exists only if $\hat{A}(0) \equiv A(1) = 0$; and that if this condition is fulfilled, the solutions are given by

$$F = \hat{F}(0)1 + \sum_{n\neq 0}(in)^{-1}\hat{A}(n)e^{inx} \tag{12.5.9}$$

the series being distributionally convergent.

An interesting special case is that in which $A = \varepsilon - 1$. Then (12.5.9) leads to the unique solution $F = E$ satisfying $\hat{E}(0) = 0$, namely,

$$E = \sum_{n\neq 0}(in)^{-1}e^{inx}. \tag{12.5.10}$$

Now the series here is boundedly convergent (in the pointwise sense), as is seen from Exercise 1.5 or 7.2.2(2). The pointwise sum is identifiable with the distributional sum by virtue of the substance of Example 12.3.2(2). Thus E is (generated by) the pointwise sum function, which can be shown to be defined by

$$E(x) = (\pi - |x|)\operatorname{sgn} x \tag{12.5.11}$$

for $|x| \leqslant \pi$.

By using the concept of convolution of distributions introduced in the next section, the solution of (12.5.7) may be expressed in terms of A (assumed to be such that $A(1) = 0$) as $F = \hat{F}(0)1 + E * A$.

For periodic distributions, E plays the role of an *elementary solution* for the differential operator D. The notion of elementary solutions of linear differential and partial differential operators is fundamental in much of the modern work in this field; see [E], Chapter 5. See also Exercise 12.16.

12.5.10. **Measures and Functions of Bounded Variation; Fourier-Stieltjes Series.** We have in 12.2.3 referred to the theorem of F. Riesz according to which each (Radon) measure F is expressible in the form

$$F(u) = \frac{1}{2\pi}\int_{-\pi}^{\pi} u(x)\, d\phi(x) \tag{12.5.12}$$

where ϕ is a suitable (F-dependent) function of bounded variation on $[-\pi, \pi]$. It is furthermore not difficult to verify that, conversely, any such function ϕ defines, via (12.5.12), a Radon measure F; complete details appear in [HS], Section 8 and [AB], Chapter 8. On computing the Fourier coefficients of F, we have from (12.5.12)

$$\hat{F}(n) = \frac{1}{2\pi}\int_{-\pi}^{\pi} e^{-inx}\, d\phi(x),$$

and an application of the formula for partial integration in Riemann-Stieltjes integrals (which in this case is easily reduced to a simple limiting process and partial summation applied to the approximating Riemann-Stieltjes sums; compare the proof of 2.3.6) shows that

$$\hat{F}(n) = \frac{1}{2\pi}[\phi(\pi) - \phi(-\pi)](-1)^n + (in)\cdot\frac{1}{2\pi}\int_{-\pi}^{\pi}\phi(x)e^{-inx}\, dx,$$

that is,

$$\hat{F}(n) = \frac{1}{2\pi}[\phi(\pi) - \phi(-\pi)]\cdot(-1)^n + in\cdot\hat{\phi}(n). \tag{12.5.13}$$

Since (12.5.10) and (12.5.11) indicate that

$$\sum_{n\neq 0}\frac{(-1)^n e^{inx}}{in} = \sum_{n\neq 0}\frac{e^{in(x+\pi)}}{in}$$

is the Fourier-Lebesgue series of a function of bounded variation, reference to (12.5.13) and the fact (to be established in a moment), that the (distributional) derivative of any function of bounded variation is a measure, will show that a trigonometric series

$$\sum_{n\in Z} c_n e^{inx} \tag{12.5.14}$$

is a Fourier-Stieltjes series in the sense defined in 12.5.2, if and only if the formally integrated series

$$\sum_{n \neq 0} (in)^{-1} c_n e^{inx} \tag{12.5.15}$$

is the Fourier-Lebesgue series of some function ϕ of bounded variation; or again if and only if the coefficients c_n are expressible as Riemann-Stieltjes integrals with respect to such a function ϕ in the following fashion:

$$c_n = \frac{1}{2\pi} \int_{-\pi}^{\pi} e^{-inx}\, d\phi(x).$$

It is this circumstance that explains most clearly the use of the qualifier "Fourier-Stieltjes" in this connection.[1]

Inasmuch as the distributional derivative of (12.5.15) is (as follows from 12.5.6) exactly the series (12.5.14) shorn of its constant term, we may infer that any Radon measure μ is expressible in the form

$$\mu = c + D\phi, \tag{12.5.16}$$

where c is a constant and ϕ is a function of bounded variation. The converse is also true.

To verify this last point, it suffices to show that, if ϕ is of bounded variation, then $\mu = D\phi$ is a Radon measure. Now, by (12.4.1),

$$D\phi(u) = -\frac{1}{2\pi} \int Du \cdot \phi\, dx.$$

To majorize the absolute value of this expression, one may either apply partial integration for Riemann-Stieltjes integrals and so obtain

$$|D\phi(u)| = \left|\frac{1}{2\pi} \int u\, d\phi\right| \leqslant \frac{1}{2\pi} \cdot \mathrm{V}(\phi) \cdot \|u\|_\infty;$$

or one can repeat the more pedestrian argument used in the proof of 2.3.6 (based on approximating the integrals by sums), which would again lead to the majorization

$$|D\phi(u)| \leqslant \frac{1}{2\pi} \cdot \mathrm{V}(\phi) \cdot \|u\|_\infty. \tag{12.5.17}$$

Comparing (12.5.17) with (12.2.2), it is seen that $D\phi$ is indeed a Radon measure.

12.6. Convolutions of Distributions

We tackle the problem of defining the convolution of any two distributions by first concentrating on the special case in which one of them is an element of $\mathbf{C}^\infty$.

[1] The function ϕ in (12.5.12) is not generally periodic; however, it may be replaced in the term $in\hat{\phi}(n)$ in (12.5.13), and in (12.5.16) and the proof (12.5.17), by one that is.

The guiding light in framing the definition in this case will be that provided by the definition which applies when the remaining factor in the convolution is a distribution that is (generated by) a function.

If $u \in \mathbf{C}^\infty$ and $f \in \mathbf{L}^1$, and if $\check{u}$ denotes the function $t \to u(-t)$ (see Volume I, p. 31), then

$$f * u(x) = \frac{1}{2\pi}\int f(y)u(x - y)\,dy$$

$$= \frac{1}{2\pi}\int f(y)\check{u}(y - x)\,dy.$$

Regarding f as a distribution, this may be written

$$f * u(x) = f(T_x\check{u}).$$

The expression on the right makes sense even if f be replaced by an arbitrary distribution. Thus we are led to frame the following definition.

12.6.1. **The Convolution $F * u$.** If $F \in \mathbf{D}$ and $u \in \mathbf{C}^\infty$, $F * u$ is defined to be the function for which

$$F * u(x) = F(T_x\check{u}). \tag{12.6.1}$$

From this it is evident that $F * u$ is bilinear in the pair (F, u).

12.6.2. If $F \in \mathbf{D}$ and $u \in \mathbf{C}^\infty$, then $F * u \in \mathbf{C}^\infty$ and

$$D(F * u) = DF * u = F * Du.$$

Similarly,

$$T_a(F * u) = T_aF * u = F * T_au.$$

Proof. The defining equation (12.6.1) gives for any $a \neq 0$:

$$a^{-1}[F * u(x + a) - F * u(x)] = F[a^{-1}(T_{x+a}\check{u} - T_x\check{u})]$$

$$= F[T_xa^{-1}(T_a\check{u} - \check{u})].$$

Now it is easily shown (see Exercise 12.2) that

$$\lim_{a\to 0} a^{-1}(T_a\check{u} - \check{u}) = -D\check{u} = (Du)^\vee$$

in the sense of $\mathbf{C}^\infty$, so that continuity of F shows that $F * u$ is differentiable, that

$$D(F * u) = F[T_x\,(Du)^\vee] = F * Du,$$

and at the same time that

$$D(F * u) = F(-T_x\,D\check{u}) = F(-DT_x\check{u}) = (DF)(T_x\check{u})$$

$$= (DF) * u.$$

Inasmuch as $Du \in \mathbf{C}^\infty$ whenever $u \in \mathbf{C}^\infty$, it follows that $F * u \in \mathbf{C}^\infty$.

The relation involving translates is proved similarly.

12.6.3. (1) If $F \in \mathbf{D}$ and $u_k \to u$ in $\mathbf{C}^\infty$, then $F * u_k \to F * u$ in $\mathbf{C}^\infty$.

(2) If $F_k \to F$ in $\mathbf{D}$ and $u \in \mathbf{C}^\infty$, then $F_k * u \to F * u$ in $\mathbf{C}^\infty$.

Proof. (1) By 12.2.5, for some number $c \geqslant 0$ and some integer m,

$$|F(v)| \leqslant c \cdot \sup_{0 \leqslant p \leqslant m} \|D^p v\|_\infty$$

for each $v \in \mathbf{C}^\infty$. Hence 12.6.2 yields

$$\begin{aligned} |D^q(F * u_k)(x) - D^q(F * u)(x)| &= |F * D^q u_k(x) - F * D^q u(x)| \\ &= |F[T_x(D^q u_k - D^q u)^\vee]| \\ &\leqslant c \cdot \sup_{0 \leqslant p \leqslant m} \|D^p T_x(D^q u_k - D^q u)\|_\infty \\ &\leqslant c \cdot \sup_{0 \leqslant p \leqslant m+q} \|D^p(u_k - u)\|_\infty, \end{aligned}$$

from which the stated result follows, the last-written expression tending to zero as $k \to \infty$.

(2) The proof of this is somewhat deeper. We may and will assume (without loss of generality) that $F = 0$ and aim to show that $F_k * u \to 0$ in $\mathbf{C}^\infty$ as $k \to \infty$. By 12.6.2, if $q \in \{0, 1, 2, \cdots\}$,

$$D^q(F_k * u) = F_k * D^q u.$$

It will therefore suffice to prove that, given any $v \in \mathbf{C}^\infty$, $f_k = F_k * v \to 0$ in $\mathbf{C}$. Now, by (12.6.1) and (12.4.1),

$$f_k(x) = F_k(T_x \check{v}),$$
$$Df_k(x) = F_k(T_x D\check{v})$$

for all indices k and all $x \in T$, the first of which shows that $f_k(x) \to 0$ as $k \to \infty$ for all $x \in T$. The crucial point now is an appeal to Appendix B.2.1(2), bearing in mind 12.1.4 and 12.2.1. This affirms that the F_k are equicontinuous, which signifies that there exists $m \in \{0, 1, 2, \cdots\}$ such that

$$\sup_k |F_k(w)| \leq m \cdot \sup_{0 \leq p \leq m} \|D^p w\|_\infty$$

for all $w \in \mathbf{C}^\infty$. In particular, for all indices k and all $x \in T$,

$$\begin{aligned} |Df_k(x)| &\leq m \cdot \sup_{0 \leq p \leq m} \|D^p T_x D\check{v}\|_\infty \\ &\leq m \cdot \sup_{0 \leq p \leq m+1} \|D^p \check{v}\|_\infty \\ &= B \end{aligned}$$

say, where B is independent of k and x. Thus the f_k are equicontinuous on T. Let $\varepsilon > 0$. Since T is compact, points $x_1, \cdots, x_r$ of T may be chosen so that every point of T is modulo 2π, within distance $2^{-1}(B+1)^{-1}\varepsilon$ of $\{x_1, \cdots, x_r\}$. Then, for all indices k, the mean value theorem shows that

$$\|f_k\|_\infty \leq B \cdot 2^{-1}(B+1)^{-1}\varepsilon + \sup\{|f_k(x_1)|, \cdots, |f_k(x_r)|\};$$

and this is at most ε for all sufficiently large k. Hence $\|f_k\|_\infty \to 0$ as $k \to \infty$, and the proof is complete.

12.6.4. For $u \in \mathbf{C}^\infty$ we have

$$\check{u} = \sum_{n \in Z} \hat{u}(n) e_{-n}$$

and

$$T_x \check{u} = \sum_{n \in Z} e^{inx} \cdot \hat{u}(n) e_{-n},$$

the series converging in $\mathbf{C}^\infty$ (see 12.1.1). Hence the definition 12.6.1 leads to the formula

$$F * u(x) = \sum_{n \in Z} \hat{F}(n) \hat{u}(n) e^{inx}, \tag{12.6.2}$$

which shows that we still have the relation

$$(F * u)\hat{}\,(n) = \hat{F}(n)\hat{u}(n) \qquad (n \in Z);$$

the series (12.6.2) is absolutely and uniformly convergent thanks to (12.1.4) and (12.5.1).

12.6.5. **Convolution of Distributions.** On the basis of 12.6.3 we are now able to define $F * G$ as a distribution for arbitrary $F, G \in \mathbf{D}$. Thus, we define

$$F * G(u) = F[(G * \check{u})^\vee] \tag{12.6.3}$$

for each $u \in \mathbf{C}^\infty$. This makes $F * G$ a linear functional on $\mathbf{C}^\infty$ whose continuity follows from 12.6.3(1).

Taking $u = e_{-n}$ in (12.6.3) and using (12.6.2) and the orthogonality relations (1.3.1), it is seen at once that

$$(F * G)\hat{}\,(n) = F[\hat{G}(n) e_{-n}] = \hat{F}(n)\hat{G}(n) \qquad (n \in Z). \tag{12.6.4}$$

This last equation shows first that if $G = u \in \mathbf{C}^\infty$, then the present definition of $F * u$ agrees with that prescribed in 12.6.1 (provided, of course, that a function in $\mathbf{C}^\infty$ is identified with the distribution it generates). In view of 12.5.4(1), equation (12.6.4) implies secondly the commutativity of convolution:

$$F * G = G * F,$$

and likewise the associativity and distributivity of convolution.

Either of (12.6.3) or (12.6.4) shows that the mapping $(F, G) \to F * G$ is bilinear from $\mathbf{D} \times \mathbf{D}$ into $\mathbf{D}$.

From 12.5.5, 12.5.6, and (12.6.4) (or by other means) it is easy to verify that

$$T_a(F * G) = T_a F * G = F * T_a G, \tag{12.6.5}$$

$$D(F * G) = DF * G = F * DG. \tag{12.6.6}$$

12.6.6. If $F_k \to F$ in $\mathbf{D}$, then $F_k * G \to F * G$ in $\mathbf{D}$.

Proof. We know from 12.6.2 that $(G * \check{u})^{\vee} \in \mathbf{C}^{\infty}$ for each $u \in \mathbf{C}^{\infty}$, and so

$$F_k * G(u) = F_k[(G * \check{u})^{\vee}] \to F[(G * \check{u})^{\vee}] = F * G(u),$$

which signifies that $F_k * G \to F * G$ in $\mathbf{D}$.

Remark. It is even true that $F_k * G_k \to F * G$ in $\mathbf{D}$ whenever $F_k \to F$ and $G_k \to G$ in the same sense, but the proof is rather more difficult. Since we shall nowhere need this stronger assertion, the proof is omitted.

12.6.7. **Examples.** (1) The Dirac measure ε is the identity element for convolution, that is, $\varepsilon * F = F$ for all $F \in \mathbf{D}$. This explains how it comes about that all convolution algebras of *functions*, large enough to contain all trigonometric polynomials, lack identity elements.

More generally, $\varepsilon_a * F = T_a F$ for $F \in \mathbf{D}$ and $a \in T$.

(2) By (12.6.6), $DF = (D\varepsilon) * F$ for every $F \in \mathbf{D}$. Thus differentiation can be expressed as convolution with the fixed distribution $D\varepsilon$.

(3) Given $A, B \in \mathbf{D}$, the equation

$$A * F = B$$

has a solution $F \in \mathbf{D}$ if and only if there is an integer $m \geqslant 0$ such that

$$|\hat{B}(n)| \leqslant \text{const } (1 + |n|)^m |\hat{A}(n)| \qquad (n \in Z),$$

in which case the solutions are of the form

$$F = F_0 + \sum_{\hat{A}(n) \neq 0} \frac{\hat{B}(n) e_n}{\hat{A}(n)},$$

where $A * F_0 = 0$, that is,

$$F_0 = \sum_{\hat{A}(n) = 0} c_n e_n,$$

where (c_n) is some tempered sequence.

These assertions follow readily from (12.6.4) and 12.5.3(2).

(4) In order that the equation

$$A * F = B$$

shall have a solution $F \in \mathbf{D}$ for an arbitrarily given $B \in \mathbf{D}$, it is necessary and sufficient that there exist a number $c > 0$ and an integer $m \geqslant 0$ such that

$$|\hat{A}(n)| \geqslant c \cdot (1 + |n|)^{-m} \qquad (n \in Z).$$

This statement is verifiable by reference to (3) and the remark that the said equation is soluble for every B if and only if it is soluble when $B = \varepsilon$ (as follows from the fact that ε is the identity for convolution, together with associativity of convolution).

12.6.8. **Definition of $\bar{F}$, $\check{F}$, and F^*.** The definitions of $\bar{f}$, $\check{f}$, and f^*, given for functions immediately prior to 2.3.1, may be extended to distributions via the formulae

$$\bar{F}(u) = \overline{F(\bar{u})}, \qquad \check{F}(u) = F(\check{u}), \qquad F^* = (\bar{F})^{\vee} = (\check{F})^{-}.$$

Then the map $F \to \check{F}$ is linear, while $F \to F^*$ and $F \to \bar{F}$ are conjugate-linear. It is moreover easily verified that $(F^*)^{\wedge} = (\hat{F})^{-}$ (compare with 2.3.1).

If μ is a measure, each of $\bar{\mu}$, $\check{\mu}$, and μ^* is a measure, and (see 12.3.8) $\|\bar{\mu}\|_1 = \|\check{\mu}\|_1 = \|\mu^*\|_1 = \|\mu\|_1$.

12.6.9. **A Glance at the Dual Situation.** Let us pause in order to bring into somewhat sharper focus the dual problem raised in Section 3.4.

The position now is that the Fourier transform $\hat{\phi}$ is defined as a distribution whenever ϕ is a tempered function on Z (see 12.5.4(2)), and the problem to be faced concerns the validity of the equation (3.4.1), namely,

$$(\phi * \psi)^{\wedge} = \hat{\phi} \cdot \hat{\psi}. \tag{12.6.7}$$

Even now, however, this heuristic equation lacks meaning if ϕ and ψ are unrestricted tempered functions on Z. Thus

(1) $\phi * \psi$ is defined only for restricted pairs ϕ, ψ;

(2) $\hat{\phi} \cdot \hat{\psi}$ is likewise defined only for restricted pairs ϕ, ψ.

Regarding (1), it would be desirable to undertake a thorough examination of the convolution process as applied to functions on Z; this might be done along the lines usually followed in the case in which the underlying group is R (see [E], 4.19 and the references cited there), but the task is not one that can be undertaken here. Regarding (2), see 12.3.5.

It must suffice for the moment for us to remark that (12.6.7) is quite easily established on a sound footing whenever ϕ and ψ belong to ℓ^1 (a case already mentioned in Section 3.4), and again whenever at least one of ϕ and ψ, say ϕ, is rapidly vanishing at infinity, that is, is such that

$$\lim_{|n| \to \infty} \phi(n) n^k = 0 \qquad (k = 1, 2, \cdots).$$

In this latter case $\hat{\phi}$ belongs to $\mathbf{C}^{\infty}$ thanks to 12.1.1, and $\hat{\phi} \cdot \hat{\psi}$ is well-defined as in 12.3.4.

Another important case is discussed in 12.11.3.

12.7 More about M and $\mathbf{L}^p$

In this section we shall set forth some further properties of **M** and $\mathbf{L}^p$ in relation to convolution and establish some properties of the sets of Fourier coefficients of measures and of functions in $\mathbf{L}^p$. In particular, we shall show that **M** is a Banach algebra under convolution (a fact to which partial reference has already been made in 3.3.2, 4.2.5 and 11.4.1(1)). The structure of this Banach algebra **M** is still far from being known in its finer detail and appears to be a problem of great complexity.

To begin with simpler things, we recall that in 12.3.8 a norm $\|\cdot\|_1$ has been defined on **M** which extends that on $\mathbf{L}^1$; this norm is just that dual to the norm on **C**, when we identify **M** with the set of all continuous linear functionals on **C** (see 12.2.9 and I, B.1.7). The explicit formula for this norm is

$$\|\mu\|_1 = \sup\{|\mu(f)| : f \in \mathbf{C}, \|f\|_\infty \leqslant 1\}. \tag{12.7.1}$$

12.7.1. **M** is complete (and hence a Banach space).

Proof. A direct proof is called for in Exercise 12.8. We here give an alternative proof based upon 12.3.9. Suppose that $(\mu_n)_{n=1}^\infty$ is a Cauchy sequence in **M**. It is then clear that

$$\sup_n \|\mu_n\|_1 < \infty.$$

By 12.3.9, therefore, a subsequence $(\mu_{n_k})_{k=1}^\infty$ may be extracted such that, for some $\mu \in \mathbf{M}$, we have

$$\lim_{k \to \infty} \mu_{n_k}(f) = \mu(f)$$

for each $f \in \mathbf{C}$. Now the Cauchy character of the original sequence signifies that to each $\varepsilon > 0$ corresponds $n(\varepsilon)$ so that

$$\|\mu_m - \mu_n\|_1 \leqslant \varepsilon \qquad \text{for all } m, n > n(\varepsilon).$$

Accordingly, by (12.7.1),

$$|\mu_m(f) - \mu_n(f)| \leqslant \varepsilon \cdot \|f\|_\infty \qquad \text{for all } m, n > n(\varepsilon),$$

for any $f \in \mathbf{C}$. Taking herein $m = n_k$ and letting $k \to \infty$, it follows that

$$|\mu(f) - \mu_n(f)| \leq \varepsilon \cdot \|f\|_\infty \qquad \text{for all } n > n(\varepsilon),$$

and hence, by reference to (12.7.1) again,

$$\|\mu - \mu_n\|_1 \leqslant \varepsilon \qquad \text{for all } n > n(\varepsilon).$$

This shows that $\lim_{n \to \infty} \mu_n = \mu$ for the normed topology and exhibits the completeness of **M**.

12.7.2. **Convolutions of Measures and Functions.** Since $\mathbf{M} \subset \mathbf{D}$, $\mu * u$ is defined for each $u \in \mathbf{C}^\infty$ as in 12.6.1. However, using 12.2.9, we can define

$\mu * f$ for any $f \in \mathbf{C}$ in a consistent fashion by setting

$$\mu * f(x) = \mu(T_x \check{f});$$

compare with (12.6.1). Since

$$\| T_x \check{f} - T_{x_0} \check{f} \|_\infty \to 0$$

as $x \to x_0$ (since f is uniformly continuous), (12.7.1) ensures that $\mu * f \in \mathbf{C}$ and that

$$\| \mu * f \|_\infty \leqslant \| \mu \|_1 \cdot \| f \|_\infty .$$

But the end is still not reached: $\mu * f$ can be defined for still more general functions f in such a way as to extend 3.1.6.

12.7.3. (1) Suppose that $1 \leqslant p \leqslant \infty$. If $\mu \in \mathbf{M}$ and $f \in \mathbf{L}^p$, then $\mu * f \in \mathbf{L}^p$ and

$$\| \mu * f \|_p \leqslant \| \mu \|_1 \cdot \| f \|_p .$$

(2) If λ, $\mu \in \mathbf{M}$, then $\lambda * \mu \in \mathbf{M}$ and

$$\| \lambda * \mu \|_1 \leqslant \| \lambda \|_1 \cdot \| \mu \|_1 .$$

Proof. (1) Suppose first that $f, g \in \mathbf{C}$. For any partition $-\pi = x_0 < x_1 < \cdots < x_n = \pi$ we have

$$\frac{1}{2\pi} \sum_{k=1}^{n} \mu * f(x_k) g(x_k)\, \Delta x_k = \frac{1}{2\pi} \sum_k \mu(T_{x_k} \check{f}) \cdot g(x_k)\, \Delta x_k$$

$$= \mu\Big[\frac{1}{2\pi} \sum_k T_{x_k} \check{f} \cdot g(x_k)\, \Delta x_k\Big]. \qquad (12.7.2)$$

If we take a sequence of such partitions for which $\max_k \Delta x_k \to 0$ then, owing to uniform continuity of $\mu * f$ and g, the initial member of (12.7.2) converges to

$$\frac{1}{2\pi} \int \mu * f(x) g(x)\, dx .$$

At the same time, because of uniform continuity of f and g, the functions

$$\frac{1}{2\pi} \sum_k T_{x_k} \check{f} \cdot g(x_k)\, \Delta x_k$$

converge uniformly to the function

$$y \to \frac{1}{2\pi} \int T_x \check{f}(y) g(x)\, dx$$

$$= \frac{1}{2\pi} \int \check{f}(y - x) g(x)\, dx = \check{f} * g(y).$$

Consequently, the last member of (12.7.2) converges to $\mu(\check{f} * g)$. We thus infer that

$$\frac{1}{2\pi}\int (\mu * f) \cdot g\, dx = \mu(\check{f} * g).$$

By 3.1.4 and (12.7.1), therefore,

$$\left|\frac{1}{2\pi}\int (\mu * f) g\, dx\right| \leqslant \|\mu\|_1 \cdot \|f * g\|_\infty \leqslant \|\mu\|_1 \cdot \|f\|_p \cdot \|g\|_{p'}.$$

The converse of Hölder's inequality (see Exercise 3.6) leads thence to the inequality

$$\|\mu * f\|_p \leqslant \|\mu\|_1 \cdot \|f\|_p \tag{12.7.3}$$

for $f \in \mathbf{C}$.

Suppose now that $f \in \mathbf{L}^p$ and $p < \infty$. One may then choose a sequence $(f_n)_{n=1}^\infty$ from $\mathbf{C}$ such that $\|f - f_n\|_p \to 0$ as $n \to \infty$. Reference to (12.7.3) confirms that $(\mu * f_n)_{n=1}^\infty$ is a Cauchy sequence in $\mathbf{L}^p$ [apply (12.7.3) with f replaced by $f_m - f_n$]. By completeness of $\mathbf{L}^p$, the functions $\mu * f_n$ therefore converge in mean in $\mathbf{L}^p$ to some $h \in \mathbf{L}^p$. However, by 12.6.6, $\mu * f_n$ converges in $\mathbf{D}$ to $\mu * f$. It follows at once that $\mu * f$ and h are the same distribution, so that $\mu * f \in \mathbf{L}^p$ and [by (12.7.3)]

$$\|\mu * f\|_p = \|h\|_p = \lim_{n\to\infty} \|\mu * f_n\|_p \leqslant \|\mu\|_1 \cdot \|f\|_p.$$

Finally, if $p = \infty$, we may choose the $f_n \in \mathbf{C}$ so that $\|f_n\|_\infty \leqslant \|f\|_\infty$ and $f_n \to f$ pointwise almost everywhere. Then (see (12.7.3))

$$\|\mu * f_n\|_\infty \leqslant \|\mu\|_1 \cdot \|f_n\|_\infty \leqslant \|\mu\|_1 \cdot \|f\|_\infty.$$

By the case $p = 1$, $\mu * f_n \to \mu * f$ in mean in $\mathbf{L}^1$, so that a suitably chosen subsequence is pointwise convergent almost everywhere. From this we conclude that $\|\mu * f\|_\infty \leqslant \|\mu\|_1 \cdot \|f\|_\infty$, so completing the proof of (1).

(2) By 12.6.5, $\lambda * \mu$ is the distribution defined by

$$\lambda * \mu(u) = \lambda[(\mu * \check{u})^\vee] \qquad (u \in \mathbf{C}^\infty).$$

By 12.6.2, $(\mu * \check{u})^\vee \in \mathbf{C}^\infty$ and, by 12.7.2,

$$\|(\mu * \check{u})^\vee\|_\infty \leqslant \|\mu\|_1 \cdot \|u\|_\infty.$$

Hence

$$|\lambda * \mu(u)| \leqslant \|\lambda\|_1 \cdot \|\mu\|_1 \cdot \|u\|_\infty,$$

and so (see 12.2.6) $\lambda * \mu \in \mathbf{M}$ and $\|\lambda * \mu\|_1 \leqslant \|\lambda\|_1 \cdot \|\mu\|_1$.

Remark. If $1 < p < \infty$ the inequality in 12.7.3(1) can be improved; see Exercise 13.5.

12.7.4. **M as an Algebra and Related Problems.** The results of Section 12.6, together with 12.7.1 and 12.7.3(2), affirm that **M** is a complex Banach algebra under convolution with ε as its identity element (see 11.4.1); it is in fact another strong contender for the title of "group algebra" (see Section 3.3). In a few respects it is easier, but in most respects more difficult, to handle than $\mathbf{L}^1$. In particular, the important problem of determining the nontrivial complex homomorphisms of $\mathbf{M}$ is much more difficult than the analogous problem for $\mathbf{L}^1$ (which was solved in 4.1.2) and cannot even now be regarded as satisfactorily solved. In outline the situation is still much as described in Chapter 3 of [Hew].

To illustrate the difficulties and the extent of existing ignorance, we recall from 4.1.2 that the only nontrivial complex homomorphisms of $\mathbf{L}^1$ are of the type $\gamma_n: f \to \hat{f}(n)$, where n ranges over Z. Each of these homomorphisms has a natural extension to $\mathbf{M}$ [see (12.6.4)], which we continue to denote by γ_n. These, however, do *not* exhaust the nontrivial complex homomorphisms of $\mathbf{M}$. There are several ways of supporting this statement; we briefly describe two of them. The first is concerned with a purely existential proof, while the second is considerably more precise and interesting.

(1) Considered as linear functionals on $\mathbf{M}$, the γ_n are equicontinuous. Applying I, B.4.2 to the sequence $(\gamma_n)_{n=1}^{\infty}$, one derives the existence of a continuous linear functional γ_∞ on $\mathbf{M}$ with the following property: given any $\varepsilon > 0$, any finite subset $\{\mu_1, \cdots, \mu_r\}$ of $\mathbf{M}$, and any integer n_0, there exists an integer $n > n_0$ for which

$$|\gamma_n(\mu_i) - \gamma_\infty(\mu_i)| < \varepsilon \qquad (i = 1, 2, \cdots, r).^1 \tag{12.7.4}$$

The relations (12.7.4) may be shown to imply that γ_∞ is a (continuous) complex homomorphism of $\mathbf{M}$, and that $\gamma_\infty \mid \mathbf{L}^1 = 0$. In particular, γ_∞ is certainly distinct from all the γ_n. Since (12.7.4) also entails that $\gamma_\infty(\varepsilon) = 1$, γ_∞ is nontrivial.

(2) Inasmuch as $\gamma_n(\mu^*) = \overline{\gamma_n(\mu)}$ for all $n \in Z$ and all $\mu \in \mathbf{M}$, any complex homomorphism γ of $\mathbf{M}$ for which $\gamma(\mu^*) \neq \overline{\gamma(\mu)}$ holds for some $\mu \in \mathbf{M}$ is necessarily nontrivial and distinct from all the γ_n. Such homomorphisms γ actually do exist, and various ways of producing them have been discussed. One method is discussed in [R], Theorem 5.3.4.

The existence of such homomorphisms γ of $\mathbf{M}$, which is usually expressed by saying that the Banach algebra $\mathbf{M}$ is *asymmetric*, is one of the most striking and most significant differences between $\mathbf{M}$ and $\mathbf{L}^1$ considered as Banach algebras. A direct corollary of this asymmetry of $\mathbf{M}$ is the existence of real-valued functions ϕ on Z such that $\phi(n) \geqslant 1$ for $n \in Z$ and ϕ is a Fourier-Stieltjes transform but ϕ^{-1} is not a Fourier-Stieltjes transform. This asymmetry and its corollary are known to hold whenever T is replaced by any nondiscrete group (see [R], Theorem 5.3.4; [HR], Vol. 2, p. 519; MR **43** # **6659**); the corollary was first noted for the case in which the underlying group is R by Wiener and Pitt in 1938.

[1] I,B.4.1 is not applicable, since $\mathbf{M}$ is not separable.

Even more surprising (not to say shocking) things can happen, as the reader will discover on looking at Section 5.4 of [R] and pp. 143–144 of [Hew]. Much of the early work on complex homomorphisms of **M** is due to the Russian mathematician Šreĭder; later elaborations are due to Hewitt [2] and Hewitt and Kakutani [1], [2] jointly. See also [Ta]; Taylor [1]–[11]; Johnson [1]–[6]; MR **38** # # 6308, 6309; **41** # **2409**.

There are some subalgebras of **M** of moderate interest for which the difficulties so far described are to some extent surmountable; see Exercises 12.51 and 16.28.

We have mentioned in Section 4.2 the problem of homomorphisms of $\mathbf{L}^1$ into itself. This, together with the problem of homomorphisms of $\mathbf{L}^1$ into **M**, has been solved for general groups by Cohen [2], [3], an essential step being the determination of all the idempotents in **M** (again for a general group). See (3) immediately below; also MR **41** # 8929; **42** # 6518; **54** # 5741.

In view of the mystery surrounding the complex homomorphisms of **M**, it is not surprising that the problem of homomorphisms of **M** raises new difficulties. One knows (see 4.2.6) which maps α of Z into $Z \cup \{\infty\}$ define homomorphisms T of $\mathbf{L}^1$ into **M** by means of the formula $(Tf)^\wedge = \hat{f} \circ \alpha$; and, as appears in Exercise 12.49, each such homomorphism can be extended into a homomorphism T' of **M** into itself such that $(T'\mu)^\wedge = \hat{\mu} \circ \alpha$ for $\mu \in \mathbf{M}$, $\hat{\mu}(\infty)$ being understood to be 0 whenever $\mu \in \mathbf{M}$. However, there exist homomorphisms of **M** that do not arise by thus extending a homomorphism of $\mathbf{L}^1$ into **M**; in fact, there exist homomorphisms $T' \neq 0$ of **M** such that $T'|\mathbf{L}^1 = 0$. One type of such homomorphisms is described in Section 3.4.1 of [R]. Alternatively, it suffices to define $T'\mu = \gamma(\mu)\iota$, where γ is any nonzero complex homomorphism of **M** such that $\gamma|\mathbf{L}^1 = 0$, and ι is any nonzero idempotent in **M**. The classification of such homomorphisms of **M** is, as far as the author is aware, still largely unsolved.

Concerning norm-decreasing homomorphisms of $\mathbf{L}^1$ into **M**, and the same problem for more general underlying groups, see Glicksberg [1] and Greenleaf [1]. (In these papers, the underlying group is not assumed to be Abelian.)

Further reading on some of the topics mentioned above: Brown and Hewitt [1], MR **22** # 9809; **36** # 6879; **37** # # 4222, 4224, 6693; **38** # # 489, 491; **44** # # 1993, 1994; **48** # # 2666, 4642; **49** # 9539; **50** # # 5359, 7950; **51** # # 3798, 6287, 8737; **52** # # 8796, 8797, 14846, 14848; **54** # # 3291, 3292, 5743, 5744, 8163; **56** # # 982, 3571, 3572, 16254, 1659; **57** # # 7034, 7037.

(3) *Helson's theorem.* In 3.1.1(d) we have remarked that the only idempotents in $\mathbf{L}^1$ are trigonometric polynomials $\sum_{n \in F} e^{inx}$, where F is a finite subset of Z. But in **M** there are many other idempotents. In more concrete terms, the problem is that of determining which subsets S of Z are such that $\sum_{n \in S} e^{inx}$ is a Fourier-Stieltjes series. The solution was given by Helson [2], [3]; and the analogous problem for general groups was solved by P. J. Cohen [3]. For details, see [R], Chapter 3.

Helson's result is very simple to state: $\sum_{n \in S} e^{inx}$ is a Fourier-Stieltjes series if and only if S differs by a finite set from some periodic subset of Z.

The "if" part of Helson's theorem is simple to prove; see Exercise 12.48. A special case of the "only if" part is discussed in Exercise 12.46. See also 16.8.4.

In 12.7.5 we shall give a much more rudimentary necessary and sufficient condition in order that a general trigonometric series $\sum_{n \in Z} c_n e^{inx}$ shall be a Fourier-Stieltjes series. This will be applied in 12.7.8 to show directly (that is, without any appeal to Helson's theorem) that in particular $\sum_{n \geqslant 0} e^{inx}$ is *not* a Fourier-Stieltjes series. This example is especially significant in connection with conjugate series (see Section 12.8).

It is a simple consequence of 9.2.4, 9.2.8 and the substance of 12.5.10 that, if $(c_n)_{n \in Z}$ is odd and $c_n \geqslant 0$ $(n \in Z,\ n > 0)$, then a necessary condition in order that

$$\sum_{n \in Z} c_n e^{inx} = \sum_{n=1}^{\infty} 2ic_n \sin nx$$

be a Fourier-Stieltjes series is that

$$\sum_{n=1}^{\infty} \frac{c_n}{n} < \infty\,;$$

in particular, if also $(c_n)_{n=0}^{\infty}$ is ultimately periodic, then the said series is a Fourier-Stieltjes series only if the periodic part is zero. Inspired by some remarks of Helson, Goes [2], [3] has proved a great deal more of the same nature concerning series $\sum_{n=1}^{\infty} c_n \cos nx$ and $\sum_{n=1}^{\infty} c_n \sin nx$ in which the coefficients exhibit ultimately periodic or almost periodic features. He shows in particular that, if $(n_k)_{k=1}^{\infty}$ is any strictly increasing sequence of positive integers, then $\sum_{k=1}^{\infty} \sin n_k x$ is never a Fourier-Stieltjes series.

(4) *Littlewood's conjecture*. Finally we mention an interesting issue relating to idempotents. Let

$$m(k) = \inf \Big\| \sum_{j=1}^{k} e^{in_j x} \Big\|_1,$$

the infimum being taken over all sets of k distinct integers $n_1, \cdots, n_k$. What is the true order of magnitude of $m(k)$ as $k \to \infty$? If the n_j are in arithmetic progression, then $m(k) \geqslant \text{const} \log k$; Littlewood conjectured that this inequality is valid in general. Davenport proved in 1960 that

$$m(k) \geqslant 1/8[\log k/\log \log k]^{1/4}$$

for all sufficiently large k. This estimate was extended to all compact Abelian groups by Hewitt and Zuckerman [2]. For some related results, see Salem [1], Uchiyama [1], Fournier [2], Pichorides [1], [2], [3], MR **40** # 6150. Just before this book went to press, McGehee, Pigno and Smith [1], [2] jointly announced a proof of a generalized Littlewood conjecture. In fact, they prove the following. Assume that $S = \{n_1 < n_2 < \cdots\}$ is a subset of Z: then

(i) (Generalized Hardy Inequality) For every S-spectral measure μ (see 15.1.1 below),

$$\sum_{k=1}^{\infty} k^{-1} |\hat{\mu}(n_k)| \leqslant 30 \| \mu \|_1\,;$$

(ii) (Generalized Littlewood Inequality) for every positive natural number N and every complex-valued sequence $(c_k)_{k=1}^{N}$ such that $|c_k| \geqslant 1$ for every $k \in \{1, 2, \cdots, N\}$,

$$\left\| \sum_{k=1}^{N} c_k e_{n_k} \right\|_1 \geqslant (30)^{-1} \log N;$$

(iii) for every complex-valued sequence $(c_k)_{k=1}^{\infty}$ such that $|c_k| \leqslant k^{-1}$ for every $k \in \{1, 2, \cdots\}$, there exists $F \in \mathbf{L}^\infty$ such that $\| F \|_\infty \leqslant 30$ and $\hat{F}(n_k) = c_k$ for every $k \in \{1, 2, \cdots\}$.

The significance of estimates of the type involved in the Littlewood Conjecture is indicated in part by their application to the study of idempotents in measure algebras, which in turn is related to the problem of homomorphisms of measure algebras (Cohen [2], [3]); see Section 4.2 and Subsection 16.8.4.

For remarks concerning the Littlewood conjecture in a more general setting, see Price [4].

12.7.5. **Criterion for Fourier-Stieltjes Series.** Let $(c_n)_{n \in Z}$ be a given sequence and put

$$s_N(x) = \sum_{|n| \leqslant N} c_n e^{inx}, \tag{12.7.5}$$

$$\sigma_N(x) = (N+1)^{-1}[s_0(x) + \cdots + s_N(x)]$$
$$= \sum_{|n| \leqslant N} \left(1 - \frac{|n|}{N+1}\right) c_n e^{inx}. \tag{12.7.6}$$

In order that $\sum_{n \in Z} c_n e^{inx}$ be a Fourier-Stieltjes series, it is necessary and sufficient that

$$\limsup_{N \to \infty} \|\sigma_N\|_1 < \infty. \tag{12.7.7}$$

Proof. If $\sum_{n \in Z} c_n e^{inx}$ is a Fourier-Stieltjes series, there exists a measure $\mu \in \mathbf{M}$ such that $c_n = \mu(n)$ for $n \in Z$. Hence (compare (5.1.6))

$$\sigma_N = \mu * F_N,$$

as follows from (12.6.4) and 12.5.4(1). Then 12.7.3 shows that

$$\|\sigma_N\|_1 \leqslant \|\mu\|_1 \cdot \|F_N\|_1 = \|\mu\|_1,$$

and (12.7.7) is visibly fulfilled.

Conversely, suppose that (12.7.7) holds. Then evidently

$$\sup_N \|\sigma_N\|_1 < \infty,$$

and 12.3.9 entails that there exists a measure $\mu \in \mathbf{M}$ and a subsequence $(\sigma_{N_k})_{k=1}^{\infty}$ such that

$$\lim_{k \to \infty} \sigma_{N_k}(f) = \mu(f)$$

for each $f \in \mathbf{C}$. Taking $f = e_{-n}$, this gives

$$\lim_{k \to \infty} \hat{\sigma}_{N_k}(n) = \hat{\mu}(n) \qquad (n \in Z).$$

The limit on the left-hand side here is just c_n, as reference to (12.7.6) will confirm. Thus $c_n = \hat{\mu}(n)$ for all $n \in Z$ and $\sum_{n \in Z} c_n e^{inx}$ is a Fourier-Stieltjes series.

Remark. Some deeper questions are mentioned in 12.7.9. See also Exercise 12.50.

12.7.6. **Criterion for Fourier Series of Class $\mathbf{L}^p$ ($p > 1$).** The notation being as in 12.7.5, suppose further that $1 < p \leqslant \infty$. In order that $\sum_{n \in Z} c_n e^{inx}$ be the Fourier series of a function in $\mathbf{L}^p$, it is necessary and sufficient that

$$\limsup_{N \to \infty} \|\sigma_N\|_p < \infty. \tag{12.7.8}$$

Proof. This proceeds on exactly the same lines as does that of 12.7.5, appealing to 12.3.10(2) in place of 12.3.9. We leave to the reader the task of filling in the details.

12.7.7. **Remark.** When $p = 1$, the analogue of 12.7.6 is false, one of several possible corrected versions being in fact 12.7.5. In the last resort this breakdown is due to the fact that $\mathbf{L}^1$ is not the dual of $\mathbf{L}^\infty$ in the same sense that $\mathbf{L}^{q'}$ is the dual of $\mathbf{L}^q$ when $1 \leqslant q < \infty$ (see I, C.1). Some other corrected versions will be discussed briefly in 12.7.9.

12.7.8. **Example.** Let us use 12.7.5 to show that $\sum_{n \geqslant 0} e^{inx}$ is not a Fourier-Stieltjes series.

Direct calculation shows that here

$$s_N(x) = (1 - e^{ix})^{-1}[1 - e^{i(N+1)x}]$$

and

$$\sigma_N(x) = (1 - e^{ix})^{-1}\{1 - [(N+1)(1 - e^{ix})]^{-1}[e^{ix} - e^{i(N+2)x}]\}.$$

Since $|1 - e^{ix}| = 2|\sin \tfrac{1}{2}x|$, it follows that for large N we have

$$\begin{aligned} |\sigma_N(x)| &\geqslant \tfrac{1}{2}\,|\operatorname{cosec} \tfrac{1}{2}x|\left[1 - \frac{1}{(N+1)|\sin \tfrac{1}{2}x|}\right] \\ &\geqslant \tfrac{1}{2} \operatorname{cosec} \tfrac{1}{2}x\left[1 - \frac{\pi}{(N+1)|x|}\right] \\ &\geqslant \tfrac{1}{4} \operatorname{cosec} \tfrac{1}{2}x \end{aligned}$$

for $2\pi/(N + 1) \leqslant x \leqslant \pi$. Hence

$$\|\sigma_N\|_1 \geqslant \frac{1}{2\pi}\int_{2\pi/(N+1)}^{\pi} \operatorname{cosec} \tfrac{1}{2}x\,dx,$$

which tends to infinity with N. Thus (12.7.7) is violated and $\sum_{n \geqslant 0} e^{inx}$ is not a Fourier-Stieltjes series.

From this example it may be inferred easily that if

$$c_n = \alpha + \beta \cdot \operatorname{sgn} n + b_n,$$

where α and β are constants, $\beta \neq 0$, and $\sum_{n \in Z} |b_n|^2 < \infty$, then $\sum_{n \in Z} c_n e^{inx}$ is not a Fourier-Stieltjes series.

An alternative argument runs thus. Bochner's theorem 9.2.8 combines with 12.7.2 or 12.7.3 to show that, if $f \in \mathbf{L}^\infty$ and $\hat{f} \notin l^1(Z)$, then sgn $\hat{f}$ is not a Fourier-Stieltjes transform. Applying this with f taken to be the function such that

$$f(x) = i \sum_{n=1}^{\infty} n^{-1} \sin nx \qquad \text{for all } x \in R$$

(or

$$f(x) = i \sum_{n=1}^{\infty} n^{-1}(\log(1 + n))^{-1} \sin nx \qquad \text{for all } x \in R),$$

using 7.2.2, and noting that

$$\operatorname{sgn} \hat{f}(n) = \operatorname{sgn} n \qquad \text{for all } n \in Z,$$

it follows that the sequence whose n-th term is

$$\tfrac{1}{2}(1 + \operatorname{sgn} \hat{f}(n)) = 1 \text{ or } 0 \quad \text{according as} \quad n > 0 \text{ or } n < 0$$

is not a Fourier-Stieltjes transform.

For many other examples, see Exercise 12.37.

Remark. Concerning maps α of Z into itself such that

$$\sum_{n \in Z} e^{i\alpha(n) + inx}$$

is a Fourier-Stieltjes series, see Edwards [13].

12.7.9. Criteria for Fourier-Lebesgue Series; the Steinhaus-Littlewood Problem. We revert to the matters mentioned in 12.7.7. Let $c = (c_n)$ be a function on Z and consider the trigonometric series

$$S : \sum_{n \in Z} c_n e^{inx} \tag{12.7.9}$$

and its partial sums

$$s_N(x) = \sum_{|n| \leqslant N} c_n e^{inx}. \tag{12.7.10}$$

As follows from 12.7.5, the condition

$$\sup_N \|s_N\|_1 < \infty \tag{12.7.11}$$

is *sufficient* in order that S be a Fourier-Stieltjes series. Helson [1] (see also [Z_1], p. 286) showed that (12.7.11) also entails that $c \in \mathbf{c}_0(Z)$. On the other hand, Mary Weiss [1] proved that (12.7.11) does *not* ensure that S is a Fourier-Lebesgue series; see also Hewitt and Zuckerman [4]. (Why such a failure is to be expected was explained in 12.7.7.) We proceed to consider briefly two developments suggested by this breakdown, the second of which constitutes a necessary and sufficient condition on (c_n) in order that S be a Fourier-Lebesgue series. (As always, the criterion is very difficult to apply to specific examples.)

(1) It seems that both Steinhaus and Littlewood are responsible for the following question: given that

$$s_N(x) \geqslant 0 \qquad \text{for all real } x \text{ and } N = 1, 2, \cdots, \tag{12.7.12}$$

does it follows that S is a Fourier-Lebesgue series?

Inasmuch as (12.7.12) implies (12.7.11), it ensures (by 12.7.5) that S is a Fourier-Stieltjes series; see also Exercise 12.25.

Steinhaus himself showed that if $\lim_{N \to \infty} s_N(x)$ exists finitely and is non-negative for every real x, then S is a Fourier-Lebesgue series; see [Ba_1], p. 244 and [Ba_2], p. 353. In this connection we remark that if

$$c_n = \begin{cases} (\log |n|)^{-1} & \text{for } |n| \geqslant 2, \\ 0 & \text{for } n = 0, \pm 1 \end{cases}$$

then $f(x) \equiv \lim_{N \to \infty} s_N(x)$ exists finitely for all $x \not\equiv 0 \bmod 2\pi$ and $f(x) \geqslant 0$ for all such x. It is moreover true (but far from trivial) that $f \in \mathbf{L}^1$ and that S is the Fourier-Lebesgue series of f; this follows from a general theorem (see [Ba_2], p. 353). From Exercise 13.1 it then appears that f does not belong to $\mathbf{L}^p$ for any $p > 1$. This shows that Steinhaus's result is in a sense the best possible.

As to the Steinhaus-Littlewood question itself, a negative answer was established very recently by Katznelson [1], who constructed an example of a series S satisfying (12.7.12) but which is not a Fourier-Lebesgue series: in this example, the measure μ for which $\hat{\mu} = c$ is in fact singular with respect to Lebesgue measure. See also Brown and Hewitt [1], [2]; MR **55** # 13160; **56** # 9184; **54** # 10365.

(2) By using 6.1.1 in conjunction with results in general integration theory referring to criteria for weak compactness for subsets of $\mathbf{L}^1$ (see [DS_1], pp. 294–295; [E], pp. 274–276), the following characterization of Fourier-Lebesgue series can be established.

Introducing the Cesàro sums

$$\sigma_N(x) = \sum_{|n| \leqslant N} \left(1 - \frac{|n|}{N + 1}\right) c_n e^{inx}, \tag{12.7.13}$$

the following four conditions are equivalent:

(a) a subsequence $(\sigma_{N_k})_{k=1}^{\infty}$ exists such that the set functions

$$E \to \int_E \sigma_{N_k}(x)\,dx \qquad (k = 1, 2, \cdots) \tag{12.7.14}$$

are uniformly (or equi-) absolutely continuous [that is, to each $\varepsilon > 0$ corresponds $\delta > 0$ such that

$$\sup_k \left| \int_E \sigma_{N_k}(x)\,dx \right| \leqslant \varepsilon$$

for all measurable sets E satisfying $m(E) < \delta$];

(b) some subsequence $(\sigma_{N_k})_{k=1}^{\infty}$ converges weakly in $\mathbf{L}^1$;

(c) some subsequence $(\sigma_{N_k})_{k=1}^{\infty}$ converges in $\mathbf{L}^1$;

(d) $(c_n)_{n\in Z} \in \mathbf{A}(Z)$.

If any one of these conditions holds, and if $f = \lim_k \sigma_{N_k}$ weakly or strongly in $\mathbf{L}^1$, then $c = \hat{f}$, $\sigma_N = \sigma_N f$, and so (by 6.1.1) $\lim_{N\to\infty} \|\sigma_N - f\|_1 = 0$.

There is also a partial analogue of the above result, due to Keogh [1], in which σ_N is everywhere replaced by s_N. If (a′), (b′), and (c′) are the statements that result from (a), (b), and (c) respectively on replacing σ_N by s_N, the analogue asserts the equivalence of (a′), (b′), and (c′), together with the fact that, if any one of these is satisfied, and if $f = \lim_k s_{N_k}$ weakly or strongly in $\mathbf{L}^1$, then $c = \hat{f}$, $s_N = s_N f$, and $\lim_{k\to\infty} \|s_{N_k} - f\|_1 = 0$.

In proving this analogue one uses, in place of 6.1.1, the fact that $s_N f \to f$ in measure as $N \to \infty$ whenever $f \in \mathbf{L}^1$. (Convergence in measure is defined in 1.2.5, and the stated result follows from the remarks in 12.10.2.)

Remark. The uniform absolute continuity of the set functions (12.7.14) can be expressed as uniform absolute continuity of the point functions

$$F_k(x) = \int_0^x \sigma_{N_k}(t)\,dt,$$

the condition being precisely that to each $\varepsilon > 0$ corresponds a number $\delta = \delta(\varepsilon) > 0$ such that

$$\sup_k \sum_{j=1}^{r} |F_k(b_j) - F_k(a_j)| \leqslant \varepsilon$$

for any finite sequence $((a_j, b_j))_{j=1}^r$ of disjoint open intervals (a_j, b_j) the sum of whose lengths does not exceed δ. In this connection see, for example, [HS], Theorem (19.53) and its proof.

(3) If G is a general (locally compact Hausdorff) group, one can define the concept of bounded Radon measure on G; see the remarks following 12.13.3, the references cited there, Exercise 12.34, and also [HR]. Denote by $\mathbf{M}(G)$ the set of bounded Radon measures on G. If G is Abelian, one can introduce the character group X of G (see 2.2.1 and [HR], Chapter 6). To each $\mu \in \mathbf{M}(G)$ corresponds the Fourier (or Fourier-Stieltjes) transform $\hat{\mu} = \mathscr{F}\mu$, namely, the bounded continuous function on X defined by

$$\hat{\mu}(\chi) = \mu(\bar{\chi}) \qquad (\chi \in X).$$

(This definition of $\hat{\mu}$ presupposes some development of the integration theory associated with μ; compare the closing remarks in 12.2.3.)

Most of the problems mentioned already in connection with $\mathbf{M} = \mathbf{M}(T)$ have their exact analogues for $\mathbf{M}(G)$. In addition, and more specifically, some of the functional analytic properties of $\mathscr{F}\mathbf{M}(G)$, considered as a normed linear space with the norm

$$\|\hat{\mu}\|^{\wedge} = \sup \{|\hat{\mu}(\chi)| : \chi \in X\},$$

have been investigated by Beurling and Hewitt (unpublished) and subsequently by Ramirez [1], [2], [3].

12.8 Hilbert's Distribution and Conjugate Series

In this section and the next we shall examine the operation of passing from a trigonometric series

$$\sum_{n \in Z} c_n e^{inx}$$

to the so-called *conjugate* (or *allied*) series

$$\sum_{n \in Z} (-i \cdot \operatorname{sgn} n) c_n e^{inx};$$

the explanation of the use of the term "conjugate" will be given shortly. By way of example, the reader will notice that, in the notation of Chapter 7, (S) is the series conjugate to (C).

Our glance will be incomplete in at least two respects. First, and as is by now customary in this book, we shall have next to nothing to say about the traditional approach to the problem of pointwise convergence or summability (except in so far as the results of Chapter 10 and the attached exercises may be applied). Secondly, although the explanation of the term "conjugate" will hint at close connections with complex variable theory, and with the study of those trigonometric series said to be of *power-series type* because they are of the form

$$\sum_{n \geq 0} c_n e^{inx},$$

there will be no space to give a full account of these connections. For these aspects we must refer the reader to [Z_1], Chapters III and IV; [Ba_2], Chapter VIII; [HaR], Sections 4.8 and 5.8; [Hel]; [Kz], Chapter III; and, for the modern viewpoint applying to a category of more general groups, to [R], Chapter 8. See also [EG], Sections 6.7 and 6.8.

For our part, we shall be interested in representing the passage from the Fourier series of a distribution to the conjugate series as the operation of convolution with a certain distribution H which is neither a function nor a measure, and with the nature of this operation in its action on the $\mathbf{L}^p$ spaces. This outlook fits in well with the discussion of multiplier operators to be discussed in Chapter 16. It will also form the basis of the proof, given in 12.10, of the mean convergence in $\mathbf{L}^p$ of the Fourier series of a function in $\mathbf{L}^p$, when $1 < p < \infty$.

The reason for the term "conjugate" is simply and adequately explained by regarding any trigonometric polynomial

$$\sum_{n \in Z} c_n e^{inx}$$

as the boundary value for $r = 1$ of the harmonic function

$$re^{ix} \to \sum_{n \in Z} c_n r^{|n|} e^{inx}.$$

For it then appears that the conjugate series represents the value for $r = 1$ of that one of the conjugate harmonic functions of re^{ix} (these conjugates being undetermined up to addition of constants; see [He], p. 55) which has a zero mean value. Some use will be made of this fact in 12.9.7.

12.8.1. **Hilbert's Distribution.** It will be plain from (12.6.4) that the series conjugate to the Fourier series of a distribution F is the Fourier series of the *conjugate* (or *allied*) distribution $\tilde{F} \equiv H * F$, where the distribution H is given by

$$H = \sum_{n \in Z} -i \cdot \operatorname{sgn} n \cdot e^{inx}; \tag{12.8.1}$$

here sgn n denotes 0 or $|n|^{-1} n$ according as $n \in Z$ is or is not equal to 0. It will appear from (12.8.7) and (12.8.8) that the operation $F \to \tilde{F}$ can be expressed in a way which makes it plainly analogous (for periodic functions) to the so-called Hilbert transform for functions of a real variable. For this reason we take the liberty of referring to H as *Hilbert's* (periodic) *distribution*. Referring to (7.1.3), we see that $s_N(H * F) = \tilde{D}_N * F$, and that $H = \lim_{N \to \infty} \tilde{D}_N$ in $\mathbf{D}$.

Starting from (12.8.1) we shall now derive some other expressions for H.

To begin with, since we have by (7.1.3) the relations

$$\tilde{D}_N(x) \equiv \sum_{|n| \leqslant N} -i \cdot \operatorname{sgn} n \cdot e^{inx} = 2 \sum_{n=1}^{N} \sin nx$$

$$= \cot \tfrac{1}{2}x - \frac{\cos (N + \frac{1}{2})x}{\sin \frac{1}{2}x},$$

we have for $u \in \mathbf{C}^\infty$

$$H(u) = \lim_{N\to\infty} \frac{1}{2\pi}\int_{-\pi}^{\pi} u(x)\left[\frac{\cos \frac{1}{2}x - \cos (N + \frac{1}{2})x}{\sin \frac{1}{2}x}\right] dx$$

$$= \lim_{N\to\infty} \frac{1}{2\pi}\int_{0}^{\pi} [u(x) - u(-x)]\left[\frac{\cos \frac{1}{2}x - \cos (N + \frac{1}{2})x}{\sin \frac{1}{2}x}\right] dx.$$

Now $[u(x) - u(-x)]/\sin \frac{1}{2}x$ is integrable over $(0, \pi)$ and so the Riemann-Lebesgue lemma 2.3.8 shows that

$$H(u) = \frac{1}{\pi}\int_{0}^{\pi} [u(x) - u(-x)] \cdot \tfrac{1}{2} \cot \tfrac{1}{2}x \, dx. \tag{12.8.2}$$

The inequality

$$|u(x) - u(-x)| \leqslant 2|x| \cdot \|Du\|_\infty$$

combines with (12.8.2) to show that

$$|H(u)| \leqslant \frac{2}{\pi}\int_{0}^{\pi} \tfrac{1}{2}x \cdot \cot \tfrac{1}{2}x \, dx \cdot \|Du\|_\infty, \tag{12.8.3}$$

so that $H \in \mathbf{D}^1$. On the other hand, the final statement in Example 12.7.8 shows that the series (12.8.1) defining H is not a Fourier-Stieltjes series, so that H is not a measure. Thus H is a distribution of order exactly 1.

From (12.8.2) we have also

$$H(u) = \lim_{\varepsilon\to 0} \frac{1}{\pi}\int_{\varepsilon}^{\pi} [u(x) - u(-x)] \cdot \tfrac{1}{2} \cot \tfrac{1}{2}x \, dx.$$

On expressing the remaining integral as the difference of those whose integrands are, respectively, $u(x)\frac{1}{2} \cot \frac{1}{2}x$ and $u(-x)\frac{1}{2} \cot \frac{1}{2}x$, and making a change of variable in the second, it appears that

$$H(u) = \lim_{\varepsilon\to 0} \frac{1}{\pi}\int_{\varepsilon \leqslant |x| \leqslant \pi} u(x) \cdot \tfrac{1}{2} \cot \tfrac{1}{2}x \, dx. \tag{12.8.4}$$

In other words, and by comparison with Example 12.3.2(5), this may be expressed by writing

$$H = \text{P.V.} \cot \tfrac{1}{2}x. \tag{12.8.5}$$

Partial integration may be applied to (12.8.4) to show that

$$H = 2 \cdot D[\log |\sin \tfrac{1}{2}x|]; \tag{12.8.6}$$

the argument is very similar to that laid out in Example 12.4.3(3) and the reader is urged to supply the details. Naturally, the differentiation involved in (12.8.6) is understood in the distributional sense (see 12.4.1).

From 12.6.1 and (12.8.4) we derive for $u \in \mathbf{C}^\infty$ the relation

$$H * u(x) = \lim_{\varepsilon\to 0} \frac{1}{2\pi}\int_{\varepsilon \leqslant |y| \leqslant \pi} u(x - y) \cot \tfrac{1}{2}y \, dy; \tag{12.8.7}$$

it is this expression that exhibits most clearly the similarity of the operation $u \to H * u$ with the Hilbert transform. Similarly, using 12.6.1 and (12.8.2), there follows the relation

$$H * u(x) = \frac{1}{\pi}\int_0^{\pi} [u(x - y) - u(x + y)]\tfrac{1}{2} \cot \tfrac{1}{2} y \, dy \tag{12.8.8}$$

for $u \in \mathbf{C}^{\infty}$.

12.8.2. The Classical Approach to the Conjugate Function.

The form of equations (12.8.4), (12.8.7), and (12.8.8) suggests strongly the examination of the pointwise limit

$$\begin{aligned} f^c(x) &\equiv \lim_{\varepsilon \to 0} H_\varepsilon * f(x) \\ &\equiv \lim_{\varepsilon \to 0} \frac{1}{2\pi} \int_{\varepsilon \leqslant |y| \leqslant \pi} f(x - y) \cot \tfrac{1}{2} y \, dy \qquad (12.8.9) \\ &\equiv \lim_{\varepsilon \to 0} \frac{1}{2\pi} \int_{\varepsilon}^{\pi} [f(x - y) - f(x + y)] \cot \tfrac{1}{2} y \, dy, \end{aligned}$$

where

$$H_\varepsilon(x) = \begin{cases} \cot \tfrac{1}{2} x & \text{for } \varepsilon \leqslant |x| \leqslant \pi, \\ 0 & \text{for } 0 \leqslant |x| < \varepsilon, \end{cases}$$

and is defined elsewhere by periodicity, and where f is free from the smoothness restrictions imposed upon u in (12.8.4), (12.8.7), and (12.8.8). The use of the principal value integrals in (12.8.9) is suggested, since it is obvious that the integrands in (12.8.9) are in general integrable over no neighborhood of the origin.

The classical theory of conjugate series and functions [for which see the references in (3) below] is expressed entirely in terms of f^c rather than our $\tilde{f} \equiv H * f$. The reader must guard against thinking that f^c and $\tilde{f}$ are always and obviously the same thing; see (3). Even for continuous functions f, $f^c(x)$ may fail to exist finitely for certain values of x; it is not trivial to show even that $f^c(x)$ exists finitely for almost all x for an arbitrary continuous f.

We shall have neither occasion nor space to discuss the existence of f^c in general. All that we shall need, and all that we shall prove, is contained in (1) and (2) immediately below. A good deal more that we shall not prove is mentioned in (3) with the main aim of clarifying the connection between f^c and $\tilde{f}$.

(1) Suppose that $f \in \mathbf{L}^1$ is such that

$$\int_0^{\pi} |f(x - y) - f(x + y)| \cot \tfrac{1}{2} y \, dy < \infty$$

for almost all x, so that for such x we have

$$f^c(x) = \frac{1}{2\pi}\int_0^\pi [f(x-y) - f(x+y)] \cot \tfrac{1}{2}y \, dy;$$

suppose further that f^c, thus defined almost everywhere, belongs to $\mathbf{L}^1$. Then $f^c = \tilde{f}$ qua distributions. These hypotheses on f are fulfilled whenever $f \in \mathbf{L}^1$ and

$$|f(x-y) - f(x+y)| \leqslant h(x)|y|^\alpha$$

for almost all x, where $h \in \mathbf{L}^1$ and $\alpha > 0$; and hence in particular whenever f is absolutely continuous and $Df \in \mathbf{L}^p$ for some $p > 1$.

Proof. The hypotheses on f ensure that the theorems of Fubini and Tonelli ([W], Theorems 4.2b and 4.2c) can be applied to compute $\hat{f^c}$. The result of this computation reads:

$$\hat{f^c}(n) = -\frac{i}{\pi}\hat{f}(n)\int_0^\pi \cot \tfrac{1}{2}y \cdot \sin ny \, dy.$$

Further computation on the remaining integral then shows that

$$\hat{f}^c(n) = -i \cdot \operatorname{sgn} n \cdot \hat{f}(n) \qquad \text{for all } n \in Z.$$

This combined, with (12.6.4) and (12.8.1), shows that $f^c = H * f \equiv \tilde{f}$ qua distributions. The details of these calculations are left to the reader (Exercise 12.18).

(2) If $f \in \mathbf{L}^2$, then

$$\lim_{\varepsilon \to 0} H_\varepsilon * f = H * f = \tilde{f}$$

in $\mathbf{L}^2$. In particular, there exists a sequence $\varepsilon_\nu \to 0$ such that

$$\tilde{f}(x) \equiv H * f(x) = \lim_{\nu \to \infty} H_{\varepsilon_\nu} * f(x)$$

for almost all x.

Proof. One can compute $\hat{H}_\varepsilon$ and verify that $\hat{H}_\varepsilon \to H$ boundedly on Z (see Exercise 12.18 again). So (12.6.4), (12.8.1), and 8.3.1 show that $H * f \in L^2$ and that $H_\varepsilon * f \to H * f$ in $\mathbf{L}^2$ as $\varepsilon \to 0$. The rest follows from the well-known fact that a mean convergent sequence contains a subsequence which converges pointwise almost everywhere (see the proof of [W], Theorem 4.5a).

Remark. From (2) it appears that $\tilde{f}(x) = f^c(x)$ a.e. for any $f \in \mathbf{L}^2$ for which f^c is known to exist almost everywhere. (This last is in fact true for every $f \in \mathbf{L}^2$; see (3).)

(3) Perhaps the simplest route toward a fairly general identification of f^c and $\tilde{f}$ lies in proving (somewhat along the lines of the methods of Chapter 6) that for any $f \in \mathbf{L}^1$ the conjugate series is Cesàro summable almost everywhere to $f^c(x)$, that is, that

$$\lim_{N \to \infty} \sigma_N \tilde{f}(x) = f^c(x)$$

for almost all x; for a proof of this, see [Z_1] p. 92; [Ba_2], p. 58. Let us assume this result.

As we shall prove in 12.9.1, $\tilde{f} \in \mathbf{L}^p$ whenever $1 < p < \infty$ and $f \in \mathbf{L}^p$. In this case, as follows from 6.1.1, $\sigma_N \tilde{f}$ converges in $\mathbf{L}^p$ to $\tilde{f}$, and it appears from the last paragraph that $\tilde{f} = f^c$ a.e. Similarly, on the basis of the results in 12.9.9, one can infer that $\tilde{f} = f^c$ a.e. whenever, more generally, $f \cdot \log^+ |f| \in \mathbf{L}^1$; see 12.9.9(1) and (2) and compare 12.10.2.

For a general $f \in \mathbf{L}^1$, f^c may fail to be integrable (a result analogous, but not trivially equivalent, to the corresponding assertion about $\tilde{f}$, which will be proved in 12.8.3); see [Z_1], p. 257; [Ba_2], pp. 95, 112. In such cases f^c does not generate a distribution in the manner described in 12.2.2, and the question of the identification of f^c and $\tilde{f}$ scarcely arises.

We mention that it was proved by Lusin and Privalov (see [Z_1], p. 131; [Z_2], p. 252; [Ba_2], p. 62; [HaR], Theorem 89; G. Weiss [1], p. 164) that $f^c(x)$ exists finitely almost everywhere for each $f \in \mathbf{L}^1$; see also the remarks in 13.10.3. The function f^c, although not necessarily integrable in Lebesgue's sense, is integrable in various generalized senses and the corresponding generalized Fourier series of f^c is indeed the series conjugate to the Fourier series of f; for details, see [Z_1], pp. 262–263 and [Ba_2], pp. 128–137. One suitable concept of integration for this purpose is the so-called B-integral (see [Z_1], pp. 262–263) and the result is then due to Kolmogorov (1928); another is the so-called A-integral (see [Ba_2], pp. 128–137), in which case the result is due to Ul'yanov (1957).

See also MR **54** # 5719.

12.8.3. Conjugates of Functions in $\mathbf{L}^2$, $\mathbf{L}^1$, and $\mathbf{C}$. It will be encouraging to begin with one of the few really simple properties of the conjugate function operator.

(1) If $f \in \mathbf{L}^2$, then $\tilde{f} \equiv H * f \in \mathbf{L}^2$ and

$$\|\tilde{f}\|_2 \leqslant \|f\|_2. \tag{12.8.10}$$

Proof. The relation (12.8.1) shows that

$$|\hat{H}(n)| \leqslant 1 \qquad \text{for all } n \in Z$$

whence the stated result follows on the basis of (12.6.4) and 8.3.1.

(2) When $\mathbf{L}^2$ is replaced by $\mathbf{L}^1$ or $\mathbf{C}$, the situation is less simple.

In the first place, Exercise 7.7 furnishes an example of a function $f \in \mathbf{L}^1$ such that $\tilde{f} \notin \mathbf{L}^1$.

Again, consider the pointwise sum-function f of the series

$$\sum_{n=2}^{\infty} \frac{\sin nx}{n \log n}. \tag{12.8.11}$$

By 7.2.2(1), f is continuous. We will show that in fact f is absolutely continuous and yet, nevertheless, $\tilde{f} \notin \mathbf{L}^\infty$.

Indeed,

$$Df = \sum_{n=2}^{\infty} \frac{\cos nx}{\log n}$$

and Exercise 7.7 and the substance of 12.4.2 affirm the absolute continuity of f. The Fourier series of $\tilde{f}$ is [see 7.2.2(3)]

$$-\sum_{n=2}^{\infty} \frac{\cos nx}{n \log n}$$

which, by 7.3.1, is pointwise convergent for $x \not\equiv 0 \pmod{2\pi}$. Since, according to (1) immediately above, the series converges in $\mathbf{L}^2$ to $\tilde{f}$, it follows that

$$\tilde{f}(x) = -\sum_{n=2}^{\infty} \frac{\cos nx}{n \log n} \tag{12.8.12}$$

pointwise almost everywhere. We proceed to estimate the pointwise sum of the series (12.8.12), concentrating on small values of $x > 0$.

For any such value of x, let $N = N(x)$ be the positive integer such that

$$N \log N \leqslant x^{-1} < (N+1) \log (N+1).$$

By partial summation and the estimate (see Exercise 1.2)

$$\left|\sum_{n=1}^{r} \cos nx\right| \leqslant \frac{A}{x},$$

it follows that

$$\left|\sum_{n=N+1}^{\infty} \frac{\cos nx}{n \log n}\right| \leqslant \frac{2A}{x(N+1) \log (N+1)} \leqslant A',$$

where A and A' are independent of x. On the other hand, for $2 \leqslant n \leqslant N$,

$$\cos nx \geqslant \cos Nx \geqslant \cos \{(\log N)^{-1}\}.$$

From (12.8.12) it now follows that, for any preassigned $\varepsilon > 0$, there exists $\delta = \delta(\varepsilon) > 0$ such that

$$-\tilde{f}(x) \geqslant (1 - \varepsilon) \log \log \frac{1}{x} \tag{12.8.13}$$

for almost all $x \in (0, \delta)$.

The inequality (12.8.13) shows that $\tilde{f} \notin \mathbf{L}^{\infty}$.

Even more striking examples of a similar nature have been constructed by Lusin and Tolstov; see [Ba_2], pp. 95–98. See also Goes [2], Section V.

A similar argument shows that the pointwise sum function

$$f(x) = \sum_{n=1}^{\infty} \frac{\sin nx}{n}, \tag{12.8.14}$$

which (see Exercise 10.8) agrees on $(0, 2\pi)$ with $(\pi - x)/2$ (and so is but very mildly discontinuous), is such that, for any given $\varepsilon > 0$, there exists a number $\delta = \delta(\varepsilon) > 0$ such that

$$-\tilde{f}(x) \geqslant (1 - \varepsilon) \log \frac{1}{x} + A_{\varepsilon} \tag{12.8.15}$$

for almost all $x \in (0, \delta)$, A_ε being independent of x. This example will be useful later [see 13.9.2(3)].

A little more generally, consider a function f with a finite number of jump discontinuities at points $a_1, \ldots, a_k$ in $[-\pi, \pi)$ and such that

$$Df = \sum_{j=1}^{k} c_j \varepsilon_{a_j} + g,$$

where the c_j are complex numbers and g is so smooth that $\tilde{g} \in \mathbf{L}^1$. Then

$$D\tilde{f} = D(H * f) = H * Df$$

$$= \sum_{j=1}^{k} c_j T_{aj} H + \tilde{g}.$$

Defining temporarily $\varphi(x) = 2 \log|\sin \frac{1}{2}x$ for all $x \in (-\pi, \pi)$ and by periodicity elsewhere, it follows from (12.8.6) that

$$\tilde{f}(x) = \sum_{j=1}^{k} c_j \varphi(x - a_j) + h(x),$$

where h is continuous. This exhibits the misbehaviour of $\tilde{f}$: in particular, if $c_j \neq 0$, $\tilde{f}$ is unbounded near a_j like

$$-\log(|x - a_j|^{-1}).$$

That the preceding examples are not in the nature of isolated freaks is shown by the next result.

(3) The set of functions $f \in \mathbf{L}^1$ (respectively, $\mathbf{C}$), for which $\tilde{f} \in \mathbf{M}$ (respectively, $\mathbf{L}^\infty$), is a meagre subset of $\mathbf{L}^1$ (respectively, $\mathbf{C}$); the complementary set is therefore everywhere dense in $\mathbf{L}^1$ (respectively, $\mathbf{C}$).

Proof. This will follow on combining the known fact that H is not a measure with the more general theorem 12.8.4 immediately following, as a preliminary to which the reader may find a glance at I, A and I, B.2 profitable.

12.8.4. Suppose that $F \in \mathbf{D}$ satisfies *either* of the following two conditions

(1) $F * f \in \mathbf{M}$ for each f in a nonmeagre subset $\mathbf{S}$ of $\mathbf{L}^1$;

(2) $F * f \in \mathbf{L}^\infty$ for each f in a nonmeagre subset $\mathbf{S}$ of $\mathbf{C}$.

Then $F \in \mathbf{M}$.

Proof. (1) Suppose first that F satisfies condition (1). We write

$$\mathbf{S} = \bigcup_{n=1}^{\infty} \mathbf{S}_n, \tag{12.8.16}$$

where $\mathbf{S}_n$ denotes the set of $f \in \mathbf{L}^1$ such that $F * f \in \mathbf{M}$ and $\|F * f\|_1 \leqslant n$ [the norm on $\mathbf{M}$ being defined as in (12.3.6)]. The first step is to show that each $\mathbf{S}_n$ is a closed subset of $\mathbf{L}^1$.

To this end, suppose that f_k $(k = 1, 2, \cdots)$ belongs to $\mathbf{S}_n$ and $f_k \to f$ in $\mathbf{L}^1$: it must be shown that $f \in \mathbf{S}_n$. Now, by definition of $\mathbf{S}_n$, $F * f_k \in \mathbf{M}$ and

$\|F * f_k\|_1 \leqslant n$ for all k. Hence, by 12.3.9, there is a subsequence $(f_{k_i})_{i=1}^{\infty}$ and $\mu \in \mathbf{M}$ such that $F * f_{k_i} \to \mu$ weakly in $\mathbf{M}$. At the same time, however, 12.6.6 affirms that $F * f_n \to F * f$ in $\mathbf{D}$. It follows that $F * f = \mu \in \mathbf{M}$. In addition, by Exercise 12.13,

$$\|F * f\|_1 = \|\mu\|_1 \leqslant \liminf_{i \to \infty} \|F * f_{k_i}\|_1 \leqslant n,$$

and so f does indeed belong to $\mathbf{S}_n$.

Knowing that $\mathbf{S}_n$ is closed in $\mathbf{L}^1$ and that $\mathbf{S}$ is nonmeagre, the relation (12.8.16) forces the conclusion that, for some n, $\mathbf{S}_n$ has interior points relative to $\mathbf{L}^1$. For this n there exist a number $\rho > 0$ and a function $f_0 \in \mathbf{L}^1$ such that $h \in \mathbf{S}_n$ whenever $h \in \mathbf{L}^1$ and $\|h - f_0\|_1 \leqslant \rho$. Then, if $f \in \mathbf{L}^1$ and $\|f\|_1 \leqslant \rho$, $h = f_0 + f \in \mathbf{S}_n$ and so

$$F * f = F * h - F * f_0$$

is seen to belong to $\mathbf{M}$ and to satisfy $\|F * f\|_1 \leqslant 2n$. It follows at once that $F * f \in \mathbf{M}$ and

$$\|F * f\|_1 \leqslant 2n\rho^{-1}\|f\|_1 \tag{12.8.17}$$

for all $f \in \mathbf{L}^1$.

For the final step, we choose an approximate identity $(f_i)_{i=1}^{\infty}$ in $\mathbf{L}^1$ satisfying $\|f_i\|_1 \leqslant 1$ and infer from (12.8.17) that

$$\|F * f_i\|_1 \leqslant 2n\rho^{-1} \tag{12.8.18}$$

for all i.

Repeating the arguments employed three paragraphs above, and using 12.3.2(3) and 12.6.7(1), it may be inferred from (12.8.18) that $F \in \mathbf{M}$.

(2) Suppose now that F satisfies condition (2). It will suffice to show that

$$\sup_i \|F * f_i\|_1 < \infty \tag{12.8.19}$$

is again true for some approximate identity $(f_i)_{i=1}^{\infty}$ in $\mathbf{L}^1$, which we may assume to satisfy $\|f_i\|_1 \leqslant 1$. For reasons that will appear shortly, we shall assume that each $f_i \in \mathbf{C}^{\infty}$.

Now, by the case $p = 1$ of 3.1.4, we have

$$|F * f * f_i(0)| \leqslant \|F * f\|_\infty \|f_i\|_1 \leqslant \|F * f\|_\infty$$

for each $f \in \mathbf{S}$. Thus

$$\sup_i |F * f * f_i(0)| < \infty \qquad (f \in \mathbf{S}). \tag{12.8.20}$$

Since each $f_i \in \mathbf{C}^{\infty}$, and since convolution is associative (see 12.6.5), 12.6.2 and 3.1.4 make it plain that

$$f \to F * f * f_i(0)$$

is, for each i, a continuous linear functional on $\mathbf{C}$. In view of (12.8.20), an application of I, B.2.1(2) leads to the existence of a number $c > 0$ such that

$$\sup_i |F * f * f_i(0)| \leqslant c\|f\|_\infty \qquad (f \in \mathbf{C}). \tag{12.8.21}$$

Reference to (12.3.6) leads at once from (12.8.21) to

$$\sup_i \|F * f_i\|_1 \leqslant c,$$

which entails (12.8.19). The proof is complete.

12.8.5. **Remarks concerning 12.8.3 and 12.8.4.** (1) There are close connections between 12.8.4 and problems concerning multiplier operators, a topic discussed in Chapter 16; see especially Section 16.3.

(2) Concerning 12.8.3(2), we remark that it is possible to exhibit specific and quite simple functions $f \in \mathbf{L}^1$ such that the function f^c (see 12.8.2) is non-integrable over any nondegenerate interval; see, for example, $[Z_1]$, p. 257.

(3) As 12.8.4 shows, the underlying reason for the existence of continuous functions f such that $\tilde{f}$ is essentially unbounded is simply that H is not a measure, that is, in view of the substance of 12.5.10, that H is not of the form

$$\text{const} + D\phi$$

with ϕ a function of bounded variation.

It is therefore interesting to note that the underlying reason for the existence of absolutely continuous functions f such that $\tilde{f}$ is essentially unbounded on every nondegenerate interval (compare 12.8.3(2)) can be shown (Exercise 16.27 and Edwards [3]) to be the circumstance that H is not of the form

$$\text{const} + Dh$$

with $h \in \mathbf{L}^\infty$. That H is not of this form follows easily from the fact that $\sum_{n=1}^\infty \cos nx/n$ is not the Fourier series of a function in $\mathbf{L}^\infty$, which in turn follows from Exercise 6.3 or from (12.8.15).

(4) We take this opportunity to mention in passing a famous and remarkable theorem due jointly to F. and M. Riesz, namely: if μ and $\tilde{\mu}$ are both measures, then $\mu \in \mathbf{L}^1$. For proofs of this, see $[Z_1]$, p. 285; $[Ba_2]$, pp. 87–92; $[R_1]$, p. 335; and [R], Section 8.2.

An equivalent formulation of the theorem asserts that if f is a function of bounded variation, and if $\tilde{f}$ is (the distribution generated by) a function of bounded variation, then f is in fact absolutely continuous; compare the Hardy–Littlewood theorem quoted in 10.6.2(7). (The equivalence of the two versions hinges upon the substance of 12.5.10; see Exercise 12.19.)

For some abstract versions of the theorem, see de Leeuw and Glicksberg [1], Lumer [1], Ahern [1], and Glicksberg [2], [3].

12.9 The Theorem of Marcel Riesz

The results stated in 12.8.3 may be extended and balanced by the following major result, due to Marcel Riesz (1927).

12.9.1. **Statement of the Theorem.** Suppose that $1 < p < \infty$. There exists a number $k_p \geqslant 1$ such that $H * f \in \mathbf{L}^p$ and

$$\|H * f\|_p \leqslant k_p \|f\|_p \tag{12.9.1}$$

for each $f \in \mathbf{L}^p$.

12.9.2. **Start of Proof.** The proof we shall give is somewhat lengthy and will be prefaced by a number of reductions and manipulations. Many other proofs are known; an entirely different one will be given in Section 13.9.

Although the following proof is intended to be complete in all details, the reader may find it of interest to glance at the sketch proof using other complex variable techniques that appears on pp. 165–167 of G. Weiss [1]. An approach to 12.9.1 based upon the study of harmonic functions appears in [R_1], pp. 345–348 (especially Exercise 17); compare 12.9.8(2). See also [Kz], pp. 68–70 and Remark (2) following 13.9.1 and 13.9.2 below.

12.9.3. For a given number $k \geqslant 0$ denote by $\mathbf{E}_k$ the set of $f \in \mathbf{L}^p$ such that $H * f \in \mathbf{L}^p$ and

$$\|H * f\|_p \leqslant k \|f\|_p. \tag{12.9.1'}$$

Our task is to show that, if $1 < p < \infty$, then some $k = k_p$ exists for which $\mathbf{E}_k = \mathbf{L}^p$. We begin by observing some simple properties of $\mathbf{E}_k$ for a given k and p.

12.9.4. (1) $\mathbf{E}_k$ is a closed subset of L^p;

(2) if $u = \operatorname{Re} f$ and $v = \operatorname{Im} f$ belong to $\mathbf{E}_k$, then $f \in \mathbf{E}_{2k}$;

(3) if f is real-valued, and if $f_+ = \sup(f, 0)$ and $f_- = \sup(-f, 0)$ belong to $\mathbf{E}_k$, then $f \in \mathbf{E}_{2k}$.

Proof. (1) Let $(f_n)_{n=1}^{\infty}$ be any sequence extracted from $\mathbf{E}_k$ such that $f_n \to f$ in mean in $\mathbf{L}^p$. By (12.9.1') and 12.3.10(2), a suitable subsequence $(H * f_{n_i})_{i=1}^{\infty}$ converges weakly in $\mathbf{L}^p$ to some $g \in \mathbf{L}^p$. On the other hand, by 12.6.6, $H * f_n \to H * f$ in $\mathbf{D}$. It follows that $H * f = g \in \mathbf{L}^p$ and, by Exercise 12.14, that

$$\begin{aligned} \|H * f\|_p = \|g\|_p &\leqslant \liminf_{i \to \infty} \|H * f_{n_i}\|_p \\ &\leqslant \liminf_{i \to \infty} k \|f_{n_i}\|_p \leqslant k \|f\|_p. \end{aligned}$$

This proves (1).

(2) By linearity of $H * f$ in the variable f, we see that $H * f = H * u + i(H * v)$. If u and v belong to $\mathbf{E}_k$, it follows that $H * f \in \mathbf{L}^p$ and that

$$\begin{aligned} \|H * f\|_p &\leqslant \|H * u\|_p + \|H * v\|_p \\ &\leqslant k(\|u\|_p + \|v\|_p) \\ &\leqslant 2k\|f\|_p, \end{aligned}$$

which proves (2).

(3) The proof is exactly similar to that of (2).

12.9.5. In order to establish 12.9.1 for a given p satisfying $1 < p < \infty$, it suffices to show that for some $k = k_p$ the inequality (12.9.1) holds for each trigonometric polynomial $f > 0$.

Proof. This follows by repeated applications of 12.9.4. Thus if (12.9.1′) holds for each trigonometric polynomial $f > 0$, 12.9.4(1) and 6.1.1 show that (12.9.1′) holds for all continuous $f > 0$. By 12.9.4(1), this inequality extends to all continuous $f \geqslant 0$. Then 12.9.4(3) and (2) show that the same is true, with $4k$ in place of k, for any complex-valued continuous f. Finally 12.9.4(1) and 6.1.1 show that (12.9.1′) is valid, with $4k$ in place of k, for any $f \in \mathbf{L}^p$.

12.9.6. If (12.9.1) holds for some p satisfying $1 < p < \infty$ and each $f \in \mathbf{L}^p$, then it also holds, with $k_{p'} = k_p$, for each $f \in \mathbf{L}^{p'}$ (where, as usual, $1/p + 1/p' = 1$).

Proof. Suppose that f and g are trigonometric polynomials. Then

$$\begin{aligned} \frac{1}{2\pi}\int (H * f)\check{g}\,dx &= (H * f) * g(0) = f * (H * g)(0) \\ &= \frac{1}{2\pi}\int \check{f}(H * g)\,dx. \end{aligned}$$

By Hölder's inequality and the main hypothesis, it follows that

$$\left|\frac{1}{2\pi}\int \check{f}(H * g)\,dx\right| \leqslant k_p\|f\|_p \cdot \|g\|_{p'}.$$

The converse of Hölder's inequality (Exercise 3.6) now entails that

$$\|H * g\|_{p'} \leqslant k_p\|g\|_{p'}$$

for all trigonometric polynomials. Since $p' < \infty$, 12.9.4(1) now serves to show that (12.9.1) is true with p' in place of p and $k_{p'} = k_p$.

Remark. The preceding result is a special case of a more general principle which will appear in 16.4.1.

12.9.7. **Final Stage of Proof.** By 12.9.5 and 12.9.6, in order to prove 12.9.1 completely, it will suffice to consider a value of p satisfying $1 < p \leqslant 2$

and establish the existence of a number k for which (12.9.1′) holds for all trigonometric polynomials $f > 0$. We now undertake this task.

Write

$$g = H * f = \sum_{n \in Z} -i \cdot \operatorname{sgn} n \cdot \hat{f}(n) e^{inx}, \tag{12.9.2}$$

$$h = f + ig. \tag{12.9.3}$$

Since f is real-valued, so that $\overline{\hat{f}(n)} = \hat{f}(-n)$ (see 2.3.1), it is easily seen that g is real-valued. The reader will also observe that

$$h = \hat{f}(0) + 2 \sum_{n > 0} \hat{f}(n) e^{inx}. \tag{12.9.4}$$

It is crucial to our proof to know *either* that

$$-\operatorname{Re} \frac{1}{2\pi} \int h^p \, dx \leqslant \text{const} \, \|f\|_p^p, \tag{12.9.5}$$

or that

$$\operatorname{Re} \frac{1}{2\pi} \int h^p \, dx \geqslant 0. \tag{12.9.5′}$$

The discussion of (12.9.5) and (12.9.5′) is deferred until 12.9.8. Meanwhile we proceed on the assumption that at least one of these inequalities is valid.

Choose $\delta = \delta_p$ so that $0 < \delta < \frac{1}{2}\pi < p\delta < \frac{1}{2}p\pi \leqslant \pi$ (recall that we are assuming that $1 < p \leqslant 2$) and put

$$\alpha = \alpha_p = \sec p\delta, \qquad \beta = \beta_p = (\sec \delta)^p (1 + |\alpha|),$$

so that $\alpha < 0$ and $\beta > 0$. We claim that

$$1 \leqslant \alpha \cdot \cos pt + \beta (\cos t)^p \qquad \text{for } |t| \leqslant \tfrac{1}{2}\pi. \tag{12.9.6}$$

Indeed, if $\delta \leqslant |t| \leqslant \frac{1}{2}\pi$, the right-hand side is not less than $\alpha \cos pt \geqslant \alpha \cos p\delta = 1$; and if $|t| \leqslant \delta$, it is not less than $\beta(\cos \delta)^p - |\alpha| = 1$.

Now $f = |h| \cos t$ where, since $f > 0$, $|t| < \frac{1}{2}\pi$. So (12.9.6) yields

$$\frac{1}{2\pi} \int |h|^p \, dx \leqslant \alpha \left(\frac{1}{2\pi}\right) \int |h|^p \cos pt \cdot dx + \beta \left(\frac{1}{2\pi}\right) \int |h|^p (\cos t)^p \, dx$$

$$= \alpha \cdot \operatorname{Re} \left(\frac{1}{2\pi}\right) \int h^p \, dx + \beta \left(\frac{1}{2\pi}\right) \int f^p \, dx,$$

provided we take that branch of the pth power which is real and positive on the positive real axis. Since $\alpha < 0$, either of (12.9.5) or (12.9.5′) leads thence to the inequality

$$\frac{1}{2\pi} \int |h|^p \, dx \leqslant \text{const} \, \|f\|_p^p.$$

Then (12.9.3) yields

$$\|g\|_p = \|-i(h - f)\|_p \leqslant \text{const } \|f\|_p,$$

which is equivalent to (12.9.1′).

12.9.8. **Return to (12.9.5) and (12.9.5′).** It thus remains only to establish one or the other of (12.9.5) or (12.9.5′). It is in fact true that

$$\frac{1}{2\pi}\int h^p\,dx = [\frac{1}{2\pi}\int h\,dx]^p, \tag{12.9.7}$$

and we shall substantiate this in a moment. Since

$$\frac{1}{2\pi}\int h\,dx = \hat{h}(0) = \hat{f}(0) = \frac{1}{2\pi}\int f\,dx,$$

(12.9.7) plainly implies both (12.9.5) and (12.9.5′). Let us therefore consider (12.9.7).

(1) There is a proof of (12.9.7), due to Helson and based on the theory of Banach algebras, which is indicated in Exercise 12.26. The above proof of the M. Riesz theorem itself is also due in part to Helson.

(2) One may also observe that h is the boundary value (on the unit circumference) of the polynomial

$$H(w) = \hat{f}(0) + 2\sum_{n>0}\hat{f}(n)w^n$$

in the complex variable w; see (12.9.4). Then, since $f > 0$ is the boundary value of the harmonic function $\operatorname{Re} H$, it follows from the maximum principle for harmonic functions that $\operatorname{Re} H(w) > 0$ for $|w| \leqslant 1$. An analytic branch of H^p may thus be defined and (12.9.7) follows at once from Cauchy's theorem applied to this branch.

(3) A third approach is as follows. It is simple to verify that

$$\frac{1}{2\pi}\int P\circ h\,dx = P[\frac{1}{2\pi}\int h\,dx] \tag{12.9.8}$$

is true for each polynomial P in one complex variable w. Now the range of h lies within some compact rectangle in the half-plane $\operatorname{Re} w > 0$ and one may there choose an analytic branch of w^p. This chosen branch can then be approximated, uniformly on this rectangle, by polynomials $P(w)$ (a special case of Runge's theorem proved in Appendix D in Volume 1; alternatively, consider the chosen branch of w^p in the disc $|w - n| < n$, where n is a sufficiently large positive integer). Then h^p is the uniform limit of the corresponding functions $P \circ h$. A limiting process on (12.9.8) leads directly to (12.9.7).

Of the arguments (1), (2), and (3), (1) and (3) may be taken over for more general groups, but (2) cannot.

The proof of 12.9.1 is entirely complete.

Remarks. For extensions of 12.9.1 and related theorems, see [R], Chapter 8; Helson [7]; MR **20** # # 4155, 5397; **41** # 4136.

Concerning the best possible value of k_p in (12.9.1), see MR **47** # 702; for the same question relating to (12.9.9), see MR **54** # 10967.

Weighted norm inequalities for the operator H have also been studied; see, for example, MR **47** # 701; **54** # 5720.

12.9.9. **Further Inequalities.** As we know from 12.8.3 and 12.8.4, 12.9.1 is false for $p = 1$ and for $p = \infty$. It can however be shown that

$$\|H * f\|_p \leqslant k_p \|f\|_1 \qquad \text{for } 0 < p < 1 \tag{12.9.9}$$

and that

$$\|H * f\|_1 \leqslant \frac{A}{2\pi} \int |f| \log^+ |f| \, dx + B \tag{12.9.10}$$

whenever f is a trigonometric polynomial. Proofs will be found in [Z_1], pp. 254–256; [Ba_2], pp. 103–122; and [R], pp. 220–221. In each of these references the proofs use the same general principles as does the preceding proof of 12.9.1. See also [Kz], p. 66. In Section 13.9 we shall discuss proofs of 12.9.1, and equations (12.9.9) and (12.9.10), depending on a general interpolation theorem due to Marcinkiewicz.

We mention also that an elegant type of proof of (12.9.1), based upon a study of rearrangements of functions and an inequality of Hardy (see [Z_1], p. 20), has been given by O'Neil and Weiss [1].

Let us temporarily assume the truth of (12.9.9) and (12.9.10) for trigonometric polynomials f and see how their range of validity can be extended.

(1) If we take any $f \in \mathbf{L}^1$ and apply (12.9.9) to the trigonometric polynomials $\sigma_N f$, noticing en route that

$$H * \sigma_N f = \sigma_N(H * f) = \sigma_N \tilde{f},$$

we obtain

$$\|\sigma_N \tilde{f}\|_p \leqslant k_p \|f\|_1 .$$

Calling on the fact that $\sigma_N \tilde{f} \to f^c$ almost everywhere (see 12.8.2(3)), it may be deduced from Fatou's lemma ([W], Theorem 4.1d) that

$$\|f^c\|_p \leqslant k_p \|f\|_1 \qquad \text{if } 0 < p < 1 \text{ and } f \in \mathbf{L}^1 . \tag{12.9.11}$$

In (2) it will be shown that $H * f \in \mathbf{L}^1$ and that (12.9.10) continues to hold, provided $f \cdot \log^+ |f| \in \mathbf{L}^1$. In this case 6.4.4 gives

$$\sigma_N \tilde{f} = \sigma_N(H * f) \to H * f \qquad \text{a.e.}$$

Hence [by 12.8.2(3) once more] $\tilde{f} \equiv H * f = f^c$ a.e. whenever $f \cdot \log^+ |f| \in \mathbf{L}^1$. This confirms an identification stated in 12.8.2(3). It shows also that (12.9.11) holds with $\tilde{f}$ in place of f^c whenever $f \cdot \log^+ |f| \in \mathbf{L}^1$, but this estimate will be bettered in (2).

(2) Starting from the assumed validity of (12.9.10) for trigonometric polynomials f, we aim to show that $H * f \in \mathbf{L}^1$ and that (12.9.10) continues to hold whenever $f \cdot \log^+ |f| \in \mathbf{L}^1$.

Proof. Suppose first that $f \in \mathbf{L}^\infty$. Then, by 12.8.2(2), $H * f \in \mathbf{L}^2 \subset \mathbf{L}^1$. That (12.9.10) holds in this case is easily seen by approximating f by the trigonometric polynomials $\sigma_N f$ which, by 6.4.4 and 6.4.7, converge boundedly and almost everywhere to f, and by making appeal to Lebesgue's theorem ([W], Theorem 4.1b).

Before proceeding to handle a general f satisfying $f \log^+ |f| \in \mathbf{L}^1$, we observe that if in (12.9.10) we replace f by αf, where α is any positive number, and then divide both sides by α, it appears that to any $\varepsilon > 0$ corresponds a number $r = r(\varepsilon) > 0$ such that

$$\|H * f\|_1 \leqslant \varepsilon + \frac{A}{2\pi} \int |f| \log^+ (r|f|)\, dx. \tag{12.9.12}$$

By virtue of what we have already established, (12.9.12) holds for any $f \in \mathbf{L}^\infty$.

Take now any f satisfying $f \log^+ |f| \in \mathbf{L}^1$ and define f_n to be equal to f at points where $|f| \leqslant n$ and to be zero elsewhere, so that $f_n \in \mathbf{L}^\infty$. Applying (12.9.12) to $f_m - f_n$ in place of f, we see that

$$\|H * f_m - H * f_n\|_1 \leqslant \varepsilon + \frac{A}{2\pi} \int |f_m - f_n| \log^+ (r|f_m - f_n|)\, dx. \tag{12.9.13}$$

Suppose that $m < n$. The integrand appearing on the right-hand side of (12.9.13) vanishes on the set E_m of points $x \in [-\pi, \pi]$ satisfying $|f(x)| \leqslant m$ and is everywhere majorized by $2|f| \log^+ (2r|f|)$, which is integrable. Since the measure of the complement $[-\pi, \pi] \backslash E_m$ tends to zero as $m \to \infty$, it follows from (12.9.13) that

$$\|H * f_m - H * f_n\|_1 \leqslant 2\varepsilon$$

provided $n > m \geqslant m(\varepsilon)$. The sequence $(H * f_n)_{n=1}^\infty$ is thus Cauchy, and therefore convergent, in $\mathbf{L}^1$. But, since $f_n \to f$ in $\mathbf{L}^1$, $H * f_n \to H * f$ distributionally (see 12.6.6). It follows that $H * f \in \mathbf{L}^1$. Finally, if (12.9.10) be written down with f_n in place of f, the passage to the limit as $n \to \infty$ will show, since $H * f_n \to H * f$ in $\mathbf{L}^1$ and since Lebesgue's theorem can be applied to the integrals on the right-hand side, that (12.9.10) continues to hold for the chosen f. The proof of (2) is thus complete.

Remark. Statement (2) is, in a sense, the best possible of its type; for the details, see [Z_1] p. 257.

(3) We remark finally that there are also integral inequalities applying to the functions conjugate to bounded functions and to continuous functions. Thus, there exists an absolute constant $\lambda_0 > 0$ such that

$$\frac{1}{2\pi}\int \exp[\lambda|\tilde{f}(x)|]\,dx < \infty \qquad \text{if } \|f\|_\infty \leqslant 1 \text{ and } \lambda < \lambda_0. \quad (12.9.14)$$

Moreover,

$$\frac{1}{2\pi}\int \exp[\lambda|\tilde{f}(x)|]\,dx < \infty \qquad \text{if } f \in \mathbf{C} \text{ and } \lambda \text{ is real}. \quad (12.9.15)$$

Complex variable proofs of these results will be found in [Z_1], pp. 254–257; alternative proofs will appear in 13.9.2. See also [Kz], p. 70.

12.10 Mean Convergence of Fourier series in $\mathbf{L}^p$ $(1 < p < \infty)$

Kolmogorov remarked in 1925 on a way of expressing $s_N f$ in terms of conjugate functions, the use of which leads painlessly from 12.9.1 to mean convergence of the Fourier series of a function in $\mathbf{L}^p$ when $1 < p < \infty$ [see relation (B) in 1.3.2].

12.10.1. **Another Theorem of Marcel Riesz.** Suppose that $1 < p < \infty$. There exists a number $k = k_p$ such that for $f \in \mathbf{L}^p$ one has

$$\|s_N f\|_p \leqslant k\|f\|_p \qquad \text{for all } N \in \{1, 2, \ldots\} \quad (12.10.1)$$

and

$$\lim_{N\to\infty} \|f - s_N f\|_p = 0. \quad (12.10.2)$$

Proof. Put $f_{\pm N}(x) = e^{\pm iNx}f(x)$. Then

$$H * f_N(x) = -i\sum_{n\in Z} \operatorname{sgn} n \cdot \hat{f}(n - N)e^{inx}$$

$$H * f_{-N}(x) = -i\sum_{n\in Z} \operatorname{sgn} n \cdot \hat{f}(n + N)e^{inx}.$$

Hence, for $N > 0$,

$$e^{-iNx}[H * f_N(x)] - e^{iNx}[H * f_{-N}(x)]$$
$$= -i\sum_{n\in Z}[\operatorname{sgn}(n + N) - \operatorname{sgn}(n - N)]\hat{f}(n)e^{inx},$$

or

$$\left.\begin{aligned} &i[e^{-iNx}[H * f_N(x)] - e^{iNx}[H * f_{-N}(x)]] \\ &\quad = 2s_{N-1}f(x) + \hat{f}(N)e^{iNx} + \hat{f}(-N)e^{-iNx} \end{aligned}\right\}. \quad (12.10.3)$$

Using 12.9.1 and 2.3.2, (12.10.1) follows directly. [The value of k appearing in (12.10.1) is not necessarily the same as that appearing in (12.9.1).]

Now (12.10.1) shows that the set, say $\mathbf{S}$, of $f \in \mathbf{L}^p$ for which (12.10.2) is true, is closed in $\mathbf{L}^p$. For suppose that $(f_n)_{n=1}^{\infty}$ is a sequence extracted from $\mathbf{S}$

which converges in $\mathbf{L}^p$ to f. Then

$$\begin{aligned}\|s_N f - f\|_p &\leqslant \|s_N f - s_N f_n\|_p + \|s_N f_n - f_n\|_p + \|f_n - f\|_p \\ &\leqslant k\|f - f_n\|_p + \|s_N f_n - f_n\|_p + \|f - f_n\|_p \\ &= (k+1)\|f - f_n\|_p + \|s_N f_n - f_n\|_p.\end{aligned}$$

Given $\varepsilon > 0$, first choose and fix $n = n(\varepsilon)$ so that

$$\|f - f_n\|_p \leqslant \frac{\varepsilon}{2(k+1)},$$

and then choose $N_0 = N_0(\varepsilon)$ so that

$$\|s_N f_n - f_n\|_p \leqslant \tfrac{1}{2}\varepsilon$$

for all $N \geqslant N_0$. This last choice is possible since $f_n \in \mathbf{S}$. It appears thus that

$$\|s_N f - f\|_p \leqslant \varepsilon$$

for all $N \geqslant N_0$, which shows that $f \in \mathbf{S}$.

Having seen that $\mathbf{S}$ is closed in $\mathbf{L}^p$, it remains only to show that $\mathbf{S}$ is everywhere dense in $\mathbf{L}^p$. But it is evident that $\mathbf{S}$ contains all trigonometric polynomials. These are everywhere dense in $\mathbf{L}^p$ by 6.1.1. The proof is thus complete.

12.10.2. **Further Inequalities.** Assuming the inequality (12.9.9) to hold for trigonometric polynomials, the equation (12.10.3) yields at once the estimate

$$\|s_N f\|_p \leqslant k_p \|f\|_1 \qquad \text{for all } p \in (0, 1) \text{ and all } N \in \{1, 2, \ldots\} \tag{12.10.4}$$

for trigonometric polynomials f. (The constant k_p need not have the same value in (12.10.4) as in (12.9.9).) The extension to any $f \in \mathbf{L}^1$ is almost immediate. (Approximate f by the trigonometric polynomials $\sigma_{N'} f$.) Then, much as in the closing stage of the proof of 12.10.1, it may be inferred that (12.10.2) holds whenever $f \in \mathbf{L}^1$ and $0 < p < 1$.

Likewise, assuming the results in 12.9.9(2), it may be inferred that

$$\|s_N f\|_1 \leqslant \frac{A}{2\pi} \int |f| \log^+ |f| \, dx + B \qquad \text{for all } N \in \{1, 2, \ldots\} \tag{12.10.5}$$

whenever $f \cdot \log^+ |f| \in \mathbf{L}^1$. A little more argument will then show that (12.10.2) holds with $p = 1$ whenever $f \cdot \log^+ |f| \in \mathbf{L}^1$; for the details, see $[Z_1]$, p. 267.

The result stated in Exercise 10.2 shows at once that (12.10.1) and (12.10.2) are false when $p = 1$ and f is suitably chosen from $\mathbf{L}^1$. (But see 12.7.9(2).)

See also MR **55** # 963.

12.10.3. **Projection of $\mathbf{L}^p$ onto $\mathbf{H}^p$.** Consider the distribution

$$P = \tfrac{1}{2}(1 + \varepsilon + iH), \tag{12.10.6}$$

which is such that

$$\hat{P}(n) = 1 \text{ if } n \geqslant 0, \qquad = 0 \text{ if } n < 0. \tag{12.10.7}$$

For any distribution F we have

$$P * F = \sum_{n \geqslant 0} \hat{F}(n) e^{inx} \tag{12.10.8}$$

Most of the results about P can be read off from those of H appearing in Section 12.8, 12.9.1, and 12.10.2. In particular, P is a distribution of order exactly 1; and the set of $f \in \mathbf{L}^1$ for which $P * f \in \mathbf{M}$ is meagre [see 12.8.3(3)].

The operation $F \to P * F$ is a *projection*, that is, it is linear and idempotent.

If we introduce the Hardy space $\mathbf{H}^p$ for $1 \leqslant p \leqslant \infty$ (see Exercise 3.9, the references cited there, and also Chapter 17 of $[\mathrm{R}_1]$), then it follows from 12.9.1 that, when $1 < p < \infty$, $f \to P * f$ is a continuous projection of $\mathbf{L}^p$ onto $\mathbf{H}^p$. This assertion is false for $p = 1$ [see 12.8.3(3)]; indeed it is known (D. J. Newman [1]) that there exists no continuous projection whatsoever of $\mathbf{L}^1$ onto $\mathbf{H}^1$.

12.11 Pseudomeasures and Their Applications

12.11.1. **Definition of Pseudomeasures; the Space P.** By a *pseudomeasure* is meant a distribution S such that $\hat{S}$ is a bounded function on Z; the terminology appears to have been coined by Kahane and Salem; see [KS], Appendices I and II; $[\mathrm{Kah}_2]$, Chapitre III; [Kz], p. 150. Thus, the distributions H and P introduced in 12.8.1 and 12.10.3, respectively, are pseudomeasures. We denote by $\mathbf{P}$ the set of pseudomeasures; $\mathbf{P}$ is a linear subspace of $\mathbf{D}$, and $\mathbf{M} \subset \mathbf{P}$ properly. $\mathbf{P}$ is also a convolution algebra.

Pseudomeasures arise quite naturally in the representation of multipliers, to be discussed in Chapter 16.

12.11.2. Let us reintroduce the linear space $\mathbf{A}$ of continuous functions f such that

$$\|f\|_{\mathbf{A}} \equiv \sum_{n \in Z} |\hat{f}(n)| < \infty; \tag{12.11.1}$$

see Section 10.6 and 11.4.17.

As has been noted in 10.6.1 and 11.4.17, $\mathbf{A}$ is a Banach space and also a Banach algebra under pointwise operations.

12.11.3. **P as a Dual Space.** One reason for the significance of $\mathbf{P}$ is that it can be identified with the set of continuous linear functionals on $\mathbf{A}$: each continuous linear functional on **A** is expressible in the form

$$f \to S(f) = \sum_{n \in Z} \hat{S}(n)\hat{f}(-n)$$

for a uniquely determined $S \in \mathbf{P}$; and, conversely, each $S \in \mathbf{P}$ generates thus a continuous linear functional on **A**. The dual norm of S is

$$\|S\|_{\mathbf{P}} \equiv \sup\{|S(f)| : f \in \mathbf{A}, \|f\|_{\mathbf{A}} \leqslant 1\},$$

which proves to be equal to

$$\|\hat{S}\|_{\infty} \equiv \sup\{|\hat{S}(n)| : n \in Z\}.$$

All this is scarcely more than a restatement of the fact that each continuous linear functional on $\ell^1(Z)$ has the form

$$\phi \to \sum_{n \in Z} \phi(n)\psi(-n)$$

for a unique $\psi \in \ell^\infty(Z)$, and conversely; see Exercise 12.32.

Since **A** is a Banach algebra, the product fS of $f \in \mathbf{A}$ and $S \in \mathbf{P}$ can be defined as a pseudomeasure by means of the relation

$$fS(g) = S(fg) \qquad (g \in \mathbf{A}); \tag{12.11.2}$$

compare the substance of 12.3.4.

If $\phi \in \ell^1(Z)$ and $\psi \in \ell^\infty(Z)$, then $\hat{\phi} \in \mathbf{A}$ and $\hat{\psi}$, the distributional sum of the series

$$\sum_{n \in Z} \psi(n)e^{inx},$$

is a pseudomeasure. Moreover, $\phi * \psi \in \ell^\infty(Z)$ and (see Exercise 12.32 again)

$$(\phi * \psi)\hat{} = \hat{\phi}\hat{\psi}. \tag{12.11.3}$$

12.11.4. **Problems Involving Pseudomeasures. Spectral Synthesis Sets.** The most fascinating problems concerning pseudomeasures arise in connection with applications of the Hahn-Banach theorem to the study of the analogue, for the group Z, of the problems discussed in Section 11.2 in relation to the groups T and R (see 11.2.2 and 11.2.5; 12.12.6; [R], Chapter 7; [KS], Chapitres IX and X and Appendice II).

In order to describe such problems, it is necessary to speak of the support of a pseudomeasure. Generally speaking, one can show that for any distribution S there exists a smallest closed set E with the following property: $S(u) = 0$ for each $u \in \mathbf{C}^\infty$ whose support $\operatorname{supp} u = \{x : u(x) \neq 0\}^-$ does not intersect E; this set E is the *support* of S, denoted by $\operatorname{supp} S$. See Exercise 12.29 and also, for the case in which S is a measure, Exercise 12.27.

A distribution S is said to be *supported by* a set F, and F is said to *support* S, if and only if $\operatorname{supp} S \subset F$.

Suppose now that S is a pseudomeasure. There is no good reason to expect the relation

$$S(f) = 0 \tag{12.11.4}$$

to be implied by

$$f \in \mathbf{A}, \quad f(E) \leqslant \{0\} \quad \operatorname{supp} S \subset E; \tag{12.11.5}$$

compare Exercise 12.30. That (12.11.4) does *not* in general follow from (12.11.5) is a corollary of the results due to Malliavin concerning spectral synthesis mentioned in 11.2.3. (The connections hinge upon the use of the Hahn-Banach theorem in the manner indicated in 12.12.6 and again in the hints to Exercise 12.32.)

Actually, as subsequent work of Kahane and Rudin has disclosed, something more specific is true, namely: there exist real-valued functions $f \in \mathbf{A}$ such that, if $\mathbf{J}_k$ denotes the closed ideal in the algebra $\mathbf{A}$ generated by f^k ($k = 1, 2, \cdots$), then $\mathbf{J}_{k+1}$ is a proper subset of $\mathbf{J}_k$ for each k. (For the details, see Exercise 12.53 and the references cited there. The statement remains true when T is replaced by any infinite compact Abelian group whatsoever.) Expressed otherwise (in dual form, in fact), this means that there exist functions $\phi \in \ell^1(Z)$ with the property that, if $\mathbf{I}_k$ denotes the closed ideal in the convolution algebra $\ell^1(Z)$ generated by the convolution power ϕ^{*k} where $k \in \{1, 2, \ldots\}$, then $\mathbf{I}_{k+1}$ is a proper subset of $\mathbf{I}_k$ for each k. The case $k = 1$ entails (via the Hahn-Banach theorem) that a pseudomeasure S exists such that

$$\hat{\phi}^2 S = 0, \qquad \hat{\phi} S \neq 0. \tag{12.11.6}$$

Since the first equation in (12.11.6) shows that $\operatorname{supp} S \subset E \equiv \hat{\phi}^{-1}(\{0\})$, (12.11.6) entails at once that (12.11.5) does not imply (12.11.4).

Other striking counterexamples have been given by Kahane and Katznelson [2].

Faced with this, two courses of investigation suggest themselves, namely:

(1) to seek special types of closed set E for which the relations (12.11.5) imply (12.11.4).

(2) to seek extra conditions upon $f \in \mathbf{A}$ which, in conjunction with (12.11.5), suffice to entail (12.11.4).

The pursuit of the aim specified in (1) amounts to the study of the analogue, for the group T of the concept of spectral synthesis set in R^m mentioned in 11.2.3. Thus, with our present approach, it is natural to define a *spectral* or *harmonic synthesis set* (= *ensemble de synthèse spectrale* ou *harmonique*) in T as a closed subset E of T such that (12.11.5) implies (12.11.4); the terminology will be explained in Subsection 12.11.6. It is equivalent to say that E is a spectral synthesis set, if and only if

(12.11.5) implies $fS = 0$. Thus if ϕ is as in (12.11.6), $E = \hat{\phi}^{-1}(\{0\})$ is a closed set which is not a spectral synthesis set.

There is complete accord between the above definition of spectral synthesis sets in T and the apparently different one used in Subsection 11.2.3 for spectral synthesis sets in R^m. In other words, it can be shown without trouble that a closed subset E of T is a spectral synthesis set (as defined in the preceding paragraph) if and only if it has the following property: the relation

$$\mathbf{I} = \mathbf{I}_E \equiv \{\phi \in \ell^1(Z) : \hat{\phi} = 0 \text{ on } E\}$$

holds for every closed ideal $\mathbf{I}$ in $\ell^1(Z)$ for which

$$Z_{\mathbf{I}} \equiv \bigcap \{\hat{\phi}^{-1}(\{0\}) : \phi \in \mathbf{I}\} = E.$$

The reader is invited to construct a proof of this equivalence, using the Hahn-Banach theorem as an intermediary; compare the arguments presented in Subsection 12.12.6.

The remarks made in Subsection 11.2.3 about spectral synthesis sets in R^m apply in the main to spectral synthesis sets in T; in particular, although conditions are known which are sufficient to ensure that a given closed set is a spectral synthesis set, a complete structural characterization of such sets appear to be extremely difficult; see 12.11.5.

As regards (2), most of the known sufficient conditions impose smoothness restrictions on f. Perhaps the simplest nontrivial sufficient condition is that f shall belong to $\mathbf{C}^1$ and that Df (as well as f itself) shall vanish on E; see Exercise 12.31. A deeper result asserts that it is sufficient that f satisfy a supplementary Lipschitz condition of order $1/2$, that is, that

$$|f(x) - f(x')| \leqslant \text{const } |x - x'|^{1/2};$$

the exponent $\frac{1}{2}$ is known to be best possible (see MR **40** # 629). For this and other similar sets of conditions, see Subsection 13.5.5 and [KS], p. 123. (These and similar questions are discussed in a more general setting by Herz [2] and Edwards [4].) Once again, no necessary and sufficient conditions are known.

12.11.5. **Some Examples and Counterexamples concerning Spectral Synthesis Sets.** For details concerning the matters touched upon lightly here, the reader should in general consult Chapter 7 of [R]; [HR], Chapter 10; [Kah$_2$], Chapitre V; and Chapitres IX and XI of [KS]; see also [Kah] and Malliavin [2]. More specific references will appear as we proceed.

(1) The simplest and oldest specific condition on a closed set E sufficient (but not necessary) to ensure that E shall be a spectral synthesis set is that

the frontier of E contains no nonvoid perfect set (see 11.2.3 and Exercise 12.52). This condition is fulfilled whenever E is countable.

There is a version of this result applying to general groups (see [R], p. 161) and another applying in the general context of commutative Banach algebras (see [Lo], p. 86 and [N], p. 226, Theorem 5). In each of these references, spectral synthesis sets in T are discussed in terms of closed ideals in $\ell^1(Z)$; see 11.2.3 and 11.2.5. See also [HR], (39.24) and (39.29).

(2) A number of examples of nonvoid perfect spectral synthesis sets are known. Thus, Herz showed in 1956 that the Cantor ternary set (see Exercise 12.44) is a spectral synthesis set. Herz's original proof has been developed and generalized; see [KS], pp. 124–125, and [R], Section 7.4 and [Kah_2], pp. 58–59. See also MR **50** # 7956; **52** # 8800.

(3) There is another method of constructing nonvoid perfect spectral synthesis sets, due originally to Kahane and Salem; see [KS], pp. 125–127. In order to explain this in a little detail, we introduce some notation, supposing in what follows that E denotes a closed subset of T.

Denote by $\mathbf{M}(E)$ [respectively, $\mathbf{P}(E)$] the set of measures μ (respectively, pseudomeasures S) such that $\operatorname{supp}\mu \subset E$ (respectively, $\operatorname{supp} S \subset E$). Denote further by $\mathbf{P}^0(E)$ the set of pseudomeasures S such that $S(f) = 0$ whenever $f \in \mathbf{A}$ vanishes on E. $\mathbf{P}^0(E)$ is weakly closed in $\mathbf{P}(E)$. It is then evident that $\mathbf{P}^0(E) \subset \mathbf{P}(E)$ and almost evident that $\mathbf{M}(E) \subset \mathbf{P}^0(E)$. (The second point depends on the remark that any continuous function which vanishes on the complement of E is the uniform limit of continuous functions, each having its support contained in the complement of E.) Thus

$$\mathbf{M}(E) \subset \mathbf{P}^0(E) \subset \mathbf{P}(E). \tag{12.11.7}$$

A moment's thought will show that E is a spectral synthesis set if and only if

$$\mathbf{P}^0(E) = \mathbf{P}(E). \tag{12.11.8}$$

A fortiori, therefore, any set E such that

$$\mathbf{M}(E) = \mathbf{P}(E) \tag{12.11.9}$$

is a spectral synthesis set. Sets E satisfying (12.11.9) are usually said to "support no true pseudomeasures." For example (see Exercise 12.33), every finite set is of this type; however, there exist countable closed sets E that do not satisfy (12.11.9); see Exercise 15.21.

The first nontrivial examples of sets supporting no true pseudomeasures were given by Kahane and Salem in 1956 and were nonvoid and perfect (see [KS], pp. 126–127). More recently, Varopoulos [2] has shown that among the sets supporting no true pseudomeasures are to be found all the so-called Kronecker sets defined in Subsection 15.7.4.

(4) On the other hand, as soon as one can exhibit (or prove the existence of) pseudomeasures $S \in \mathbf{P}(E)$ and functions $f \in \mathbf{A}$ vanishing on E and satisfying $S(f) \neq 0$, it will become certain that E is *not* a spectral synthesis set.

This principle is valid for quite general groups and lies behind Schwartz's proof that the unit sphere $E = \{x \in R^3 : |x| = 1\}$ is not a spectral synthesis set in the group R^3 (see 11.2.3): in this case it suffices to take for S the distributional derivative $\partial\mu/\partial x_1$, where μ is the measure obtained by distributing a unit total mass uniformly over E (see [R], pp. 165–166). In this connection we remark that it is known ([R], p. 172) that each of the sets $\{x \in R^3 : |x| \leqslant 1\}$ and $\{x \in R^3 : |x| \geqslant 1\}$ is a spectral synthesis set in R^3; it follows that the intersection of two spectral synthesis sets, and the frontier of a spectral synthesis set, may fail to be such a set. See also [HR], (40.19); Varopoulos [3]; MR **20** # 7186; **48** # 2671; **51** # 13592; **55** # 8699.

The closing remarks of the last paragraph prompt the question: is the union of two spectral synthesis sets always a spectral synthesis set? It is not difficult to see that the union of two *disjoint* spectral synthesis sets is a set of the same nature, but the general case is unsolved. See also Varopoulos [3] and Drury [1], [2].

12.11.6. **Spectral Synthesis in $\ell^\infty(Z)$.** In order to explain briefly the term "spectral synthesis set" applied to certain closed subsets of $R/2\pi Z$, it seems best to consider the spectral (or harmonic) analysis and synthesis problems for the group Z.

For the group T, these problems have been introduced in Subsection 2.2.1 and solved in the course of Sections 11.1 and 11.2. For the space $\mathbf{L}^\infty$ (with its weak topology), the analysis problem is that of determining which (bounded continuous) characters $e_n (n \in Z)$ of T belong to $\bar{\mathbf{V}}_f^\infty$, the weakly closed invariant subspace of $\mathbf{L}^\infty$ generated by f. The answer is contained in 11.1.1, namely, $e_n \in \bar{\mathbf{V}}_f^\infty$ if and only if $n \in \operatorname{supp} \hat{f}$. The synthesis problem is that of the recapture of f from these e_n, in the sense that f shall be the weak limit in $\mathbf{L}^\infty$ of linear combinations of characters $e_n \in \bar{\mathbf{V}}_f^\infty$; the possibility of doing this is recorded in Remark (1) following 11.2.1.

In the case of T the same is true when we replace $\mathbf{L}^\infty$ by $\mathbf{L}^p$ $(1 \leqslant p < \infty)$ or $\mathbf{C}$.

Let us now turn to the case of the discrete group Z, whose bounded continuous characters are the functions $e_x : n \to e^{ixn}$ $(x \in T)$; see Subsection 2.5.4 and Exercise 2.3. In this case we cannot discuss the analysis and synthesis problems for $\ell^p(Z)$ $(1 \leqslant p < \infty)$ or $\mathbf{c}_0(Z)$, since no character e_x belongs to any of these spaces. There remains the space $\ell^\infty(Z)$, which contains each e_x. It will be necessary to consider $\ell^\infty(Z)$ with its so-called weak topology generated by $\ell^1(Z)$ (see I, B.4.2): this is the weakest topology on $\ell^\infty(Z)$ such that, for each $\psi \in \ell^1(Z)$, the function $\phi \to \sum_{n \in Z} \phi(n)\psi(n)$ is continuous in the said topology. (Thus, a sequence or net (ϕ_i) of elements of $\ell^\infty(Z)$ converges weakly to an element ϕ of $\ell^\infty(Z)$ if and only if

$$\lim_i \sum_{n \in Z} \phi_i(n)\psi(n) = \sum_{n \in Z} \phi(n)\psi(n)$$

for each $\psi \in \ell^1(Z)$.)

The spectral analysis problem for $\ell^\infty(Z)$ is as follows: given $\phi \in \ell^\infty(Z)$, which characters e_x belong to $\overline{\mathbf{V}}_\phi^\infty$, the weakly closed invariant subspace of $\ell^\infty(Z)$ generated by ϕ? By arguing much as in Subsection 12.12.6, the answer is seen to take the form: $e_x \in \overline{\mathbf{V}}_\phi^\infty$ if and only if $x \in -E$, where $E = \operatorname{supp} \hat{\phi}$. (The minus sign is a consequence of the way we chose to define $\hat{\phi}$ in equation (2.5.1) and 12.5.4(2); it has no sinister significance.) As for the synthesis problem, it can be shown (again by arguments similar to those to be used in Subsection 12.12.6) that ϕ is the weak limit in $\ell^\infty(Z)$ of linear combinations of characters $e_x \in \overline{\mathbf{V}}_\phi^\infty$ if and only if $f\hat{\phi} = 0$ for each $f \in \mathbf{A}$ which vanishes on $-E$. In turn, this will be true for *all* $\phi \in \ell^\infty(Z)$ for which $\operatorname{supp} \hat{\phi} \subset -E$, if and only if $-E$ is a spectral synthesis set in T, which is trivially equivalent to saying that E itself is a spectral synthesis set in T.

It is now evident why the term "spectral (or harmonic) synthesis set" was selected: put very crudely, all those functions in $\ell^\infty(Z)$, which *ought* to be synthesizable from the characters e_x $(x \in E)$, are *in fact* so synthesizable if and only if E is a spectral synthesis set in T.

At the same time, the results recalled above concerning spectral analysis and synthesis in $\mathbf{L}^\infty$ signify in particular that *every* subset of the discrete group Z is (or may be regarded as) a spectral synthesis set in Z.

Many of the problems concerned with spectral analysis and synthesis stem from a study of the case in which the underlying group is R, given by Beurling [3]. An exposition of generalizations and analogues of Beurling's work, applying to general groups, is to be found in Herz [2] and the references cited there; see also MR **41** # 5893; **50** # 5366; **55** # 3685. For an especially interesting chapter in this story, see Koosis [1].

12.12 Capacities and Beurling's Problem

The aim in this section is to explain the concept of capacity referred to in 10.4.6, to apply it to the study of Beurling's problem mentioned in 11.2.5, to indicate its kinship with ideas in potential theory, and to provide a guide to further reading.

It is not possible to provide detailed proofs of *all* the necessary results about capacities, nor even to mention specifically *all* the concepts of capacity which have shown themselves to be relevant to harmonic analysis and trigonometric series; see the references cited in 12.12.7(5).

In 12.12.1–12.12.5 we assemble as quickly as possible enough information about capacity to make an application to Beurling's problem, which is discussed in 12.12.6. In 12.12.7 appear some diverse remarks of a general nature.

The notion of support of a distribution (see 12.11.4) plays a vital role throughout this section.

Save in 12.12.7(1), we assume that the number α satisfies $0 < \alpha < 1$.

12.12.1. **The Kernels K_α.** For each $\alpha \in (0, 1)$ we choose a corresponding *kernel of order* α, namely, the function

$$K_\alpha(x) = |\sin \tfrac{1}{2}x|^{-\alpha}. \tag{12.12.1}$$

Each such kernel can be used to develop a corresponding potential theory in which K_α takes over the role played in Newtonian potential theory by the Newtonian kernel $|x|^{-1}$ on R^3 (where $|x|$ here denotes the Euclidean length of the vector $x \in R^3$); see 12.12.7(5). In working toward our immediate objective, we shall naturally use harmonic analysis much more than is traditional in Newtonian theory (although modern trends in the latter field also invoke harmonic analysis to a fair degree). Thus we shall be especially interested in the Fourier transform of K_α.

The essential facts are that

$$\hat{K}_\alpha(n) > 0 \qquad (n \in Z) \tag{12.12.2}$$

and that

$$\hat{K}_\alpha(n) \sim C_\alpha |n|^{\alpha - 1} \qquad \text{as } |n| \to \infty, \tag{12.12.3}$$

where C_α is a positive number. The reader should experience no great difficulty in proving (12.12.2) and (12.12.3); see [KS], pp. 32–33, 39–40.

12.12.2. **The α-energy of a Distribution.** If A denotes any distribution, its *α-energy* is defined to be

$$E_\alpha(A) = \sum_{n \in Z} \hat{K}_\alpha(n) |\hat{A}(n)|^2 \quad (\leqslant \infty). \tag{12.12.4}$$

Using (12.12.2) and Bochner's theorem 9.2.8, it is not difficult to verify that $E_\alpha(A)$ is finite if and only if the distribution $K_\alpha * A * A^*$ is equal distributionally to a continuous function f, in which case $E_\alpha(A) = f(0)$. This criterion could therefore be used to define the α-energy of A.

In view of (12.12.3) it appears that A has finite α-energy if and only if

$$\sum_{n \neq 0} |n|^{\alpha - 1} |\hat{A}(n)|^2 < \infty. \tag{12.12.5}$$

12.12.3. **The α-capacity of Closed Sets.** If E is a closed subset of T, its α-capacity $c_\alpha(E)$ is defined by the relation

$$c_\alpha(E) = \frac{1}{\inf E_\alpha(\mu)} \tag{12.12.6}$$

the infimum being taken with respect to all positive measures μ having total mass $\mu(1) = 1$ and with support $\operatorname{supp} \mu \subset E$; it is to be understood that $\inf\{\infty\} = \inf \varnothing = \infty$ and that $1/\infty = 0$. Recall also that a measure μ is positive if and only if $\mu(f) \geqslant 0$ for each nonnegative continuous function f (compare Exercise 12.7); for this it is enough that $\mu(f) \geqslant 0$ for each nonnegative $f \in \mathbf{C}^\infty$.

In particular, $c_\alpha(E) = 0$ if and only if E supports no nonzero positive measure μ such that

$$\sum_{n \neq 0} |n|^{\alpha - 1} |\hat{\mu}(n)|^2 < \infty .$$

It is plain that $c_\alpha(E) = 0$ implies that $c_\beta(E) = 0$ whenever $\alpha < \beta < 1$.

It is equally clear that $c_\alpha(E) > 0$ whenever E has positive measure. The converse is false: given α, $0 < \alpha < 1$, there exist perfect sets E of measure zero and positive α-capacity (see [Ba_2], p. 406).

12.12.4. **Criterion for Positive α-capacity.** By definition, in order that a closed set E shall have positive α-capacity, it is *necessary* (and sufficient) that E shall support a nonzero positive measure having finite α-energy.

It is true and entirely unexpected that a *sufficient* condition is that E shall support a nonzero *distribution* having finite α-energy.

A proof of somewhat more than this, which depends on delving a little more deeply into the potential-theoretic development of the notion of α-energy, is suggested in Exercise 12.42.

12.12.5. **Capacitary Dimension.** For a closed set $E \subset T$ we define the *capacitary dimension* of E, denoted by cap. dim. E, to be the supremum of numbers α such that $0 < \alpha < 1$ and $c_\alpha(E) > 0$.

It is a fact, which need not detain us at all, that cap. dim. E is numerically equal to the so-called Hausdorff dimension of E; see [KS], p. 34 Théorème I.

We shall have need of two results concerning capacitary dimension, one of which is a very simple deduction from 12.12.4 and will be proved here; the other is more difficult and we must refer the reader to [KS], p. 106, Théorème IV for its proof.

(1) If cap. dim. $E = \alpha$, $0 < \alpha < 1$, and $0 < q < 2/\alpha$, then E supports no pseudomeasures $\sigma \neq 0$ for which $\hat{\sigma} \in \ell^q(Z)$. (The conclusion holds, indeed, even if σ be assumed to be merely a distribution.)

(2) If $0 < \alpha < 1$ and $q > 2/\alpha$, there exist a closed set E satisfying cap. dim. $E = \alpha$ and a nonzero positive measure μ supported by E for which $\hat{\mu} \in \ell^q(Z)$.

Proof of (1). We may suppose that $q > 2$ and choose α' so that $c_{\alpha'}(E) = 0$ and $q < 2/\alpha'$.

If $\operatorname{supp} \sigma \subset E$ and $\hat{\sigma} \in \ell^q(Z)$, a simple application of Hölder's inequality for series shows that $E_{\alpha'}(\sigma) < \infty$. Since $c_{\alpha'}(E) = 0$ and $\operatorname{supp} \sigma \subset E$, 12.12.4 entails that $\sigma = 0$.

Remarks. If $0 < \alpha < 1$, there exist closed sets E with cap. dim. $E = \alpha$ which support nonzero positive measures μ for which $\hat{\mu}(n) = O(|n|^{-\frac{1}{2}\beta})$ for any preassigned $\beta < \alpha$; and closed sets E of measure zero exist which support nonzero positive measures μ for which $\hat{\mu}(n) = O(|n|^{\varepsilon - \frac{1}{2}})$ for any preassigned $\varepsilon > 0$; see [Z_2], p. 146. The best such results are due to Ivašev-Mousatov; see [KS], pp. 100–111 (and Hewitt and Zuckerman [3]; Hewitt and Ritter [2], [3]; Brown [1], [2] for extensions to more general groups). See also MR **37** # 3277.

The computation of the α-capacity, or even of the capacitary dimension, of a given set is seldom easy. However, for the purposes of examples, whole classes of sets (obtained in a fashion rather like that which leads to the famous Cantor ternary set; see Exercise 12.44) have been defined and some information about their capacitary dimensions accumulated; see [KS], Chapitres I, II, and III.

So, for example, given any integer $\nu \geqslant 2$ and any number ξ satisfying $0 < \xi < 1/\nu$, one can construct perfect nowhere dense sets E (the Cantor set corresponding to $\nu = 2$, $\xi = \frac{1}{3}$) having measure zero and for which cap. dim. $E = \log \nu / \log (1/\xi)$; see [KS], pp. 16–17, 34. In particular, Cantor's set E has capacitary dimension equal to $\alpha_0 = \log 2/\log 3$ and α_0-capacity equal to zero.

Incidentally, among such sets E one finds many sets of uniqueness (see 12.12.8 and [KS], p. 59).

12.12.6. **Application to Beurling's Problem.** This problem, already mentioned in 11.2.5, can be formulated in the following terms: suppose that $\phi \in \ell^1(Z)$ and that $E = \hat{\phi}^{-1}(\{0\})$; under what conditions upon E does the closed invariant subspace $\overline{\mathbf{V}}_\phi^p$ of $\ell^p(Z)$ generated by ϕ coincide with $\ell^p(Z)$? (The notation $\overline{\mathbf{V}}_\phi^p$ is suggested by that introduced in Section 11.1.)

As we shall see, one of the most interesting cases of this problem is almost (but not quite) completely solved in terms of the capacitary dimension of E.

We discuss three cases, according to the value of p involved.

(1) The case $p = 1$. The complete solution is contained in Wiener's closure of translations theorem for the group Z (see 11.2.5 and Exercise 12.32): $\overline{\mathbf{V}}_\phi^1 = \ell^1(Z)$ if and only if $E = \varnothing$.

(2) The case $2 \leqslant p \leqslant \infty$. (If $p = \infty$, it is understood that $\ell^\infty(Z)$ is taken with its weak topology; see the introductory remarks to Chapter 11 and I, B.4.2.)

According to the Hahn-Banach theorem (I, B.5.2), $\overline{\mathbf{V}}_\phi^p = \ell^p(Z)$ if and only if the unique $\psi \in \ell^{p'}(Z)$ satisfying

$$\phi * \psi = 0 \tag{12.12.7}$$

is $\psi = 0$. In this application of the Hahn-Banach theorem, we are identifying the dual of $\ell^p(Z)$ with $\ell^{p'}(Z)$ (a discrete analogue of I, C.1; see [E], Exercise 1.2 and, for $p = \infty$, a special case of Theorem 8.1.1; and Exercise 12.32 below.)

Since $\phi \in \ell^1(Z)$, (12.12.7) is equivalent to

$$\hat{\phi}\hat{\psi} = 0, \tag{12.12.8}$$

$\hat{\psi}$ being at any rate a pseudomeasure and $\hat{\phi}$ a member of $\mathbf{A}$; see (12.11.3).

However, since $1 \leqslant p' \leqslant 2$, more can be said about $\hat{\psi}$: as will be seen in 13.4.1, $\hat{\psi}$ is actually a function in $\mathbf{L}^p$; if $p = \infty$, it is evident that $\hat{\psi} \in \mathbf{C}$.

It follows that the condition that E be null suffices to ensure that $\overline{\mathbf{V}}_\phi^p = \ell^p(Z)$; that this condition is also necessary when $p = 2$; that a necessary condition for any of the specified values of p is that E be nowhere dense; and that this last condition is also sufficient when $p = \infty$.

Nothing more precise seems to be known for $2 < p < \infty$.

(3) The case $1 < p < 2$. This is the case in which the capacitary dimension of E enters into the discussion. We will give two results.

(a) In order that $\overline{\mathbf{V}}_\phi^p = \ell^p(Z)$, it suffices that

$$\text{cap. dim. } E < 2/p'. \tag{12.12.9}$$

(b) If $1 > \alpha > 2/p'$, there exists $\phi \in \ell^1(Z)$ such that cap. dim. $E = \alpha$ and $\overline{\mathbf{V}}_\phi^p \neq \ell^p(Z)$.

Proof of (a). We start by using the Hahn-Banach theorem exactly as in (2) above. From (12.12.8), it appears that $\sigma \equiv \hat{\psi}$ satisfies supp $\sigma \subset E$. Moreover, $\hat{\sigma} = \psi \in \ell^{p'}(Z)$. So (12.12.9) and 12.12.5(1) combine to show that $\sigma = 0$ and therefore $\psi = 0$. By virtue of the Hahn-Banach theorem, (a) is thus established.

Proof of (b). If $\alpha > 2/p'$, 12.12.5(2) affirms the existence of a closed subset E of T satisfying cap. dim. $E = \alpha$ and a nonzero positive measure μ supported by E and such that $\psi \equiv \hat{\mu} \in \ell^{p'}(Z)$. Now it is not difficult to construct a $\phi \in \ell^1(Z)$ such that $\hat{\phi}^{-1}(\{0\}) = E$; compare the proof of Lemma (6.1) in Edwards [4]. Then, since μ is a measure, $\hat{\phi}\mu = 0$, which is equivalent to (12.12.8). Since $\mu \neq 0$, ψ is not the zero element of $\ell^{p'}(Z)$. Thus (12.12.7) shows that $\overline{\mathbf{V}}_\phi^p$ cannot coincide with $\ell^p(Z)$.

12.12.7. **Further Remarks about Capacity.** In this subsection we collect a few remarks about capacity and give references for further reading.

(1) *The cases $\alpha = 0, 1$: logarithmic capacity.* Hitherto it has been assumed that $0 < \alpha < 1$.

Corresponding formally to the case $\alpha = 0$ is the logarithmic kernel

$$K_0(x) = \log(|\sin \tfrac{1}{2} x|^{-1});$$

the associated concept of capacity is termed *logarithmic capacity*.

The formulae (12.12.2) and (12.12.3) remain valid for $\alpha = 0$, and it follows that a closed set E for which $c_\alpha(E) > 0$ for some α satisfying $0 < \alpha < 1$ has positive logarithmic capacity. The converse is false; compare the closing remark in 12.12.3.

Corresponding to the case $\alpha = 1$ one must take as kernel the Dirac measure ε: this leads to a concept of capacity that is essentially equivalent to invariant measure and is of little interest in the present context.

(2) *Interior and exterior capacities.* In search of greater flexibility, the concept of capacity, so far defined only for closed sets, can be extended to more general sets. The first step is the introduction of the so-called interior and exterior capacities.

The *interior α-capacity* of an arbitrary set E is by definition

$$c_{\alpha *}(E) = \sup c_\alpha(F), \tag{12.12.10}$$

the supremum being taken with respect to all closed sets $F \subset E$. Thus $c_{\alpha *}(E) = c_\alpha(E)$ if E is closed.

The *exterior α-capacity* of an arbitrary set E is defined to be

$$c_\alpha^*(E) = \inf c_{\alpha *}(U), \tag{12.12.11}$$

the infimum being taken relative to all open sets U containing E.

Before saying any more about these set functions, we observe a further connection with trigonometric series.

(3) *Capacity and convergence of trigonometric series.* Reverting to the topics discussed in 10.4.6, it can be shown that if $(c_n)_{n\in Z}$ is such that

$$\sum_{n\in Z} |n|^{1-\alpha}|c_n|^2 < \infty,$$

then the (Borel) set of points of divergence of the series

$$\sum_{n\in Z} c_n e^{inx}$$

has interior α-capacity zero. This is a special case of a result applying to more general notions of capacity; see [Ba$_1$], p. 411. The original result of Beurling referred to in 10.4.6 concerns the case of logarithmic capacity ($\alpha = 0$). There is a converse assertion, also due to Beurling. For all this, see also [KS], p. 41–47.

(4) *Capacities as set functions.* The interior and exterior capacities just defined are plainly nonnegative and increasing set functions, in which respect they are like measures. There the similarity ends, however: interior and exterior capacities are not even finitely additive on simple types of sets. However, it is not difficult to show that c_α^* is countably subadditive for arbitrary sets.

On the contrary, to show that $c_{\alpha *}(E) = c_\alpha^*(E)$ for a reasonably wide class of sets is much more difficult than is the corresponding problem for measures. The satisfactory solution is due to Choquet, who showed that $c_{\alpha *}(E) = c_\alpha^*(E)$

is true for a wide class of sets E containing at least all Borel sets. This result dates from 1952 and was the outcome of a profound and original investigation of nonadditive set functions carried out by Choquet.

A set E such that $c_{\alpha *}(E) = c_\alpha^*(E)$ is said to be *α-capacitable* and its *α-capacity* $c_\alpha(E)$ is the common value of $c_{\alpha *}(E)$ and $c_\alpha^*(E)$.

(5) *Relations with potential theory.* The "capacity" terminology belonged originally to potential theory, and its use in the present context is due to the fact that in Newtonian potential theory (to take the classic example) one may follow very similar steps. In place of K_α one uses the Newtonian kernel $K(x) = |x|^{-1}$ on R^3, where $|x|$ here denotes the Euclidean length of the point x of R^3, in terms of which the self-energy of the distribution of matter represented by the measure μ on R^3 is expressed by the integral

$$E(\mu) = \iint K(x - y)\, d\mu(x)\, d\bar{\mu}(y). \tag{12.12.12}$$

The use of energy considerations in Newtonian potential theory dates back to Gauss and was rejuvenated in modern times by Henri Cartan. The reader will observe that the expression (12.12.12) involves viewing a measure as a set function; see the remarks in 12.2.3 and Exercises 12.38–12.42. On replacing E_α by E, the definitions in 12.12.3 and (2) above lead to the *Newtonian capacities* of subsets of R^3.

The formal similarity is apparently complete, but we must indicate one point that has to be checked with care, namely, it has to be verified that for all positive measures μ on T one has the equality

$$\iint K_\alpha(x - y)\, d\mu(x)\, d\mu(y) = \sum_{n \in Z} \hat{K}_\alpha(n)|\hat{\mu}(n)|^2. \tag{12.12.13}$$

This is not altogether trivial; see Exercise 12.40. (The analogous problem for Newtonian potentials was investigated by Deny.)

The reader who wishes to look into the details of the relations with potential theory may consult the long article by Ohtsuka [1] and the references cited there (especially the items listed as Cartan [5], [6], Fuglede [1] and Deny [1], [2] in Ohtsuka's bibliography). Sad to say, the present writer knows of no account in book form of modern potential theory. For more about connections between capacity and harmonic analysis, see [Ba_1], pp. 398 ff., [KS], Chapitres III, IV, VIII and (for general groups) Herz [2], Section 3, and the references cited in these works.

12.12.8. **Sets of Multiplicity and Sets of Uniqueness.** Other concepts of smallness of sets, somewhat similar to that expressed by vanishing capacity, play a central role in the theory of trigonometric series.

Consider again a closed subset E of T. It is known that to assert the existence of at least one nonzero pseudomeasure σ with support contained in E and satisfying

$$\lim_{|n| \to \infty} \hat{\sigma}(n) = 0 \tag{12.12.14}$$

is necessary and sufficient in order that E be a so-called *set of multiplicity in the wide sense*, or simply a *set of multiplicity*, that is, that there shall exist a trigonometric series which converges to zero at all points not in E and whose coefficients are not identically vanishing; see [Z_1], pp. 344–347; [Ba_2], p. 366; [Kah_2], p. 44; and [KS], p. 54; and compare with the closing remarks in 12.12.5.

In this case, contrary to what the substance of 12.12.4 may lead one to expect, if in (12.12.14) one were to demand that σ must be a positive measure (or a measure at all, positive or not), a different class of sets E would result, namely, the class of *sets of multiplicity in the strict sense*. In other words, there exist closed sets of multiplicity that are not sets of multiplicity in the strict sense; see [KS], p. 57.

A set that is not a set of multiplicity is termed a *set of uniqueness* (in the strict sense); and a set that is not a set of multiplicity in the strict sense is termed a *set of uniqueness in the wide sense*.

Cantor and Young (1870, 1908) showed that any countable set is a set of uniqueness. It is virtually obvious that any set of positive measure is a set of multiplicity in the strict sense. The advent of the theory of Lebesgue measure and integration brought with it the feeling that all null sets should be sets of uniqueness, but this expectation was shattered by Men'shov (1916).

For details concerning all these matters, see [Z_1], Chapter IX; [Ba_2], Chapter XIV; [KS], Chapitres V and VI. See also Kahane and Mandelbrot [1]; MR **35** # 3379; **40** # 631; **51** # 11016.

12.13 The Dual Form of Bochner's Theorem

In this section we shall temporarily turn aside from our main pursuit in order to apply something of what has been learned in this chapter to formulate and prove the form of Bochner's theorem about positive definite functions that is applicable to the group Z. The form of the theorem applicable to the circle group T has been dealt with in Chapter 9.

12.13.1. Positive Definite Functions on Z. A complex-valued function ϕ on Z is said to be *positive definite* if and only if

$$\sum_{m\in Z}\sum_{n\in Z}\phi(m-n)\bar{c}_m c_n \geqslant 0 \tag{12.13.1}$$

for each sequence $(c_n)_{n\in Z}$ of complex numbers having a finite support (that is, such that $c_n = 0$ for all but a finite set of $n \in Z$). This definition should be compared with (9.2.1) and (9.2.2).

It follows readily from (12.13.1) that

$$\phi(-n) = \overline{\phi(n)}, \qquad |\phi(n)| \leqslant \phi(0) \qquad (n \in Z); \tag{12.13.2}$$

in particular, each positive definite ϕ belongs to $\ell^\infty(Z)$.

12.13.2. **Relation with Fourier-Stieltjes Transforms.** Let us consider a function ϕ on Z of the form $\phi = \hat{\mu}$, where $\mu \in \mathbf{M}$. In this case a simple calculation shows that

$$\sum \sum \phi(m - n)\bar{c}_m c_n = \mu(|t|^2), \tag{12.13.3}$$

where t is the trigonometric polynomial defined by

$$t(x) = \sum_{n \in Z} c_n e^{inx}.$$

From this we infer at once that ϕ is positive definite whenever the measure μ is *positive* in the sense (compare Exercise 12.7) that $\mu(f) \geqslant 0$ for any nonnegative $f \in \mathbf{C}$.

On the other hand, if ϕ is positive definite, (12.13.3) shows that $\mu(|t|^2) \geqslant 0$ for every trigonometric polynomial t, whence it follows (see 12.2.3 and Exercise 2.18) that μ is a positive measure.

To sum up, we find that a Fourier-Stieltjes transform $\hat{\mu}$ is positive definite if and only if the measure μ is positive. If the measure μ is (generated by) a continuous function f, then the transform $\hat{f}$ is positive definite if and only if f is nonnegative.

We are now ready to state and prove the appropriate Bochner representation theorem.

12.13.3. **The Bochner-type Theorem.** The positive definite functions ϕ on Z are precisely the functions of the form $\phi = \hat{\mu}$ for some positive measure $\mu \in \mathbf{M}$.

Remarks. That this is indeed an exact analogue of the representation formula (9.2.3) for positive definite functions on the circle group T hinges upon the following remarks.

By a (*bounded Radon*) *measure* on Z will be meant a continuous linear functional Λ on $\mathbf{c}_0(Z)$ (compare the substance of 12.2.3 together with Exercise 4.45 of [E] or p. 364 of [HS]), and it is quite easy to show that to each such measure Λ on Z corresponds a unique function $\lambda \in \ell^1(Z)$ such that

$$\Lambda(\phi) = \sum_{n \in Z} \lambda(n)\phi(n)$$

for all $\phi \in \mathbf{c}_0(Z)$. Furthermore, the measure Λ is *positive*, in the sense that $\Lambda(\phi) \geqslant 0$ for any nonnegative function $\phi \in \mathbf{c}_0(Z)$, if and only if the corresponding function λ is nonnegative. Bearing this in mind, the Bochner theorem 9.2.8 for the group T asserts that any continuous positive definite function f on the circle group is equal to the Fourier transform $\hat{\Lambda}$ of some positive (bounded Radon) measure Λ on Z.

To this we might add that on the compact group T all Radon measures are automatically bounded (because of compactness); and that on the discrete group Z all functions are continuous (because of discreteness).

12.13.4. **Proof of 12.13.3.** In view of 12.13.2 it remains only to show that, if ϕ is a positive definite function on Z, then $\phi = \hat{\mu}$ for some measure $\mu \in \mathbf{M}$. To this end, we choose any approximate identity $(k_r)_{r=1}^{\infty}$ in $\mathbf{L}^1$ such that $k_r \geqslant 0$, $\|k_r\|_1 \leqslant 1$, and $\hat{k}_r \in \ell^1(Z)$ for each r, and we consider the functions

$$\phi_r = \hat{k}_r \phi. \tag{12.13.4}$$

Let us first verify that each ϕ_r is positive definite. We have in fact

$$\sum \sum \phi_r(m - n)\bar{c}_m c_n = \frac{1}{2\pi} \int k_r(x) \{\sum \sum \phi(m - n)\overline{c_m e^{imx}} c_n e^{inx}\} \, dx,$$

all sums appearing being over a finite range. The last integral appearing is nonnegative since $k_r \geqslant 0$ and $\{\dots\} \geqslant 0$, so that ϕ_r is indeed positive definite.

Besides this, $\phi_r \in \ell^1(Z)$ because $\hat{k}_r \in \ell^1(Z)$ and $\phi \in \ell^\infty(Z)$. We can therefore write

$$\phi_r = \hat{f}_r, \tag{12.13.5}$$

where

$$f_r(x) = \sum_{n \in Z} \phi_r(n) e^{inx}$$

is a continuous function on the circle group. Since ϕ_r is positive definite, 12.13.2 shows that f_r is nonnegative. Consequently,

$$\begin{aligned} \|f_r\|_1 &= \hat{f}_r(0) = \phi_r(0) = \hat{k}_r(0)\phi(0) \\ &\leqslant \|k_r\|_1 \cdot \phi(0) \leqslant \phi(0). \end{aligned}$$

Applying 12.3.9, it follows that there exists a subsequence $(f_{r_s})_{s=1}^{\infty}$ such that the measures generated by the functions f_{r_s} converge weakly in $\mathbf{M}$ to a measure $\mu \in \mathbf{M}$. Then, by (12.13.5), we have

$$\hat{\mu} = \lim_{s \to \infty} \hat{f}_{r_s} = \lim_{s \to \infty} \phi_{r_s} \tag{12.13.6}$$

pointwise on Z. But (12.13.4) combines with 3.2.4 to show that

$$\lim_{r \to \infty} \phi_r = \phi \tag{12.13.7}$$

pointwise on Z. A comparison of (12.13.6) and (12.13.7) shows that $\phi = \hat{\mu}$. That μ is positive, follows either from 12.13.2, or from its construction as the weak limit in $\mathbf{M}$ of the positive measures generated by the nonnegative functions f_{r_s}. The proof is thus complete.

12.13.5. **Product of Positive Definite Functions on Z.** It follows at once from 12.13.3 that the pointwise product of two positive definite functions on Z is again positive definite (compare Exercise 9.4).

12.13.6. **A Line to Pursue.** We end this section with what it is hoped will be a leading question and an invitation to the reader to provide his own answer thereto.

Suppose that $\mathbf{H}$ denotes a Hilbert space and U a unitary endomorphism of $\mathbf{H}$. Assign to each $\mathbf{h} \in \mathbf{H}$ the complex-valued function $\phi_{\mathbf{h}}$ on Z defined by $\phi_{\mathbf{h}}(n) = (U^n\mathbf{h}, \mathbf{h})$, where $(\cdot, \cdot)$ denotes the scalar (or inner) product in $\mathbf{H}$. It is simple to verify that $\phi_{\mathbf{h}}$ is positive definite. The question is: what results if to $\phi_{\mathbf{h}}$ one applies the Bochner theorem 12.13.3? Can you relate the result to the so-called spectral resolution theorem for U and to an operational calculus for U?

EXERCISES

12.1. Supply a proof of the equivalence of (12.1.2) and (12.1.3).

12.2. Prove that if $u \in \mathbf{C}^\infty$, then

$$\mathbf{C}^\infty - \lim_{a \to a_0} T_a u = T_{a_0} u$$

and

$$\mathbf{C}^\infty - \lim_{a \to 0} \frac{T_{-a}u - u}{a} = Du.$$

12.3. Is it true that

(1) $\mathbf{C}^\infty - \lim_{n \to \infty} \sin nx/n^{10} = 0$?

(2) $\mathbf{C}^\infty - \lim_{n \to \infty} \cos nx/\exp[(\log n)^{3/2}] = 0$?

Give your reasons.

12.4. Define $u(x) = 0$ for $x \equiv 0 \pmod{2\pi}$ and $u(x) = \exp(-\operatorname{cosec}^2 \tfrac{1}{2}x)$ otherwise. Show that $u \in \mathbf{C}^\infty$ and that $D^n u(0) = 0$ for $n = 0, 1, 2, \cdots$. This shows that $\mathbf{C}^\infty$ contains many nonanalytic functions.

12.5. (1) Suppose n is an integer, $n > 2$, and define $u_n(x)$ to be 0 if $x = 0$ or if $2\pi/n \leqslant |x| \leqslant \pi$, to be $\exp(-\operatorname{cosec}^2 \tfrac{1}{2}nx)$ if $0 < |x| < 2\pi/n$, and elsewhere so as to be periodic. Show that $u_n \in \mathbf{C}^\infty$ and that $(2\pi)^{-1} \int u_n \, dx \equiv c_n > 0$. Verify that the $v_n = c_n^{-1} \cdot u_n$ form an approximate identity.

(2) Deduce from (1) that to each $f \in \mathbf{C}$ and each neighborhood V of the support $\overline{\{x : f(x) \neq 0\}}$ of f there corresponds at least one sequence $(u_n)_{n=1}^\infty$ extracted from $\mathbf{C}^\infty$ satisfying the following conditions:

(a) $\|u_n\|_\infty \leqslant \|f\|_\infty$ for all n;

(b) u_n vanishes outside V for all n;

(c) $\lim_{n \to \infty} u_n = f$ uniformly.

Show similarly that the conclusion stands when $\mathbf{C}$ is replaced by $\mathbf{L}^\infty$, provided condition (c) is replaced by

(c′) $\lim_{n \to \infty} u_n(x) = f(x)$ a.e.

Note: We know that there exist approximate identities formed of trigonometric polynomials. The essential feature of the v_n is that they vanish outside smaller and smaller neighborhoods of 0 in T. This property is crucial for the local study of distributions. The results in (2) have already proved to be useful in 12.3.2(5); see also Exercises 12.29 and 12.30.

12.6. Verify in detail the statements made in 12.1.2 and 12.1.5.

12.7. A distribution F is said to be *positive* if and only if $F(u) \geqslant 0$ for any nonnegative real-valued $u \in \mathbf{C}^\infty$. Show that any positive distribution is a measure.

12.8. Give a proof of 12.7.1 without using 12.3.9.

12.9. Prove the impossibility of multiplying measures in the fashion described in 12.3.5.

Hints: Use Exercise 12.7 to show that $\alpha\beta \in \mathbf{M}$ whenever α and β are positive measures. The next and crucial step is to show that there exists a number $c > 0$ such that

$$\|\alpha\beta\|_1 \leqslant c\|\alpha\|_1\|\beta\|_1$$

for any two positive measures α and β. Assuming the contrary, show that there would exist positive measures α_n and β_n ($n = 1, 2, \cdots$) such that $\|\alpha_n\|_1 = \|\beta_n\|_1 = 1$ and

$$\|\alpha_n\beta_n\|_1 > n^5.$$

Consider the product $\alpha\beta$, where

$$\alpha = \sum_{n=1}^{\infty} n^{-2}\alpha_n, \qquad \beta = \sum_{n=1}^{\infty} n^{-2}\beta_n,$$

in order to reach a contradiction, observing that the hypotheses made in 12.3.5 ensure that $\alpha\beta$ is a positive measure such that $\alpha\beta \geqslant n^{-4}\alpha_n\beta_n$ for all n.

12.10. Show that if μ is the measure generated by a function $f \in \mathbf{L}^1$, then $\|\mu\|_1 = \|f\|_1$ (see 12.3.8).

12.11. Prove that if F is a distribution such that $DF = 0$, then F is a constant function.

12.12. Suppose that F is a distribution and that $DF = f \in \mathbf{L}^1$. Show that $\hat{f}(0) = 0$, and that F is the distribution generated by the absolutely continuous periodic function $c + \int_0^x f(y)\,dy$, c being a suitably chosen constant.

12.13. Show that if $\mu_n \to \mu$ weakly in $\mathbf{M}$ (see 12.3.9), then

$$\|\mu\|_1 \leqslant \liminf_{n\to\infty} \|\mu_n\|_1.$$

12.14. Show that if $f_n \to f$ weakly in $\mathbf{L}^p$, where $1 \leqslant p \leqslant \infty$ (see 12.3.10), then

$$\|f\|_p \leqslant \liminf_{n\to\infty} \|f_n\|_p.$$

12.15. Show that if $A(1) = \hat{A}(0) = 0$, then the solution of equation (12.4.2) is

$$F = \hat{F}(0) + \sum_{n \neq 0} (in)^{-1}\hat{A}(n)e^{inx}.$$

12.16. Show that the equation $D^2F + F = \varepsilon$ has no distributional solutions F.

If $P(D)$ is a linear differential operator with constant coefficients and $A \in \mathbf{D}$ is given, discuss the solubility and solutions of the equation $P(D)F = A$, where the unknown F is to belong to $\mathbf{D}$.

Remark. In case the reader finds the above conclusion puzzling, it should perhaps be stressed that in this book we speak only of distributions on the group T. There are, of course, distributions F on the group R which satisfy $D^2F + F = \varepsilon$. It can be shown that distributions on T correspond to distributions on R which are periodic; it is this additional requirement of periodicity that is incompatible with the given differential equation.

12.17. If $(K_n)_{n=1}^{\infty}$ is an approximate identity in $\mathbf{L}^1$ and $\mu \in \mathbf{M}$, show that $K_n * \mu \to \mu$ weakly in $\mathbf{M}$.

12.18. Verify in detail the computations referred to in 12.8.2(1) and (2).

12.19. Assuming the theorem of F. and M. Riesz cited in 12.8.5(4), show that if f and $\tilde{f}$ are each (equal almost everywhere to) functions of bounded variation, then each is (equal almost everywhere to) an absolutely continuous function.

Hint: Use the substance of 12.5.10.

12.20. Writing $\tilde{f} = H * f$, where $f \in \mathbf{L}^1$, and using the notations introduced in 7.1.1, verify that

$$s_N\tilde{f}(x) = -\frac{1}{2\pi}\int_0^{\pi} [f(x + y) - f(x - y)]\tilde{D}_N(y)\,dy.$$

Show that

$$\int_0^{\pi} \tilde{D}_N(y)\,dy \sim 2 \log N$$

for large N, and conclude that if $f \in \mathbf{L}^1$ and

$$f(x + y) - f(x - y) = d + \varepsilon(y),$$

where $\varepsilon(y) \to 0$ as $y \to 0$. Then

$$s_N\tilde{f}(x) \sim -\frac{d}{\pi} \log N$$

as $N \to \infty$.

12.21. Show that if S is a pseudomeasure (see Section 12.11), then $S = c + Df$, where c is a constant and $f \in \mathbf{L}^2$.

12.22. Suppose that $F \in \mathbf{D}$ is such that

$$\sup_N \|s_N F\|_{\infty} < \infty.$$

Prove that $F \in \mathbf{L}^\infty$ and that

$$\|F\|_\infty \leqslant \sup_N \|s_N F\|_\infty .$$

12.23. Prove that a measure $\mu \in \mathbf{M}$ such that $\|T_a\mu - \mu\|_1 \to 0$ as $a \to 0$ is (generated by) a function in $\mathbf{L}^1$.

Remark. Numerous stronger results are known; see Edwards [2] and the remarks following Exercise 3.5. See also MR **52** # 1161.

Hint: Show that if $\|T_a\mu - \mu\|_1 \leqslant \varepsilon$ for $|a| \leqslant \delta$, then $\|\mu * f - \mu\|_1 \leqslant \varepsilon$ for any $f \in \mathbf{L}^1$, $f \geqslant 0$, $(1/2\pi)\int f\,dx = 1$, $f = 0$ outside $|x| \leqslant \delta$ modulo 2π. Choose a sequence of such f corresponding to $\delta \to 0$ and use the completeness of $\mathbf{L}^1$.

12.24. Prove the assertion made in Remark 8.5.5(2).

12.25. Suppose that $(c_n)_{n\in Z}$ is a complex-valued sequence such that, for some $f \in \mathbf{L}^1$ and all $N = 1, 2, \cdots,$

$$s_N(x) \equiv \sum_{|n|\leqslant N} c_n e^{inx} \geqslant f(x)$$

for almost all x. Show that there exists a measure $\mu \in \mathbf{M}$ such that $c_n = \hat{\mu}(n)$ ($n \in Z$). Compare 12.7.9(1).

Hint: Use 12.7.5.

12.26. Let $\mathbf{B}$ denote the set of $h \in \mathbf{C}$ such that $\hat{h}(n) = 0$ for all integers $n < 0$.

(1) Verify that $\mathbf{B}$, when taken with pointwise operations and with the norm induced on it by that on $\mathbf{C}$, is a Banach algebra satisfying the conditions (a) to (c) in 11.4.1.

(2) Show that if $h \in \mathbf{B}$ and $\operatorname{Re} h(x) > 0$ for all real x, then $\operatorname{Re}\gamma(h) > 0$ for $\gamma \in \Gamma(\mathbf{B})$.

(3) Deduce that if $h \in \mathbf{B}$ is as in (2), and if $p > 0$, then

$$(h^p)^\frown(0) = (\hat{h}(0))^p,$$

where the pth power denotes the branch that is positive on the positive real axis.

Remark. This is the argument referred to in 12.9.8(1).

Hints: For (2) use the Hahn-Banach theorem (I, B.5.2) to show that there exists a Radon measure μ such that $\mu(h) = \gamma(h)$ for all $h \in B$ and $\|\mu\|_1 = \|\gamma\| = 1$. Use the relations $\mu(1) = \gamma(1) = 1 = \|\mu\|_1$ to show that $\mu \geqslant 0$. For (3), use (2), 11.4.10 and 11.4.15 with Φ equal to the said branch of λ^p.

12.27. Suppose that $\mu \in \mathbf{M}$ and let

$$\mathbf{I} = \{f \in \mathbf{C} : \mu(fg) = 0 \quad \text{for all } g \in \mathbf{C}\}.$$

Show that $\mathbf{I}$ is a closed ideal in $\mathbf{C} = \mathbf{C}(T)$ (see 11.4.1).

Put E for the set of common zeros of elements of $\mathbf{I}$ and regard E as a closed subset T. Let U be the complement of E in T. By using Exercise

11.17, show that U is precisely the set of points x of T with the following property: there exists a neighborhood U_x of x in T such that $\mu(f) = 0$ for every $f \in \mathbf{C}$ which vanishes outside U_x.

Remarks. This last property is more briefly expressed by saying that the measure μ vanishes on the neighborhood U_x, and it appears that μ also vanishes in the same sense on the open set U. The closed set E is the *support* of the measure μ (compare the substance of 12.11.4).

12.28. Formulate and prove an analogue of the result in Exercise 11.18 applying to the case in which $\mathbf{A}$ is replaced throughout by $\mathbf{C}^\infty$.

12.29. As a matter of definition, a distribution S is said to vanish on a given open subset Ω of T(in symbols $S = 0$ on Ω) if and only if $S(u) = 0$ for every $u \in \mathbf{C}^\infty$ satisfying supp $u \subset \Omega$. Show that if (Ω_α) is a family of open subsets of T, and if $S = 0$ on Ω_α for each α, then $S = 0$ on $\bigcup_\alpha \Omega_\alpha$.

Deduce that there exists a largest open subset Ω of T on which a given distribution S vanishes. The complement $T \backslash \Omega$ is the *support* of S, denoted by supp S; see 12.11.4.

Show also that if $S \in \mathbf{D}^m$ and $S = 0$ on an open subset Ω of T, then $S(u) = 0$ for each $u \in \mathbf{C}^m$ satisfying supp $u \subset \Omega$.

12.30. Let $S \in \mathbf{D}^m$ and write E = supp S. Show that $S(u) = 0$ whenever $u \in \mathbf{C}^m$ and $D^p u = 0$ on E for $0 \leqslant p \leqslant m$.

Remark. Despite the preceding result, it is *not* generally true that $\lim_{k \to \infty} S(u_k) = 0$ for any sequence $(u_k)_{k=1}^\infty$ such that $D^p u_k \to 0$ uniformly on E as $k \to \infty$ for $p = 0, 1, 2, \cdots$.

Hints: Reduce the problem to showing that there exists a sequence $(u_j)_{j=1}^\infty$ of elements of $\mathbf{C}^m$ such that each u_j coincides with u on some open set containing E and $u_j \to 0$ in $\mathbf{C}^m$. To construct (u_j) proceed as follows ($\dot{x}$ denotes the coset modulo $2\pi Z$ of $x \in R$): introduce the metric

$$d(\dot{x}, \dot{y}) = \inf_{n \in Z} |x - y + 2n\pi|$$

on T, in terms of which define E_δ as the set of points within distance δ of E. Then

$$\varepsilon(\delta) \equiv \sup \{|D^p u(\dot{x})| : \dot{x} \in E_\delta, 0 \leqslant p \leqslant m\}$$

tends to 0 with δ. Using the relation

$$D^p u(\dot{x}) = \int_{x_0}^{x} D^{p+1} u(t)\, dt;$$

where $x_0 \in \dot{x}_0 \in E$, show that

$$|D^p u(\dot{x})| \leqslant \delta^{m-p} \varepsilon(\delta) \tag{1}$$

for $\dot{x} \in E_\delta$ and $0 \leqslant p \leqslant m$.

Construct functions $w_\delta \in \mathbf{C}^\infty$ so that

$$\left.\begin{aligned} w_\delta(\dot{x}) &= 1 \text{ on } E_{\frac{1}{4}\delta}, = 0 \text{ outside } E_\delta, \\ |D^p w_\delta| &\leqslant A_p \delta^{-p}; \end{aligned}\right\} \tag{2}$$

this may be done by setting $w_\delta = k_\delta * v_n$, where $k_\delta \in \mathbf{C}$, $0 \leqslant k_\delta \leqslant 1$, $k_\delta = 1$ on $E_{\frac{1}{2}\delta}$, $k_\delta = 0$ outside $E_{\frac{3}{4}\delta}$, and where v_n is as in Exercise 12.5 with n chosen so that $2\pi/n \leqslant \frac{1}{4}\delta$; k_δ is easily constructed in terms of the function

$$\dot{x} \to d(\dot{x}, E) \equiv \inf_{\dot{y} \in E} d(\dot{x}, \dot{y}).$$

Now put $u_j = u \cdot w_{\delta_j}$, where δ_j is any sequence tending to 0. Use (1) and (2) to verify that $D^p u_j \to 0$ uniformly as $j \to \infty$, provided $0 \leqslant p \leqslant m$.

12.31. Let S be a pseudomeasure whose support is denoted by E (see 12.11.4 and Exercise 12.29). Show that the relation $fS = 0$ holds whenever $f \in \mathbf{C}^1$ and $f = Df = 0$ on E.

Hint: Use Exercises 12.21 and 12.30.

12.32. Let $\phi_0 \in \ell^1(Z)$ and $E = \hat{\phi}_0^{-1}(\{0\}) \subset T$. Prove that $\bar{\mathbf{V}}^1_{\phi 0}$ contains every $\phi \in \ell^1(Z)$ such that $\hat{\phi} \in \mathbf{C}^1$ and $\hat{\phi} = D\hat{\phi} = 0$ on E; and that $\bar{\mathbf{V}}^1_{\phi_0} = \ell^1(Z)$ if E is void. (The notations are as in 12.12.6. Compare 11.2.5.)

Hints: First verify that any continuous linear functional on $\ell^1(Z)$ is expressible as

$$\theta \to \sum_{n \in Z} \theta(n)\psi(-n)$$

for some $\psi \in \ell^\infty(Z)$. Next, by applying the Hahn-Banach theorem (I, B.5.2), reduce the problem to showing that, if $\psi \in \ell^\infty(Z)$ satisfies $\phi_0 * \psi = 0$, then $\phi * \psi = 0$. Introduce the pseudomeasure $S = \hat{\psi}$, verify (12.11.3), and so conclude that it suffices to establish the implication

$$\hat{\phi}_0 S = 0 \Rightarrow \hat{\phi} S = 0.$$

Finally, show that the hypothesis here entails that $\operatorname{supp} S \subset E$ and then apply the preceding exercise.

12.33. Let S be a distribution whose support is a finite subset of T, say $\{a_1, \ldots, a_k\}$. Prove that there exists an integer $m \geqslant 0$ and complex numbers $c_{jp}(j = 1, 2, \ldots, k; p = 0, 1, \ldots, m)$ such that

$$S = \sum_{j=1}^{k} \sum_{p=0}^{m} c_{jp} D^p \varepsilon_{a_j}.$$

Deduce that a pseudomeasure with a finite support is a measure.

Hints: Suppose $S \in \mathbf{D}^m$. Apply Exercise 12.30 in combination with the following simple lemma (which should be proved): if $\mathbf{L}$ is a linear space and $\ell, \ell_1, \cdots, \ell_r$ are linear functionals on $\mathbf{L}$ such that $\ell(y) = 0$ whenever $y \in \mathbf{L}$ and $\ell_1(y) = \cdots = \ell_r(y) = 0$, then ℓ is a linear combination of $\ell_1, \cdots, \ell_r$. For an alternative proof, see MR **37** # 6752.

12.34. (1) Suppose that ϕ_r $(r = 1, 2, \cdots)$ and ϕ are functions on Z such that

(a) ϕ_r is positive definite $(r = 1, 2, \cdots)$;
(b) $m \equiv \sup_r \phi_r(0) < \infty$;
(c) $\lim_{r\to\infty} \phi_r(n) = \phi(n)$ for $n \in Z$.

Show that there exist positive measures $\mu_r \in \mathbf{M}$ $(r = 1, 2, \cdots)$ and $\mu \in \mathbf{M}$ such that $\phi_r = \hat{\mu}_r$ $(r = 1, 2, \cdots)$, $\phi = \hat{\mu}$ and $\lim_{r\to\infty} \mu_r = \mu$ weakly in $\mathbf{M}$ (see 12.3.9).

(2) State and prove an analogue of (1) for the case in which Z is replaced by T.

Remarks. The analogue of (1) for the case in which Z is replaced by R is Lévy's so-called *continuity theorem*. Positive definite functions on R have close connections with probability theory.

A *bounded* (*Radon*) *measure* on R may be defined as a continuous linear functional on the Banach space $\mathbf{C}_0(R)$ composed of the continuous functions on R which tend to zero at infinity, the norm on $\mathbf{C}_0(R)$ being defined by

$$\|f\| = \sup\{|f(x)| : x \in R\};$$

see 12.2.9 and the remarks following 12.13.3. Such a measure, μ, is said to be *positive* if and only if $\mu(f) \geqslant 0$ whenever $f \in \mathbf{C}_0(R)$ is real and nonnegative-valued.

A bounded positive Radon measure μ on R such that $\mu(1) = 1$ is termed a *probability measure* on R. The Fourier transform of μ, namely, the function

$$\hat{\mu}: \xi \to \mu(e^{-i\xi x}) \qquad (\xi \in R),$$

is in probability theory usually termed the *characteristic function* of the probability distribution defined by μ.

As has been stated in Section 9.4, there is a version of the Bochner theorem valid for positive definite functions f on R: it asserts that the continuous positive definite functions f on R satisfying $f(0) = 1$ are precisely the Fourier transforms (that is, characteristic functions) of probability measures on R.

12.35. A distribution F (on T) is said to be *positive definite* if and only if

$$F(u * u^*) \geqslant 0$$

for each $u \in \mathbf{C}^\infty$ (compare 9.2.1). Show that this is so if and only if $\hat{F} \geqslant 0$, and deduce that F is positive definite if and only if

$$F = \sum_{n\in Z} c_n e^{inx},$$

where $(c_n)_{n\in Z}$ is a nonnegative tempered sequence.

Remarks. We have mentioned in Section 9.4 some of the many extensions of Bochner's theorem 9.2.8; the present exercise falls into this category. If it in turn is generalized to distributions on the group R^m, a great profusion of possibilities arise. The interested reader should consult [GV], Chapter II, where diverse modifications of the concept of positive definiteness are also discussed.

12.36. The Riesz-Markov-Kakutani theorem mentioned in 12.2.3 has as corollaries the following three statements.

(1) To each measure $\mu \in \mathbf{M}$ corresponds a positive measure $|\mu| \in \mathbf{M}$ such that $|\mu(f)| \leqslant |\mu|(|f|)$ for each bounded complex-valued Borel-measurable function f.

(2) If f and f_n $(n = 1, 2, \cdots)$ are complex-valued Borel-measurable functions such that

$$\sup_{n,x} |f_n(x)| < \infty, \qquad \lim_{n \to \infty} f_n = f \text{ pointwise},$$

then

$$\lim_{n \to \infty} \mu(f_n) = \mu(f)$$

for each measure $\mu \in \mathbf{M}$.

(3) The set of points x for which $|\mu|(\chi_{\{x\}}) > 0$ is countable, μ denoting a given measure and $\chi_{\{x\}}$ denoting the characteristic function of $\{x\}$.

Making use of these results, show that, if ϕ is a complex-valued function on Z which is a Fourier-Stieltjes transform, and if

$$S_{N,p}(x) = N^{-1} \sum_{p < n \leqslant p+N} \phi(n) e^{inx}$$

for $N = 1, 2, \cdots$ and p real, then

$$\lim_{N \to \infty} \sup_{p \in R} |S_{N,p}(x)| = 0 \tag{4}$$

for all real values of x save perhaps those belonging to a countable set.

Show that (4) remains true whenever ϕ is the limit, uniformly on Z, of Fourier-Stieltjes transforms.

12.37. Let F be a real-valued function defined on some real interval (a, ∞) and having the following properties:

(1) the set E of real numbers x, such that $F^{-1}(\{x\})$ is unbounded above, is uncountable;

(2) for any $c > 0$,

$$\sup_{T < t \leqslant T+c} |F(t) - F(T)| = o\left(\frac{1}{T}\right) \qquad \text{as } T \to \infty.$$

By using the preceding exercise, show that any function ϕ on Z, such that

$$\phi(n) = \exp\{inF(n)\}$$

for $n \in Z$ and $n > a$, is not the limit, uniformly on Z, of Fourier-Stieltjes transforms.

12.38. Suppose that E is a closed subset of T, that $0 < \alpha < 1$, and that $c_\alpha(E) > 0$. Prove that there exists a unique positive measure μ such that $\mu(1) = 1$, $\operatorname{supp} \mu \subset E$, and

$$c_\alpha(E) = \frac{1}{E_\alpha(\mu)}.$$

Remarks. This measure μ is termed the *equilibrium measure on* (or *for*) E, mainly because of the physical significance of potential theory in the Newtonian case. Equilibrium measures play an essential role in the proof of the nontrivial assertion in 12.12.4; see Exercise 12.42.

Hints: Verify that

(1) $E_\alpha(\frac{1}{2}(A + B)) < \frac{1}{2}E_\alpha(A) + \frac{1}{2}E_\alpha(B)$ for any two distinct distributions A and B having finite α-energy;

(2) $E_\alpha(A) \leqslant \liminf_{k\to\infty} E_\alpha(A_k)$ whenever $A_k \to A$ in $\mathbf{D}$.

Then use 12.3.9.

12.39. Suppose that E and α are as in the preceding exercise. Denote by $\mathscr{E}$ the set of measures having finite α-energy, by $\mathscr{E}_E$ the set of measures in $\mathscr{E}$ supported by E, and by $\mathscr{E}^+$ and $\mathscr{E}_E^+$ the set of positive measures in $\mathscr{E}$ and in $\mathscr{E}_E$, respectively. In $\mathscr{E}$ define the inner (or scalar) product

$$(\mu|\nu) = \sum_{n\in Z} \hat{K}_\alpha(n)\hat{\mu}(n)\overline{\hat{\nu}(n)}$$

and associated norm $\|\mu\| = E_\alpha(\mu)^{1/2}$.

Which (if any) of $\mathscr{E}$, $\mathscr{E}_E$, $\mathscr{E}^+$, $\mathscr{E}_E^+$ is complete for the above norm?

Can the projection method, explained for the case of $\mathbf{L}^2$ in Exercise 8.14, be adapted to establish the conclusion of the preceding exercise? If so, give the details.

Note: In the following three exercises it is necessary to assume the integration theory associated with a Radon measure; compare the remarks in 12.2.3; [E], Chapter 4; and [HS], Chapter III. Moreover, $0 < \alpha < 1$, E is a closed subset of T, $K = K_\alpha$, $c(E) = c_\alpha(E)$, μ_E is the equilibrium measure on E (see Exercise 12.38); and in general we drop the suffix "α".

12.40. If A and B are distributions having finite energy, we define the inner product

$$(A|B) = \sum_{n\in Z} \hat{K}(n)\hat{A}(n)\overline{\hat{B}(n)},$$

the corresponding norm being $\|A\| = (A|A)^{1/2}$, the square root of the energy of A; compare Exercise 12.39.

Prove that if μ and ν are positive measures having finite energy, then

$$(\mu|\nu) = \int U^\mu \, d\nu,$$

where $U^\mu(x) = \int K(x - y)\, d\mu(y)$ is a Borel-measurable function belonging to $\mathbf{L}^1$ which generates the distribution $K * \mu$.

Remark. U^μ is termed the (α-) *potential* of μ; it can be defined as a nonnegative extended real-valued function by the above integral for *any* positive measure μ and is easily seen then to be lower semicontinuous.

Hint: See [KS], p. 35, Proposition 3.

12.41. Suppose that $c(E) > 0$ and that μ_E is defined as in Exercise 12.38. Prove that

$$U^{\mu_E}(x) = \frac{1}{c(E)}$$

at almost all points x of E.

Hint: See [KS], pp. 36–37, where more refined results are established.

12.42. Suppose that $c(E) > 0$. Prove that

$$\|A\|^2 \geqslant \frac{1}{c(E)}$$

for any distribution A such that $\operatorname{supp} A \subset E$ and $\hat{A}(0) = 1$.

What can be said if $c(E) = 0$?

Hints: Let E_n denote the set of points of T at distance at most $1/n$ (with respect to the metric introduced in the hints to Exercise 12.30) from E; put $\mu_n = \mu_{E_n}$. Show that

$$(\mu_n|A) = \lim_{N \to \infty} (\mu_n \mid A * v_N),$$

where the v_N are as in Exercise 12.5, and use the preceding exercise to deduce that $(\mu_n|A) = \|\mu_n\|^2$ and so that $\|A\| \geqslant \|\mu_n\|$. Using 12.3.9, extract a subsequence (μ_{n_k}) converging weakly in $\mathbf{M}$ to a positive measure μ' such that $\mu'(1) = 1$ and $\operatorname{supp} \mu' \subset E$; apply (2) of the hints to Exercise 12.38.

For the second part, show that $c(E_n) \to 0$ if $c(E) = 0$.

12.43. Let $(\mu_k)_{k=1}^\infty$ be a sequence of measures such that

$$\sup_k \|\mu_k\|_1 < \infty$$

and

$$\phi = \lim_{k \to \infty} \hat{\mu}_k$$

exists pointwise on Z. Show that there exists a measure μ such that $\lim_{k \to \infty} \mu_k = \mu$ weakly in $\mathbf{M}$ and $\hat{\mu} = \phi$.

Formulate and prove an analogous assertion applying when $\mathbf{M}$ is replaced throughout by $\mathbf{L}^p$ where $1 < p \leqslant \infty$.

12.44. Consider *Cantor's ternary set* E, formed as follows: from $[0, 2\pi]$ delete the open middle third, leaving the set E_1; from each component interval of E_1 delete the open middle third, leaving E_2; and so on indefinitely; define $E = \bigcap_{k=1}^\infty E_k$. Verify that $m(E) = 0$. (It is not difficult to show that E is uncountable, perfect, and nowhere dense.)

Construct *Lebesgue's singular function* ℓ as follows: define ℓ_k so that $\ell_k(0) = 0$, $\ell_k(2\pi) = 1$, and ℓ_k increases linearly by amount 2^{-k} on each component interval of E_k. The function ℓ_k is increasing and absolutely continuous. The functions ℓ_k converge uniformly as $k \to \infty$ to a continuous increasing function ℓ such that $\ell(0) = 0$, $\ell(2\pi) = 1$. (It can be shown that $\ell'(x) = 0$ a.e.)

Let λ_k be the measure defined by

$$\lambda_k(u) = \int_0^{2\pi} u(x)\ell_k'(x)\,dx \qquad (u \in \mathbf{C}).$$

Using the preceding exercise, show that $\lim_{k\to\infty} \lambda_k = \lambda$ weakly in $\mathbf{M}$, where λ is a positive measure for which

$$\hat{\lambda}(n) = \exp(-i\pi n) \prod_{j=1}^{\infty} \cos\frac{2\pi n}{3^j} \qquad (n \in Z)$$

and $\operatorname{supp}\lambda \subset E$. Verify in particular that $\hat{\lambda}(\pm 3^k)$ is equal to a nonzero number independent of $k = 1, 2, 3, \cdots$, so that $\hat{\lambda}(n) \neq o(1)$ as $|n| \to \infty$. (The associated set-function measure m_λ is easily seen to be continuous, in the sense that $m_\lambda(J) \to 0$ as $m(J) \to 0$ for subintervals J of $[0, 2\pi]$.)

Finally, define $f(x) = 2\pi\ell(x) - x$ for $0 \leqslant x \leqslant 2\pi$ and by periodicity elsewhere. Show that f is continuous and of bounded variation, and that $\hat{f}(n) \neq o(1/|n|)$ as $|n| \to \infty$ (compare the Remarks following 2.3.6).

12.45. Consider $\mathbf{M}$ as a Banach algebra (see 11.4.1).

Prove that any ideal in $\mathbf{M}$ is a translation-invariant subspace of $\mathbf{M}$.

Let $\mathbf{V}$ be the closed translation-invariant subspace of $\mathbf{M}$ generated by the Dirac measure ε. Show that $\mathbf{V}$ is not an ideal in $\mathbf{M}$. (See 11.1.3(2).)

12.46. Let $\mu \neq 0$ be an idempotent element of $\mathbf{M}$ satisfying $\|\mu\|_1 \leqslant 1$. Show that $\|\mu\|_1 = 1$ and that $\hat{\mu} = \chi_S$, where S is a coset modulo some subgroup of Z (see 12.7.4(3)).

Hints: Show first that $\|\mu\|_1 = 1$. Choose $n_0 \in Z$ such that $\lambda = e_{n_0}\mu$ satisfies $\hat{\lambda}(0) = 1$. Show that $\lambda \geqslant 0$. Put $S_0 = \{n \in Z : \hat{\lambda}(n) = 1\} = \operatorname{supp}\hat{\lambda}$. Verify the inequality

$$|\hat{\lambda}(m) - \hat{\lambda}(n)|^2 \leqslant 2\hat{\lambda}(0) \cdot \operatorname{Re}\{\hat{\lambda}(0) - \hat{\lambda}(m-n)\} \qquad (m, n \in Z)$$

and deduce that S_0 is a subgroup of Z.

12.47. Let μ be an idempotent element of $\mathbf{M}$ satisfying $\|\mu\|_1 > 1$. Prove that $\|\mu\|_1 \geqslant \frac{1}{2}\sqrt{5}$.

Remark. It is apparently unknown whether $\frac{1}{2}\sqrt{5} = 1.118\cdots$ is the best-possible constant in the above statement.

Hints: The proof of the preceding exercise shows that

$$S = \{n \in Z : \hat{\mu}(n) = 1\} = \operatorname{supp}\hat{\mu}$$

is not a coset. Show that this entails the existence of $n_1, n_2, n_3 \in S$ such that $n_1 + n_2 - n_3 \notin S$. To estimate $\|\mu\|_1$, consider $\mu(f)$, where

$$f = 2e_{-n_1}(1 + e_{n_1 - n_3}) + e_{-n_2}(1 - e_{n_3 - n_1})$$
$$= 2e_{-n_1} + 2e_{-n_3} + e_{-n_2} - e_{-(n_1 + n_2 - n_3)}.$$

12.48. Any periodic subset P of Z is expressible in the form

$$P = A + kZ \equiv \{a + kn : n \in Z, a \in A\}$$

where k is a positive integer and A is a (possibly void) subset of

$$\{0, 1, \cdots, k - 1\}.$$

Use this remark to write down an explicit closed expression for the measure whose Fourier transform is the characteristic function of P. (Compare 12.7.4(3) and see 16.8.4(2).)

12.49. Let $\mathbf{E}$ denote $\mathbf{L}^p$ $(1 < p \leqslant \infty)$ or $\mathbf{M}$, and let T be a homomorphism of $\mathbf{L}^1$ into $\mathbf{E}$ (each being regarded as a convolution algebra). Using the arguments of 4.2.2, show that there exists a map α of Z into $Z \cup \{\infty\}$ such that $(Tf)^\wedge = \hat{f} \circ \alpha$ for $f \in \mathbf{L}^1$, and deduce that T is continuous from $\mathbf{L}^1$ into $\mathbf{E}$.

Show that T can be extended into a homomorphism T' of $\mathbf{M}$ into $\mathbf{E}$ such that $(T'\mu) = \hat{\mu} \circ \alpha$ for $\mu \in \mathbf{M}$, $\hat{\mu}(\infty)$ being interpreted to be 0, and $\|T'\| \leqslant \|T\|$.

Hence (or otherwise) determine all the homomorphisms of $\mathbf{L}^1$ into $\mathbf{L}^p$, where $1 < p \leqslant \infty$. See also Subsection 15.3.6.

Remark. The extension T' of T plays a useful role in the study of homomorphisms as set out in Chapter 4 of [R].

Hints: Use the closed graph theorem (I, B.3.3) to establish continuity. For the rest, use Exercise 12.43. A different type of proof for the case $\mathbf{E} = \mathbf{M}$ appears on p. 83 of [R].

12.50. Let ϕ be a complex-valued function on Z, and let m be a non-negative real number. Prove that the following assertions are equivalent:

(1) $|\sum c_r \phi(n_r)| \leqslant m \|\sum c_r e_{n_r}\|_\infty$ for all trigonometric polynomials $\sum c_r e_{n_r}$;

(2) $\phi = \hat{\mu}$ for some $\mu \in \mathbf{M}$ satisfying $\|\mu\|_1 \leqslant m$.

Remarks. There is an analogue of this assertion valid for general groups and due to Eberlein; see [R], Theorem 1.9.1 and the references cited there. The special case applying to the case in which T and Z interchange their roles reads as follows: a continuous complex-valued function f on T belongs to $\mathbf{A}$ and satisfies $\|f\|_{\mathbf{A}} \leqslant m$, if and only if

$$\left|\sum c_r f(x_r)\right| \leqslant m \cdot \sup_{n \in Z} \left|\sum c_r e^{ix_r n}\right|$$

for all finite sequences (c_r) and (x_r) of complex numbers c_r and points x_r of T. See also MR **36** # 3065.

Hints: Consider the linear functional γ defined on the normed linear subspace **T** of **C** by the formula $\gamma(t) = \sum_{n \in Z} \phi(n)\hat{t}(n)$. Recall 12.2.9.

12.51. Denote by **Q** the set of all measures $\mu \in \mathbf{M}$ expressible in the form

$$\mu = \sum_{k=1}^{\infty} c_k \varepsilon_{a_k} + f \equiv \delta + f, \tag{1}$$

where $f \in \mathbf{L}^1$, $(a_k)_{k=1}^{\infty}$ is a sequence of distinct points of T, and $(c_k)_{k=1}^{\infty}$ is a sequence of complex numbers satisfying $\sum_{k=1}^{\infty} |c_k| < \infty$, each of f, $(a_k)_{k=1}^{\infty}$ and $(c_k)_{k=1}^{\infty}$ possibly depending upon μ. (**Q** is in fact the closed subalgebra of **M** generated by $\{\varepsilon_a : a \in T\} \cup \mathbf{L}^1$.) Measures of the form δ are termed *discrete* (or sometimes *atomic* or *purely discontinuous*); we will denote by $\mathbf{M}_d$ the set of discrete measures.

Verify that **Q**, regarded as a subalgebra of **M**, is a Banach algebra of the type described in 11.4.1, and that

$$\|\mu\|_1 = \sum_{k=1}^{\infty} |c_k| + \|f\|_1$$

whenever $\mu \in \mathbf{Q}$ is given by (1).

Show that each $\gamma \in \Gamma(\mathbf{Q})$ (see 11.4.9) takes one of two forms, namely:

(a) $\gamma(\mu) = \hat{\delta}(\chi) \equiv \sum_{k=1}^{\infty} c_k \overline{\chi(a_k)}$, where χ is a bounded (but not necessarily continuous) character of T;

(b) $\gamma(\mu) = \gamma_n(\mu) \equiv \hat{\mu}(n)$, where $n \in Z$.

By using the general version of Kronecker's theorem (see the Remarks following Exercise 2.2), deduce that the measure μ given by (1) is inversible in **Q** provided

$$\inf_{n \in Z} |\hat{\delta}(n)| > 0, \qquad \hat{\mu}(n) \neq 0 \ (n \in Z). \tag{2}$$

Remarks. **Q** does not exhaust **M**, comprising in fact precisely those measures whose continuous singular part vanishes; see [HS], Section 19, especially p. 337. The continuous singular measures are the ones that present all the difficulty in studying complex homomorphisms of **M**; see 12.7.4.

In view of the almost periodicity of $\hat{\delta}$ on Z (see Subsection 2.5.4), it turns out that (2) is actually equivalent to

$$\inf_{n \in Z} |\hat{\mu}(n)| > 0. \tag{3}$$

Since it is trivial that (3) is necessary in order that μ be inversible in **M**, either of (2) or (3) is necessary and sufficient in order that μ be inversible in **Q** or in **M**.

Hints: In discussing a given $\gamma \in \Gamma(\mathbf{Q})$, define the function χ on T by $\chi(a) = \gamma(\varepsilon_{-a})$. Discuss separately the cases in which $\gamma \mid L^1 = 0$ and $\gamma \mid L^1 \neq 0$, using 4.1.2 in the later case.

12.52. Let E be a closed subset of T whose frontier F contains no nonvoid perfect set. Prove that E is a spectral synthesis set.

Hints: Suppose that $S \in \mathbf{P}(E)$ and that $f \in \mathbf{A}$ vanishes on E; it has to be shown that $fS = 0$. Putting $S_1 = fS$, use Exercise 12.31 to show first that $E_1 \equiv \operatorname{supp} S_1$ is a subset of F. Assuming that $S_1 \neq 0$, E_1 must have at least one isolated point, say a. Without loss of generality, assume that $a = 0$. Consider the functions u_ε constructed in Exercise 10.26 and show that $\|u_\varepsilon S_1\|_{\mathbf{P}} \to 0$ with ε. Verify that, on the other hand, $u_\varepsilon S_1$ is independent of ε if ε is small enough, and show that this leads to a contradiction of the relation $0 \in E_1 = \operatorname{supp} S_1$.

12.53. This exercise provides a general basis for one half of the Malliavin-Kahane-Rudin non-synthesis theorem mentioned in 11.2.3 in connection with the algebra $\mathbf{L}^1(R^m)$ and again in 12.11.4 for the algebra $\mathbf{A} = \mathbf{A}(T)$; see [Kz], pp. 231–232; [HR], (42.15); [Kah$_2$], pp. 63–64, 68. The other half of the programme consists of proving the existence of f and σ as prescribed below for certain algebras $\mathbf{B}$. This is in itself fairly complicated; see [R], Section 7.6.4; [Kz], pp. 233–235; [HR], (42.16)–(42.19); [Kah$_2$], pp. 68–72; MR, **39** # **4611**.

In what follows, $\mathbf{B}$ denotes a commutative Banach algebra with unit e (as in 11.4.1). If $\mathbf{I}$ is an ideal in $\mathbf{B}$, the *zero-set* or *hull* of $\mathbf{I}$ is defined to be the set

$$Z(\mathbf{I}) = \{\gamma \in \Gamma(\mathbf{B}) : \gamma(f) = 0 \quad \text{for all } f \in \mathbf{I}\};$$

the notation is a natural extension of that used in 11.4.3. For every ideal $\mathbf{I}$ in $\mathbf{B}$, $Z(\mathbf{I})$ is a closed subset of $\Gamma(\mathbf{B})$ (see 11.4.18).

If E is a subset of $\Gamma(\mathbf{B})$, the *kernel* of E is defined to be the set

$$\mathbf{I}(E) = \{f \in \mathbf{B} : \gamma(f) = 0 \quad \text{for all } \gamma \in E\};$$

this is always a closed ideal in $\mathbf{B}$. For regular algebras $\mathbf{B}$ (see [Kz], p. 223), $Z(\mathbf{I}(E)) = E$ for every closed subset E of $\Gamma(\mathbf{B})$.

A closed subset E of $\Gamma(\mathbf{B})$ is said to be a spectral *synthesis set* (or *spectral set*) for $\mathbf{B}$, if and only if $\mathbf{I}(E)$ is the *unique* closed ideal $\mathbf{I}$ in $\mathbf{B}$ such that $Z(\mathbf{I}) = E$.

One says that *spectral synthesis holds* in $\mathbf{B}$, if and only if every closed subset E of $\Gamma(\mathbf{B})$ is a spectral set for $\mathbf{B}$; this is so, if and only if

$$\mathbf{I} = \mathbf{I}(Z(\mathbf{I}))$$

for every closed ideal $\mathbf{I}$ in $\mathbf{B}$.

Denote by $\mathbf{B}'$ the topological dual of $\mathbf{B}$, the duality being indicated by $\langle\ ,\ \rangle$.

Assume that $f \in \mathbf{B}$, $\sigma \in \mathbf{B}'$ and $\sigma \neq 0$. Define for all $t \in R$,

$$C(t) = \|\exp(itf) \cdot \sigma\|_{\mathbf{B}'} \tag{1}$$

and assume further that $N \in \{1, 2, \ldots\}$ is such that

$$\int_R |t|^N C(t)\, dt < \infty. \tag{2}$$

For every $a_0 \in R$, define $f_0 = a_0 e + f$; and, for every natural number n, define $\mathbf{I}_n$ to be the closed ideal in $\mathbf{B}$ generated by f_0^n. Prove that there exists $a_0 \in R$ such that $\mathbf{I}_j \neq \mathbf{I}_k$ for all $j, k \in \{1, \ldots, N+1\}$ such that $j \neq k$. (Since $Z(\mathbf{I}_n) = Z(\mathbf{I}_1) = E$, say, for all positive natural numbers n, it follows that E is not a spectral set for $\mathbf{B}$.)

Hints: Proceed in the following stages.

(a) For all $h \in \mathbf{B}$ and all $\sigma \in \mathbf{B}'$, define $h \cdot \sigma \in \mathbf{B}'$ by

$$\langle g, h \cdot \sigma \rangle = \langle hg, \sigma \rangle \quad \text{for all } g \in \mathbf{B}.$$

Verify that

$$\|h \cdot \sigma\|_{\mathbf{B}'} \leqslant \|h\|_{\mathbf{B}} \cdot \|\sigma\|_{\mathbf{B}'}.$$

(b) For all $f \in \mathbf{B}$ and all $t \in R$, define $e_t = \exp(itf) \in \mathbf{B}$ by

$$e_t = \sum_{k=0}^{\infty} (it)^k f^k / k!.$$

Prove that the function $t \to e_t$ is uniformly continuous from R into $\mathbf{B}$ (cf. [HR], (42.14)). Conclude from (2) that

$$\lim_{|t| \to \infty} |t|^N C(t) = 0. \tag{3}$$

Prove also that, for all $t \in R$,

$$\frac{d}{dt} e_t = \lim_{\delta \to 0,\, \delta \neq 0} \delta^{-1}(e_{t+\delta} - e_t) = ife_t. \tag{4}$$

(c) One may assume that $\langle e, \sigma \rangle \neq 0$. (If not, replace σ by $h \cdot \sigma$, where h is suitably chosen in $\mathbf{B}$, and use (a).)

(d) Define $\Phi(t) = \langle e, e_t \cdot \sigma \rangle$ for all $t \in R$. Prove that Φ is continuous on R and

$$|\Phi(t)| \leqslant C(t) \quad \text{for all } t \in R. \tag{5}$$

Since $\Phi(0) = \langle e, \sigma \rangle \neq 0$ (see (c)), the Fourier transform

$$\hat{\Phi}(s) = \int_R \Phi(t) e^{-ist}\, dt$$

is non-vanishing for some real s, say for $s = -a_0$. On replacing f by $f_0 = a_0 + f$, one may assume that

$$\hat{\Phi}(0) \neq 0;$$

that is,

$$\int_R \langle e, e_t \cdot \sigma \rangle\, dt \neq 0. \tag{6}$$

(e) Suppose $p \in \{0, 1, \ldots, N\}$. Using (2) and (b), prove that there exists $\tau_p \in \mathbf{B}'$ such that

$$\langle g, \tau_p \rangle = \int_R (it)^p \cdot \langle g, e_t \cdot \sigma \rangle \, dt \quad \text{for all } g \in \mathbf{B}.$$

(f) Assume that $q \in \{0, 1, 2, \ldots\}$ and $g \in \mathbf{B}$. Define

$$J_{p,q} = \langle gf^q, \tau_p \rangle = \int_R (it)^p \cdot \langle gf^q, e_t \cdot \sigma \rangle \, dt.$$

Use (3) and (4) and partial integration to prove that

$$J_{p,q} = \begin{cases} -p \cdot J_{p-1,q-1} \text{ for all } p \in \{1, 2, \ldots, N\} \text{ and all } q \in \{1, 2, \ldots\} \\ 0 \text{ for } p = 0 \text{ and all } q \in \{1, 2, \ldots\}. \end{cases} \tag{7}$$

(g) Deduce from (7) that, if $p \in \{0, 1, \ldots, N\}$, then $J_{p,p+1} = 0$ for all $g \in \mathbf{B}$; hence that τ_p annihilates I_{p+1}. On the other hand, taking $p = q \in \{0, 1, \ldots, N\}$ and $g = e$, deduce that

$$\langle f^p, \tau_p \rangle = J_{p,p} = (-1)^p p! \int_R \langle e, e_t \cdot \sigma \rangle \, dt \neq 0,$$

the last step by (6). Conclude that f^p does not belong to $\mathbf{I}_{p+1}$

12.54. Prove that, if m is a positive integer, S a distribution, and $DS \in \mathbf{D}^m$, then $S \in \mathbf{D}^{m-1}$. Deduce that, if F is a distribution and $F \in \mathbf{D}^m \backslash \mathbf{D}^{m-1}$, then $DF \in \mathbf{D}^{m+1} \backslash \mathbf{D}^m$.

Remark. This result was suggested to me by Dr Jo Ward.

Hints: Look again at 12.4.3(2) and 12.4.6.

CHAPTER 13

Interpolation Theorems

This chapter is devoted to the proofs and some of the applications of the theorems of Riesz-Thorin and of Marcinkiewicz, each of which is concerned with operators T defined on subsets of Lebesgue spaces (constructed over fairly general measure spaces) and taking values in similar such spaces. Only relatively simple versions of the theorems are treated. Even so, some of the proofs are fairly complex and one aim has been to present all the important details.

Section 13.1 collects some preliminaries concerning measure spaces. The treatment of the Riesz-Thorin theorem and its applications occupies Sections 13.2–13.6, while Sections 13.7–13.11 deal with Marcinkiewicz' theorem and its applications. Among the latter is to be found the promised alternative approach to the study of the conjugate operator $f \to H * f = \tilde{f}$ (see Sections 12.8 and 12.9).

Suggestions for reading on further developments will be found in 13.4.2 and 13.8.2.

Any reader who is perturbed by the introduction of general measure spaces is advised to concentrate on the two concrete instances described in 13.1.3 and interpret all the general definitions and concepts in these special cases, which suffice for all the important applications made in this book. See also the remarks in 13.2.3(5).

13.1 Measure Spaces

13.1.1. **Some Definitions.** By a *measure space* is meant a triplet $(X, \mathscr{M}, \mu)$ in which X denotes a set, $\mathscr{M}$ a σ-algebra of subsets of X (see [HS], p. 4), and μ a function on $\mathscr{M}$ with values in $[0, \infty]$ such that

$$\mu(\varnothing) = 0$$

and

$$\mu(M) = \sum_{n=1}^{\infty} \mu(M_n)$$

whenever M is the countable disjoint union of sets $M_n \in \mathscr{M}$ $(n = 1, 2, \cdots)$.

In connection with the second equation, a sum composed of nonnegative terms, at least one of which is ∞, is interpreted to be ∞; and the sum of a divergent series of nonnegative numbers is likewise interpreted to be ∞.

A subset M of X is said to be *μ-finite* if $M \in \mathscr{M}$ and $\mu(M) < \infty$. The measure space is $(X, \mathscr{M}, \mu)$ is termed *σ-finite* if X can be expressed as the union of a countable sequence of μ-finite sets. *All the measure spaces we shall need to consider are σ-finite*, and we shall therefore make this a standing hypothesis. (The requirement of σ-finiteness is not essential at all points; especially is this the case when one considers the measure spaces associated with Radon measures, which is the situation prevailing in harmonic analysis.)

Nor will anything be lost by assuming, as we shall henceforth, that *all measure spaces $(X, \mathscr{M}, \mu)$ are complete*: this means (compare [HS], p. 155) that, if M is a *μ-null set* (that is, if $M \in \mathscr{M}$ and $\mu(M) = 0$), then so too is any subset of M.

13.1.2. **The Spaces $\mathbf{L}^p(X, \mathscr{M}, \mu)$.** As is indicated in Chapter 6 of [W], or in greater detail in Sections 12 and 13 of [HS], and in 3.4 and 3.5 of [AB], one can associate with any measure space $(X, \mathscr{M}, \mu)$ a Lebesgue-like integration theory. In particular, there is an associated concept of measurable function (the members of $\mathscr{M}$ playing the role of measurable sets); and one can construct the associated Lebesgue spaces $\mathbf{L}^p(\mu) = \mathbf{L}^p(X, \mathscr{M}, \mu)$ for $0 < p \leqslant \infty$. (The reader will notice that the space here denoted by $\mathbf{L}^p$ is symbolized $\mathfrak{L}_p$ in [HS].)

If f is a given function on X, the *μ-equivalence class* of f is the set of all functions on X that agree μ-a.e. with f, that is, which agree with f save perhaps on a μ-null subset of X.

For an arbitrary measurable function f on X we define the symbol $\|f\|_{\mathbf{L}^p(X, \mathscr{M}, \mu)}$ or, more briefly, $\|f\|_{p,\mu}$ to mean

$$\{\int_X |f|^p \, d\mu\}^{1/p} \qquad \text{if } 0 < p < \infty$$

and

$$\operatorname*{ess\,sup}_X |f| \qquad \text{if } p = \infty.$$

Then (compare Exercise 8.17 and Section 13.7) $\|f\|_{p,\mu}$ may be ∞. $\mathbf{L}^p(X, \mathscr{M}, \mu)$ consists precisely of those complex-valued measurable functions f on X for which $\|f\|_{p,\mu} < \infty$. If we identify functions belonging to any one μ-equivalence class, then $(f_1, f_2) \to \|f_1 - f_2\|_{p,\mu}$ (or $\|f_1 - f_2\|_{p,\mu}^p$ if $0 < p < 1$) appears as a metric on $\mathbf{L}^p(X, \mathscr{M}, \mu)$ and makes the latter into a complete metric space. If also $1 \leqslant p \leqslant \infty$, then $\| \cdot \|_{p,\mu}$ is a norm on $\mathbf{L}^p(X, \mathscr{M}, \mu)$ and makes the latter into a Banach space. For details, see [W], pp. 68–72 and [HS], Section 13.

13.1.3. **Examples of Measure Spaces.** For the applications we have in mind there are essentially only two types of measure space required, namely:

(1) The case in which X is a real interval, $\mathscr{M}$ is the set of Lebesgue measurable subsets of X and μ is a multiple of the restriction of Lebesgue measure to $\mathscr{M}$. The standard case for future reference is that in which X is the interval $[-\pi, \pi)$ (or indeed any chosen interval of length 2π) and μ is $(2\pi)^{-1}$ times the restriction of Lebesgue measure to the Lebesgue-measurable subsets of the chosen interval, the associated Lebesgue space being what in this book has been, and will continue to be, denoted by $\mathbf{L}^p$.

(2) The case in which X is Z, $\mathscr{M}$ comprises all subsets of Z and μ is the so-called *counting measure*: $\mu(M) =$ the cardinal number of M whenever M is finite, $\mu(M) = \infty$ whenever M is infinite. A slightly more general situation is that in which X and $\mathscr{M}$ retain the same meaning while μ is defined by

$$\mu(M) = \sum_{n \in M} c_n, \tag{13.1.1}$$

where $(c_n)_{n \in Z}$ is a fixed nonnegative real-valued sequence defined on Z. Here, as in a similar context in 13.1.1, it is to be understood that $\mu(M) = \infty$ whenever M is infinite and such that the series on the right-hand side of (13.1.1) is divergent. The aforesaid counting measure arises when $c_n = 1$ for all $n \in Z$.

In case μ is specified by (13.1.1), $\mathbf{L}^p(X, \mathscr{M}, \mu)$ comprises exactly those complex-valued functions ϕ on Z such that

$$\sum_{n \in Z} c_n |\phi(n)|^p < \infty$$

if $p < \infty$, or such that

$$\sup_{n \in Z} \{|\phi(n)| : n \in Z,\ c_n > 0\} < \infty$$

if $p = \infty$, two functions being identified if they agree at all points $n \in Z$ for which $c_n > 0$. In particular, if μ is the counting measure, $\mathbf{L}^p(X, \mathscr{M}, \mu)$ is just $\ell^p(Z)$ (as defined in 2.2.5).

Occasionally [Example 13.2.3(2) provides an instance], it is necessary to speak of counting measures on sets other than Z. Accordingly, it should be observed that the preceding remarks about counting measures and measures of the type (13.1.1) apply equally when Z is replaced by any subset thereof, or by any countable set whatsoever. In fact, similar considerations apply when Z is replaced by an absolutely arbitrary set S, except that the counting measure on S will be σ-finite (see 13.1.1) if and only if S is countable.

13.1.4. **Simple Functions.** Given a measure space $(X, \mathscr{M}, \mu)$, a complex-valued function f on X is termed $(X, \mathscr{M}, \mu)$-*simple* (or just *simple*, if the measure space is understood from the context) if it is expressible in the form

$$f = \sum_{k=1}^{n} c_k \chi_{M_k}, \tag{13.1.2}$$

where the c_k are complex numbers, the M_k are μ-finite subsets of X, and χ_M is used to denote the characteristic function of the subset M of X.

It is easy to see that a simple f may always be represented in the form (13.1.2) wherein, furthermore, the M_k are disjoint.

A simple function f belongs to $\mathbf{L}^p(\mu)$ for every p satisfying $0 < p \leqslant \infty$. Moreover, the set of simple functions is everywhere dense in $\mathbf{L}^p(\mu)$, provided either that $p < \infty$ or, if $p = \infty$, that X is μ-finite. (See [W], p. 93 and [HS], p. 187.)

13.1.5. **Another Converse of Hölder's Inequality.** For the proof of the Riesz-Thorin theorem we shall need the following variant of the converse of Hölder's inequality (compare Exercise 3.6).

Suppose that $(X, \mathscr{M}, \mu)$ is a measure space and that f is a complex-valued, measurable function on X which is known to be integrable over each μ-finite subset of X, that is, that

$$\int_M |f| \, d\mu \equiv \int_X |f| \chi_M \, d\mu < \infty$$

for each μ-finite subset M of X. Then, if $1 \leqslant p \leqslant \infty$, we have

$$\|f\|_{p,\mu} = \sup \left| \int_X fg \, d\mu \right|, \tag{13.1.3}$$

the supremum being taken with respect to all simple functions g satisfying $\|g\|_{p',\mu} \leqslant 1$. (As usual, p' is defined by the relation $1/p + 1/p' = 1$, together with the convention that $p' = \infty$ if $p = 1$ and $p' = 1$ if $p = \infty$.) Our hypotheses on f ensure that fg is integrable for every simple function g.

It is to be remarked that, if f is assumed a priori to belong to

$$\mathbf{L}^p(\mu) = L^p(X, \mathscr{M}, \mu),$$

the assertion is contained in Theorem (15.1) of [HS]. It is essential that we dispense with this assumption, and additional argument is needed.

Proof. The cases $p = \infty$ and $p = 1$ are especially simple and demand no further explanation. Assume therefore that $1 < p < \infty$.

Let m denote the supremum appearing in (13.1.3). If $\|f\|_{p,\mu}$ is finite, Hölder's inequality shows that

$$m \leqslant \|f\|_{p,\mu};$$

the same is vacuously true if $\|f\|_{p,\mu}$ is infinite. It thus suffices to show that

$$\|f\|_{p,\mu} \leqslant m. \tag{13.1.4}$$

Now if g is measurable, vanishes outside a μ-finite set M, and $\|g\|_{p',\mu} \leqslant 1$, then g can be expressed as the pointwise limit of a sequence g_k $(k = 1, 2, \cdots)$ of simple functions vanishing outside M and satisfying $\|g_k\|_{p',\mu} \leqslant 1$. It then follows from Lebesgue's convergence theorem that m, be it finite or infinite, is unaltered if in its definition we allow g to range over all measurable functions vanishing outside a μ-finite set and subject to $\|g\|_{p',\mu} \leqslant 1$. Then, on replacing g by $g \cdot \operatorname{sgn} f$, where

$$(\operatorname{sgn} f)(x) = \begin{cases} \dfrac{\overline{f(x)}}{|f(x)|} & \text{if } f(x) \neq 0 \\ 0 & \text{otherwise}, \end{cases}$$

we infer that

$$m \geqslant \sup \int_X |f| g \, d\mu, \tag{13.1.5}$$

where now g ranges over all nonnegative measurable functions vanishing outside some μ-finite set and satisfying $\|g\|_{p',\mu} \leqslant 1$. The monotone convergence theorem, together with the assumed σ-finiteness of the measure space $(X, \mathscr{M}, \mu)$, shows next that the inequality (13.1.5) remains undisturbed if the competing g's are freed from the demand that they vanish outside some μ-finite set.

This being so, let $f_r = \inf(|f|, r\chi_{M_r})$ for $r = 1, 2, \cdots$, where $M_r \in \mathscr{M}$, $M_r \subset M_{r+1}$, $\mu(M_r) < \infty$, and $\bigcup_{r=1}^{\infty} M_r = X$. Then $f_r \in \mathbf{L}^p(\mu)$ and it is clear from the final version of (13.1.5) that

$$\int_X f_r g \, d\mu \leqslant m$$

for any nonnegative measurable function g satisfying $\|g\|_{p',\mu} \leqslant 1$. So, by Theorem (15.1) of [HS],

$$\|f_r\|_{p,\mu} \leqslant m. \tag{13.1.6}$$

Since $f_r \uparrow |f|$ as $r \uparrow \infty$, the monotone convergence theorem combines with (13.1.6) to yield (13.1.4) and so completes the proof.

13.2 Operators of Type (p, q)

13.2.1. The Concept of Type. Let $(X, \mathscr{M}, \mu)$ and $(Y, \mathscr{N}, \nu)$ be measure spaces as described in 13.1.1, and let $\mathfrak{D}$ denote a linear subspace of

$$\mathbf{L}^p(\mu) = \mathbf{L}^p(X, \mathscr{M}, \mu).$$

We shall in this section be concerned with a linear operator T defined on $\mathfrak{D}$ and taking values in the space of complex-valued ν-measurable functions on Y which are ν-integrable over ν-finite sets (or in the set of ν-equivalence classes of such functions).

Given two exponents p and q from the range $[1, \infty]$, T is said to be of *type* (p, q) [or, more precisely, of *strong type* (p, q)] if and only if there exists a number $m \geqslant 0$ such that

$$\|Tf\|_{q,\nu} \leqslant m \|f\|_{p,\mu} \tag{13.2.1}$$

for $f \in \mathfrak{T}$. The norms appearing in (13.2.1) are to be interpreted in the fashion explained in 13.1.2. The least admissible value of m in (13.2.1) is denoted by $\|T\|_{p,q}$ or $\|T\|_{p,q;\mu,\nu}$ and is termed the (*strong*) (p, q)-*norm* of T. If T is not of type (p, q) we shall write $\|T\|_{p,q} = \infty$. These definitions apply even if T is not linear (see Section 13.7) but we shall in this section consider only the case in which T is linear. They apply also when $p > 0$, $q > 0$; but, unless the contrary is explicitly stated, we assume that $p \geqslant 1$ and $q \geqslant 1$.

From 13.1.5 it follows that in all cases

$$\|T\|_{p,q} = \sup \left| \int_Y Tf \cdot g \, d\nu \right|, \tag{13.2.2}$$

the supremum being taken with respect to those $f \in \mathfrak{T}$ satisfying $\|f\|_{p,\mu} \leqslant 1$ and those $(Y, \mathcal{N}, \nu)$-simple functions g satisfying $\|g\|_{q',\nu} \leqslant 1$. Because of linearity of T, one may replace the inequality signs in the preceding sentence by equality signs. It is also true that

$$\|T\|_{p,q} = \sup \|Tf\|_{q,\nu}, \tag{13.2.3}$$

the supremum being taken relative to those $f \in \mathfrak{T}$ satisfying $\|f\|_{p,\mu} \leqslant 1$ (or $\|f\|_{p,\mu} = 1$).

These remarks make it plain that T is of type (p, q) if and only if it maps the normed linear subspace $\mathfrak{T}$ of $\mathbf{L}^p(\mu)$ into the Banach space $\mathbf{L}^q(\nu) = \mathbf{L}^q(Y, \mathcal{N}, \nu)$ in a continuous fashion.

In applications it often arises that the heuristic form of T is given in advance and that there is a considerable freedom of choice in $\mathfrak{T}$: T may, for example, be initially defined on $\mathbf{L}^2(\mu)$, in which case $\mathfrak{T}$ might be selected to be $\mathbf{L}^2(\mu)$ itself, or to be the space of $(X, \mathcal{M}, \mu)$-simple functions. However we will show that if, as is usually the case, $\mathfrak{T}$ is everywhere dense in $\mathbf{L}^p(\mu)$, then the type classification of T is largely independent of this ambiguity, depending in fact solely on the possibility or otherwise of extending T into a continuous linear operator from the whole of $\mathbf{L}^p(\mu)$ into $\mathbf{L}^q(\nu)$.

13.2.2. **Extension Theorem.** Suppose that T and $\mathfrak{T}$ are as in 13.2.1 and that T is of type (p, q). Suppose also that $\mathfrak{T}$ is everywhere dense in $\mathbf{L}^p(\mu)$. Then T can be uniquely extended into a continuous linear map of $\mathbf{L}^p(\mu)$ into $\mathbf{L}^q(\nu)$ which satisfies

$$\|Tf\|_{q,\nu} \leqslant \|T\|_{p,q} \cdot \|f\|_{p,\mu} \tag{13.2.4}$$

for all $f \in \mathbf{L}^p(\mu)$.

Proof. If such an extension exists, it is unique on account of its continuity and the assumed denseness of $\mathfrak{T}$ in $\mathbf{L}^p(\mu)$.

An extension of the desired type is obtainable in the following way. Given $f \in \mathbf{L}^p(\mu)$, chose a sequence $(f_n)_{n=1}^{\infty}$ of elements of $\mathfrak{T}$ such that

$$\lim_{n \to \infty} \|f - f_n\|_{p,\mu} = 0. \tag{13.2.5}$$

Taking any finite value of m not less than $\|T\|_{p,q}$, (13.2.1) gives

$$\|Tf_n - Tf_{n'}\|_{q,\nu} \leqslant m \, \|f_n - f_{n'}\|_{p,\mu}$$

for all n and n', so that (13.2.5) indicates that the sequence $(Tf_n)_{n=1}^{\infty}$ is Cauchy in $\mathbf{L}^q(\nu)$. Furthermore, (13.2.1) entails that the limit of $(Tf_n)_{n=1}^{\infty}$ is independent of which sequence $(f_n)_{n=1}^{\infty}$ is chosen, provided only that it is subject to (13.2.5). Denote the limit by $T_1 f$. It is then clear from the preceding sentence that T_1 maps $\mathbf{L}^p(\mu)$ linearly into $\mathbf{L}^q(\nu)$, and that the restriction of T_1 to $\mathfrak{T}$ is none other than T. Finally, by (13.2.1) and (13.2.5), the definition of $T_1 f$ shows that

$$\begin{aligned} \|T_1 f\|_{q,\nu} &= \lim_{n\to\infty} \|Tf_n\|_{q,\nu} \\ &\leqslant \liminf_{n\to\infty} m \, \|f_n\|_{p,\mu} \\ &= m \, \|f\|_{p,\mu}. \end{aligned} \tag{13.2.6}$$

Since m is arbitrary save for the restriction

$$m \geqslant \|T\|_{p,q},$$

(13.2.4) follows from (13.2.6).

13.2.3. **Some Examples.** (1) *The Fourier transformation.* Take $(X, \mathscr{M}, \mu)$ as in 13.1.3(1), and $(Y, \mathscr{N}, \nu)$ as in 13.1.3(2) with ν the counting measure on $Y = Z$. The operator T is taken to be the Fourier transformation:

$$Tf = \hat{f}$$

for $f \in \mathbf{L}^1(\mu) \equiv \mathbf{L}^1$. It is evident that T maps $\mathbf{L}^1$ into $\mathbf{L}^\infty(\mu) \equiv \ell^\infty(Z)$. Moreover, since the ν-finite sets are just the finite sets, the hypotheses of 13.2.1 are fulfilled.

That T is of type $(1, \infty)$ is the content of 2.3.2.

On the other hand (8.2.2) says exactly that $T|\mathbf{L}^2(\mu)$ is of type (2, 2).

In particular, therefore, the restriction of T to the μ-simple functions is simultaneously of types $(1, \infty)$ and (2, 2). This, combined with the Riesz-Thorin theorem, will permit us to make further type-statements about T; see Section 13.4.

(2) *The moment operator.* Here we take $X = [0, 1]$, $\mathscr{M}$ the set of Lebesgue-measurable subsets of X, μ the restriction of Lebesgue measure to $\mathscr{M}$, $Y = \{0, 1, 2, \cdots\}$, $\mathscr{N}$ the set of all subsets of Y, and ν the counting measure on Y. Let S be the so-called *moment operator* defined on $\mathbf{L}^1(\mu)$ by $Sf = f^{\#}$, where

$$f^{\#}(n) = \int_0^1 f(x)x^n \, dx \qquad (n = 0, 1, 2, \cdots).$$

Once again the hypotheses of 13.2.1 are satisfied.

It is not difficult to show (see Exercise 13.7) that S is a one-to-one linear map of $\mathbf{L}^1(\mu)$ into $\mathbf{c}_0(Y) \subset \mathbf{L}^\infty(\nu)$, and that S is of type $(1, \infty)$. (The definition of $\mathbf{c}_0(Y)$ is exactly analogous to that of $\mathbf{c}_0(Z)$ given in 2.2.5.)

It is also true, but not trivial, that $S \mid \mathbf{L}^2(\mu)$ is of type (2, 2): this is in fact the content of a famous inequality due to Hilbert (see [HLP], pp. 212 and 226).

Once more, an application of the Riesz-Thorin theorem leads to further information about S; see Exercise 13.8.

(3) *The conjugate function operator.* Take $(X, \mathscr{M}, \mu) = (Y, \mathscr{N}, \nu)$ as in 13.1.3(1), so that $\mathbf{L}^p(\mu) = \mathbf{L}^p(\nu) = \mathbf{L}^p$. Consider the transformation $T: f \to \tilde{f} = H * f$ defined in Section 12.8. We know that Tf is defined distributionally whenever $f \in \mathbf{L}^1$. M. Riesz' theorem 12.9.1 asserts that $T|\mathbf{L}^p$ is of type (p, p) whenever $1 < p < \infty$. On the other hand, from 12.8.4 it appears that T is not of type (1, 1) nor of type (∞, ∞). More about the nature of T acting on $\mathbf{L}^1$ will appear in Section 13.9 in connection with the concept of "weak type" of operators.

(4) Further examples appear in Section 13.6 and Exercises 13.5 and 13.6.

(5) This is a convenient point at which to interject some remarks addressed to the reader who chooses to limit the general Theorems 13.4.1 and 13.8.1 to versions possessing a degree of abstraction just adequate to cover the essential applications made in this book.

As has been said in the introductory material to this chapter, it is for this purpose enough to be prepared to meet cases in which each of $(X, \mathscr{M}, \mu)$ and $(Y, \mathscr{N}, \nu)$ is one of the measure spaces described in 13.1.3.

The operators which are likewise essential to subsequent applications are all expressible (or reducible without loss of essential scope) to one of two forms. Those to which Theorem 13.4.1 will be applied are of the form

$$Tf(y) = \lim_{N \to \infty} T_N f(y), \tag{13.2.7}$$

while Theorem 13.8.1 will be applied also to certain operators of the form

$$Tf(y) = \sup_{N \geqslant 1} |T_N f(y)|; \tag{13.2.8}$$

the domain of T may without loss be taken to a subset of $\mathbf{L}^1(X, \mathscr{M}, \mu)$ and to be such that earlier results in this book guarantee that (13.2.7) holds pointwise ν-a.e. and in mean with various exponents. In either case one can choose the T_N $(N = 1, 2, \cdots)$ to be of the form

$$T_N f(y) = \int_X K_N(x, y) f(x)\, d\mu(x), \tag{13.2.9}$$

where K_N is bounded, measurable in the pair (x, y) (see [HS], p. 379), and such that

$$\int_X |K_N(x, y)|^p\, d\mu(x) < \infty$$

for $0 < p < \infty$. (Different choices of the T_N can, of course, lead to the same T; we do not always indicate explicitly an admissible choice of the T_N

satisfying all these conditions.) It is then very simple to show that $T_N f$ is a bounded ν-measurable function on Y whenever f belongs to $\mathbf{L}^q(X, \mathscr{M}, \mu)$ for some q satisfying $1 \leqslant q \leqslant \infty$.

For operators of the form (13.2.8), however, we shall not be able to guarantee a priori that Tf is finite-valued (everywhere or even ν-a.e.). The opening remarks in Section 13.7 will refer specifically to this state of affairs.

13.2.4. **Preamble to the Riesz-Thorin Theorem.** In terms of the "type" language, it is very simple to state roughly the aims of the Riesz-Thorin theorem. This theorem asserts that if an operator T is simultaneously of types (p_0, q_0) and (p_1, q_1), where $1 \leqslant p_j, q_j \leqslant \infty$ for $j = 0, 1$, then T is also of type (p, q) for certain "intermediate" pairs (p, q) and that for such pairs, $\|T\|_{p,q}$ can be majorized in a certain way. This explains why the theorem is often described as an *interpolation theorem*.

Before we can embark on the proof of the Riesz-Thorin theorem one more auxiliary is required, this time from complex variable theory.

13.3 The Three Lines Theorem

The three lines theorem, a simple result in complex variable theory, is the major tool used in the proof we give of the Riesz-Thorin theorem.

Throughout this section V denotes the vertical strip in the complex ζ-plane defined by

$$\zeta = \xi + i\eta, \qquad 0 \leqslant \xi \leqslant 1, \qquad \eta \in R. \tag{13.3.1}$$

We are concerned with a function F defined, bounded, and continuous on the strip V and analytic interior to V. The crucial result is as follows.

13.3.1. With the above notations and assumptions, put

$$M_\xi = \sup\{|F(\xi + i\eta)| : \eta \in R\} \qquad (0 \leqslant \xi \leqslant 1). \tag{13.3.2}$$

Then, for $0 \leqslant \xi \leqslant 1$,

$$M_\xi \leqslant M_0^{1-\xi} M_1^{\xi}. \tag{13.3.3}$$

Proof. Since it is plainly enough to show that (13.3.3) holds when M_0 and M_1 are replaced by arbitrary fixed numbers exceeding M_0 and M_1, respectively, we may assume that M_0 and M_1 are positive.

Then, by considering the function $F(\zeta)/M_0^{1-\zeta}M_1^{\zeta}$ in place of F, it is seen to be enough to deal with the case in which $M_0 = M_1 \leqslant 1$. In other words, we wish to show that from the assumptions

$$|F(i\eta)| \leqslant 1, \qquad |F(1 + i\eta)| \leqslant 1 \qquad (\eta \in R) \tag{13.3.4}$$

it follows that

$$|F(\xi + i\eta)| \leqslant 1 \qquad (0 \leqslant \xi \leqslant 1, \eta \in R). \tag{13.3.5}$$

To this end we consider first the case in which

$$\lim_{|\eta| \to \infty} F(\xi + i\eta) = 0 \tag{13.3.6}$$

holds uniformly for $0 \leqslant \xi \leqslant 1$. In this case the desired conclusion (13.3.5) follows directly on applying the maximum principle to a rectangle $0 \leqslant \xi \leqslant 1$, $|\eta| \leqslant \gamma$, where γ is chosen so large that

$$|F(\xi \pm i\gamma)| \leqslant 1$$

for $0 \leqslant \xi \leqslant 1$, the choice of such a number γ being possible by virtue of (13.3.6) holding uniformly for $0 \leqslant \xi \leqslant 1$.

In general, we apply the special case just established to each of the functions

$$F_n(\zeta) \coloneqq F(\zeta) \cdot \exp n^{-1}(\zeta^2 - 1) \qquad (n = 1, 2, \cdots).$$

Each F_n satisfies the condition (13.3.6) previously and temporarily imposed upon F and

$$|F_n(\xi + i\eta)| \leqslant 1$$

for $\xi = 0$ or 1 by virtue of (13.3.4). Accordingly, (13.3.5) holds with F_n in place in F, and (13.3.5) itself follows thence on letting $n \to \infty$.

13.3.2. **Remark.** Almost any textbook on complex function theory will contain a discussion of many extensions and generalizations of 13.3.1, which we leave in the simple and unadorned version directly useful in Section 13.4.

13.4 The Riesz-Thorin Theorem

13.4.1. (Riesz-Thorin) Let $(X, \mathscr{M}, \mu)$ and $(Y, \mathscr{N}, \nu)$ be measure spaces, as in 13.1.1, and let T be a linear operator defined for all μ-simple functions and taking values in the set of functions on Y which are ν-integrable over each ν-finite set[1]. Suppose that T is simultaneously of types (p_j, q_j) $(j = 0, 1)$ where $1 \leqslant p_j, q_j \leqslant \infty$. Then, for any exponent pair (p, q) of the form

$$\frac{1}{p} = \frac{1-t}{p_0} + \frac{t}{p_1}, \qquad \frac{1}{q} = \frac{1-t}{q_0} + \frac{t}{q_1}, \qquad 0 \leqslant t \leqslant 1,$$

T is of type (p, q) and

$$\|T\|_{p,q} \leqslant \|T\|_{p_0,q_0}^{1-t} \cdot \|T\|_{p_1,q_1}^{t}. \tag{13.4.1}$$

In particular, if $p < \infty$, or if X is μ-finite, T can be extended so as to map $\mathbf{L}^p(\mu)$ continuously into $\mathbf{L}^q(\nu)$.

Remarks. (1) The preceding result may be expressed in the following way: the function $\log \|T\|_{1/\alpha, 1/\beta}$ on the rectangle $0 \leqslant \alpha, \beta \leqslant 1$ is convex on any segment joining two points at which it is finite. This

[1] Or in the set of ν-equivalence classes of such functions.

explains why the Riesz-Thorin theorem is often described as a *convexity theorem*.

(2) If $p_j < \infty$ for $j \in \{0, 1\}$, and if X is a locally compact Hausdorff space and μ a positive Radon measure on X, one may replace μ-simple functions by linear combinations of characteristic functions of relatively compact μ-measurable subsets of X.

Proof. Once (13.4.1) is established, the final sentence will follow at once from 13.2.2.

To handle (13.4.1) we write $\alpha_j = 1/p_j$, $\beta_j = 1/q_j$ for $j = 0, 1$ and $\alpha = 1/p$, $\beta = 1/q$, making the conventions that $1/0 = \infty$ and $1/\infty = 0$. Define also

$$\alpha(\zeta) = (1 - \zeta)\alpha_0 + \zeta\alpha_1, \qquad \beta(\zeta) = (1 - \zeta)\beta_0 + \zeta\beta_1$$

for all complex ζ. Thus $\alpha(j) = \alpha_j$ and $\beta(j) = \beta_j$ for $j = 0, 1$ and $\alpha(t) = \alpha$ and $\beta(t) = \beta$.

Let f be a μ-simple function. By 13.1.5,

$$\|Tf\|_{q,\nu} = \sup \left| \int_Y Tf \cdot g \, d\nu \right|, \tag{13.4.2}$$

the supremum being taken with respect to all ν-simple functions g satisfying $\|g\|_{q',\nu} = 1$.

In order to establish (13.4.1), it will thus suffice to choose and fix a μ-simple function f satisfying $\|f\|_{p,\mu} = 1$, say

$$f = \sum_k a_k \chi_{A_k},$$

the A_k being pairwise disjoint, and a ν-simple function g satisfying $\|g\|_{q',\nu} = 1$, say

$$g = \sum_h b_h \chi_{B_h},$$

the B_h being pairwise disjoint and show that the integral

$$I = \int_Y Tf \cdot g \, d\nu$$

satisfies the inequality

$$|I| \leqslant \|T\|_{p_0,q_0}^{1-t} \cdot \|T\|_{p_1,q_1}^{t}. \tag{13.4.3}$$

In the above expressions for f and g, each A_k is a μ-finite set, each B_h is a ν-finite set, the sums are finite, and we may assume that $a_k \neq 0$ and $b_h \neq 0$.

In proving (13.4.3) we write

$$f = |f|e^{iu}, \qquad g = |g|e^{iv},$$

where u and v are real-valued and μ-measurable and ν-measurable, respectively. At this point we consider two cases in turn.

(1) Suppose first that $\alpha > 0$ and $\beta < 1$. In this case we introduce the functions f_ζ and g_ζ defined by

$$f_\zeta = |f|^{\alpha(\zeta)/\alpha} e^{iu}, \qquad g_\zeta = |g|^{(1-\beta(\zeta))/(1-\beta)} e^{iv}. \tag{13.4.4}$$

Define further

$$F(\zeta) = \int_Y Tf_\zeta \cdot g_\zeta \, d\nu, \tag{13.4.5}$$

in terms of which

$$I = F(t). \tag{13.4.6}$$

The linearity of T leads to the formula

$$F(\zeta) = \sum_{k,h} |a_k|^{\alpha(\zeta)/\alpha} |b_h|^{(1-\beta(\zeta))/(1-\beta)} \int_Y T\chi_{A_k} \cdot \chi_{B_h} e^{i(u+v)} \, d\nu,$$

which makes it plain that F is an entire analytic function which is bounded on the strip V. We aim to apply 13.3.1.

If $\xi \equiv \operatorname{Re} \zeta = 0$, then $\operatorname{Re} \alpha(\zeta) = \alpha_0$ and Hölder's inequality gives from (13.4.5) and the assumptions about T

$$\begin{aligned} |F(i\eta)| &\leqslant \|Tf_{i\eta}\|_{1/\beta_0,\nu} \cdot \|g_{i\eta}\|_{1/(1-\beta_0),\nu} \\ &\leqslant \|T\|_{1/\alpha_0,1/\beta_0} \cdot \|f_{i\eta}\|_{1/\alpha_0,\mu} \cdot \|g_{i\eta}\|_{1/(1-\beta_0),\nu}. \end{aligned} \tag{13.4.7}$$

But from (13.4.4) it follows by direct calculation that

$$\|f_{i\eta}\|_{1/\alpha_0,\mu} = \|f\|_{1/\alpha,\mu}^{\alpha_0/\alpha} = 1$$

and that

$$\|g_{i\eta}\|_{1/(1-\beta_0),\nu} = \|g\|_{1/(1-\beta),\nu}^{(1-\beta_0)/(1-\beta)} = 1,$$

since $1/\alpha = p$ and $1/(1-\beta) = q'$ and

$$\|f\|_{p,\mu} = \|g\|_{q',\nu} = 1.$$

Consequently (13.4.7) shows that

$$M_0 \equiv \sup\{|F(i\eta)| : \eta \in R\} \leqslant \|T\|_{1/\alpha_0,1/\beta_0} = \|T\|_{p_0,q_0}. \tag{13.4.8}$$

An exactly similar argument shows that

$$M_1 \equiv \sup\{|F(1+i\eta)| : \eta \in R\} \leqslant \|T\|_{p_1,q_1}. \tag{13.4.9}$$

On applying 13.3.1, equations (13.4.6), (13.4.8), and (13.4.9) lead to (13.4.3), and the proof is complete in this case.

(2) Of the excluded cases, that in which $\alpha = 0$ and $\beta = 1$ leaves nothing to be proved.

If $\beta = 1$, in which case $\beta_0 = \beta_1 = 1$, and $\alpha > 0$, f_ζ is defined as in (13.4.4) while g_ζ is defined to be g. The proof then proceeds as before.

An entirely similar modification applies if $\alpha = 0$, in which case $\alpha_0 = \alpha_1 = 0$, and $\beta < 1$.

13.4.2. Comments on the Riesz-Thorin Theorem. (1) In its original form, as given by M. Riesz in 1926, the measure spaces involved had finite sets X and Y and counting measures μ and ν, the functions (finite sequences) allowed were real-valued, and convexity was asserted only for exponent pairs (p, q) subject to the additional restriction $1 \leqslant p \leqslant q \leqslant \infty$. Moreover, the result was expressed in terms of bilinear functionals rather than linear operators. (This corresponds to the introduction of the integral $\int_Y Tf \cdot g \, d\nu$ in the preceding proof and represents a standard possible variation in form of the theorem.) See [HLP], pp. 214–220, where complex-valued sequences are admitted but the proof really uses only real-variable methods, and the conclusion is again the restricted form of convexity. This restricted version is, however, adequate for several important applications.

Thorin's major contribution was the use of complex-variable methods, together with some simplifications in the proof, leading to unrestricted convexity (as formulated in 13.4.1); see Thorin [1], [2].

The proof given here is based on that appearing in [Z_2], pp. 95–96. It plainly depends crucially on using complex-valued functions. For a somewhat more general discussion, see [DS_1], pp. 520–526.

(2) Various important extensions of the theorem are now known.

Some of these apply to spaces $\mathbf{L}^p$ with $0 < p < \infty$ and some to the Hardy spaces $\mathbf{H}^p$ for the same range of values of p (see Exercise 3.9); for the details see [Z_2], Chapter XII, where convexity theorems for multilinear operators are also considered. (An application of such a convexity theorem will be mentioned briefly at the end of 16.4.9.) Other types of extension are given by O'Niel [2], Campanato and Murthy [1].

We also direct the reader's attention to extensions given by Stein and Weiss [2] in which the measures μ and ν, as well as the exponents p and q, are allowed to vary in a certain way. In this treatment 13.4.1 is formulated for operators T which are not necessarily linear but merely "sublinear." Stronger versions of 13.4.1 are also proved; cf. [Moz], Chapter 1.

(3) As we shall see in 13.8.3, (13.4.1) does not express all that is known to follow from hypotheses like those in 13.4.1. Moreover, a conclusion similar to, but a little weaker than, (13.4.1) derives from weaker hypotheses; see Theorem 13.8.1.

(4) Elaborate studies of interpolation problems and techniques, which are in some senses abstract versions of the Riesz-Thorin theorem, have been made. See, for example, Calderón [1] and the references cited there, Stampacchia [1]; Stein [2]; Schechter [1]; Peetre [1], [2], [3]; Coifmann, Cwikel, Rochberg, Sagher and Weiss [1]. See also MR **37** # 1951.

13.5 The Theorem of Hausdorff-Young

As a first application of the Riesz-Thorin theorem, we derive a result that constitutes a partial extension of Parseval's formula (8.2.2) and the Riesz-Fischer theorem 8.3.1. The result we obtain appears in two mutually dual assertions.

13.5.1. (Hausdorff-Young) Suppose that $1 \leqslant p \leqslant 2$.

(1) If $f \in \mathbf{L}^p$ then

$$\|\hat{f}\|_{p'} \leqslant \|f\|_p .$$

(2) If $\phi \in \ell^p(Z)$, then the series

$$\sum_{n \in Z} \phi(n) e^{inx}$$

converges in $\mathbf{L}^{p'}$ to a function $\hat{\phi}$ such that

$$\|\hat{\phi}\|_{p'} \leqslant \|\phi\|_p .$$

Proof. (1) This will follow from applying 13.4.1 and 13.2.2 to the situation described in 13.2.3(1). With the notation used there, 2.3.2 and (8.2.2) show that, if we restrict T to the μ-simple functions, then

$$\|T\|_{1,\infty} \leqslant 1, \qquad \|T\|_{2,2} = 1. \tag{13.5.1}$$

So, by 13.4.1,

$$\|T\|_{p,q} \leqslant 1$$

whenever

$$\frac{1}{p} = \frac{1-t}{1} + \frac{t}{2}, \qquad \frac{1}{q} = \frac{1-t}{\infty} + \frac{t}{2}$$

for some t satisfying $0 \leqslant t \leqslant 1$. These requirements signify that $1 \leqslant p \leqslant 2$ and $q = p'$. To derive (1), it now suffices to apply 13.2.2.

(2) Here we interchange the roles of the measure spaces $(X, \mathscr{M}, \mu)$ and $(Y, \mathscr{N}, \nu)$ used in the proof of (1). The ν-simple functions ϕ are precisely those with finite supports. For such ϕ we define

$$T\phi = \sum_{n \in Z} \phi(n) e_n ,$$

a trigonometric polynomial. It is evident that T is of type $(1, \infty)$ and of type $(2, 2)$, and that (13.5.1) holds. [The assertion concerning $\|T\|_{2,2}$ follows from Exercise 1.7(1).] By 13.4.1 and 13.2.2 it follows, as in (1), that T can be continuously extended so as to map $\mathbf{L}^p(\nu) \equiv \ell^p(Z)$ into $\mathbf{L}^{p'}(\mu) \equiv \mathbf{L}^{p'}$ whenever $1 \leqslant p \leqslant 2$, and that

$$\|T\|_{p,p'} \leqslant 1.$$

The continuity of this extension of T shows at once that, for any $\phi \in \ell^p(Z)$, the series appearing in (2) converges in $\mathbf{L}^{p'}$ in accordance with the relation

$T\phi = \lim_{N\to\infty} T\phi_N$ in $\mathbf{L}^{p'}$, where

$$\phi_N(n) = \begin{cases} \phi(n) & \text{if } |n| \leqslant N, \\ 0 & \text{otherwise}, \end{cases}$$

so that $\lim_{N\to\infty} \phi_N = \phi$ in $\ell^p(Z)$.

13.5.2. **Remarks.** The theorem was proved by W. H. Young in 1912–1913 for the special values of p of the form $2k/(2k-1)$, where k is a positive integer. For general values of p in the interval $[1, 2]$, the result is due originally to Hausdorff (1923). F. Riesz showed that the result was true for expansions in terms of a general orthonormal system of functions u_n satisfying $|u_n| \leqslant 1$ a.e.; this result also dates from 1923; see [Z_2], p. 102.

The idea of deriving 13.5.1 and F. Riesz's extension thereof from the convexity theorem is itself due to M. Riesz (1926).

The Hausdorff-Young theorem, like the $\mathbf{L}^2$-theory of Chapter 8, can be extended to Hausdorff locally compact Abelian groups in general. (That the measure spaces involved are no longer σ-finite in general leads to no insuperable difficulties.) See [R], Chapter 1; [E], Sections 10.3 and 10.4; [We], Chapitre VI; [HR], (31.22). See also 13.6.3 below.

For other cases of inequalities of the type $\|\hat{f}\|_q < \infty$, see Prohorenko [1].

For further results, see Edwards [14]; Williams [1]; Fournier [1]; MR **49** # 9518; **51** # 1243; **52** # 8788; **55** ## 8689a,b; **56** # 953; **57** # 10366.

13.5.3. **Best-possible Nature of the Hausdorff-Young Theorem.** There are various senses in which 13.5.1 is the best possible of its type.

(1) In 13.5.1(1), the exponent p' cannot be replaced by anything smaller; that is, if $q < p'$ is given, there exist functions $f \in \mathbf{L}^p$ such that $\hat{f} \notin \ell^q(Z)$.

Indeed, as has been noted in 8.3.2, the breakdown is rather dramatic when $p = 2$. For general values of $p \in [1, \infty)$ we observe that Exercise 7.8 shows that, for any $\delta > 0$, there exists a function $f \in \mathbf{L}^p$ such that $\hat{f}(n) = |n|^{-(1/p' + \delta)}$ for $n \neq 0$. Since $q < p'$, δ can be chosen so small that $q(1/p' + \delta) \leqslant 1$, in which case it is evident that $\hat{f} \notin \ell^q(Z)$.

For the same purpose one might use the periodic function f for which $f(x)$ is $|x|^{\alpha - 1}$ or 0 according as $|x| \leqslant 1$ or $1 < |x| \leqslant \pi$, where $0 < \alpha < 1$; for this function it is easily verified that $\lim_{|n|\to\infty} |n|^{\alpha}\hat{f}(n)$ exists and is nonzero, so that $f \in \mathbf{L}^p$ if and only if $(1-\alpha)p < 1$ and $\hat{f} \in \ell^q(Z)$ if and only if $\alpha q > 1$.

This last example serves to show also that, in 13.5.1(2), the exponent p' cannot be replaced by anything larger.

See also Brown [1]; Hewitt and Ritter [1]; Edwards, Hewitt and Ritter [1].

(2) Both 13.5.1(1) and 13.5.1(2) become false when $p > 2$: this will be established in two ways in Sections 14.4 and 15.4, respectively.

(3) If $1 \leqslant p < 2$, $\mathscr{F}\mathbf{L}^p \equiv \{\hat{f} : f \in \mathbf{L}^p\}$ is a proper subset of $\ell^{p'}(Z)$ (whereas the two sets are identical when $p = 2$); compare the next paragraph and see also Section 15.4.

From (2) immediately above it follows that, dually, $\mathscr{F}\ell^p \equiv \{\hat{\phi} : \phi \in \ell^p(Z)\}$ is a proper subset of $\mathbf{L}^{p'}$, whenever $1 \leqslant p < 2$. (For otherwise it would appear that $\hat{g} \in \ell^p(Z)$ whenever $g \in \mathbf{L}^{p'}$, which is contrary to (2) since $p' > 2$.)

13.5.4. **Cases of Equality in 13.5.1** It is possible (see [Z_2], p. 105) to show that equality occurs in 13.5.1(1) if and only if $f(x) = \text{const}\, e^{inx}$ for some $n \in Z$, and in 13.5.1(2) if and only if ϕ vanishes for all but at most one element of Z.

In each case the "if" assertion is trivial, but the "only if" assertion is not so. The result is due to Hardy and Littlewood (1926).

A discussion applying to general groups (compare 13.5.2) is due to Hewitt and Hirschman [1].

13.5.5. **An Application.** The Hausdorff-Young theorem will in this subsection be employed to prove a result stated in Subsection 12.11.4 in connection with equalities of the type $fS = 0$, where $f \in \mathbf{A}$ and $S \in \mathbf{P}$.

Throughout the present subsection it will be supposed that $1 \leqslant q \leqslant \infty$; that S is a pseudomeasure of the form $S = \hat{\phi}$, where $\phi \in \ell^q(Z)$ (so that, if $q = \infty$, S may be an arbitrary pseudomeasure); that E denotes a closed subset of T containing supp S; and that $f \in \mathbf{A}$, so that $f = \hat{\psi}$ for some $\psi \in \ell^1(Z)$. It will be assumed furthermore that f vanishes on E and that, for some sequence $(\varepsilon_j)_{j=1}^{\infty}$ of positive numbers tending to zero,

$$f(x) = O(\varepsilon_j^{(1/2)-(1/q)}) \tag{13.5.2}$$

uniformly for x at periodic distance (see Exercise 12.30) at most ε_j from E. Our aim is to show that under these conditions

$$fS = 0. \tag{13.5.3}$$

Before commencing the proof we observe that if $1 \leqslant q \leqslant 2$, the condition (13.5.2) becomes void and may be dropped entirely; in this case, too, the proof presents no trouble since, by 13.5.1, S is a function in $\mathbf{L}^{q'}$. In case $q = \infty$, $1/q$ is to be interpreted as zero: this case covers the statement made in Subsection 12.11.4.

As a final preliminary, it is to be observed that, by (12.11.3), the equation (13.5.3) is equivalent to

$$\psi * \phi = 0, \tag{13.5.4}$$

which is what we shall in fact establish.

Proof of (13.5.3). For brevity we shall write c in place of $\frac{1}{2} - 1/q$, and we shall make a legitimate change of stance by regarding E as a closed periodic set of real numbers. In view of the preliminary remarks, we may and will assume that $2 < q \leqslant \infty$. Throughout the computations, B will denote various numbers depending at most upon q and S; the value denoted by B is not necessarily the same at each appearance.

Let A_j and U_j denote the periodic sets

$$(-\tfrac{1}{2}\varepsilon_j, \tfrac{1}{2}\varepsilon_j) \pmod{2\pi} \qquad \text{and} \qquad (-\varepsilon_j, \varepsilon_j) \pmod{2\pi},$$

respectively, so that (13.5.2) holds for $x \in E + U_j$. Define further the functions

$$k_j = \frac{(\chi_{A_j} * \chi_{A_j})}{(\varepsilon_j/2\pi)^2}$$

and $\kappa_j = \hat{k}_j$, and note that $\kappa_j \in \ell^1(Z)$. It is very simple to verify that $0 \leqslant k_j \leqslant B\varepsilon_j^{-1}$, and that k_j vanishes outside U_j. Consequently,

$$\|k_j\|_s \leqslant B\varepsilon_j^{1/s-1} \qquad (1 \leqslant s \leqslant \infty).$$

In particular, taking $s = 2q/(q-2) > 2$, 13.5.1 can be applied to yield

$$\|\kappa_j\|_s \leqslant \|k_j\|_{s'} \leqslant B\varepsilon_j^{-c}. \tag{13.5.5}$$

Now the k_j form an approximate identity in $\mathbf{L}^1$, so that $\lim_{j\to\infty} \kappa_j = 1$ boundedly on Z, and therefore $\psi * \phi = \lim_{j\to\infty} \psi * (\kappa_j\phi)$ pointwise on Z. This may also be written:

$$\psi * \phi(n) = \lim_{j\to\infty} \frac{1}{2\pi} \int e^{-inx} f(x) S_j(x)\, dx \tag{13.5.6}$$

for $n \in Z$, where $S_j = (\kappa_j\phi)^\wedge = k_j * S$. Since S_j is easily verified to vanish outside $E + U_j$, and since $f = 0$ on E, (13.5.6) and (13.5.2) combine to yield the majorizations

$$\begin{aligned} |\psi * \phi(n)| &\leqslant \liminf_{j\to\infty} \frac{1}{2\pi} \int_{D_j} |fS_j|\, dx \\ &\leqslant \liminf_{j\to\infty} O(\varepsilon_j{}^c) \cdot \frac{1}{2\pi} \int_{D_j} |S_j|\, dx, \end{aligned} \tag{13.5.7}$$

where we have written D_j for the intersection of $(E + U_j)\backslash E$ with $(0, 2\pi)$.

Now the Cauchy-Schwarz inequality gives

$$\frac{1}{2\pi} \int_{D_j} |S_j|\, dx \leqslant (2\pi)^{-1/2} m(D_j)^{1/2} \|S_j\|_2, \tag{13.5.8}$$

where $m(D_j)$ denotes the Lebesgue measure of D_j. Furthermore, by Parseval's formula, Hölder's inequality for sums, and (13.5.5),

$$\begin{aligned} \|S_j\|_2 &= \|\kappa_j\phi\|_2 \leqslant \|\kappa_j\|_s \cdot \|\phi\|_q \\ &\leqslant B\varepsilon_j{}^{-c}. \end{aligned} \tag{13.5.9}$$

Accordingly, (13.5.7), (13.5.8), and (13.5.9) together show that

$$\begin{aligned} |\psi * \phi(n)| &\leqslant \liminf_{j\to\infty} O(\varepsilon_j{}^c) \cdot m(D_j)^{1/2} \cdot B\varepsilon_j^{-c} \\ &= \liminf_{j\to\infty} O[m(D_j)^{1/2}]. \end{aligned} \tag{13.5.10}$$

Since E is closed, the intersection of the D_j is void; so countable additivity of Lebesgue measure guarantees that $m(D_j) \to 0$ as $j \to \infty$, and (13.5.4) follows from (13.5.10).

Remarks. (1) If one assumes E to be such that $m(D_j) \to 0$ as $j \to \infty$ with some preassigned degree of rapidity, one may correspondingly relax the condition (13.5.2).

In any case, in place of (13.5.2) one might assume that

$$\{\int_{D_j} |f|^2 \, dx\}^{1/2} = o(\varepsilon_j^{1/2 - 1/q}), \tag{13.5.11}$$

which in turn is satisfied if

$$\{\int_{D_j} |f|^p \, dx\}^{1/p} = O(\varepsilon_j^{1/2 - 1/q}) \tag{13.5.12}$$

for some $p > 2$.

(2) Somewhat similar procedures can be applied to general groups; see Herz [2], p. 210. See also MR **35** # 7081.

(3) In case $q = \infty$, the resulting exponent $\frac{1}{2}$ is (13.5.2) is best possible; see MR **40** # 629.

13.6 An Inequality of W. H. Young

We shall now apply 13.4.1 to the proof of a result about convolutions forecast in Remark (2) following 3.1.6.

13.6.1. (W. H. Young) Suppose that

$$1 \leqslant p \leqslant \infty, \quad 1 \leqslant q \leqslant \infty, \quad \frac{1}{r} = \frac{1}{p} + \frac{1}{q} - 1 \geqslant 0. \tag{13.6.1}$$

Then $f * g \in \mathbf{L}^r$ and

$$\|f * g\|_r \leqslant \|f\|_p \cdot \|g\|_q \tag{13.6.2}$$

whenever $f \in \mathbf{L}^p$ and $g \in \mathbf{L}^q$.

Proof. We take $(X, \mathscr{M}, \mu) = (Y, \mathscr{N}, \nu)$ as in 13.1.3(1). Fix $f \in \mathbf{L}^p$ and let T be the linear operator defined for all simple functions g by $Tg = f * g$. By 3.1.6, T is of type $(1, p)$ and

$$\|T\|_{1,p} \leqslant \|f\|_p.$$

By 3.1.4, T is of type (p', ∞) and

$$\|T\|_{p', \infty} \leqslant \|f\|_p.$$

An application of 13.4.1 shows that

$$\|T\|_{q,r} \leqslant \|f\|_p$$

whenever there exists a number t in $[0, 1]$ such that

$$\frac{1}{q} = \frac{1-t}{1} + \frac{t}{p'}, \qquad \frac{1}{r} = \frac{1-t}{p} + \frac{t}{\infty};$$

and these requirements are equivalent to (13.6.1).

The simple functions are everywhere dense in $\mathbf{L}^q$ and 13.2.2 shows that T can be continuously extended so as to map $\mathbf{L}^q$ into $\mathbf{L}^r$ in such a way that

$$\|Tf\|_r \leqslant \|f\|_p \cdot \|g\|_q.$$

Consequently it remains only to make sure that, for any $g \in \mathbf{L}^q$, the extension of T is such that Tg and $f * g$ agree almost everywhere. However, if we take a sequence $(g_n)_{n=1}^{\infty}$ of simple functions converging to g in $\mathbf{L}^q$, then $Tg = \lim_{n\to\infty} Tg_n$ in $\mathbf{L}^r$ and $Tg_n = f * g_n$ by the initial definition of T. On the other hand, (3.1.2) shows that $\lim_{n\to\infty} f * g_n = f * g$ in $\mathbf{L}^1$. The desired identification follows at once and the proof is complete.

13.6.2. There are more general results of a similar sort; see, for example, [E], Theorem 9.5.1, where a proof quite independent of the convexity theorem is given. The inequality extends to convolutions over quite general groups; for the case of the group Z, see Exercise 13.4. For extensions in a somewhat different direction, see O'Niel [1].

13.6.3. **Best possible constants.** Both 13.5.1(1) and 13.6.1 have analogues for the groups R^n. It was for long unknown what were the best possible values of the constants C_p and $C_{p,q,r}$ in the inequalities

$$\|\hat{f}\|_{p'} \leqslant C_p \|f\|_p$$

$$\|f * g\|_r \leqslant C_{p,q,r} \|f\|_p \cdot \|g\|_q$$

in the case of R^n. The answer was provided by Beckner (see MR **52** # # 6316, 6317), namely

$$C_p = (p^{1/p}(p')^{-1/p'})^{n/2}$$

$$C_{p,q,r} = C_p \cdot C_q \cdot C_{r'}.$$

Regarding a similar question in relation to (13.6.2), see MR **57** # 1021. See also MR **57** # # 10358, 10366.

13.7 Operators of Weak Type

The remainder of this chapter is concerned with the proof and applications of the second of the two interpolation theorems on our program, namely, the theorem of Marcinkiewicz. The present section sets out some definitions and concepts in terms of which Marcinkiewicz's theorem is customarily expressed, together with some simple preliminary results and some examples.

Some of the operators of weak type we shall wish to consider (the majorant operators σ^* and s^* are examples) are such that it is not a priori obvious that they transform finite-valued functions into functions which are either finite-valued or finite-valued almost everywhere. As a result of this, it is convenient to allow into the discussion extended real-valued functions. (The introduction of such functions is a convenience only and could be avoided in various ways and at the expense of enough circumlocution; on balance, it seems a good deal simpler to introduce them.) We begin by recalling very rapidly a few basic facts and conventions concerning such functions.

By definition (compare [HS], pp. 54–55), the *extended real number system* is the set $R^{\#} = R \cup \{-\infty, \infty\}$, obtained by adjoining to R two new elements $-\infty$ and ∞, and endowed with that linear order which extends the usual order on R and which makes $-\infty$ and ∞ the least and greatest elements, respectively, of $R^{\#}$. The supremum (respectively, infimum) of any subset of R which is unbounded above (respectively, below) relative to the usual order of R will thus be ∞ (respectively, $-\infty$). An *extended real-valued function* on a set X is simply a function on X with values in $R^{\#}$.

The manipulation of extended real numbers and extended real-valued functions will be governed by the following rules, in which a denotes an arbitrary real number:

$$\infty + a = a + \infty = \infty, \quad -\infty + a = a + (-\infty) = a - \infty = -\infty,$$

$$\infty + \infty = \infty, \quad (-\infty) + (-\infty) = -\infty,$$

$$-(\infty) = -\infty, \quad -(-\infty) = \infty,$$

$$\infty \cdot a = a \cdot \infty = \infty, \quad (-\infty) \cdot a = a \cdot (-\infty) = -\infty \qquad \text{if } a > 0,$$

$$\infty \cdot a = a \cdot \infty = -\infty, \quad (-\infty) \cdot a = a \cdot (-\infty) = \infty \qquad \text{if } a < 0,$$

$$\infty \cdot \infty = \infty, \quad \infty \cdot 0 = 0 \cdot \infty = (-\infty) \cdot 0 = 0 \cdot (-\infty) = 0,$$

$$|\infty| = |-\infty| = \infty, \quad \infty^a = \infty \qquad \text{if } a > 0.$$

We do not define the expressions $\infty + (-\infty)$, $(-\infty) + \infty$, $\infty \cdot (-\infty)$, $(-\infty) \cdot \infty$, $(-\infty) \cdot (-\infty)$.

In many cases where extended real-valued functions have to be handled, they are nonnegative-valued; in such cases the above conventions appear more intuitive than in the general situation.

Concerning the definition and properties of $\int_X f \, d\mu$ for extended real- and complex-valued μ-measurable functions f on X, see again [HS], Section 12. In particular, if f is an extended real-valued μ-measurable function on a measure space $(X, \mathcal{M}, \mu)$ (see [HS], p. 149), we define $\int_X |f|\, d\mu$ to be

$$\sup \sum_{k=1}^{n} \inf f(M_k)\mu(M_k),$$

the supremum being taken with respect to all partitions $(M_k)_{k=1}^{n}$ of X in which $M_k \in \mathcal{M}$ for all $k \in \{1, 2, \cdots, n\}$; see [HS], Section 12 for details. In terms of this convention, $\| f \|_{p, \mu}$ is defined to be

$$\{\int_X |f|^p \, d\mu\}^{1/p}$$

whenever $0 < p < \infty$. The definition of $\|f\|_{\infty,\mu}$ remains almost the same as when f is finite-valued: it is the smallest (extended real) number m such that $|f(x)| \leqslant m$ is true μ-almost everywhere on X.

It is also convenient to modify the definition of $\mathbf{L}^p(X, \mathcal{M}, \mu)$ given in 13.1.2 so as to include in $\mathbf{L}^p(X, \mathcal{M}, \mu)$ those μ-measurable extended real-valued functions f for which $\|f\|_{p,\mu} < \infty$. This enlargement is in reality rather trifling insofar as the relation $\|f\|_{p,\mu} < \infty$ entails that $|f(x)| < \infty$ for μ-almost all $x \in X$ (compare [HS], pp. 154, 169–170) and, as far as integration with respect to μ is concerned, f can therefore be replaced by a function which is everywhere finite-valued.

We now turn from generalities to particularities. In what follows the term "function" means "extended real- or complex-valued function."

13.7.1. **Truncation of Functions.** Suppose that f is a function on a set X and that $a > 0$ is a real number. The *a-truncation* $f_{1,a}$ of f is the function on X defined by the specification:

$$f_{1,a}(x) = \begin{cases} f(x) & \text{if } |f(x)| \leqslant a, \\ af(x)/|f(x)| & \text{otherwise;} \end{cases}$$

here we make the special convention that $\pm\infty/\infty = \pm 1$. Thus

$$|f_{1,a}| = \min(|f|, a).$$

We define further

$$f_{2,a} = f - f_{1,a}.$$

Notice that $f_{1,a}$ is always finite-valued, and that the relation

$$f = f_{1,a} + f_{2,a}$$

is universally valid.

13.7.2. **Distribution Functions.** Suppose that $(X, \mathcal{M}, \mu)$ is a measure space and that f is a μ-measurable function on X. The *distribution function* D_f or D_f^{μ} is the nonnegative extended real-valued function on $(0, \infty)$ defined as follows:

$$D_f^{\mu}(t) = \mu(\{x \in X : |f(x)| > t\}). \tag{13.7.1}$$

(This use of the term "distribution" has nothing whatsoever to do with that in Chapter 12.) The reader will observe that $D_f{}^\mu = D_{|f|}{}^\mu$, that $D_f{}^\mu$ depends only on the μ-equivalence class of f, and that $D_f{}^\mu(t) \leqslant \mu(X)$.

We shall often write

$$E_f(t) = \{x \in X : |f(x)| > t\},$$

so that

$$D_f{}^\mu(t) = \mu(E_f(t)).$$

It is not difficult to see that $D_f{}^\mu$ is decreasing (in the wide sense) and continuous on the right, that is,

$$D_f{}^\mu(t) = D_f{}^\mu(t+0) \equiv \lim_{t' \downarrow t} D_f{}^\mu(t')$$

for $0 < t < \infty$. Monotonicity ensures that $D_f{}^\mu$ is Lebesgue-measurable on $(0, \infty)$.

If f is a μ-measurable function on X and $a > 0$, the following relations hold:

$$\left.\begin{aligned} D^\mu_{f_{1,a}}(t) &= D^\mu_f(t) && \text{if } 0 < t < a, \\ D^\mu_{f_{1,a}}(t) &= 0 && \text{if } \quad t \geqslant a, \\ D^\mu_{f_{2,a}}(t) &= D^\mu_f(t+a) && \text{if } \quad t > 0. \end{aligned}\right\} \tag{13.7.2}$$

The verification is left to the reader.

13.7.3. **Integrals in Terms of Distribution Functions.** For us the main significance of the distribution function stems from the fact that it enables us to express the integral of a nonnegative (extended real-valued) μ-measurable function on X in terms of a Lebesgue integral over $(0, \infty)$. The appropriate formula reads

$$\int_X f^p \, d\mu = \int_0^\infty pt^{p-1} D_f{}^\mu(t) \, dt, \tag{13.7.3}$$

where $0 < p < \infty$, it being understood that $\infty^p = \infty$. For a proof in the case where f is real-valued, see [HS], Corollary (21.72), where (13.7.3) is shown to be an almost immediate consequence of an appropriate form of the Fubini theorem. The general case may be deduced from this by considering the real-valued functions $f_n = \inf(f, n)$, noting that $D_{f_n}{}^\mu(t) \uparrow D_f{}^\mu(t)$, and using the theorem on termwise integration of monotone-increasing sequences of measurable functions [[HS], Theorem (12.22)].

The remarks in 13.7.2 show that (13.7.3) holds with $|f|^p$ in place of f^p whenever f is any (complex- or extended real-valued) μ-measurable function f on X.

We remark in passing that (13.7.3) could be used quite effectively to *define* integrals with respect to μ.

13.7.4. **Quasilinear Operators.** Suppose that $(X, \mathscr{M}, \mu)$ and $(Y, \mathscr{N}, \nu)$ are measure spaces. Suppose further that T is an operator whose domain $\mathfrak{D}$ is a set of μ-measurable complex-valued functions on X. The range of T is assumed to lie in the set of ν-measurable complex-valued functions on Y, or in the set of ν-measurable nonnegative extended real-valued functions on Y, or in the corresponding sets of ν-equivalence classes of functions (see 13.1.2 and 13.2.1). (Quite often the operators T we need to consider can be assigned a domain whose elements are μ-equivalence classes of functions, but to assume this is an unnecessary luxury.)

The operator T is said to be *quasilinear* (with constant κ) if $f_1 + f_2 \in \mathfrak{D}$ and

$$|T(f_1 + f_2)| \leqslant \kappa(|Tf_1| + |Tf_2|) \tag{13.7.4}$$

for ν-almost all points of Y whenever $f_1, f_2 \in \mathfrak{D}$. (In case Tf_1 and Tf_2 are ν-equivalence classes, $|Tf_1| + |Tf_2|$ is of course the ν-equivalence class of any function $g_1 + g_2$ where g_k $(k = 1, 2)$ is a function chosen from the equivalence class $|Tf_k|$; and (13.7.4) is then understood to signify that the same inequality holds at ν-almost all points between functions chosen from the appropriate ν-equivalence classes.)

For example, any linear operator T whose domain and range are as specified above is quasilinear with a constant $\kappa \leqslant 1$.

Again, if $(X, \mathscr{M}, \mu) = (Y, \mathscr{N}, \nu)$ are as in Example 13.1.3(1), the majorant operators σ^* and s^* defined in (6.4.10) and (6.4.14), respectively, have domain $\mathbf{L}^1 = \mathbf{L}^1(X, \mathscr{M}, \mu)$ and range in the set of nonnegative extended real-valued ν-measurable functions on Y. It is very simple to verify that they are each quasilinear with constant $\kappa \leqslant 1$. They are not linear operators.

Other illustrations appear in Example 13.7.6(2).

13.7.5. **Operators of Weak Type.** Suppose again that $(X, \mathscr{M}, \mu)$ and $(Y, \mathscr{N}, \nu)$ are measure spaces and that p and q are exponents chosen from the interval $[1, \infty]$. Suppose also that T is an operator whose domain $\mathfrak{D}$ and range are as described in 13.7.4; T may or may not be quasilinear, however.

If $q < \infty$, the operator T is said to be of *weak type* (p, q) on $\mathfrak{D}$ if there exists a number $A \geqslant 0$ such that

$$\nu(E_{Tf}(t)) \leqslant \left(\frac{A\|f\|_{p,\mu}}{t}\right)^q \tag{13.7.5}$$

for all $f \in \mathfrak{D}$ and all $t > 0$, where (compare 13.7.2)

$$E_{Tf}(t) = \{y \in Y : |Tf(y)| > t\}.$$

To cover the excluded case in which $q = \infty$, it is agreed that T will be said to be of *weak type* (p, ∞) on $\mathfrak{D}$ if and only if it satisfies a (strong) type (p, ∞) inequality (see Section 13.2) on its domain of definition, that is, if and only if there exists a number $A \geqslant 0$ such that

$$\|Tf\|_{\infty,\nu} \leqslant A\|f\|_{p,\mu} \tag{13.7.6}$$

for all $f \in \mathfrak{D}$.

In (13.7.5) and (13.7.6), $\|f\|_p$ is defined as in 13.1.2, so that $\|f\|_p$ may be ∞; in this case $\infty^a = \infty$ for any $a > 0$. (See the conventions listed at the beginning of this section.)

In either case the smallest admissible value of A is termed the *weak* (p, q)-*norm* of T, despite the fact that (apart from the case in which $q = \infty$ and all due precautions are taken over the addition of operators) the function of T so defined has not the properties of a norm (see I, B.1.2). This abuse of the term "norm" appears to be customary in this connection.

The inequality (13.7.5) may be written

$$D_{Tf}{}^{\nu}(t) \leqslant \left(\frac{A\|f\|_{p,\mu}}{t}\right)^q. \tag{13.7.7}$$

Most frequently the hypothesis that T is of weak type (p, q) $(q < \infty)$ is brought into play by employing (13.7.7) in conjunction with (13.7.3) applied with (Tf, q, ν) in place of (f, p, μ); see the proof of 13.8.1.

It is clear that the above definitions of the concept of weak type (p, q) can be formulated if either or both of p and q lie in the interval $(0, 1)$, but this extension appears to be of relatively little interest.

13.7.6. **Examples of Operators of Weak Type.** (1) If T is of type (p, q) on a domain $\mathfrak{D}$, then it is of weak type (p, q) on $\mathfrak{D}$.

This is a matter of definition if $q = \infty$. If $q < \infty$, one has by hypothesis for each $f \in \mathfrak{D}$:

$$\|Tf\|_{q,\nu} \equiv \{\int_Y |Tf|^q \, d\nu\}^{1/q} \leqslant \|T\|_{p,q}\|f\|_{p,\mu},$$

confirming that for any $t > 0$

$$\{t^q \cdot D_{Tf}{}^{\nu}(t)\}^{1/q} \leqslant \|T\|_{p,q}\|f\|_{p,\mu}$$

and therefore

$$D_{Tf}{}^{\nu}(t) \leqslant \frac{\|T\|_{p,q}^q\|f\|_{p,\mu}^q}{t^q},$$

which shows that T is of weak type (p, q) on $\mathfrak{D}$ and that the weak (p, q)-norm of T does not exceed the (p, q)-norm $\|T\|_{p,q}$.

Thus every operator of type (p, q) is ipso facto an example of an operator of weak type (p, q); see 13.2.1, Section 13.6, and Exercises 13.5 and 13.6. The converse is false, even for operators defined and linear on certain subspaces of $\mathbf{L}^p(\mu)$ and having ranges in $\mathbf{L}^q(\nu)$; but see Exercise 13.16.

We shall in Sections 13.9 and 13.10 see that the conjugate function operator $f \to \tilde{f}$ and the majorant operator $f \to \sigma^* f$ are of weak type $(1, 1)$ on suitably chosen domains, even though they are not of type (1, 1) on those domains. Another such example will appear in Section 13.11. The theorem of Kolmogorov cited in 10.3.4 shows that the majorant operator $f \to s^*f$ is not of weak type $(1, q)$ on $\mathbf{L}^1$ for any $q > 0$.

In view of Carleson's result cited in 10.4.5(3), together with Stein's theorem 16.2.8, the operator s^* is of weak type (2, 2) on $\mathbf{L}^2$.

(2) *The Hardy-Littlewood maximal operator.* This operator is perhaps the forerunner of all others as an example of an operator of weak type (1, 1).

Here we suppose that $(X, \mathscr{M}, \mu) = (Y, \mathscr{N}, \nu)$, X denoting the real axis R, $\mathscr{M}$ the set of Lebesgue-measurable subsets of R, and μ the Lebesgue measure on R.

For any measurable function f that is integrable over $(-a, a)$ for every $a > 0$, define

$$f^r(x) = \sup_{s>0} s^{-1} \int_x^{x+s} |f(y)|\, dy,$$

$$f^\ell(x) = \sup_{s>0} s^{-1} \int_{x-s}^{x} |f(y)|\, dy.$$

It can be shown ([HS], Lemma (21.75)) that if we write for $j = \ell, r$ and $t > 0$

$$E_t^j = \{x \in R : f^j(x) > t\},$$

then

$$\nu(E_t^j) \leqslant t^{-1} \int_{E_t^j} |f|\, d\mu. \tag{13.7.8}$$

The *maximal operator* $M^j: f \to f^j$ is clearly quasilinear, and (13.7.8) evidently entails that M^j is of weak type (1, 1) on $\mathbf{L}^1(\mu)$.

However, consideration of the function

$$f(x) = \begin{cases} x^{-1}(\log x)^{-2} & \text{for } 0 < x < \tfrac{1}{2} \\ 0 & \text{elsewhere}, \end{cases}$$

for which $f^\ell(x) \geqslant x^{-1}|\log x|^{-1}$ for small $x > 0$, shows that $f \to f^\ell$ is not of type (1, 1) on $\mathbf{L}^1(\mu)$. Similarly, $f \to f^r$ is not of type (1, 1) on $\mathbf{L}^1(\mu)$.

Similar statements apply to the maximal operator

$$M^\Delta : f \to f^\Delta \equiv \sup(f^\ell, f^r).$$

It is shown in [HS], Theorem (21.76) that M^j $(j = \ell, r, \Delta)$ is of type (p, p) on $\mathbf{L}^p(\mu)$ for $1 < p < \infty$. This, together with certain other inequalities established in [HS], Theorem (21.80), will follow, on the basis of Marcinkiewicz's theorem 13.8.1 and the results in 13.8.3, from the statement that each of these operators is of weak type (1, 1) on $\mathbf{L}^1(\mu)$ and (what is quite evident) that each is also of type (∞, ∞) on $\mathbf{L}^\infty(\mu)$. The proofs in [HS] are, on the contrary, direct and lead to specific estimates for the constants involved in the inequalities. See also [Kz], pp. 74–76.

The keystone in all these results is the assertion that M^j is of weak type (1, 1) on $\mathbf{L}^1(\mu)$. It is therefore interesting to note that this is itself a consequence of the general theorem of Stein recorded in 16.2.8 and the Lebesgue theorem on the differentiability almost everywhere of an indefinite integral which is quoted in 6.4.2; see Exercise 16.14.

The Hardy-Littlewood maximal operators have been defined and studied in connection with more general measure spaces; see, for example, Smith [1], Shimogaki [1], Rauch [1] and MR **35** # 6788. There will be no place in this book for these extensions; but see also 13.10.3 below.

13.7.7. ***Remark.*** There is an analogue of 13.2.2 applying to linear operators of weak type (p, q), but which appears to be of somewhat peripheral interest; see Exercise 13.19.

13.8 The Marcinkiewicz Interpolation Theorem

In the Riesz-Thorin convexity theorem 13.4.1, we have already encountered one result that enables us to "interpolate" properties of a functional operator T on the basis of given "extreme" properties of T. There, continuity of type (p, q) of T was deduced, for certain pairs (p, q) intermediate to two pairs (p_j, q_j) $(j = 0, 1)$, from continuity of types (p_j, q_j) $(j = 0, 1)$ of T. It is now time to prove another famous interpolation theorem, which asserts that the same conclusion applies to somewhat different intermediate pairs on the basis of weak-type continuity for the pairs (p_j, q_j), the latter being subject to special relations that are not needed in the Riesz-Thorin theorem but which are essential for this second theorem (compare 13.8.1 and 13.4.1, where no such conditions appear).

13.8.1. **Marcinkiewicz' Interpolation Theorem.** Suppose that $(X, \mathscr{M}, \mu)$ and $(Y, \mathscr{N}, \nu)$ are measure spaces and that (p_0, q_0) and (p_1, q_1) are two exponent pairs having the following properties:

$$\left.\begin{aligned} 1 \leqslant p_j \leqslant q_j \leqslant \infty \qquad (j = 0, 1), \\ q_0 \neq q_1. \end{aligned}\right\} \tag{13.8.1}$$

Let T be a quasilinear operator [with constant κ; see (13.7.4)], whose domain $\mathfrak{D}$ is a linear subspace of $\mathbf{L}^{p_0}(X, \mathscr{M}, \mu) \cap \mathbf{L}^{p_1}(X, \mathscr{M}, \mu)$ which is closed under the formation of truncations (see 13.7.1), and whose range is as described in 13.7.4. Suppose further that T is simultaneously of weak types (p_0, q_0) and (p_1, q_1) with weak norms A_0 and A_1, respectively. Suppose finally that

$$\frac{1}{p} = \frac{t}{p_0} + \frac{(1-t)}{p_1}, \qquad \frac{1}{q} = \frac{t}{q_0} + \frac{(1-t)}{q_1}, \qquad 0 < t < 1. \tag{13.8.2}$$

The conclusion is that T is of (strong) type (p, q) on $\mathfrak{D}$, that is, that there exists a number $A = A(t, p_0, q_0, p_1, q_1, A_0, A_1)$ such that

$$\|Tf\|_{q,\nu} \leqslant A\|f\|_{p,\mu} \tag{13.8.3}$$

for $f \in \mathfrak{D}$.

Proof. There are several cases to be considered separately. Throughout the proof the symbols $A_2, A_3, A_4, \cdots$ are used to denote nonnegative numbers depending at most on T, μ, ν, t, p_0, q_0, p_1, q_1, A_0, and A_1.

(1) We shall begin with the case in which q_0 and q_1 are finite. Of this there is one subcase which can be dismissed rather quickly, namely, the case in which $p_0 = p_1 = p$.

In this subcase we may suppose without loss of generality that $q_0 < q < q_1$. If $a > 0$, (13.7.3) gives for $f \in \mathfrak{D}$

$$\|Tf\|_{q,\nu}^q = q\int_0^\infty b^{q-1}D_{Tf}^{\nu}(b)\,db$$
$$= q\left\{\int_0^a + \int_a^\infty\right\}.$$

In the first integral we majorize $D_{Tf}^{\nu}(b)$ by using (13.7.5) with p and q replaced by p_0 and q_0, respectively; the second integral is treated likewise, using p_1 and q_1 in place of p and q. As a result it appears that

$$\|Tf\|_{q,\nu}^q \leqslant \text{const}\,\{a^{q-q_0}\|f\|_{p,\mu}^{q_0} + a^{q-q_1}\|f\|_{p,\mu}^{q_1}\},$$

since $q_0 < q < q_1$ by hypothesis. If $\|f\|_{p,\mu} > 0$, and if we choose $a = \|f\|_{p,\mu}$, the desired result (13.8.3) appears; and if $\|f\|_{p,\mu} = 0$, Tf is null and (13.8.3) is trivial.

Having disposed of this subcase, we suppose henceforth that $p_0 \neq p_1$ and that in fact $p_0 < p < p_1$. (The proof to follow breaks down if $p_0 = p_1$, so the subcase just examined demands separate treatment.) Assume also to begin with that $q_0 < q_1$.

Suppose that $f \in \mathfrak{T}$. The substance of 13.7.1 and (13.7.4), coupled with the hypotheses in 13.8.1, show that for any $b > 0$

$$E_{Tf}(b) \subset E_{Tf_{1,a}}(\tfrac{1}{2}\kappa^{-1}b) \cup E_{Tf_{2,a}}(\tfrac{1}{2}\kappa^{-1}b)$$

and therefore

$$D_{Tf}^{\nu}(b) \leqslant D_{Tf_{1,a}}^{\nu}(\tfrac{1}{2}\kappa^{-1}b) + D_{Tf_{2,a}}^{\nu}(\tfrac{1}{2}\kappa^{-1}b). \tag{13.8.4}$$

It must be stressed that in (13.8.4) one is at liberty to choose a and b as positive-valued functions each of the other. This freedom will play a vital role in the subsequent proof.

Since T is of weak types (p_0, q_0) and (p_1, q_1), (13.8.4) shows that

$$D_{Tf}^{\nu}(b) \leqslant A_2(b^{-q_1}\|f_{1,a}\|_{p_1,\mu}^{q_1} + b^{-q_0}\|f_{2,a}\|_{p_0,\mu}^{q_0}).$$

The reader should experience no trouble in verifying that, if a be chosen to be a positive monotone function $a(b)$ of $b \in (0, \infty)$, then $\|f_{k,a}\|_{p_j}$, $(j = 0, 1;$ $k = 1, 2)$ is a monotone, and therefore measurable, function of $b \in (0, \infty)$.

So an application of (13.7.2) and (13.7.3) leads to

$$\begin{aligned}\|Tf\|_{q,\nu}^{q} &= q\int_0^\infty b^{q-1}D_{Tf}^{\nu}(b)\,db \\ &\leqslant qA_2\int_0^\infty \{b^{q-q_1-1}\|f_{1,a(b)}\|_{p_1,\mu}^{q_1} + b^{q-q_0-1}\|f_{2,a(b)}\|_{p_0,\mu}^{q_0}\}\,db \\ &\leqslant A_3\{\int_0^\infty b^{q-q_1-1}[\int_0^{a(b)} D_f^{\mu}(c)c^{p_1-1}\,dc]^{q_1/p_1}\,db \\ &\quad + \int_0^\infty b^{q-q_0-1}[\int_{a(b)}^\infty D_f^{\mu}(c)(c-a)^{p_0-1}\,dc]^{q_0/p_0}\,db\} \\ &\equiv A_3(I_1+I_2). \end{aligned} \tag{13.8.5}$$

The crucial step is to choose a as a function $a(b)$ of $b \in (0, \infty)$ with properties to be described as we go along. From this point on, it is to be assumed that $a(b)$ is a positive strictly monotone function of $b \in (0, \infty)$, the inverse of which is denoted by $b = b(a)$.

We proceed to estimate I_1 and I_2, taking first the case in which $q_0 < q_1$ and assuming that $a(b)$ is strictly increasing.

In considering I_1, it will be convenient to introduce the measure λ on $R_+ = (0, \infty)$ defined by

$$\lambda(E) = \int_E b^{q-q_1-1}\,db$$

for Lebesgue-measurable subsets E of R_+. Then $I_1^{p_1/q_1}$ is the $\mathbf{L}^{q_1/p_1}(R_+, \lambda)$-norm of the function

$$b \to \int_0^{a(b)} D_f^{\mu}(c)c^{p_1-1}\,dc,$$

which is λ-integrable over every λ-finite measurable set; recall that $q_1/p_1 \geqslant 1$. So, in view of 13.1.5, $I_1^{p_1/q_1}$ is equal to

$$\sup\left|\int_0^\infty b^{q-q_1-1}g(b)\{\int_0^{a(b)} D_f^{\mu}(c)c^{p_1-1}\,dc\}\,db\right| \equiv \sup|J|,$$

the supremum being taken relative to all λ-simple functions g satisfying

$$\int_0^\infty |g(b)|^{(q_1/p_1)'}b^{q-q_1-1}\,db \leqslant 1.\text{[1]} \tag{13.8.6}$$

It is moreover clear that in this case it suffices to take the supremum relative to those functions g of the prescribed type which are nonnegative, for which functions g it is evident that $J \geqslant 0$.

Now

$$J = \int_0^\infty\{\int_0^\infty b^{q-q_1-1}g(b)\chi_{(0,a(b))}(c)D_f^{\mu}(c)c^{p_1-1}\,dc\}\,db$$

[1] We assume that $q_1/p_1 > 1$; otherwise, (13.8.6) and the following derivation of (13.8.7) need slight and obvious modifications.

and the Fubini theorem (in the form in which it appears in [HS], Theorem (21.12)), together with the fact that $a(b)$ is strictly increasing, shows that

$$\begin{aligned} J &= \int_0^\infty \{\int_0^\infty b^{q-q_1-1} g(b)\chi_{(b(c),\infty)}(b) D_f{}^\mu(c) c^{p_1-1}\, db\}\, dc \\ &= \int_0^\infty c^{p_1-1} D_f{}^\mu(c)\{\int_{b(c)}^\infty b^{q-q_1-1} g(b)\, db\}\, dc \\ &\leqslant \int_0^\infty c^{p_1-1} D_f{}^\mu(c)\{[\int_{b(c)}^\infty b^{q-q_1-1}\, db]^{p_1/q_1}[\int_{b(c)}^\infty g(b)^{(q_1/p_1)'} b^{q-q_1-1}\, db]^{1/(q_1/p_1)'}\}\, dc \\ &\leqslant \int_0^\infty c^{p_1-1}\, D_f{}^\mu(c)\{\int_{b(c)}^\infty b^{q-q_1-1}\, db\}^{p_1/q_1}\, dc, \end{aligned} \tag{13.8.7}$$

the penultimate step being an application of Hölder's inequality for integrals with respect to the measure λ restricted to $(b(c), \infty)$, and the last step using (13.8.6). Since $q_0 < q_1$, $q < q_1$ and

$$\{\int_{b(c)}^\infty b^{q-q_1-1}\, db\}^{p_1/q_1} = b(c)^{(q-q_1)p_1/q_1}(q_1 - q)^{-p_1/q_1}.$$

Thus (13.8.7) and the relation $I_1^{p_1/q_1} = \sup J$ show that

$$I_1 \leqslant A_4\{\int_0^\infty c^{p_1-1} D_f{}^\mu(c) b(c)^{(q-q_1)p_1/q_1}\, dc\}^{q_1/p_1}. \tag{13.8.8}$$

Turning to the consideration of I_2, it is to be observed first that, since $(c - a)^{p_0-1} \leqslant c^{p_0-1}$,

$$I_2 \leqslant I_2' \equiv \int_0^\infty b^{q-q_0-1}\{\int_{a(b)}^\infty c^{p_0-1} D_f{}^\mu(c)\, dc\}^{q_0/p_0}\, db.$$

The method used above to majorize $I_1^{p_1/q_1}$ can now be adapted so as to majorize $I_2'^{p_0/q_0}$, which, by 13.1.5, is equal to

$$\sup \int_0^\infty b^{q-q_0-1} h(b)\{\int_{a(b)}^\infty c^{p_0-1} D_f{}^\mu(c)\, dc\}\, db,$$

the supremum being taken relative to a suitable set of nonnegative measurable functions h on $(0, \infty)$ satisfying (compare the footnote to page 159)

$$\int_0^\infty h(b)^{(q_0/p_0)'} b^{q-q_0-1}\, db \leqslant 1.$$

In the course of the argument, use is made of the fact that

$$\chi_{(a(b),\infty)}(c) = \chi_{(0,b(c))}(b)$$

and of the relation $q > q_0$. The outcome of this procedure is the estimate

$$I_2 \leqslant I_2' \leqslant A_5\{\int_0^\infty c^{p_0-1} D_f{}^\mu(c) b(c)^{(q-q_0)p_0/q_0}\, dc\}^{q_0/p_0}. \tag{13.8.9}$$

From (13.8.5), (13.8.8), and (13.8.9) it follows that

$$\|Tf\|_{q,\nu}^{q} \leqslant A_6[\{\int_0^\infty c^{p_1-1}D_f^{\mu}(c)b(c)^{(q-q_1)p_1/q_1}\,dc\}^{q_1/p_1}$$

$$+\{\int_0^\infty c^{p_0-1}D_f^{\mu}(c)b(c)^{(q-q_0)p_0/q_0}\,dc\}^{q_0/p_0}]. \qquad (13.8.10)$$

The aim being to obtain integrands on the right-hand side of (13.8.10) which (compare (13.7.3)) are each of the form const $c^{p-1}D_f^{\mu}(c)$, we write tentatively

$$b(a) = (K^{-1}a)^{\rho},$$

where it is hoped to choose $K > 0$ and $\rho > 0$ so as to achieve this aim. The restrictions $K > 0$ and $\rho > 0$ will ensure that $b(a)$ is a positive and strictly monotone function of a, as has been deposed earlier.

If success is to be achieved at all, the only possible choice of ρ is already discernible on looking at the first integrand in (13.8.10) and is that for which $p_1 - 1 + \rho(q - q_1)p_1/q_1 = p - 1$, that is, that for which

$$\rho(q/q_1 - 1) = p/p_1 - 1. \qquad (13.8.11)$$

Since $q/q_1 < 1$ and $p/p_1 < 1$, this choice renders $\rho > 0$. Direct calculations, using (13.8.2), show that this choice of ρ arranges that

$$p_0 - 1 + \rho(q - q_0)p_0/q_0 = p - 1,$$

thereby taking care of the second integrand in (13.8.10). Consequently, (13.8.10) becomes

$$\|Tf\|_{q,\nu}^{q} \leqslant A_7[K^{q_1(p_1-p)/p_1}\|f\|_{p,\mu}^{q_1p/p_1}$$

$$+ K^{q_0(p_0-p)/p_0}\|f\|_{p,\mu}^{q_0p/p_0}]. \qquad (13.8.12)$$

The quantity K is now to be chosen so as to achieve the desired qth power of $\|f\|_{p,\mu}$ on the right-hand side of (13.8.12). Assuming, as we may, that $\|f\|_{p,\mu} > 0$, this object is achieved by taking

$$K = \|f\|_{p,\mu}^{\beta},$$

where β is determined by the relation

$$(1 - p/p_1)\,\beta = \frac{q}{q_1} - \frac{p}{p_1}.$$

Then, indeed, straightforward calculations show that (13.8.12) reduces to

$$\|Tf\|_{q,\nu}^{q} \leqslant 2A_7\|f\|_{p,\mu}^{q},$$

and our goal is achieved provided $q_0 < q_1$.

Let us now consider what happens when $q_0 > q_1$. A careful examination of the preceding proof shows that it is only in the step from (13.8.7) to (13.8.8), and at the corresponding point in the estimate of I_2, that the assumption $q_0 < q_1$ is essential. To counter this difficulty we observe that if, when

$q_0 > q_1$, we assume that $a(b)$ is strictly *decreasing* (instead of increasing), then the integral $\int_{b(c)}^{\infty}$ in (13.8.7) must (providentially) be replaced by $\int_0^{b(c)}$. As a consequence, (13.8.8) remains intact. A similar escape takes place at the corresponding point in the estimation of I_2. Thus (13.8.10) stands unblemished. The proof then closes much as before: this time we require that $\rho < 0$, and this is ensured by (13.8.11) since now $q/q_1 > 1$ and $p/p_1 < 1$. The reader is urged strongly to write out the details of what has just been sketched in brief; see Exercise 13.10.

This completes the proof in case (1).

(2) The case in which $p_0 = p_1 = p$, $q_0 < \infty$, $q_1 = \infty$: for this, see Exercise 13.11.

(3) The case in which $1 \leqslant p_0 \leqslant q_0 < \infty$, $p_1 = q_1 = \infty$: for this, see Exercise 13.12.

(4) The case in which $1 \leqslant p_0 \leqslant q_0 < \infty$, $p_0 < p_1 < \infty$, $q_1 = \infty$.

Applying (13.8.4) and making a change of variable, we shall have, if $a = a(b)$ is any positive and strictly monotone function of $b \in (0, \infty)$,

$$\|Tf\|_{q,\nu}^q \leqslant A_8\{\int_0^{\infty} b^{q-1} D^{\nu}_{Tf_{1,a(b)}}(b)\, db + \int_0^{\infty} b^{q-1} D^{\nu}_{Tf_{2,a(b)}}(b)\, db\}, \quad (13.8.13)$$

On the other hand, by the weak type hypotheses on T,

$$\|Tf_{1,a(b)\infty,\nu} \leqslant A_1\|f_{1,a(b)}\|_{p_1,\mu}.$$

In view of (13.7.2), (13.7.3), and the fact that $p_1 > p$, we have

$$\begin{aligned}\|Tf_{1,a(b)}\|_{\infty,\nu} &\leqslant A_1\{\int_0^{a(b)} p_1 t^{p_1-1} D_f^{\mu}(t)\, dt\}^{1/p_1}\\ &\leqslant A_9 a(b)^{1-p/p_1}\{\int_0^{\infty} pt^{p-1} D_f^{\mu}(t)\, dt\}^{1/p_1}.\end{aligned}$$

So, by (13.7.3) once again,

$$\|Tf_{1,a(b)}\|_{\infty,\nu} \leqslant A_9 a(b)^{1-p/p_1}\|f\|_{p,\mu}^{p/p_1} \quad (13.8.14)$$

Our hope is to choose a and b as positive and strictly monotone functions each of the other in such a way that

$$A_9 a(b)^{1-p/p_1}\|f\|_{p,\mu}^{p/p_1} \leqslant b. \quad (13.8.15)$$

For, if this is possible, (13.8.14) shows that $D^{\nu}_{Tf_{1,a(b)}}(b)$ will vanish for all $b > 0$ and the first term on the right-hand side of (13.8.13) will also vanish.

If, at the same time as satisfying (13.8.15), we can choose $a(b)$ to be strictly increasing, the estimate of I_2 in (1) (for the case in which $p_0 \neq p_1$) will proceed exactly as before and will yield, in place of (13.8.10), the inequality

$$\|Tf\|_{q,\nu}^q \leqslant A_{10}\{\int_0^{\infty} c^{p_0-1} D_f^{\mu}(c) b(c)^{(q-q_0)p_0/q_0}\, dc\}^{q_0/p_0}. \quad (13.8.16)$$

Since, as before, we wish to obtain an integrand of the form const $c^{p-1}D_f^{\mu}(c)$, we again choose tentatively

$$b(a) = (K^{-1}a)^{\rho},$$

and hope to dispose of $K > 0$ and $\rho > 0$ advantageously.

The only possible successful choice of ρ is given by

$$p_0 - 1 + \frac{\rho(q - q_0)p_0}{q_0} = p - 1,$$

that is, by

$$\rho(q/q_0 - 1) = p/p_0 - 1,$$

which, in view of (13.8.2) and the supposition that $q_1 = \infty$, gives

$$\rho = 1 - p/p_1$$

This choice of ρ does indeed make $\rho > 0$; and direct calculations show that it leads from (13.8.16) to

$$\|Tf\|_{q,\nu}^{q} \leqslant A_{11}K^{(p_0 - p)q_0/p_0}\|f\|_{p,\mu}^{q_0 p/p_0}. \tag{13.8.17}$$

In order to satisfy (13.8.15) and to obtain $\|f\|_{q,\mu}^{q}$ on the right-hand side of (13.8.17), we write $K = K_1K_2$, where

$$K_1 = \|f\|_{p,\mu}^{\beta}$$

and β is to be chosen to obtain $\|f\|_{p,\mu}^{q}$ on the right-hand side of (13.8.17), and where $K_2 > 0$ is to be chosen (if possible) so as to accommodate (13.8.15).

For β the only possible choice is such that

$$\frac{\beta(p_0 - p)q_0}{p_0} + \frac{q_0 p}{p_0} = q,$$

which signifies that $\rho(1 - \beta) = 1$. Then, since $\rho = 1 - p/p_1$ and $\beta\rho = -p/p_1$, (13.8.15) is seen to be accommodated by choosing $K_2 > 0$ so that

$$A_9 \leqslant K_2^{p/p_1 - 1},$$

a choice that is possible since $p/p_1 - 1 \neq 0$.

This completes the discussion of case (4).

(5) The final case, in which $1 \leqslant p_1 \leqslant q_1 < \infty$, $p_0 < p_1 < \infty$ and $q_0 = \infty$, is left for the reader's attention in Exercise 13.13.

13.8.2. **Remarks.** (1) It will be clear to the reader that the restrictions $p_j \leqslant q_j$ $(j = 0, 1)$, $q_0 \neq q_1$ are necessary in the preceding proof of 13.8.1; it will further be apparent that one cannot by modifying the proof dispose of the condition $q_0 \neq q_1$ (since otherwise the case $p_0 = p_1$, $q_0 = q_1$ of 13.8.1 would imply that any quasilinear operator of weak type (p, q) is also of strong type (p, q), which is shown to be false in Example 13.7.6(2), and also by 12.8.3(2) in conjunction with 13.9.1).

A more important question is whether one can suppose that $p_j > q_j$ $(j = 0, 1)$ and still obtain a valid theorem. That the answer is negative has been shown by Richard A. Hunt [1].

(2) An outline of a proof of 13.8.1 for the case in which q_0 and q_1 are finite has been given by Hunt and Weiss [2]. Calderón (lecture notes) and Hunt [1] have also extended the theorem to the case of an operator T acting between pairs of Lorentz spaces (which generalize the $\mathbf{L}^p$-spaces). See also Oklander [1] and Stampacchia [1].

(3) In the preceding proof, which is essentially an expanded and annotated version of that appearing on pp. 112–115 of [Z_2], no attention has been paid to the precise form of the functional dependence of A on the p's, q's, A_0, A_1, and t. There is no great difficulty in gaining some precision, though the question of maximum precision is another matter. The interested reader may seek an estimate as an exercise, and also refer to [Z_2], p. 114, where it is shown that, if $p_0 \geqslant p_1$ and q_0 and q_1 are finite, then

$$\|T\|_{p,q} \leqslant KA_0^tA_1^{1-t},$$

where

$$K^q = (2\kappa)^q q\left\{\frac{(p_0/p)^{q_0/p_0}}{|q - q_0|} + \frac{(p_1/p)^{q_1/p_1}}{|q - q_1|}\right\}.$$

It is important to note that, if T is a quasilinear operator of *strong* types (p_j, q_j) $(j = 0, 1)$, such estimates of $A = \|T\|_{p,q}$ as *do* result from a simple-minded examination of the preceding proof of 13.8.1 *do*, in some cases, tend to infinity as $t \downarrow 0$ or as $t \uparrow 1$. The appropriate version of 13.4.1 shows, however, that the *best* estimate of $\|T\|_{p,q}$ remains bounded. This shows that simple-mindedness is not enough, and that 13.8.1 is not a complete substitute for 13.4.1. For further comments in this vein, see [Z_2], pp. 115–116.

(4) Although some applications of the Marcinkiewicz theorem to harmonic analyis appear later in this chapter, the reader may wish to consult also G. Weiss' article [1]. Concerning the role of the same theorem in the study of singular integral equations, see Calderón [2].

A more restricted version of the Marcinkiewicz theorem, together with some of its applications, appear on pp. 1166–1184 of [DS_2]. (However, the proof of the theorem itself presented by Dunford and Schwartz contains several errors.) See also [EG], Appendix A.

(5) For generalised versions of the Marcinkiewicz theorem, see MR **37** # 4601. Multilinear versions of the theorem are given in MR **38** # 6346. Vector-valued versions are dealt with in MR **52** # 1162.

(6) A stronger form of the theorem has been given by Stein and Weiss [2] (see also [Moz], Chapter 1), the essential differences being that one

assumes that $1 \leqslant p_j \leqslant q_j < \infty$; that $p_0 \neq p_1$ and $q_0 \neq q_1$; and that in place of weak type conditions one assumes restricted weak type conditions; the conclusion being that T is of type (p, q) whenever (13.8.2) holds. To say that T is of *restricted weak type* (p, q), signifies that (13.7.5) holds for all f which are characteristic functions of μ-finite subsets of X.

13.8.3. **Further Inequalities.** Partly for their own interest, and partly for subsequent application in Sections 13.9 and 13.10, we propose to establish some inequalities subsidiary to those implied in Marcinkiewicz's theorem 13.8.1.

It will be assumed that $T: f \to f^{\#}$ is quasilinear and has as domain a linear subspace $\mathfrak{D}$ of $\mathbf{L}^1(X, \mathscr{M}, \mu) \cap \mathbf{L}^a(X, \mathscr{M}, \mu)$, which is stable under multiplication by characteristic functions of μ-measurable sets and under truncation. The range of T is to lie in the space of ν-measurable functions (or in the space of equivalence classes thereof). It is further assumed that $\nu(Y) < \infty$.

The major hypotheses are that T is of weak types $(1, 1)$ and (a, a) on $\mathfrak{D}$, where $1 < a \leqslant \infty$, that is, that

$$D^{\nu}_{f^{\#}}(t) = \nu(E^{\#}(t)) \leqslant \frac{A\|f\|_{1,\mu}}{t}, \tag{13.8.18}$$

and either

$$D^{\nu}_{f^{\#}}(t) = \nu(E^{\#}(t)) \leqslant \frac{B\|f\|^a_{a,\mu}}{t^a} \tag{13.8.19}$$

if $a < \infty$, or

$$\|f^{\#}\|_{\infty,\nu} \leqslant B\|f\|_{\infty,\mu} \tag{13.8.20}$$

if $a = \infty$, in each case for $f \in \mathfrak{D}$ and $t > 0$. In these inequalities we have written

$$E^{\#}(t) = E^{\#}_f(t) = \{y \in Y : |f^{\#}(y)| > t\},$$

just as subsequently we shall write

$$E(t) = E_f(t) = \{x \in X : |f(x)| > t\}.$$

The conclusions are that

$$\|f^{\#}\|_{1,\nu} \leqslant A' + B' \int |f| \log^+ |f| \, d\mu \qquad \text{if } \mu(X) < \infty, \tag{13.8.21}$$

and that

$$\|f^{\#}\|_{p,\nu} \leqslant A_p \|f\|_{1,\mu} \qquad \text{if } 0 < p < 1, \tag{13.8.22}$$

in each case for $f \in \mathfrak{D}$; A' and B' denote numbers that may depend on T and the measure spaces involved, A_p a number that may depend on these things and on p as well, but A', B', and A_p are independent of f. These remarks apply also to A'' and B'', which appear shortly.

Details of the proofs will be given for the case in which $1 < a < \infty$. If $a = \infty$ one can either make simple modifications in the proof exhibited (compare the modifications in the proof of 13.8.1 to handle the case $p_1 = q_1 = \infty$), or one can use Theorem 13.8.1 so as to reduce oneself to the case in which $a < \infty$.

Proof of (13.8.21). We denote by $\alpha = \alpha(t)$ a positive-valued function of $t > 0$ to be specified later and write

$$g(x) = \begin{cases} f(x) & \text{if } |f(x)| > \alpha, \\ 0 & \text{otherwise}, \end{cases}$$

so that $g \in \mathfrak{D}$ whenever $f \in \mathfrak{D}$. Accordingly

$$f = g + h,$$

where $h \in \mathfrak{D}$ satisfies $|h| \leqslant \alpha$. By quasilinearity of T (compare (13.7.4)),

$$|f^{\#}| \leqslant \kappa(|g^{\#}| + |h^{\#}|)$$

and so

$$E_f^{\#}(t) \subset E_g^{\#}\left(\frac{t}{2\kappa}\right) \cup E_h^{\#}\left(\frac{t}{2\kappa}\right).$$

By (13.8.18) and (13.8.19), therefore,

$$\begin{aligned} D_{f^{\#}}^{\nu}(t) &\leqslant D_{g^{\#}}^{\nu}\left(\frac{t}{2\kappa}\right) + D_{h^{\#}}^{\nu}\left(\frac{t}{2\kappa}\right) \\ &\leqslant \frac{2\kappa A\|g\|_{1,\mu}}{t} + \frac{(2\kappa)^a B\|h\|_{a,\mu}^a}{t^a} \\ &= 2\kappa A t^{-1}\int_{E(\alpha)} |f|\, d\mu + (2\kappa)^a B t^{-a}\int_{X \setminus E(\alpha)} |f|^a\, d\mu. \end{aligned}$$

Hence

$$D_{f^{\#}}^{\nu}(t) \leqslant 2\kappa A t^{-1}\int_{E(\alpha)} |f|\, d\mu + (2\kappa)^a B \alpha^{a-1} t^{-a}\|f\|_{1,\mu}. \qquad (13.8.23)$$

Now

$$\begin{aligned} \|f^{\#}\|_{1,\nu} &= \int_Y |f^{\#}|\, d\nu = \int_0^{\infty} D_{f^{\#}}^{\nu}(t)\, dt = \int_0^1 + \int_1^{\infty} \\ &\leqslant \nu(Y) + \int_1^{\infty} D_{f^{\#}}^{\nu}(t)\, dt, \end{aligned}$$

so that it will suffice to show that $\int_1^{\infty} D_{f^{\#}}^{\nu}(t)\, dt$ is majorized by an expression of the type appearing on the right-hand side of (13.8.21).

To this end we choose any r satisfying $0 < r < 1$ and take $\alpha = t^r$, so that $\alpha^{a-1}t^{-a} = t^{-s}$, where $s > 1$. Then (13.8.23) yields

$$D_{f^{\#}}^{\nu}(t) \leqslant 2\kappa A t^{-1}\int_{E(t^r)} |f|\, d\mu + (2\kappa)^a B t^{-s}\|f\|_{1,\mu}. \qquad (13.8.24)$$

To $\int_1^\infty D_{f\#}^\nu(t)\,dt$ the second term on the right-hand side of (13.8.24) contributes a term majorized by

$$(2\kappa)^a B\|f\|_{1,\mu}\cdot\int_1^\infty t^{-s}\,dt = B'\|f\|_{1,\mu}.$$

Inasmuch as $u \leqslant C + C'u\log^+ u$ for $u \geqslant 0$, while $\mu(X) < \infty$, this last term is majorized by an expression of the type appearing on the right-hand side of (13.8.21). The remaining problem is therefore to verify that

$$\int_1^\infty t^{-1}\{\int_{E(t^r)} |f(x)|\,d\mu(x)\}\,dt \leqslant A'' + B''\int_X |f|\log^+|f|\,d\mu. \qquad (13.8.25)$$

However, it is a simple matter to show that the function

$$(x, t) \to \chi_{E(t^r)}(x)|f(x)|$$

is measurable in the pair (x, t), as a consequence of which the theorems of Fubini and Tonelli show that the left-hand member of (13.8.25) is equal to

$$\int_X |f(x)|\{\int_1^\infty t^{-1}\chi_{E(t^r)}(x)\,dt\}\,d\mu(x),$$

in which the inner integral has the value

$$\log^+\{|f(x)|^{1/r}\}$$

owing to the fact that $\chi_{E(t^r)}(x)$ is 1 or 0 according as $t^r < |f(x)|$ or $t^r \geqslant |f(x)|$. Since $\log^+\{u^c\} = c\cdot\log^+ u$ for $u > 0$, $c > 0$, we see that the left-hand member of (13.8.25) is in fact equal to

$$\left(\frac{1}{r}\right)\int_X |f|\log^+|f|\,d\mu,$$

so that (13.8.25) is certainly valid.

Proof of (13.8.22). We have in this case to show that

$$\|f\#\|_{p,\nu}^p = \int_0^\infty pt^{p-1}D_{f\#}^\nu(t)\,dt \leqslant A_p\|f\|_{1,\mu}^p, \qquad (13.8.26)$$

provided $0 < p < 1$.

Let $\varepsilon > 0$. Then

$$\int_0^\varepsilon pt^{p-1}D_{f\#}^\nu(t)\,dt \leqslant \nu(Y)\varepsilon^p. \qquad (13.8.27)$$

Also, taking $\alpha(t) = t$ in (13.8.23), it appears that

$$\int_\varepsilon^\infty pt^{p-1} D^\nu_{f\#}(t)\,dt \leqslant 2\kappa Ap \int_\varepsilon^\infty t^{p-2}\{\int_{E(t)} |f(x)|\,d\mu(x)\}\,dt$$

$$+ (2\kappa)^a Bp\|f\|_{1,\mu} \int_\varepsilon^\infty t^{p-2}\,dt. \qquad (13.8.28)$$

Since $p < 1$, the second term on the right of (13.8.28) is equal to

$$\frac{(2\kappa)^a Bp\|f\|_{1,\mu}\varepsilon^{p-1}}{1-p}.$$

By the theorems of Fubini and Tonelli, the first term is equal to

$$2\kappa Ap \int_X |f(x)|\{\int_\varepsilon^\infty \chi_{E(t)}(x) t^{p-2}\,dt\}\,d\mu(x),$$

the inner integral in which is easily seen to equal

$$\frac{\max\{0, \varepsilon^{p-1} - |f(x)|^{p-1}\}}{1-p} \leqslant \frac{\varepsilon^{p-1}}{1-p}$$

because $\chi_{E(t)}(x)$ is 1 or 0 according as $t < |f(x)|$ or $t \geqslant |f(x)|$. Thus (13.8.28) and (13.8.27) lead to

$$\|f^\#\|^p_{p,\nu} \leqslant \nu(Y)\varepsilon^p + \frac{(2\kappa A + 2^a\kappa^a B)p\varepsilon^{p-1}\|f\|_{1,\mu}}{1-p},$$

from which (13.8.26) follows on letting $\varepsilon \downarrow \|f\|_{1,\mu}$.

13.8.4. **Remarks.** (1) Arguments very similar to those appearing in 13.8.3 will show that, if T is of weak types (1, 1) and (∞, ∞) on $\mathfrak{T}$, then it is of type (p, p) on $\mathfrak{T}$ for $1 < p < \infty$. (This is, of course, a special case of Theorem 13.8.1.) The details, written out for the case of the Hardy-Littlewood maximal operator (see Example 13.7.6(2)) but really quite general in scope, appear in the proof of Theorem (21.76) in [HS].

Numerous further inequalities of the same general type appear in Exercise 13.22; see also [Z_2], pp. 116–121.

(2) A significant class of operators T satisfying the conditions mentioned in Remark (1) immediately above has been exhibited by Dunford and Schwartz (see [DS_1], pp. 668–684) in connection with ergodic theory.

They begin with a linear operator S with domain $\mathbf{L}^1(X, \mathscr{M}, \mu) \cap \mathbf{L}^\infty(X, \mathscr{M}, \mu)$ and range in the space of μ-measurable functions (or in the space of equivalence classes of such functions), S being assumed to be of types (1, 1) and (∞, ∞) and to have associated norms

$$\|S\|_{1,1} \leqslant 1, \qquad \|S\|_{\infty,\infty} \leqslant 1;$$

such operators have since been christened *Dunford-Schwartz operators*. From S one constructs its iterates S^n ($n = 0, 1, 2, \cdots$; $S^0 = I$, the identity operator), their arithmetic means

$$A_N = N^{-1} \sum_{n=0}^{N-1} S^n,$$

and finally the majorant operator A^* defined by

$$A^*f(x) = \sup_N |A_N f(x)|.$$

It is clear that A^* is sublinear and of type (∞, ∞). Dunford and Schwartz show (what is by no means obvious) that A^* is of weak type (1, 1), so that the inequalities (13.8.21) and (13.8.22) hold with A^*f in place of $f\#$. (They prove these inequalities directly and with specific estimates of the constants A', B', and A_p.)

It is also shown that the sequence $(A_N f)_{N=1}^{\infty}$ converges pointwise almost everywhere if $f \in \mathbf{L}^p(X, \mathscr{M}, \mu)$ and $1 \leqslant p < \infty$, and that it converges in $\mathbf{L}^p(X, \mathscr{M}, \mu)$ if $f \in \mathbf{L}^p(X, \mathscr{M}, \mu)$ and $1 < p < \infty$. These statements constitute the so-called *pointwise* and *mean ergodic theorems*, respectively (see [DS_1], *loc. cit.*).

13.9 Application to Conjugate Functions

It is now time to redeem the promise, made in 12.9.9, to provide an alternative approach to the proofs of the inequalities (12.9.1), (12.9.9), and (12.9.10). This is to be done by applying Marcinkiewicz's theorem 13.8.1 and the results in 13.8.3 to the conjugate function operator $T: f \to \tilde{f} = H * f$ introduced in Section 12.8. As will become clear in 13.9.2, the crucial step in this program is the proof that T, regarded as an operator with domain and range in $\mathbf{L}^2$, is of weak type (1, 1). The required proof is quite troublesome and occupies the major portion of the present section.

Before embarking on this task, a comment is in order. The reader may at first be puzzled why we do not regard T as an operator with domain $\mathbf{L}^1$ and prove that it is of weak type (1, 1) on this domain. The reason why we cannot follow this course is a consequence of our divergence from the traditional treatment, discussed in 12.8.2. The traditional account works with the operator $T': f \to f^c$ with domain $\mathbf{L}^1$ and includes a proof of the fact that T' is of weak type (1, 1); see [Z_1], p. 134 and [Ba_2], p. 113. Now f^c is the pointwise limit almost everywhere of $H_\varepsilon * f$, whereas our $\tilde{f}$ is the distributional limit of $H_\varepsilon * f$, in each case as $\varepsilon \downarrow 0$. As we have seen in 12.8.2(2), f^c and $\tilde{f}$ may be identified if (for example) $f \in \mathbf{L}^2$. The identification is not, however, possible for a general $f \in \mathbf{L}^1$. Indeed, it was seen in 12.8.3 that, for a general $f \in \mathbf{L}^1$, the distribution $\tilde{f}$ is not (generated by) a function at all. Thus we cannot even hope to prove that T is of weak type (1, 1) on the domain $\mathbf{L}^1$. Instead, we must of necessity restrict its domain to a set of functions f for which we can be sure that $\tilde{f}$ is (generated by and identifiable with) a function, and which is still large enough for us to derive the inequalities (12.9.1), (12.9.9), and (12.9.10). The domain $\mathbf{L}^2$ is amply large enough to fit the bill and is, moreover, convenient in other respects. (Smaller domains, such as $\mathbf{L}^\infty$, would in fact suffice.)

After this preamble we proceed to state and prove the key result of this section.

13.9.1. **The Main Inequality.** There exists an absolute constant $A > 0$ such that, for all $f \in \mathbf{L}^2$ and all $\lambda > 0$,

$$m(\{x \in [-\pi, \pi) : |\tilde{f}(x)| > \lambda\}) \leqslant A\lambda^{-1} \int_{-\pi}^{\pi} |f(t)|\, dt, \tag{13.9.1}$$

where m denotes Lebesgue measure. That is to say, the operator $f \to \tilde{f}$, with domain $\mathbf{L}^2$ and range in $\mathbf{L}^2$, is of weak type (1, 1).

Proof. This will be broken into several steps, in the course of which it will sometimes be convenient to write $|E|$ in place of $m(E)$. The symbols $A, A', A_0, A_1, \cdots$ will be used to denote positive absolute constants.

Since the operator in question is linear, a little thought will convince the reader that it is enough to prove (13.9.1) for nonnegative functions $f \in \mathbf{L}^2$ satisfying

$$\int_{-\pi}^{\pi} |f(t)|\, dt = 1,$$

which restrictions will be assumed henceforth.

(1) We begin by transforming the problem. By 12.8.2(2), $\tilde{f} = \lim_{\varepsilon \to 0} H_\varepsilon * f$ in $\mathbf{L}^2$, and also $\tilde{f} = \lim_{n\to\infty} H_{\varepsilon_n} * f$ pointwise almost everywhere for a suitable (possibly f-dependent) sequence of positive numbers ε_n tending to 0. Write $\tilde{f}_n = H_{\varepsilon_n} * f$, so that

$$\begin{aligned} \tilde{f}_n(x) &= \frac{1}{2\pi} \int_{\varepsilon_n \leqslant |y| \leqslant \pi} f(x-y) \cot \tfrac{1}{2} y\, dy \\ &= -\frac{1}{2\pi} \left\{ \int_{-\pi}^{x-\varepsilon_n} + \int_{x+\varepsilon_n}^{\pi} \right\} \frac{f(t)\, dt}{\tan \frac{1}{2}(t-x)}. \end{aligned} \tag{13.9.2}$$

As x and t range separately over $[-\pi, \pi)$, $t - x$ ranges over $(-2\pi, 2\pi)$. The function $\cot \frac{1}{2}(t - x)$ has singularities to concern us at $t - x = -2\pi$, 0 and 2π, and these plainly present the potentially important features in (13.9.2). Accordingly, we examine, in place of the functions $\tilde{f}_n$, the functions h_n defined on $[-\pi, \pi)$ as follows:

$$h_n(x) = \left\{ \int_{-\pi}^{x-\varepsilon_n} + \int_{x+\varepsilon_n}^{\pi} \right\} \frac{f(t)\, dt}{t - x},$$

$$+ \int_{-\pi}^{\pi} f(t)\{(t - x - 2\pi)^{-1} + (t - x + 2\pi)^{-1}\}\, dt \tag{13.9.3}$$

$$= \left\{ \int_{-3\pi}^{x-\varepsilon_n} + \int_{x+\varepsilon_n}^{3\pi} \right\} \frac{f(t)\, dt}{t - x}. \tag{13.9.4}$$

A little manipulation shows that

$$-2\pi\tilde{f}_n(x) = 2h_n(x) + j_n(x), \tag{13.9.5}$$

where

$$j_n(x) = \left\{\int_{-\pi}^{x-\varepsilon_n} + \int_{x+\varepsilon_n}^{\pi}\right\} f(t)g(\tfrac{1}{2}(t - x))\,dt$$

$$- \int_{x-\varepsilon_n}^{x+\varepsilon_n} f(t)\{[\tfrac{1}{2}(t - x) - \pi]^{-1} + [\tfrac{1}{2}(t - x) + \pi]^{-1}\}\,dt,$$

the function g being defined on $[-\pi, \pi)$ by continuous extension of the formula

$$g(u) = \cot u - u^{-1} - (u - \pi)^{-1} - (u + \pi)^{-1} \qquad (0 < |u| < \pi)$$

and being bounded and continuous, say $|g| \leqslant A_0$. It is evident that each j_n is continuous.

Defining the function j by the formula

$$j(x) = \int_{-\pi}^{\pi} f(t)g(\tfrac{1}{2}(t - x))\,dt,$$

it is easily shown that $j_n \to j$ uniformly and so, a fortiori, in $\mathbf{L}^2(-\pi, \pi)$. Therefore, since $\tilde{f}_n \to \tilde{f}_n$ pointwise almost everywhere, equation (13.9.5) shows that the sequence (h_n) converges in $\mathbf{L}^2(-\pi, \pi)$ and pointwise almost everywhere on $[-\pi, \pi)$ to a limit which we denote by h.

It is clear that

$$|j(x)| \leqslant A_0 \int_{-\pi}^{\pi} |f(t)|\,dt = A_0. \tag{13.9.6}$$

Moreover, (13.9.5) shows that

$$-2\pi\tilde{f}(x) = 2h(x) + j(x) \tag{13.9.7}$$

for almost all $x \in [-\pi, \pi)$.

By altering the functions concerned on null sets, we may and will assume it arranged that $\tilde{f}_n \to \tilde{f}$, $h_n \to h$, and $j_n \to j$ pointwise *everywhere* on $[-\pi, \pi)$, so that (13.9.7) also holds at all points $x \in [-\pi, \pi)$. Such changes alter no distribution functions and no integrals.

(2) Let us next justify the transference of attention from $\tilde{f}$ to h by showing that, in order to prove 13.9.1, it suffices to show that there exist absolute constants $A' > 0$ and $\lambda_0 > 0$ such that

$$m(\{x \in [-\pi, \pi) : |h(x)| > \lambda\}) \leqslant A'/\lambda \tag{13.9.8}$$

for all $\lambda > \lambda_0$ and all nonnegative $f \in \mathbf{L}^2$ satisfying

$$\int_{-\pi}^{\pi} |f(t)|\,dt = 1.$$

To see this, write $c = A_0/(2\pi)$, so that

$$\left|\left(\frac{1}{2\pi}\right)j(x)\right| \leqslant c.$$

If $\lambda > \lambda_1 = \max(\lambda_0, 2c)$, the relation $|\tilde{f}(x)| > \lambda$ combines with (13.9.7) to entail that

$$|h(x)| = |\pi\{\tilde{f}(x) + \left(\frac{1}{2\pi}\right)j(x)\}| > \pi(\lambda - c) > \frac{1}{2}\pi\lambda > \lambda.$$

For any such λ, the assumed inequality (13.9.8) yields

$$m(\{x \in [-\pi, \pi) : |\tilde{f}(x)| > \lambda\}) \leqslant A'/\lambda.$$

On the other hand, if $0 < \lambda \leqslant \lambda_1$, we have trivially

$$m(\{x \in [-\pi, \pi) : |\tilde{f}(x)| > \lambda\}) \leqslant 2\pi \leqslant \frac{2\pi\lambda_1}{\lambda}.$$

Thus (13.9.1) will hold for all $\lambda > 0$, provided A is replaced by

$$\max(A', 2\pi\lambda_1),$$

which is another absolute constant.

(3) For subsequent use in the proof of (13.9.8), we record the inequality

$$\|h\|_2 \leqslant A_1\|f\|_2, \tag{13.9.9}$$

which follows from (13.9.6), (13.9.7), and (12.8.10).

(4) The proof of (13.9.8) may now begin. Let λ and μ be positive numbers; μ will later be chosen depending upon λ.

Define functions ϕ and r as follows:

$$\phi(x) = \begin{cases} f(x) & \text{if } 0 \leqslant f(x) \leqslant \mu, \\ 0 & \text{if } f(x) > \mu, \end{cases}$$

and $r = f - \phi \geqslant 0$. These are periodic functions. It is clear that

$$\alpha \equiv \int_{-\pi}^{\pi} r(t)\,dt \leqslant 1.$$

Introduce also the functions ψ and θ defined by

$$\psi(x) = \int_{-\pi}^{x} f(t)\,dt,$$

$$\theta(x) = \psi(x) - \tfrac{1}{2}\mu(x + \pi)$$

for $x \in [-\pi, \pi)$ and by periodicity for other values of x. Then $\psi(-\pi) = \theta(-\pi) = 0$ and $\psi(\pi - 0) = \alpha \leqslant 1$. Moreover, $\theta(\pi - 0) = \psi(\pi - 0) - \mu\pi$. From here on it is to be supposed that $\mu > \pi^{-1}$, so that $\theta(\pi - 0) < 0$.

This being so, θ has a largest zero, say x_0, in $[-\pi, \pi)$. Consider the set of points $\xi \in (x_0, \pi)$ for which $\theta(\xi) < \sup_{\xi \leqslant x < \pi} \theta(x)$; since θ is continuous, this set is open. The union of this set with $(-\pi, x_0)$ is an open set $\Omega \subset (-\pi, \pi)$. We assume that $f(x) > \mu$ on a set of positive measure, thus ensuring that the pointwise derivative $\theta'(x)$ is positive on a set of positive measure so that Ω is nonvoid. (In the contrary case we should have $\phi = f$ a.e. and, in the notation of stages (5)–(8) below, h will agree with $h_{(1)}$; (13.9.26) will follow directly from (13.9.15) and will suffice to yield (13.9.8) with $A' = A_3$.)

Let $\Omega = \bigcup \Omega_i$ be the expression of Ω as the union of a countable family of nonvoid disjoint open intervals $\Omega_i = (a_i, b_i)$. It is easy to see (Exercise 13.14) that $\theta(a_i) = \theta(b_i - 0)$ for all i. (We write $\theta(b_i - 0)$ rather than $\theta(b_i)$, since the relation $b_i = \pi$ may obtain and θ is not necessarily continuous at π.) From this it follows that

$$\psi(b_i - 0) - \tfrac{1}{2}\mu(b_i + \pi) = \psi(a_i) - \tfrac{1}{2}\mu(a_i + \pi),$$

and hence that

$$\int_{a_i}^{b_i} r(t)\,dt = \psi(b_i - 0) - \psi(a_i) = \tfrac{1}{2}\mu|\Omega_i|, \tag{13.9.10}$$

and

$$\int_{\Omega} r(t)\,dt = \tfrac{1}{2}\mu \sum |\Omega_i|. \tag{13.9.11}$$

From the definitions of r and θ, it is easily seen that almost all points $t \in [-\pi, \pi)$ for which $r(t) > 0$ belong to Ω (since at almost all such points the derivative $\theta'(t)$ exists and is positive). Because of this,

$$\alpha = \int_{-\pi}^{\pi} r(t)\,dt = \int_{\Omega} r(t)\,dt \tag{13.9.12}$$

and

$$\int_{[-\pi,\pi)\setminus\Omega} r(t)\,dt = 0.$$

From (13.9.11) and (13.9.12) we deduce that

$$|\Omega| = \sum |\Omega_i| = \frac{2\alpha}{\mu} \leqslant \frac{2}{\mu}. \tag{13.9.13}$$

It is necessary to introduce one further auxiliary function, namely the function Φ defined on $[-\pi, \pi)$ to equal $\tfrac{1}{2}\mu\chi_\Omega$ and defined elsewhere so as to be periodic. For this function (13.9.10) gives

$$\int_{a_i}^{b_i} \Phi(t)\,dt = \tfrac{1}{2}\mu|\Omega_i| = \int_{a_i}^{b_i} r(t)\,dt. \tag{13.9.14}$$

(5) The intervals Ω_i are now to be enlarged into open intervals Ω_i' having the following properties:

(a) Ω_i' is concentric with Ω_i;
(b) $\sum |\Omega_i'| \leqslant 6 \sum |\Omega_i|$;

(c) $|\Omega_i'| > 5|\Omega_i|$;

(d) $\lim_{i\to\infty} \dfrac{|\Omega_i'|}{|\Omega_i|} = \infty$.

The condition (d) is to be regarded as void if the decomposition $\Omega = \bigcup \Omega_i$ is finite; and in the contrary case it is supposed that the index i ranges over all positive integers. The intervals Ω_i' may, of course, fail to be disjoint, and some of them may project beyond $[-\pi, \pi)$. It is left to the reader to verify that it is possible to choose the intervals Ω_i' so as to fulfill the above conditions.

We write

$$S_{i,k} = \Omega_i + 2k\pi, \quad S'_{i,k} = \Omega_i' + 2k\pi, \quad S = \bigcup_{i,k} S_{i,k}, \qquad S' = \bigcup_{i,k} S'_{i,k},$$

where k ranges over Z.

By (13.9.4) we have for $x \in [-\pi, \pi)$:

$$h_n(x) = \left\{\int_{-3\pi}^{x-\varepsilon_n} + \int_{x+\varepsilon_n}^{3\pi}\right\} \frac{f(t)\,dt}{t - x}$$

$$= h_{1,n}(x) + h_{2,n}(x) + h_{3,n}(x)$$

say, where $h_{1,n}$, $h_{2,n}$, and $h_{3,n}$ are related to ϕ, Φ, and $r - \Phi$, respectively, exactly as h_n is related to f. For instance,

$$h_{1,n}(x) = \left\{\int_{-3\pi}^{x-\varepsilon_n} + \int_{x+\varepsilon_n}^{3\pi}\right\} \frac{\phi(t)\,dt}{t - x}.$$

To the reader is bequeathed the task (see Exercise 13.15) of showing that there exist functions $h_{(j)} \in \mathbf{L}^2(-\pi, \pi)$ such that

$$\lim_{n\to\infty} \|h_{(j)} - h_{j,n}\|_2 = 0$$

for $j = 1, 2, 3$. By extracting subsequences and renaming, if necessary, we may and will suppose that in addition

$$\lim_{n\to\infty} h_{j,n}(x) = h_{(j)}(x)$$

for almost all $x \in [-\pi, \pi)$.

Since $0 \leqslant \phi \leqslant \mu$, it follows on replacing f by ϕ in (13.9.9) that

$$\|h_{(1)}\|_2^2 \leqslant A_1^2 \int_{-\pi}^{\pi} \phi^2(t)\,dt \leqslant A_1^2\mu \int_{-\pi}^{\pi} \phi(t)\,dt$$

$$\leqslant A_1^2\mu \int_{-\pi}^{\pi} f(t)\,dt = A_1^2\mu. \tag{13.9.15}$$

From the same source, taken in conjunction with (13.9.13), it appears that

$$\|h_{(2)}\|_2^2 \leqslant A_1^2 \int_{-\pi}^{\pi} \Phi^2(t)\,dt = A_1^2(\tfrac{1}{2}\mu)^2|\Omega|$$

$$\leqslant \tfrac{1}{2}A_1^2\mu. \tag{13.9.16}$$

A suitable majorization of $h_{(3)}$ is more troublesome and will occupy the next two stages of the proof.

(6) As a first step towards the majorization of $h_{(3)}$ we will show that, for almost all $x \in [-\pi, \pi)\backslash S'$,

$$h_{(3)}(x) = \lim_{n\to\infty} \left\{ \sum{}' \int_{S_{i,k}} \frac{\{r(t) - \Phi(t)\}\,dt}{t - x} + \sum{}'' \int_{S_{i,k}} \frac{\{r(t) - \Phi(t)\}\,dt}{t - x} \right\}, \tag{13.9.17}$$

where $\sum'$ indicates a summation over those pairs (i, k) such that

$$S_{i,k} \subset [-3\pi, x - \varepsilon_n]$$

and $\sum''$ a summation over those pairs (i, k) such that $S_{i,k} \subset [x + \varepsilon_n, 3\pi]$.

Suppose indeed that $x \in [-\pi, \pi)\backslash S'$. Since $r(t)$ and $\Phi(t)$ vanish for almost all t outside S, it is clear that

$$h_{3,n}(x) = \left[\int_{[-3\pi, x-\varepsilon_n]\cap S} + \int_{[x+\varepsilon_n, 3\pi]\cap S}\right] \frac{\{r(t) - \Phi(t)\}\,dt}{t - x}.$$

Directing attention to the first integral on the right-hand side, a little further thought will show that, owing to the periodic structure of S,

$$\int_{[-3\pi, x-\varepsilon_n]\cap S} \frac{\{r(t) - \Phi(t)\}\,dt}{t - x} = \sum{}' + \int_{[-3\pi, x-\varepsilon_n]\cap S_{i_0,k_0}} \frac{\{r(t) - \Phi(t)\}\,dt}{t - x}, \tag{13.9.18}$$

where $(i_0, k_0) = (i_0(n), k_0(n))$ is that index pair, if any such exists, such that S_{i_0,k_0} contains $x - \varepsilon_n$. It is almost obvious that, for sufficiently large n, k_0 is independent of n.

To estimate the integral J appearing on the right-hand side of (13.9.18) we use the fact that, since r and Φ are nonnegative, there exist points t_1 and t_2 of S_{i_0,k_0} for which, if we write E for the range of integration,

$$\int_E \frac{r(t)\,dt}{(t - x)} = (t_1 - x)^{-1} \int_E r(t)\,dt$$

and

$$\int_E \frac{\Phi(t)\,dt}{t - x} = (t_2 - x)^{-1} \int_E \Phi(t)\,dt.$$

Consequently;

$$\begin{aligned} |J| &\leqslant |t_1 - x|^{-1} \int_{S_{i_0,k_0}} r(t)\,dt + |t_2 - x|^{-1} \int_{S_{i_0,k_0}} \Phi(t)\,dt \\ &= \tfrac{1}{2}\mu|t_1 - x|^{-1}|\Omega_{i_0}| + \tfrac{1}{2}\mu|t_2 - x|^{-1}|\Omega_{i_0}|, \end{aligned}$$

the last step following by use of (13.9.14) together with periodicity. Since $x \notin S'_{i_0,k_0}$, $|t_s - x| \geqslant \frac{1}{2}(|\Omega'_{i_0}| - |\Omega_{i_0}|)$ for $s = 1, 2$, and therefore

$$|J| \leqslant 2\mu\left(\frac{|\Omega'_{i_0}|}{|\Omega_{i_0}|} - 1\right)^{-1} \tag{13.9.19}$$

Consider what happens as $n \to \infty$. If, for sufficiently large n, there exist no intervals $S_{i,k}$ containing $x - \varepsilon_n$, J vanishes for all sufficiently large n. In the contrary case, the decomposition $\Omega = \bigcup \Omega_i$ must be infinite and i ranges over all positive integers. In this case, let (n_j) be the sequence of values of n for which $x - \varepsilon_{n_j} \in S_{i_0(n_j),k_0(n_j)}$ for some pair of indices $i_0(n_j)$, $k_0(n_j)$. For sufficiently large j, $k_0(n_j)$ is constantly 0 or -1, while $i_0(n_j) \to \infty$ as $j \to \infty$. Thus (13.9.19) and property (d) in (5) show that J is zero, if n is distinct from all the n_j, or is majorized by a quantity tending to zero as n ranges through the values n_j. In either case, therefore,

$$J \to 0 \qquad \text{as } n \to \infty.$$

A similar argument can be applied to the integrals involving right-hand intervals $[x + \varepsilon_n, 3\pi]$, and (13.9.17) follows on combining the two arguments.

(7) Having established (13.9.17), we proceed to estimate the integrals

$$J_{i,k}(x) = \int_{S_{i,k}} \frac{\{r(t) - \Phi(t)\}\, dt}{t - x}$$

for $x \in [-\pi, \pi)\backslash S'$ and any pair (i, k) such that $S_{i,k} \subset [-3\pi, 3\pi]$. An appraisal of

$$\int_{[-\pi,\pi)\backslash S'} |J_{i,k}(x)|\, dx$$

follows and leads finally to a crucial majorization of

$$\int_{[-\pi,\pi)\backslash S'} |h_{(3)}(x)|\, dx.$$

Repeating an argument used in (6), we find that there exist points τ_1 and τ_2 of $S_{i,k}$ for which

$$J_{i,k}(x) = \{(\tau_1 - x)^{-1} - (\tau_2 - x)^{-1}\} \cdot \tfrac{1}{2}\mu|S_{i,k}|;$$

as before, (13.9.14) is being called into play here. If we denote by $c_{i,k}$ the midpoint of $S_{i,k}$, and recall that $x \notin S'_{i,k}$ and that $|S'_{i,k}| > 5|S_{i,k}|$, we obtain

$$|J_{i,k}(x)| \leqslant A_2\mu|S_{i,k}|^2(c_{i,k} - x)^{-2}. \tag{13.9.20}$$

On the other hand,

$$\int_{[-\pi,\pi)\backslash S'} \frac{|S_{i,k}|^2\,dx}{(c_{i,k}-x)^2} \leqslant |S_{i,k}|^2 \int_{R\backslash S'_{i,k}} (c_{i,k}-x)^{-2}\,dx$$
$$\leqslant \frac{4|S_{i,k}|^2}{|S'_{i,k}|}$$
$$\leqslant |S_{i,k}|,$$

where the last step depends upon (c) in (5). Thus (13.9.20) leads to

$$I_{i,k} \equiv \int_{[-\pi,\pi)\backslash S'} |J_{i,k}(x)|\,dx \leqslant A_2\mu|S_{i,k}|. \tag{13.9.21}$$

By Fatou's lemma and (13.9.17) it follows directly that

$$\int_{[-\pi,\pi)\backslash S'} |h_{(3)}(x)|\,dx \leqslant \liminf_{n\to\infty} \{\textstyle\sum' I_{i,k} + \sum'' I_{i,k}\}.$$

Now, in the sum $\sum'$ the index k must be either -1 or 0, and in $\sum''$ it must be either 0 or 1. Therefore (13.9.21) yields

$$\int_{[-\pi,\pi)\backslash S'} |h_{(3)}(x)|\,dx \leqslant A_2\mu \sum_i (|S_{i,-1}| + |S_{i,0}| + |S_{i,1}|)$$
$$= 3A_2\mu \sum_i |\Omega_i| \leqslant 6A_2,$$

where the last step makes use of (13.9.13) and the fact that $S_{i,k}$ is a translate of Ω_i. Thus

$$\int_{[-\pi,\pi)\backslash S'} |h_{(3)}(x)|\,dx \leqslant 6A_2. \tag{13.9.22}$$

(8) For this final stage of the proof, the number $\lambda > 0$ is supposed given. Denote by E_q $(q = 1, 2, 3)$ the set of points $x \in [-\pi, \pi)$ for which $|h_{(q)}(x)| > \lambda/3$, and by E the set of points $x \in [-\pi, \pi)$ for which $|h(x)| > \lambda$. The obvious relation $h = h_{(1)} + h_{(2)} + h_{(3)}$ a.e. entails that $|E| \leqslant |E_1| + |E_2| + |E_3|$.

If $E'_3 = E_3 \cap ([-\pi, \pi)\backslash S')$, (13.9.22) yields

$$\int_{E'_3} |h_{(3)}(x)|\,dx \leqslant 6A_2,$$

whence it follows that

$$m(E'_3) \leqslant 18A_2/\lambda. \tag{13.9.23}$$

On the other hand,

$$m(E_3) \leqslant m(E'_3) + m([-\pi, \pi) \cap S'). \tag{13.9.24}$$

It has already been supposed (see (4)) that $\mu > \pi^{-1}$: we now suppose that $\mu > 24\pi^{-1}$. Then (13.9.13) and property (b) in (5) show that

$$\sum |\Omega_i'| \leqslant 6 \sum |\Omega_i| \leqslant \frac{12}{\mu} < \frac{1}{2}\pi.$$

The relation $|\Omega_i'| < \frac{1}{2}\pi$ shows that the only $S_{i,k}'$ which make a nonzero contribution to $m([-\pi, \pi) \cap S')$ are to be found among those for which $k = -1, 0$ or 1. Therefore

$$m([-\pi, \pi) \cap S') \leqslant 3 \sum |\Omega_i'| \leqslant 18 \sum |\Omega_i| \leqslant 36/\mu$$

and so, by (13.9.23) and (13.9.24),

$$m(E_3) \leqslant 18A_2/\lambda + 36/\mu, \tag{13.9.25}$$

provided $\mu > 24\pi^{-1}$.

Concerning E_1, (13.9.15) yields

$$\left(\frac{\lambda}{3}\right)^2 m(E_1) \leqslant A_1{}^2\mu,$$

so that

$$m(E_1) \leqslant \frac{A_3\mu}{\lambda^2}. \tag{13.9.26}$$

Similarly, (13.9.16) entails that

$$m(E_2) \leqslant \frac{A_4\mu}{\lambda^2}. \tag{13.9.27}$$

On combining (13.9.25), (13.9.26), and (13.9.27), and taking $\mu = \lambda > 24\pi^{-1}$, we obtain

$$m(E) \leqslant \frac{A}{\lambda}.$$

Thus (13.9.8) is established for $\lambda > 24\pi^{-1}$ and the proof is complete.

Remarks. (1) It is perhaps worthwhile to comment on the introduction of the functions ψ and θ and the set Ω. The reader should observe that θ is absolutely continuous on $[-\pi, \pi)$ with a derivative existing almost everywhere and equal to $r - \frac{1}{2}\mu$. We know that $r(x) = 0$ or $r(x) > \mu$ for all x. Roughly speaking, $r(x) > \mu$ if and only if θ is *increasing* at the point x; the set Ω is an *open* set consisting of those points where θ is increasing, together with a subset, say F, of the set of points in $[-\pi, \pi)$ where θ is nonincreasing. It turns out that the set F is of negligible importance when μ is sufficiently large.

(2) The preceding proof of (13.9.1) is bare-handed and entirely "real variable" in nature. A good deal of economy can be achieved by using the elements of complex variable theory, in particular the basic

properties of harmonic functions. We sketch a proof based on such principles; see [Kz], p. 66.

Since $T: f \to \tilde{f}$ is linear and continuous on $\mathbf{L}^2$ (see 12.8.3(1)), reference to Exercise 13.18 shows that it is enough to establish (13.9.1) in case f is a strictly positive-valued trigonometric polynomial satisfying $\|f\|_1 = 1$. Taking principal branches of arg and log, consider the function H_λ defined for all complex z satisfying $\operatorname{Re} z > 0$ by

$$\begin{aligned} H_\lambda(z) &= 1 + \pi^{-1} \cdot \arg\,((z - i\lambda)(z + i\lambda)^{-1}) \\ &= 1 + \pi^{-1} \cdot \operatorname{Im}\,(\log\,((z - i\lambda)(z + i\lambda)^{-1})). \end{aligned}$$

The function H_λ is clearly harmonic and nonnegative on the halfplane defined by $\operatorname{Re} z > 0$; moreover,

(i) $H_\lambda(z) \geqslant \frac{1}{2}$ for all z such that $|z| \geqslant \lambda$ and $\operatorname{Re} z > 0$.

Extend f and $\tilde{f}$ from the unit circumference to the complex plane by defining

$$f(re^{i\theta}) = \sum_{n \in Z} r^{|n|} \hat{f}(n) e^{in\theta},$$

$$\tilde{f}(re^{i\theta}) = -i \sum_{n \in Z} (\operatorname{sgn} n) r^{|n|} \hat{f}(n) e^{in\theta}$$

for all real $r \geqslant 0$ and all real θ. Then

$$\begin{aligned} f(re^{i\theta}) + i\tilde{f}(re^{i\theta}) &= \sum_{n \in Z} r^{|n|}(1 + \operatorname{sgn} n) \hat{f}(n) e^{in\theta} \\ &= \hat{f}(0) + 2 \sum_{n \geqslant 1} r^{|n|} \hat{f}(n) e^{in\theta} \end{aligned}$$

is an entire function of $z = re^{i\theta}$. Also, the extended f is harmonic and (by the Maximum Principle for harmonic functions) is strictly positive throughout a neighbourhood of the closed unit disc D in the complex plane. It follows that the function $z \to H_\lambda(f(z) + i\tilde{f}(z))$ is harmonic on this neighbourhood. By the mean value theorem for harmonic functions it follows that

$$\text{(ii)}\quad (2\pi)^{-1} \int_0^{2\pi} H_\lambda(f(e^{it}) + i\tilde{f}(e^{it}))\, dt = H_\lambda(f(0)) = H_\lambda(1)$$

$$= 1 - 2\pi^{-1} \tan^{-1} \lambda < 2\pi^{-1}\lambda^{-1}.$$

From (i) and (ii) it is easily deduced that the measure of the set of all $t \in [0, 2\pi]$ such that $|f(e^{it})| > \lambda$ is at most $8\lambda^{-1}$, as required.

13.9.2. **Proof of the Results in Section 12.9.** Knowing from Theorem 13.9.1 that $T: f \to \tilde{f} = H * f$ is of weak type (1, 1) on $\mathbf{L}^\infty$, say, together with the (much more evident) property that T is of type (2, 2) on $\mathbf{L}^2$ [see 12.8.2(2)], the major results of Section 12.9 can now be derived by applying what has been learned in Section 13.8.

(1) To begin with, Marcinkiewicz's interpolation theorem 13.8.1 affirms that T is of type (p, p) on $\mathbf{L}^\infty$ whenever $1 < p \leqslant 2$, so that a number k_p exists for which

$$\|H * f\|_p \leqslant k_p \|f\|_p \tag{12.9.1}$$

holds for $1 < p \leqslant 2$ and $f \in \mathbf{L}^\infty$. The extension to general $f \in \mathbf{L}^p$ follows from 12.9.4(1); and, exactly as in 12.9.6, it is seen that the inequality remains true for $2 \leqslant p < \infty$. Thus we obtain M. Riesz's theorem 12.9.1.

Furthermore, an appeal to the results in 13.8.3 now yields the inequalities

$$\|H * f\|_p \leqslant k_p \|f\|_1 \qquad \text{for } 0 < p < 1 \tag{12.9.9}$$

and

$$\|H * f\|_1 \leqslant \frac{A}{2\pi} \int |f| \log^+ |f| \, dx + B \tag{12.9.10}$$

for $f \in \mathbf{L}^\infty$. Then, as has been seen in 12.9.9, the first of these inequalities continues to hold whenever $f \in \mathbf{L}^1$ and $H * f$ is replaced by f^c (and in its given form whenever $f \log^+ |f| \in \mathbf{L}^1$), and the second whenever $f \log^+ |f| \in \mathbf{L}^1$.

(2) We turn next to the proof of (12.9.14) and (12.9.15). For this purpose we need to estimate $\|T\|_{p,p}$ for large values of p.

For $p \in (1, 2)$ the result stated in 13.8.2(3) shows, on taking $\kappa = 1$, $p_0 = q_0 = 2$, $p_1 = q_1 = 1$, $A_0 = 1$ [see (12.8.10)] and $A_1 = A$ (as in 13.9.1), that

$$\|T\|_{p,p} \leqslant 2p^{1/p} \left\{ \frac{2}{p(2 - p)} + \frac{1}{p(p - 1)} \right\}^{1/p} A^{(2-p)/p}.$$

To infer a corresponding estimate for values of p greater than 2, we use 12.9.6 (or 16.4.1). In this way it appears that there exists an absolute constant B such that

$$\|\tilde{f}\|_p \leqslant Bp \|f\|_p \qquad (p \geqslant 2). \tag{13.9.28}$$

For some remarks concerning this type of estimate, see [Ba_2], p. 107.

Suppose now that $f \in \mathbf{L}^\infty$ and $\|f\|_\infty \leqslant 1$. Then, for $\lambda > 0$,

$$\frac{1}{2\pi} \int \exp [\lambda |\tilde{f}|] \, dx = \sum_{p=0}^{\infty} \frac{\lambda^p \|\tilde{f}\|_p^p}{p!}. \tag{13.9.29}$$

Using the relations $\|\tilde{f}\|_1 \leqslant \|\tilde{f}\|_2 \leqslant \|f\|_2$ and $\|f\|_p \leqslant \|f\|_\infty \leqslant 1$, together with (13.9.28), it is easily seen that the series on the right-hand side of (13.9.29) converges to a finite sum, provided $\lambda < (eB)^{-1}$. This proves (12.9.14) with $\lambda_0 = (eB)^{-1}$.

(3) It should be added at this point that the complex variable proof of (12.9.14) given in [Z_1], p. 257, shows that it is possible to take $\lambda_0 = \frac{1}{2}\pi$. That this is the best-possible value of λ_0 follows at once on considering the

function f defined by (12.8.14), whose conjugate $\tilde{f}$ satisfies (12.8.15) for almost all small positive values of x.

(4) As for (12.9.15), suppose that $f \in \mathbf{C}$ and $\varepsilon > 0$ are given. Choose a trigonometric polynomial g such that $\|f - g\|_\infty \leqslant \varepsilon$ (see 2.4.4). Then (12.9.14) and the linearity of T show that

$$\frac{1}{2\pi} \int \exp [\lambda|\tilde{f} - \tilde{g}|]\, dx < \infty$$

provided $\varepsilon\lambda < \lambda_0$. Now $\tilde{g}$ is also a trigonometric polynomial, and

$$\exp [\lambda|\tilde{f}|] \leqslant \exp [\lambda|\tilde{g}|] \cdot \exp [\lambda|\tilde{f} - \tilde{g}|].$$

Therefore

$$\frac{1}{2\pi} \int \exp [\lambda|\tilde{f}|]\, dx < \infty,$$

again provided $\varepsilon\lambda < \lambda_0$. Since $\lambda_0 > 0$ and ε may be chosen arbitrarily small, (12.9.15) follows.

Remark. From (12.9.15) it follows easily that

$$\|\tilde{f}\|_p = o(p) \qquad \text{as } p \to \infty \tag{13.9.30}$$

for each $f \in \mathbf{C}$. Of course, this does not in itself imply the existence of a function $\varepsilon_p \to 0$ as $p \to \infty$ such that for $p \geqslant 2$

$$\|\tilde{f}\|_p \leqslant p\varepsilon_p \|f\|_p \tag{13.9.31}$$

for each $f \in \mathbf{C}$, ε_p being independent of f. Indeed, no relation (13.9.31) can hold (with an f-independent $\varepsilon_p \to 0$ as $p \to \infty$) for each $f \in \mathbf{C}$: if it did, the same would continue to hold for each $f \in \mathbf{L}^\infty$, which would entail that exp $[\lambda|\tilde{f}|]$ would be integrable for any $f \in \mathbf{L}^\infty$ and any λ. The discussion in (3) immediately above has shown this to be false.

As a consequence of this, the relation

$$\|T\|_{p,p} \neq o\{(p - 1)^{-1}\} \qquad \text{as } p \to 1 + 0$$

is easily seen to follow.

See also Koizumi [2].

13.9.3. **Remark on the Hilbert Transform.** If one replaces the group T by R, the analogue of the mapping $f \to \tilde{f}$ is the so-called *Hilbert transform*

$$f \to (x^{-1}) * f = \int_{-\infty}^{\infty} \frac{f(y)\, dy}{x - y},$$

the integral being interpreted as a principal value. This, together with similar operators involving singular integrals and functions of several real variables, have been the object of an enormous amount of study by, among others, Titchmarsh, Kober, Koizumi, and Calderòn and Zygmund. Many of the developments are extremely closely linked with the ideas of type and of weak type of operators. See Calderón and Zygmund [1], [2], Calderón [2], Cordes [1], Koizumi [1] and the references cited there, O'Neil and Weiss [1]; [EG], Section 6.7; Coifman and Weiss [1]; Coifman, Rochberg and Weiss [1]; MR **56** # 16143.

13.10 Concerning σ^*f and s^*f

We return to the topics mentioned in 6.4.7 and utilize the results established in Section 13.8 in order to establish the inequalities

$$\|\sigma^*f\|_p \leqslant A_p\|f\|_p \quad \text{if } f \in \mathbf{L}^p, \quad 1 < p \leqslant \infty, \tag{6.4.11}$$

$$\|\sigma^*f\|_p \leqslant A_p\|f\|_1 \quad \text{if } f \in \mathbf{L}^1, \quad 0 < p < 1, \tag{6.4.12}$$

$$\|\sigma^*f\|_1 \leqslant \frac{A}{2\pi}\int |f| \log^+ |f|\, dx + B \quad \text{if } f \log^+ |f| \in \mathbf{L}^1. \tag{6.4.13}$$

The reasoning proceeds as follows.

13.10.1. **The Nature of σ^*.** The operator σ^* is sublinear and, by (6.4.9),

$$\sigma^* \text{ is of type } (\infty, \infty) \text{ on } \mathbf{L}^\infty. \tag{13.10.1}$$

The statement

$$\sigma^* \text{ is of weak type } (1, 1) \text{ on } \mathbf{L}^\infty \tag{13.10.2}$$

can be established in several ways. (If σ^* were linear, (13.10.1) would imply that σ^* is of type (1, 1), and nothing would remain in doubt[1]; but σ^* is *not* linear and in fact σ^* is *not* of type (1, 1).)

One way of proving (13.10.2) is to use the properties of the Hardy-Littlewood maximal operator mentioned in Example 13.7.6(2) in conjunction with some additional arguments of an elementary nature; see Exercise 13.17.

Alternatively, one can make appeal to 6.4.4 in combination with a powerful and general theorem of Stein cited in 16.2.8.; compare Exercise 16.14.

13.10.2. **Deduction of (6.4.11) to (6.4.13), (10.4.9), and (10.4.10).** Knowing that (13.10.1) and (13.10.2) hold, the Marcinkiewicz theorem 13.8.1 yields (6.4.11) for $f \in \mathbf{L}^\infty$. On the other hand, (6.4.12) and (6.4.13) appear as special cases of (13.8.22) and (13.8.21), again provided $f \in \mathbf{L}^\infty$. The extension to the more general functions f specified is quite simple.

Since it is evident that $\sigma^*f \leqslant \sigma^*|f|$, we may assume that f is real and nonnegative. For any such f, if we take any sequence of nonnegative functions f_n $(n = 1, 2, \cdots)$ such that $f_n \leqslant f_{n+1} \uparrow f$, we have (by the monotone convergence theorem for integrals)

[1] The necessary argument would be similar to that appearing in 16.4.1.

$$\begin{aligned}\sigma^*f &= \sup_N F_N * f \\ &= \sup_N \sup_n F_N * f_n \\ &= \sup_n \sup_N F_N * f_n \\ &= \sup_n \sigma^* f_n;\end{aligned}$$

and here $\sigma^*f_n \leqslant \sigma^*f_{n+1}$. Taking $f_n = \inf(f, n) \in L^\infty$, the validity of any one of (6.4.11), (6.4.12), or (6.4.13) for each of the bounded functions f_n leads (by the monotone convergence theorem for integrals once again) to the same inequality for f. These inequalities are thus seen to be valid for the specified classes of functions f.

Using 16.2.8 again, if one grants the convergence almost everywhere of the Fourier series of each $f \in \mathbf{L}^p$ for some fixed p satisfying $1 < p \leqslant 2$ [see 10.4.5(3) for the case $p = 2$], one could derive the inequalities (10.4.9) and (10.4.10). In this connection, see Exercise 13.16.

13.10.3. **Other Maximal Operators; Singular Integrals.** The operators σ^* and M^j ($j = \ell, r, \Delta$) [see Example 13.7.6(2) and cf. Exercise 14.23] are instances of what one might term "maximal" or "majorant" operators of the type

$$f \to K^*f \equiv \sup_{r \geqslant 1} |K_r * f|,$$

where $(K_r)_{r=1}^\infty$ is a sequence of well-behaved functions. As has been mentioned earlier (see 6.4.7 and Section 6.6), the study of such maximal operators has been made in the context of general (Hausdorff locally compact) groups in Edwards and Hewitt [1] and, with special success for compact groups, by Stein [1]; see also the closing remarks in Example 13.7.6(2). The results obtained in this way assert that, under suitable conditions on the K_r, K^* will enjoy the properties asserted of σ^* in (6.4.11) to (6.4.13); see Exercise 13.17.

It would nevertheless be a mistake to infer that this automatically disposes of all problems associated with the pointwise summability of Fourier series on more general groups, the fact being that the summability methods one may wish to study do not always satisfy the required hypotheses placed on the corresponding functions K_r. This happens as soon as one passes from T to its powers T^m, that is, as soon as one passes to the consideration of multiple Fourier series of periodic functions of several real variables. In particular, the (unrestricted) Cesàro summability of such series does *not* behave in quite the expected way. For the details, see [Z_2], Chapter XVII. It is shown there that the correct analogue of (6.4.12) requires the replacement of $\|f\|_1$ by

$$\left(\frac{1}{2\pi}\right)^m \int |f|(\log^+ |f|)^{m-1}\, dx,$$

and that of (6.4.13) requires the replacement of

$$\frac{1}{2\pi}\int |f| \log^+ |f|\, dx$$

by

$$\left(\frac{1}{2\pi}\right)^m \int |f|(\log^+ |f|)^m \, dx,$$

where we have written

$$\left(\frac{1}{2\pi}\right)^m \int \cdots \, dx$$

to denote the invariant integral on the group T^m. The difference can be traced back to necessary changes in the form of the Hardy-Littlewood maximal theorem in m dimensions, and this in turn is due at least in part to the topological properties of the underlying group.

One motivation for expending effort on attempts to prove that maximal operators K^* satisfy strong- or weak-type inequalities on certain domains is the assistance deriving from any such conclusion toward proving the existence and finiteness almost everywhere of the pointwise limit of the sequence $(K_r * f)_{r=1}^{\infty}$; see Exercise 13.26. Limits of such sequences are usually rather loosely described as "singular integrals."

As an example, consider the convolution operators introduced in 12.8.2, namely,

$$H_\varepsilon * f(x) = \frac{1}{2\pi} \int_{\varepsilon \leqslant |y| \leqslant \pi} f(x - y) \cot \frac{1}{2} y \, dy,$$

where $0 < \varepsilon < \pi$. The proof of 13.9.1 can be elaborated so as to yield the conclusion that the associated maximal operator $H^*: f \to \sup_{0 < \varepsilon < \pi} |H_\varepsilon * f|$ is of weak type (1, 1) on $\mathbf{L}^1$ (compare Koizumi [1], I, p. 171; [Kz], p. 76). Since, as has been seen in 12.8.2, the pointwise limit $f^c(x) = \lim_{\varepsilon \to 0} H_\varepsilon * f(x)$ exists finitely for all x whenever $f \in \mathbf{C}^\infty$ (say), it can be inferred from Exercise 13.26 that the limit $f^c(x)$ exists finitely for almost all x whenever $f \in \mathbf{L}^1$, a result due originally to Lusin and Privalov (see the end of 12.8.2). (The fact that we have in this case a continuous parameter $\varepsilon \to 0$ in place of an integer parameter $r \to \infty$ causes no trouble whatsoever.)

Statements about mean convergence as $\varepsilon \to 0$ of the transforms $H_\varepsilon * f$ also follow quite easily from the above property of H^*; see Exercise 13.27.

For further reading relating to maximal operators and singular integrals, see [EG]; [HR], Section 44; [St]; [Kz], pp. 66 ff.; Gilbert [3]; MR **35** # 6788; **37** ## 5731, 6144, 6704, 6687; **38** ## 575, 576, 2268, 3466; **39** ## 4709, 4711; **40** # 799; **50** # 10670; **52** # 1162; **53** # 1143; **54** ## 844, 3290, 5720, 5721, 5736, 8133a,b,c, 8155, 13452; **55** # 3670, 6096; **56** ## 959, 960, 6259, 6260, 6261, 6266; **57** # 10340.

13.11 Theorems of Hardy and Littlewood, Marcinkiewicz and Zygmund

Section 13.5 has produced one sort of valid extension of the Parseval formula (8.2.2) and the Riesz-Fischer theorem 8.3.1. This section is devoted to some more results of the same type.

13.11.1. (Hardy-Littlewood) (1) If $1 < p \leqslant 2$ and $f \in \mathbf{L}^p$, then

$$\{\sum_{n \in Z} (1 + |n|)^{p-2} |\hat{f}(n)|^p\}^{1/p} \leqslant A_p \|f\|_p. \tag{13.11.1}$$

(2) If $2 \leqslant q < \infty$ and ϕ is a function on Z satisfying

$$\sum_{n \in Z} (1 + |n|)^{q-2} |\phi(n)|^q < \infty, \tag{13.11.2}$$

then the trigonometric polynomials

$$\hat{\phi}_N(x) = \sum_{|n| \leqslant N} \phi(n) e^{inx} \tag{13.11.3}$$

converge in $\mathbf{L}^q$ as $N \to \infty$ to a function f satisfying $\hat{f} = \phi$ and

$$\|f\|_q \leqslant A_{q'} \{\sum_{n \in Z} (1 + |n|)^{q-2} |\phi(n)|^q\}^{1/q}. \tag{13.11.4}$$

Remark. The function f appearing in statement (2) plainly provides a sensible interpretation of $\hat{\phi}$; see 2.5.1 and 8.3.3.

Proof. We derive (1) from an application of the Marcinkiewicz interpolation theorem in a suitable setting, and deduce (2) from (1).

(1) Here we take $(X, \mathscr{M}, \mu)$ as in Example 13.1.3(1), and $(Y, \mathscr{N}, \nu)$ as in Example 13.1.3(2) with

$$c_n = (1 + |n|)^{-2}.$$

Furthermore, take T to be defined by the formula

$$Tf(n) = n\hat{f}(n)$$

for, say, μ-simple functions f.

Parseval's formula (8.2.2) shows at once that T is of type (2, 2). We will show that T is of weak type (1, 1). Indeed, if $f^{\#} = Tf$ and $t > 0$, then

$$D^{\nu}_{f^{\#}}(t) = \nu(\{|f^{\#}(n)| > t\}) = \sum (1 + |n|)^{-2}$$

summed over those integers n for which $|n\hat{f}(n)| > t$. This last inequality implies that $|n| > t/\|f\|_1$, so that

$$D^{\nu}_{f^{\#}}(t) \leqslant \sum_{|n| > t/\|f\|_1} (1 + |n|)^{-2} \leqslant \frac{2\|f\|_1}{t},$$

which shows that T is of weak type (1, 1). [It should be noticed that Exercise 3.14 shows that T is *not* of type (1, 1).]

Appeal to the Marcinkiewicz interpolation theorem 13.8.1 shows that T is of type (p, p) whenever $1 < p < 2$; as we have noted, the same is true for $p = 2$. This leads to the inequality (13.11.1) for μ-simple f. The extension to a general $f \in \mathbf{L}^p$ is carried out in a routine manner by approximating f in $\mathbf{L}^p$ by a sequence of μ-simple functions; see 13.2.2.

(2) Write $p = q'$, so that $1 < p \leqslant 2$. For brevity write w_n in place of $(1 + |n|)^{q-2}|\phi(n)|^q$. Suppose that $g \in \mathbf{L}^p$ and that $N < N'$ are positive integers. On applying Hölder's inequality and (1) it is seen that

$$\begin{aligned}
\left|\frac{1}{2\pi}\int \hat{\phi}_N g\,dx\right| &= \left|\sum_{|n|\leqslant N} \phi(n)\hat{g}(-n)\right| \\
&= \left|\sum_{|n|\leqslant N} (1 + |n|)^{1-2/q}\phi(n) \cdot (1 + |n|)^{1-2/p}\hat{g}(-n)\right| \\
&\leqslant \{\sum_{|n|\leqslant N} w_n\}^{1/q} \cdot \{\sum_{|n|\leqslant N} (1 + |n|)^{p-2}|\hat{g}(-n)|^p\}^{1/p} \\
&\leqslant \{\sum_{|n|\leqslant N} w_n\}^{1/q} \cdot A_p\|g\|_p. \qquad (13.11.5)
\end{aligned}$$

Similarly,

$$\left|\frac{1}{2\pi}\int (\hat{\phi}_N - \hat{\phi}_{N'})g\,dx\right| \leqslant \{\sum_{N<|n|\leqslant N'} w_n\}^{1/q} \cdot A_p\|g\|_p. \qquad (13.11.6)$$

On using 13.1.5, the inequalities (13.11.5) and (13.11.6) lead respectively to the estimates

$$\|\hat{\phi}_N\|_q \leqslant \{\sum_{|n|\leqslant N} w_n\}^{1/q} \cdot A_p \qquad (13.11.7)$$

and

$$\|\hat{\phi}_N - \hat{\phi}_{N'}\|_q \leqslant \{\sum_{N<|n|\leqslant N'} w_n\}^{1/q} \cdot A_p. \qquad (13.11.8)$$

The last inequality combined with (13.11.2) shows that the sequence $(\hat{\phi}_N)_{N=1}^{\infty}$ is Cauchy and therefore convergent in $\mathbf{L}^q$; let f be its limit. Then, by mean convergence,

$$\hat{f}(n) = \lim_{N\to\infty} (\hat{\phi}_N)\hat{\ }(n),$$

which is easily seen to equal $\phi(n)$. Moreover, the defining relation

$$f = \lim_{N\to\infty} \hat{\phi}_N \qquad \text{in } \mathbf{L}^q$$

and the inequality (13.11.7) show that (13.11.4) holds and so complete the proof.

13.11.2. **Remarks.** (1) The result 13.11.1 was extended by Paley to expansions in terms of orthonormal systems of functions u_n which are uniformly bounded; see [Z_2], p. 121 and compare the remarks in 13.5.2. Yet further generality was achieved by Marcinkiewicz and Zygmund [1]. In Section 3 of Stein and Weiss [1], it is shown that this generalized version can be deduced very neatly from their version of the Riesz-Thorin theorem [mentioned in 13.4.2(2)]. Compare also with Hörmander [1], Theorem 1.10 and Gaudry [5], Theorem 5.1.

For a variant of the Hardy-Littlewood-Paley theorem involving rearrangements of Fourier coefficients, see [Z_2], p. 123.

(2) A more general version of 13.11.1 appears in Exercise 13.9. See also Exercises 13.1, 13.9, 13.21, 13.23, Askey and Wainger [1] and Izumi [2].

13.11.3. **The Dual Version of 13.11.1.** It is possible to dualize 13.11.1, that is, to formulate versions in which T and Z interchange their roles. Before making any statement of this sort, we remark that in 13.11.1 the expression $(1 + |n|)$ might everywhere be replaced by a number of other positive functions ω on Z subject to the condition

$$\sum_{\{\omega(n) > s\}} \omega(n)^{-2} \leqslant \frac{A}{s} \qquad (s > 0),$$

which would be used to ensure that the operator $T\colon f \to \omega\hat{f}$ is of weak type (1, 1).

This suggests a formulation of the dual result in terms of a measurable function ω on T such that

$$\omega(x) > 0 \text{ a.e.}, \qquad \frac{1}{2\pi}\int_{\{\omega(x) > s\}} \omega(x)^{-2}\,dx \leqslant \frac{A}{s} \qquad (s > 0). \tag{13.11.9}$$

We shall regard ω as a function on R having period 2π, and make use of the measure space $(Y, \mathscr{N}, \nu)$ in which Y is the interval $[-\pi, \pi)$, $\mathscr{N}$ is the collection of Lebesgue measurable subsets of Y, and ν is the measure defined by

$$\nu(E) = \frac{1}{2\pi}\int_E \omega(x)^{-2}\,dx \qquad (E \in \mathscr{N}).$$

It is easily verified that

$$\int_Y h\,d\nu = \frac{1}{2\pi}\int h(x)\omega(x)^{-2}\,dx$$

for any nonnegative Lebesgue measurable periodic function h on R.

13.11.4. (Marcinkiewicz-Zygmund) Let ω be as explained in 13.11.3.

(1) If $1 < p \leqslant 2$ and $\phi \in \ell^p(Z)$, then

$$\{\frac{1}{2\pi}\int \omega^{p-2}|\hat{\phi}|^p\,dx\}^{1/p} \leqslant A_p\|\phi\|_p. \tag{13.11.10}$$

(2) If $2 \leqslant q < \infty$ and f is measurable and $\int \omega^{q-2}|f|^q \, dx < \infty$, then $f \in \mathbf{L}^1$ and

$$\|\hat{f}\|_q \leqslant A_{q'} \{\frac{1}{2\pi} \int \omega^{q-2}|f|^q \, dx\}^{1/q}. \tag{13.11.11}$$

Proof. This is very similar to that of 13.11.1 and we shall be brief.

(1) We aim to apply the Marcinkiewicz interpolation theorem 13.8.1, taking $(X, \mathscr{M}, \mu)$ as in Example 13.1.3(2) with μ the counting measure on Z, $(Y, \mathscr{N}, \nu)$ as described in 13.11.3, and the operator T defined by

$$T\phi(x) = \omega(x)\hat{\phi}(x)$$

for μ-simple functions ϕ on Z.

Parseval's formula shows that T is of type (2, 2). The condition (13.11.9) ensures that T is of weak type (1, 1), since the set E_t of points x at which $|T\phi(x)| > t$ is contained in the set at which $\omega(x) > t/\|\phi\|_1$, and so

$$\nu(E_t) = \frac{1}{2\pi} \int_{E_t} \omega^{-2} \, dx \leqslant \frac{1}{2\pi} \int_{\{\omega(x) > t/\|\phi\|_1\}} \omega(x)^{-2} \, dx$$
$$\leqslant \frac{A\|\phi\|_1}{t}.$$

It follows, via Marcinkiewicz's theorem 13.8.1, that T is of type (p, p) whenever $1 < p \leqslant 2$, which leads at once to (13.11.10) for μ-simple ϕ. The extension to arbitrary $\phi \in \ell^p(Z)$ is routine and is left to the reader.

(2) Write $p = q'$. If $f \in \mathbf{L}^1$ and

$$J \equiv \{\frac{1}{2\pi} \int \omega^{q-2}|f|^q \, dx\}^{1/q} < \infty,$$

we have from Hölder's inequality and (1)

$$|\frac{1}{2\pi} \int f\hat{\phi} \, dx| = |\frac{1}{2\pi} \int \omega^{1-2/q} f \cdot \omega^{1-2/p}\hat{\phi} \, dx|$$
$$\leqslant J \cdot \{\frac{1}{2\pi} \int \omega^{p-2}|\hat{\phi}|^p \, dx\}^{1/p}$$
$$\leqslant J \cdot A_p\|\phi\|_p.$$

This shows that the assignment

$$\phi \to \frac{1}{2\pi} \int f\hat{\phi} \, dx$$

is a continuous linear functional on $\ell^p(Z)$. From the discrete analogue of I, C.1 [see [E], Exercise 1.2 and Theorem 4.16.1; [HS], Theorem (15.12)] it follows that there exists a function $\psi \in \ell^{p'}(Z) = \ell^q(Z)$ such that

$$\frac{1}{2\pi}\int f\hat{\phi}\,dx = \sum_{n\in Z}\phi(n)\psi(-n) \tag{13.11.12}$$

for all $\phi \in \ell^p(Z)$ and

$$\|\psi\|_p \leqslant A_p J. \tag{13.11.13}$$

From (13.11.12) it follows that $\hat{f} = \psi$, so that (13.11.13) yields (13.11.11).

Thus (13.11.11) holds whenever $f \in \mathbf{L}^1$ and $J < \infty$. Supposing now that f is nonnegative and measurable and that $J < \infty$, an application of (13.11.11) to $f_r = \inf(f, r) \in \mathbf{L}^1$, followed by a limiting process as $r \to \infty$, shows that

$$\|f_r\|_1 = |\hat{f}_r(0)| \leqslant \|f_r\|_q \leqslant A_{q'} \cdot J$$

and therefore that $f \in \mathbf{L}^1$. On replacing f by $|f|$, we see that $f \in \mathbf{L}^1$ whenever $J < \infty$. The proof is therefore complete.

13.11.5. **Remarks.** In 13.11.4 one may take

$$\omega(x) = |x|^\alpha \qquad \text{for } |x| \leqslant \pi$$

whenever $\alpha \leqslant 1$. The case $\alpha = 1$ is due to Marcinkiewicz and Zygmund [1] and is treated otherwise in Theorems (3.8) and (3.9) of Stein and Weiss [1].

EXERCISES

13.1. Show that $\sum_{n\neq 0} |\hat{f}(n)/n| < \infty$ whenever $f \in \mathbf{L}^p$ for some $p > 1$.

Remarks. Since (see Exercise 7.7) $\sum_{n=2}^{\infty} \cos nx/\log n$ is the Fourier series of an integrable function, it follows that $\sum_{n\neq 0} \hat{f}(n)/|n|$ diverges for suitably chosen $f \in \mathbf{L}^1$.

The stated result should be compared with Exercises 7.9, 13.21, and 13.23.

13.2. Assume that f is absolutely continuous and that $Df \in \mathbf{L}^p$, where $1 < p \leqslant 2$. Show that (with the notation introduced in 10.6.1) $f \in \mathbf{A}$ and that

$$\|f\|_{\mathbf{A}} \leqslant |\hat{f}(0)| + \{2\zeta(p)\}^{1/p}\|Df\|_p,$$

where $\zeta(s) = \sum_{n=1}^{\infty} n^{-s}$ for $\operatorname{Re} s > 1$.

Remark. There are analogous results applying when f is assumed merely to be a distribution (in the sense of Chapter 12) and Df is replaced by $D^m f$ (m a positive integer). These are simple periodic versions of results, named collectively *Sobolev's lemma*, which apply to (not necessarily periodic) distributions defined on domains in a Euclidean space R^n.

13.3. Assume that $0 < a < \pi$ and that f is defined and absolutely continuous on $[-a, a]$, that $1 < p \leqslant 2$, and that

$$I = \{\frac{1}{2\pi}\int_{-a}^{a} |Df|^p\,dx\}^{1/p}, \qquad M = \sup_{|x| \leqslant a} |f(x)|.$$

Show that there exists a function $g \in \mathbf{A}$ (see 10.6.1) such that $g(x) = f(x)$ for $|x| \leqslant a$ and

$$\|g\|_{\mathbf{A}} \leqslant \alpha_p M + \beta_p I,$$

where α_p and β_p are nonnegative numbers depending only on p.

Remarks. Estimates of the stated type are often useful in approximation theory; see, for example, Domar [2]. The given estimate can be varied by ringing changes on the construction of g.

Hint: Define g as a suitable periodic extension of f and use Exercise 13.2.

13.4. State and prove an analogue of 13.6.1 for functions on Z.

13.5. Let $\mu \in \mathbf{M}$ and define

$$B_p(\mu) = \begin{cases} \|\mu\|_1^{(2/p)-1}\|\hat{\mu}\|_\infty^{2-(2/p)} & \text{if } 1 \leqslant p \leqslant 2 \\ B_{p'}(\mu) & \text{if } 2 < p \leqslant \infty. \end{cases}$$

Prove that for $1 \leqslant p \leqslant \infty$

$$\|\mu * f\|_p \leqslant B_p(\mu)\|f\|_p$$

whenever $f \in \mathbf{L}^p$.

Hint: Consider first the case $1 \leqslant p \leqslant 2$, using the case $p = 1$ of 12.7.3, the Parseval formula, and the Riesz-Thorin theorem. For the case $2 < p \leqslant \infty$, appeal to Hölder's inequality and its converse (compare 16.4.1).

13.6. We revert to the matters discussed in Section 4.2. Let α be a map of Z into $Z \cup \{\infty\}$ such that $\alpha^{-1}(\{n\})$ is finite for each $n \in Z$. Suppose that T is the operator defined for trigonometric polynomials f by the formula

$$Tf = \sum_{n \in Z} \hat{f}(\alpha(n))e_n;$$

compare equation (4.2.5). Suppose further that $1 \leqslant p_1 < p_2 \leqslant \infty$ and that there exist numbers m_1 and m_2 such that

$$\|Tf\|_{p_i} \leqslant m_i\|f\|_{p_i}$$

for $i = 1, 2$ and all trigonometric polynomials f. Prove that, if p satisfies $p_1 \leqslant p \leqslant p_2$, T can be extended into a continuous homomorphism of $\mathbf{L}^p$ into itself.

Hint: The reader should pay especially close attention to the case in which $p = p_2 = \infty$, in which connection 12.3.10(2) will be useful.

13.7. The notation is as in 13.2.3(2). Show that S is a one-to-one linear map of $\mathbf{L}^1(\mu)$ into $\mathbf{c}_0(Y)$ which is of type $(1, \infty)$.

13.8. The notation is as in 13.2.3(2).

(1) Assuming that $S|\mathbf{L}^2(\mu)$ is of type $(2, 2)$, prove that it is of type (p, p') whenever $1 \leqslant p \leqslant 2$.

(2) Show by example that, if $p > 2$, then $S|\mathbf{L}^p(\mu)$ is not of type (p, p').

13.9. (1) Suppose that $1 < p \leqslant r \leqslant p'$ (so that $1 < p \leqslant 2$). Show that there exists a number $A = A_p > 0$ such that

$$\|(1 + |n|)^{1/p' - 1/r}\hat{f}\|_r \leqslant A\|f\|_p$$

for each $f \in \mathbf{L}^p$.

(2) Suppose that $q' \leqslant s \leqslant q < \infty$ (so that $2 \leqslant q < \infty$) and that ϕ is a complex-valued function on Z such that $(1 + |n|)^{1 - 1/q - 1/s}\phi \in \ell^s(Z)$. Show that there exists a function $g \in \mathbf{L}^q$ such that $\hat{g} = \phi$ and

$$\|g\|_q \leqslant A_{q'}\|(1 + |n|)^{1 - 1/q - 1/s}\phi\|_s.$$

Remarks. Results (1) and (2) generalize 13.11.1(1) and 13.11.1(2), respectively. Compare with [Z_2], p. 126.

There is an analogue applying in the case in which the group T is replaced by R; see Hörmander [1], Corollary 1.6.

The stated results are of interest and use in connection with multipliers; see 16.4.6(3).

Hints: For (1), use 13.5.1 and 13.11.1. Derive (2) from (1) by mimicking the argument used in the text to deduce 13.11.1(2) from 13.11.1(1).

13.10. Write out the details of the subcase $p_0 < p < p_1$, $\infty > q_0 > q_1$ of Theorem 13.8.1 [see the end of stage (1) of the proof given in the text].

13.11. Supply the proof of 13.8.1 in case (2) listed in the text.

Hints: Here $p_0 = p_1 = p$, $q_0 < \infty$, $q_1 = \infty$. Suppose $a > 0$ and write

$$\|Tf\|_{q,\nu}^q = q\int_0^a b^{q-1}D_{Tf}{}^{\nu}(b)\,db + q\int_a^{\infty} b^{q-1}D_{Tf}{}^{\nu}(b)\,db.$$

Let $a \downarrow A_1\|f\|_{p,\mu}$.

13.12. Supply the proof of 13.8.1 in case (3) listed in the text.

Hints: Here $1 \leqslant p_0 \leqslant q_0 < \infty$, $p_1 = q_1 = \infty$. Suppose $a > 0$; then

$$\|Tf\|_{q,\nu}^q \leqslant \text{const}\,\{\int_0^{\infty} b^{q-1}D_{Tf_{1,a}}^{\nu}(b)\,db + \int_0^{\infty} b^{q-1}D_{Tf_{2,a}}^{\nu}(b)\,db\}$$

Choose $a = a(b) = A_1^{-1}b$, and estimate

$$\int_0^{\infty} b^{q-1}D_{Tf_{2,a(b)}}^{\nu}(b)\,db$$

as in stage (1) of the proof in the text.

13.13. Supply the proof of 13.8.1 in case (5) listed in the text.

Hint: Modify the proof of case (4) given in the text, using as a guide the changes made in (1) to handle the case in which $q_0 > q_1$.

13.14. Let θ and $\Omega = \bigcup_i \Omega_i = \bigcup_i (a_i, b_i)$ be as in 13.9.1(4). Show that $\theta(a_i) = \theta(b_i - 0)$ for all i.

13.15. Verify the assertion made in part (5) of the proof of 13.9.1, namely, if $h_{j,n}$ are the functions there appearing, then there exist functions

$$h_{(j)} \in \mathbf{L}^2(-\pi, \pi)$$

such that $h_{j,n} \to h_{(j)}$ in $\mathbf{L}^2(-\pi, \pi)$ for $j = 1, 2, 3$.

Hint: Consider, for example, $h_{1,n}$. Observe that $h_{1,n}$ bears the same relationship to ϕ as h_n does to f.

13.16. (1) Let $(X, \mathscr{M}, \mu)$ be a measure space and f an extended real- or complex-valued μ-measurable function on X such that, for some $p > 0$, $A > 0$,

$$D_f^{\mu}(t) \leqslant (At^{-1})^p \qquad (t > 0).$$

Prove that if $0 < q < p$, then

$$\int_S |f|^q \, d\mu \leqslant \{\mu(S) + \frac{q}{p - q}\} A^q$$

for any μ-finite set $S \subset X$.

(2) Let $(X, \mathscr{M}, \mu)$, $(Y, \mathscr{N}, \nu)$ and $T: f \to f^{\#}$ be as in 13.7.5, Y being ν-finite and the domain $\mathfrak{D}$ of T having the property that $cf \in \mathfrak{D}$ whenever $f \in \mathfrak{D}$ and c is a positive number. Suppose that T is of weak type (p, q), where $0 < p \leqslant \infty$ and $0 < q \leqslant \infty$, and that $|T(cf)| \geqslant c|Tf|$ ν-a.e. for each number $c > 0$ and each $f \in \mathfrak{D}$. Prove that T satisfies a (strong) type (p, r) inequality on $\mathfrak{D}$ whenever $r < q$.

13.17. Let (K_i) be an arbitrary family of functions in $\mathbf{L}^\infty$ such that $|K_i| \leqslant H_i$ a.e. on $(-\pi, \pi)$, where the functions H_i are absolutely continuous (but not necessarily periodic) and

$$A \equiv \sup_i \frac{1}{2\pi} \int_{-\pi}^{\pi} H_i(x)\, dx < \infty,$$

$$B \equiv \sup_i \frac{1}{2\pi} \int_{-\pi}^{\pi} |xH_i'(x)|\, dx < \infty.$$

Define, for $f \in \mathbf{L}^1$,

$$K^*f(x) = \sup_i |K_i * f(x)|.$$

Show that

$$K^*f(x) \leqslant (A + 2B) f^{\Delta}(x),$$

where f^{Δ} is defined as in Example 13.7.6(2).

Show also that the same inequality holds almost everywhere if (K_i) is a countable family of functions in $\mathbf{L}^1$, the other hypotheses remaining as before.

Apply this procedure to σ^*f, taking $K_N = F_N$ for $N = 0, 1, 2, \cdots$ and

$$H_N(x) = CN(1 + N|x|)^{-2}$$

for $|x| < \pi$, C being a suitable constant, and thus conclude from the results stated in Example 13.7.6(2) that σ^* is of weak type (1, 1) on $\mathbf{L}^1$.

13.18. Let $(X, \mathscr{M}, \mu)$ be a measure space and f_n ($n = 1, 2, \cdots$) and f extended real- or complex-valued μ-measurable functions on X that are finite almost everywhere. Suppose that

$$\lim_{n\to\infty} f_n = f$$

in measure, that is, for any $t > 0$,

$$\lim_{n\to 0} D^{\mu}_{f-f_n}(t) = 0.$$

(Notice that $f - f_n$ is defined μ-a.e.; its definition may be completed in any desired fashion.) Prove that

$$D_f{}^{\mu}(t) \leqslant \inf_{\varepsilon>0} \liminf_{n\to\infty} D_{f_n}{}^{\mu}(t - \varepsilon)$$

for any $t > 0$.

Note that the hypotheses are satisfied if $f_n \to f$ in $\mathbf{L}^p(X, \mathscr{M}, \mu)$ for any $p > 0$.

13.19. Let $(X, \mathscr{M}, \mu)$ and $(Y, \mathscr{N}, \nu)$ be measure spaces. Denote by $\mathscr{F} = \mathscr{F}(Y, \mathscr{N}, \nu)$ the set of ν-equivalence classes of ν-measurable extended real- or complex-valued functions on Y which are finite valued ν-almost everywhere. Suppose that Y is expressible as the union of disjoint ν-finite sets Y_k ($k = 1, 2, \cdots$) satisfying $\nu(Y_k) > 0$. Verify that $\mathscr{F}$ is a complete metric space when the distance is defined by

$$d(g_1, g_2) = \sum_{k=1}^{\infty} \frac{1}{k^{-2}\nu(Y_k)} \int_{Y_k} \frac{|g_1 - g_2|\, d\nu}{1 + |g_1 - g_2|}.$$

(The convergence of a sequence of functions in this metric is the same as convergence in measure on each ν-finite set; see the preceding exercise.)

Suppose that T is a linear operator with domain a linear subspace $\mathfrak{T}$ of $\mathbf{L}^p(X, \mathscr{M}, \mu)$ and range in $\mathscr{F}$ which is of weak type (p, q). Show that T has a unique extension to the closure $\bar{\mathfrak{T}}$ of $\mathfrak{T}$ in $\mathbf{L}^p(X, \mathscr{M}, \mu)$ which is continuous from $\bar{\mathfrak{T}}$ into $\mathscr{F}$, and that this extension is also of weak type (p, q).

Hints: Show that a sequence of functions which converges in measure has a subsequence which converges pointwise almost everywhere. Use the preceding exercise.

13.20. Show that $\mathbf{L}^p * \mathbf{L}^p \subset \mathbf{L}^{p/(2-p)}$ if $1 \leqslant p < 2$, and that $\mathbf{L}^p * \mathbf{L}^p \subset \mathbf{A}$ if $p \geqslant 2$.

Prove also that

$$\sum_{n\in Z} |\hat{h}(n)|^{p/(2p-2)} < \infty$$

whenever $h \in \mathbf{L}^p * \mathbf{L}^p$ and $1 < p \leqslant 2$.

Remarks. These results show that $\mathbf{L}^p * \mathbf{L}^p$ is a proper subset of $\mathbf{L}^p$ whenever $p > 1$; see 7.5.3, Section 8.4, and 15.3.4.

13.21. Prove that

$$\sum_{n \in Z} (1 + |n|)^{-1} |\hat{f}(n)| \leqslant A + \frac{B}{2\pi} \int |f| \log^+ |f| \, dx$$

whenever the integral appearing on the right is finite, A and B denoting suitably chosen positive numbers. Compare with Exercises 13.1 and 13.23.

Hint: Apply (13.8.21) to a suitably chosen particular case.

13.22. The notations and assumptions are as in 13.8.3, and we now assume also that $\mu(X) < \infty$, $\nu(Y) < \infty$, and $1 < a < \infty$.

Suppose that Ψ is defined on $[0, \infty)$, vanishes on $[0, 1]$, is elsewhere positive and increasing, and

$$\Psi(2t) = O(\Psi(t)) \qquad \text{as } t \to \infty.$$

Define

$$\Phi(t) = t \int_0^t s^{-2} \Psi(s) \, ds \qquad (t \geqslant 0)$$

and suppose that

$$\int_t^\infty \Phi(s) \, ds/s^{a+1} = O\{\Phi(t)/t^a\} \qquad \text{as } t \to \infty.$$

Prove that there exist (f-independent) numbers A and B such that

$$\int_Y \Psi(|f^\#|) \, d\nu \leqslant A + B \int_X \Phi(|f|) \, d\mu$$

for each $f \in \mathfrak{D}$.

Remarks. One may take $\Psi(t) = t$ for $t > 1$, in which case $\Phi(t) = t \log^+ t$ and one recovers (13.8.21). One may also take $\Psi(t) = t(\log^+ t)^{p-1}$ for $t > 1$, where $p > 0$, and then $\Phi(t) = p^{-1} t (\log^+ t)^p$.

Hints: Write $c = \max(2\kappa, 1)$ and show first that

$$\int_Y \Psi(|f^\#|) \, d\nu \leqslant K + \sum_{j=0}^{\infty} \eta_j \delta_j, \tag{1}$$

where K denotes an f-independent number (not necessarily the same at each appearance), η_j is the ν-measure of the set where $|f^\#| > c2^j$, and

$$\delta_j = \Psi(c2^{j+1}) - \Psi(c2^j).$$

For a fixed j, split f into $f_1 + f_2$, where f_1 equals f or 0 according as $|f| \leqslant 2^j$ or not. Apply (13.8.18) and (13.8.19) to f_2 and f_1, respectively, in order to derive the inequality

$$n_j \leqslant K\{2^{-aj} \sum_{i=0}^{j} 2^{ai} \varepsilon_i + 2^{-j} \sum_{i=j+1}^{\infty} 2^i \varepsilon_i\}, \tag{2}$$

where ε_0 is the μ-measure of the set where $|f| \leqslant 1$ and ε_i is the μ-measure of the set where $2^{i-1} < |f| \leqslant 2^i$ for $i = 1, 2, \cdots$. Substitute (2) into (1) and obtain

$$\int_Y \Psi(|f^\#|)\, d\nu \leqslant K + K(S_1 + S_2),$$

where

$$S_1 = \sum_{i=0}^{\infty} 2^{ai}\varepsilon_i \sum_{j=i}^{\infty} \delta_j 2^{-aj},$$

$$S_2 = \sum_{i=1}^{\infty} 2^{i}\varepsilon_i \sum_{j=0}^{i-1} \delta_j 2^{-j}.$$

Show that each of S_1 and S_2 is majorized by an expression of the desired form.

For more details, see [Z_2], pp. 117–118.

13.23. Suppose that $p > 0$. Show that there exist f-independent numbers A and B such that

$$\sum_{n \in Z} \frac{|\hat{f}(n)|\{\log(2 + |n\hat{f}(n)|)\}^{p-1}}{1 + |n|} \leqslant A + \frac{B}{2\pi}\int |f|(\log^+ |f|)^p\, dx$$

whenever the integral appearing on the right is finite.

Deduce that

$$\sum_{n \in Z} \frac{|\hat{f}(n)|\{\log(2 + |n|)\}^{p-1}}{1 + |n|}$$

is majorized by a similar expression under the same hypotheses on f.

Remark. This result generalizes that in Exercise 13.21.

Hints: For the first part, apply the preceding exercise to a suitably chosen special case.

Derive the second part from the first by splitting the sum in question into two parts according as $|n|^{-1/2} \leqslant |\hat{f}(n)| \leqslant |n|^{1/2}$ or not.

13.24. Suppose that $2 \leqslant p \leqslant \infty$ and that F is a complex-valued function of a complex variable that is expressible in the form

$$F(z) = az + b\bar{z} + |z|^{2/p'}c(z),$$

where a and b are complex numbers and the function c is bounded on some neighborhood of the origin.

Show that to each $f \in \mathbf{L}^p$ corresponds $g \in \mathbf{L}^p$ such that $\hat{g} = F \circ \hat{f}$.

Remark. This is the simple half of Rider's result cited in 10.6.3(2); it is due to Rudin.

13.25. State and prove an analogue of the result asserted in the preceding exercise that is applicable when $1 < p \leqslant 2$.

13.26. Let $(X, \mathscr{M}, \mu)$ and $(Y, \mathscr{N}, \nu)$ be measure spaces and K_r ($r = 1, 2, \cdots$) bounded and measurable functions on the product measure space

$(X \times Y, \mathscr{M} \times \mathscr{N}, \mu \times \nu)$ (see [HS], p. 379; the condition of boundedness on each K_r could be relaxed). For each r, $T_r f$ is defined by

$$T_r f(y) = \int_X K_r(x, y) f(x)\, d\mu(x)$$

for $f \in \mathbf{L}^1(X, \mathscr{M}, \mu)$ and $y \in Y$. Define also the maximal operator

$$T^* f(y) = \sup_r |T_r f(y)|.$$

Suppose the two following conditions are fulfilled:

(1) T^* is of weak type (p, q) on $\mathbf{L}^p(X, \mathscr{M}, \mu)$, where $1 \leqslant p \leqslant \infty$, $0 < q \leqslant \infty$;

(2) there is an everywhere-dense subset $\mathfrak{D}$ of the set of real-valued functions in $\mathbf{L}^p(X, \mathscr{M}, \mu)$ such that $\lim_{r \to \infty} T_r f(y)$ exists finitely for ν-almost all $y \in Y$ whenever $f \in \mathfrak{D}$.

Show that $\lim_{r \to \infty} T_r f(y)$ exists finitely for ν-almost all $y \in Y$ whenever $f \in \mathbf{L}^p(X, \mathscr{M}, \mu)$.

Remarks. (i) Special cases of this result appear as partial converses of 16.2.8.

(ii) It may be proved that, if in (2) it is assumed that, for every $f \in \mathfrak{D}$, $\lim_{r \to \infty} T_r f(y) = 0$ for ν-almost all $y \in Y$, then the same is true for every $f \in \mathbf{L}^p$.

Hints: Show that it suffices to handle the case in which the K_r and f are real-valued. Given $\varepsilon > 0$, express any $f \in \mathbf{L}^p(X, \mathscr{M}, \mu)$ in the form $f = f_1 + f_2$, where $f_1 \in \mathfrak{D}$ and $\|f_2\|_{p,\mu} \leqslant \varepsilon$. Observe that

$$\left|\limsup_{r \to \infty} T_r f - \liminf_{r \to \infty} T_r f\right| \leqslant 2T^* f_2.$$

13.27. Assume that the maximal operator $H^*: f \to \sup_{0<\varepsilon<\pi} |H_\varepsilon * f|$ is of weak type (1, 1) on $\mathbf{L}^1$ (see 13.10.3). Deduce that

$$\sup_{0<\varepsilon<\pi} \|H_\varepsilon\|_{p,p} < \infty$$

for $1 < p \leqslant 2$, and then that the same is true for $2 \leqslant p < \infty$ (compare 12.9.6 or 16.4.1). Conclude that

$$\lim_{\varepsilon \to 0} H_\varepsilon * f = Hf \equiv \tilde{f} \qquad \text{in } \mathbf{L}^p$$

whenever $f \in \mathbf{L}^p$ and $1 < p < \infty$.

CHAPTER 14

Changing Signs of Fourier Coefficients

In this chapter we shall be concerned with some remarkable facts concerning not one Fourier series

$$\sum_{n \in Z} \hat{f}(n) e^{inx},$$

but rather "most" series

$$\sum_{n \in Z} \pm \hat{f}(n) e^{inx}$$

of the family obtained by making random changes of sign in the coefficients of the original series. It turns out that the behaviour of "most" members of such a family depends solely on the convergence or divergence of the series

$$\sum_{n \in Z} | \hat{f}(n) |^2 ;$$

if this series converges, then "most" members of the family are, in particular, Fourier series of functions in $\mathbf{L}^p$ for every $p < \infty$; while, if this series diverges, "most" members of the family fail to be Fourier-Lebesgue (or even Fourier-Stieltjes) series at all. We shall concentrate principally on the good behaviour resulting from the assumed convergence of $\sum | \hat{f}(n) |^2$; results pertaining to the case in which $\sum | \hat{f}(n) |^2 = \infty$ are mentioned only briefly in 14.2.3 and 14.3.5.

The technique we use for handling such a family, which at the same time gives a precise meaning to the term "most", stems from replacing it by

$$\sum_{n \in Z} \omega(n) \hat{f}(n) e^{inx},$$

where ω is any function from Z to $\{-1, +1\}$. This leads us to consider the set $\mathscr{C}$ of all such functions ω and we impose upon $\mathscr{C}$ the natural structure of a compact Abelian topological group and the associated normalised Haar measure. We interpret "most" to mean "belonging to a set of Haar measure one", and carry out some elementary harmonic analysis on the group $\mathscr{C}$. In the presence of the assumption $\sum |\hat{f}(n)|^2 < \infty$ we shall be able to perform a quick mental somersault and consider the series

$$\sum_{n \in Z} \omega(n) \hat{f}(n) e^{inx}$$

as defining a function belonging to either $\mathbf{L}^2(T)$ or $\mathbf{L}^2(\mathscr{C})$. This will provide the results we need to study the above family of Fourier series.

Thus the first part of this chapter is taken up with establishing properties of series which are the counterpart over $\mathscr{C}$ of Fourier series over T, namely, infinite complex linear combinations

$$\sum_{\zeta \in \mathscr{C}^\wedge} c_\zeta \zeta,$$

where $\mathscr{C}^\wedge$ denotes the group of characters of $\mathscr{C}$. These so-called Walsh-Fourier series include the families above as special cases in which the coefficients c_ζ are nonzero only for those ζ belonging to a certain lacunary (or thin) subset $\mathscr{R}$ of $\mathscr{C}^\wedge$. These lacunary series

$$\sum_{\zeta \in \mathscr{R}} c_\zeta \zeta$$

are called Rademacher series and results about them established in this chapter provide both an introduction to, and tools for the study of, concepts introduced in Chapter 15.

It should be pointed out that our approach to Rademacher series was not, historically, the first, in which Rademacher series appear instead as series of functions defined on [0, 1]. For a treatment of the classical approach we refer the reader to the first edition of this book, and for the connection between the two approaches to Exercise 14.16 or to Appendix C of [EG].

In what follows, Z_+ will denote the set of nonnegative integers. Further, for every set X, 1^X will denote the constant function with domain X and value 1.

14.1 Harmonic Analysis on the Cantor Group

14.1.1. **The Cantor Group.** We denote by $\mathscr{C}$ the set of all functions from the integers into $\{-1, +1\}$, and write its elements $\omega = (\omega(n))_{n \in Z}$. Under the operation of pointwise product, $\mathscr{C}$ is an Abelian group with identity the constant function 1^Z, and each element is its own inverse: $\omega^2 = 1^Z$, for every $\omega \in \mathscr{C}$.

If $\{-1, +1\}$ is given the discrete topology, it becomes a compact Hausdorff space. Endowing $\mathscr{C} \equiv \{-1, +1\}^Z$ with the product topology, Tychonoff's theorem ensures that $\mathscr{C}$ too becomes compact and Hausdorff.

More precisely, a subbasis for the topology on $\mathscr{C}$ consists of all sets of the form $\{\omega \in \mathscr{C} : \omega(n) = \alpha\}$, where $\alpha \in \{-1, +1\}$ and $n \in Z$. Consequently an open set in $\mathscr{C}$ is a countable union of sets of the form

$$\{\omega \in \mathscr{C} : \omega(n) = \alpha(n) \text{ for all } n \in \Phi\},$$

where Φ is a finite subset of Z and $\alpha \in \{-1, +1\}^{\Phi}$. A sequence $(\omega_n)_{n=1}^{\infty}$ converges in $\mathscr{C}$ to 1^Z if and only if for every $N \in Z_+$ there exists an $n_0 \in Z_+$ such that $\omega_n(j) = 1$ for all $j \in Z$ such that $|j| \leqslant N$ and all $n \in Z_+$ such that $n \geqslant n_0$. (The reader reluctant to appeal to Tychonoff's theorem may show directly that, with these open sets, $\mathscr{C}$ is compact and Hausdorff.)

The group operation $(\omega, \phi) \to \omega\phi$ (pointwise product) is continuous from the product space $\mathscr{C} \times \mathscr{C}$ to $\mathscr{C}$; in other words, $\mathscr{C}$ shares with T the property of being a compact Abelian (Hausdorff) topological group. We call it the *Cantor group*. (This name derives from the fact that, being a perfect, nowhere dense, totally disconnected metrisable topological space, $\mathscr{C}$ is homeomorphic to the Cantor ternary set (defined in Exercise 12.44). For a construction of such a homeomorphism, see Exercise 14.16. Sometimes $\mathscr{C}$ is termed the *Walsh group*; see Fine [1].)

14.1.2. For every $N \in Z_+$, define

$$\mathscr{C}_N = \{\omega \in \mathscr{C} : \omega(n) = 1 \text{ for all } n \in Z \text{ such that } |n| \leqslant N\}.$$

Observe that for every such N, $\mathscr{C}_N$ is a subgroup of $\mathscr{C}$ which has 2^{2N+1} cosets. By definition of the product topology, every $\mathscr{C}_N$ is both open and closed; and, for every $\omega \in \mathscr{C}$, the family $(\omega\mathscr{C}_N)_{N=0}^{\infty}$ forms a basis for the neighbourhoods of ω.

14.1.3. **The Dual of $\mathscr{C}$.** We now identify the dual group of $\mathscr{C}$ (see Volume 1, p. 20): it consists of all characters (continuous group homomorphisms) from $\mathscr{C}$ into T. Obvious characters are provided by the projection (that is, evaluation) functions. Prompted by tradition, for every $n \in Z$, we call the projection $\rho_n : \mathscr{C} \to \{-1, +1\}$ given by $\rho_n(\omega) = \omega(n)$, the *nth Rademacher character*, and we write $\mathscr{R}$ to denote the set $\{\rho_n : n \in Z\}$. The characters of $\mathscr{C}$ are describable as follows.

(1) If Φ is a nonvoid finite subset of Z then $\prod_{n \in \Phi} \rho_n$ is a character of $\mathscr{C}$.

(2) Conversely, if ζ is a character of $\mathscr{C}$, then there is a unique function a from Z to $\{0, 1\}$ with finite support and satisfying

$$\zeta(\omega) = \prod_{n \in Z} \omega(n)^{a(n)} \quad \text{for all } \omega \in \mathscr{C}. \tag{14.1.1}$$

Proof. (1) is an easy exercise. For (2) suppose that ζ is a character of $\mathscr{C}$ and that $\omega \in \mathscr{C}$. For every $N \in Z_+$ denote by ω_N the truncate of ω defined by $\omega_N(j) = \omega(j)$ for all $j \in Z$ such that $|j| \leqslant N$ and $\omega_N(j) = 1$ for all other $j \in Z$. Observe that $\omega = \lim_{N \to \infty} \omega_N$ in $\mathscr{C}$. Hence, by continuity of ζ,

$$\zeta(\omega) = \lim_{N \to \infty} \zeta(\omega_N). \tag{14.1.2}$$

Now, for all $i \in Z$, we write ξ_i for the element of $\mathscr{C}$ which maps i to $\omega(i)$ and all other integers to 1. Then

$$\omega_N = \xi_{-N} \cdot \xi_{-(N-1)} \cdots \xi_0 \cdots \xi_{N-1} \cdot \xi_N$$

so that

$$\zeta(\omega_N) = \zeta(\xi_{-N}) \cdots \zeta(\xi_0) \cdots \zeta(\xi_N).$$

But ξ_i has order 2 hence so does $\zeta(\xi_i)$, whence $\zeta(\xi_i) \in \{-1, +1\}$, for every i. Choose $a(i) \in \{0, 1\}$ so that

$$\zeta(\xi_i) = \omega(i)^{a(i)} \qquad \text{for every } \omega \in \mathscr{C}.$$

Then

$$\zeta(\omega_N) = \omega(-N)^{a(-N)} \cdots \omega(0)^{a(0)} \cdots \omega(N)^{a(N)}.$$

Since, by (14.1.2), $\lim_{N \to \infty} \zeta(\omega_N)$ exists for $\omega \in \mathscr{C}$, it follows that the $a(j)$'s must eventually equal zero. This proves (14.1.1), and uniqueness of the function a is clear.

We write $\mathscr{C}^\wedge$ for the group of characters of $\mathscr{C}$ under pointwise product, and denote its identity $1^{\mathscr{C}}$. Traditionally, members of $\mathscr{C}^\wedge$ are viewed as functions on $[0, 1]$. The Rademacher characters are called Rademacher functions and elements of $\mathscr{C}^\wedge$ termed Walsh functions; see Exercise 14.16.

The above proof justifies identifying the dual of $\{-1, +1\}$ with $\{0, 1\}$ under the exponential map, and identifying the dual of the product $\{-1, +1\}^Z$ with the weak direct sum $(\{0, 1\}^Z)^*$ (the set of elements of $\{0, 1\}^Z$ which have finite supports), under the action (14.1.1). Observe that multiplication of characters is the same as addition modulo 2 of their corresponding sequences in the weak direct sum.

The appropriate topology on $\mathscr{C}^\wedge$ is discrete; so $\mathscr{C}^\wedge$ is again an Abelian topological group (see Volume 1, p. 20).

Property (2) above shows that the set $\mathscr{R}$ of Rademacher characters generates $\mathscr{C}^\wedge$; moreover $\mathscr{R}$ is an *independent set*, in the sense that:

$$\text{if } \Phi \subset Z \text{ is nonvoid and finite, if } a \in Z^\Phi, \text{ and if}$$
$$\prod_{n \in \Phi} \rho_n^{a(n)} = 1^{\mathscr{C}}, \text{ then } \rho_n^{a(n)} = 1^{\mathscr{C}} \text{ for every } n \in \Phi. \tag{14.1.3}$$

For a representation, due essentially to Paley, of $\mathscr{C}^\wedge$ as the set Z_+, see Exercise 14.17.

14.1.4. Corresponding to the subgroups $\mathscr{C}_N$ of $\mathscr{C}$, we define

$$\mathscr{A}_N = \{\zeta \in \mathscr{C}^\wedge : \zeta(\omega) = 1 \text{ for all } \omega \in \mathscr{C}_N\}, \tag{14.1.4}$$

and note that

$$\mathscr{A}_N = \{\zeta \in \mathscr{C}^\wedge : \zeta(j) = 1 \text{ for all } j \in Z \text{ such that } |j| > N\}. \tag{14.1.5}$$

We observe that $\mathscr{A}_N$ is the subgroup of $\mathscr{C}^\wedge$ generated by $\{\rho_j\colon |j| \leqslant N\}$ and therefore possesses 2^{2N+1} elements. In general terms (see [HR], (23.23)), $\mathscr{A}_N$ is called the *annihilator* of $\mathscr{C}_N$. In Exercise 14.3 the reader is asked to prove that, as topological groups,

$$(\mathscr{C}/\mathscr{C}_N)^\wedge \text{ is isomorphic to } \mathscr{A}_N,$$

and

$$\mathscr{C}_N \text{ is isomorphic to } \mathscr{C}^\wedge/\mathscr{A}_N.$$

14.1.5. For every function f on $\mathscr{C}$ and every $\omega \in \mathscr{C}$, the *ω-translate of f* is defined to be the function $T_\omega f$ with domain $\mathscr{C}$, such that

$$(T_\omega f)(\phi) = f(\omega\phi) \qquad \text{for all } \phi \in \mathscr{C},$$

(recall that $\omega^{-1} = \omega$ for all $\omega \in \mathscr{C}$).

There is an important operation on $\mathscr{C}$ which is fundamental to the study of measure-preserving transformations on $\mathscr{C}$, but for which we have no need: the *left shift* $\sigma\colon \mathscr{C} \to \mathscr{C}$ is defined by

$$\sigma(\omega)(n) = \omega(n+1) \qquad \text{for all } \omega \in \mathscr{C} \text{ and all } n \in Z.$$

All the Rademacher characters are expressible in terms of any chosen one, together with iterates of σ and σ^{-1}:

$$\rho_{n+m}(\omega) = \rho_n(\sigma^m(\omega)) \qquad \text{for all } \omega \in \mathscr{C}.$$

14.1.6. **Integration on $\mathscr{C}$.** Just as an invariant integral on T was crucial to our analysis of functions on T, so we are now greatly aided by the use of an invariant integral over $\mathscr{C}$ (see 2.2.2). Our idea is to define an invariant integral I on the space $\mathbf{C}(\mathscr{C})$ of all continuous complex-valued functions on $\mathscr{C}$, by considering the obvious normalised invariant integral I_N on the finite group $\mathscr{C}/\mathscr{C}_N$, and letting N tend to infinity. Indeed if $f \in \mathbf{C}(\mathscr{C})$ and f is constant on the cosets of $\mathscr{C}_N$ (and so may be regarded as a continuous function on $\mathscr{C}/\mathscr{C}_N$), then the normalised invariance of I_N forces the equality

$$I_N(f) = 2^{-(2N+1)} \sum f(\omega), \tag{14.1.6}$$

where the sum runs over any subset of $\mathscr{C}$ comprising precisely one ω from each coset of $\mathscr{C}_N$ in $\mathscr{C}$. If we define

$$\mathbf{S}(\mathscr{C}) \equiv \{f \in \mathbf{C}(\mathscr{C}) : \text{for some } N \in Z_+, f \text{ is constant on the cosets of } \mathscr{C}_N\}, \tag{14.1.7}$$

then for $f \in \mathbf{S}(\mathscr{C})$, the sequence $(I_N(f))_{N=0}^{\infty}$ is eventually constant and we define $I(f)$ to equal this constant. Aiming to extend this definition of I to all of $\mathbf{C}(\mathscr{C})$, we first prove that $\mathbf{S}(\mathscr{C})$ is uniformly dense in $\mathbf{C}(\mathscr{C})$. This is easy, if the reader is prepared to tap the depths of the Stone-Weierstrass theorem (see [SMA], pp. 30–87); however, since there is an elementary proof (and since we shall need the result again in 14.9), we prove it now from first principles.

Proof. Suppose $f \in \mathbf{C}(\mathscr{C})$ and $\varepsilon > 0$. Since for every $N \in Z_+$ the characteristic function of every coset of $\mathscr{C}_N$ is continuous, it suffices to prove that $N \in Z_+$ can be chosen so that, on every coset of $\mathscr{C}_N$, the variation of f is, in absolute value, at most ε. The existence of such an $N \in Z_+$ now follows from the continuity of f combined with the facts that $\mathscr{C}$ is compact and the $\mathscr{C}_N$ form a basis for the neighbourhoods of 1^Z.

So far we have defined I on $\mathbf{S}(\mathscr{C})$, and it is easy to see that I is linear, positive, translation invariant and of norm 1 (see 2.2.2). To extend I to $\mathbf{C}(\mathscr{C})$ it now suffices to observe that whenever $(f_N)_{N=0}^{\infty}$ is a sequence in $\mathbf{S}(\mathscr{C})$ which is Cauchy with respect to the supremum norm,

$$|I(f_N) - I(f_M)| = |I(f_N - f_M)| \leqslant \|f_N - f_M\|_\infty$$

and hence that there is a complex number $I(f)$ towards which the sequence $(I(f_N))_{N \in Z_+}$ converges.

The resulting invariant integral I is again positive. Since also $I(1^C) = 1$, I may justifiably be christened *the invariant* (or *Haar*) *integral on* $\mathscr{C}$. If f is integrable on $\mathscr{C}$, we shall often denote the number $I(f)$ by $\int_{\mathscr{C}} f\, d\lambda$ or $\int_{\mathscr{C}} f(\omega)\, d\lambda(\omega)$, wherein λ is the appropriate measure on $\mathscr{C}$. (Integration theory over $\mathscr{C}$ is an instance of general integration theory which may be found, for example, in [E] and/or [E_1].) Observe that, since $\mathscr{C}_N$ has 2^{2N+1} cosets in $\mathscr{C}$, $\lambda(\chi_{\mathscr{C}_N}) = 2^{-(2N+1)}$.

We write $\mathbf{L}^p(\mathscr{C})$ for the space of equivalence classes (under equality λ-almost everywhere) of measurable functions f on $\mathscr{C}$ such that

$$\|f\|_p \equiv \left\{\int_{\mathscr{C}} |f|^p\, d\lambda\right\}^{1/p} < \infty. \tag{14.1.8}$$

14.1.7. **The Orthogonality Relations.** These state:

$$\int_{\mathscr{C}} 1^{\mathscr{C}}\, d\lambda = 1 \tag{14.1.9}$$

$$\int_{\mathscr{C}} \zeta\, d\lambda = 0 \qquad \text{for every } \zeta \in \mathscr{C}^\wedge \backslash \{1^{\mathscr{C}}\}, \tag{14.1.10}$$

and are provable by the general argument appearing in 2.2.3. The reader may find it interesting to construct an alternative proof which uses the special structure of $\mathscr{C}$, $\mathscr{C}^\wedge$ and I.

14.1.8. **The Walsh-Fourier Transform.** For much the same reasons as for T (see Chapter 1, especially 1.3.3), we now define the Fourier transform over $\mathscr{C}$, traditionally termed the *Walsh-Fourier transform*. If $f \in \mathbf{L}^1(\mathscr{C})$ then the complex-valued function $\hat{f}$ with domain $\mathscr{C}^\wedge$ is defined by

$$\hat{f}(\zeta) = \int_{\mathscr{C}} f\bar{\zeta}\, d\lambda = \int_{\mathscr{C}} f\zeta\, d\lambda \qquad \text{for all } \zeta \in \mathscr{C}^\wedge. \tag{14.1.11}$$

It is easy to see that:

(1) The Walsh-Fourier transform $f \to \hat{f}$ is linear and for all $f \in \mathbf{L}^1(\mathscr{C})$ and $\zeta \in \mathscr{C}^\wedge$, $|\hat{f}(\zeta)| \leqslant \|f\|_1$. Moreover $(\bar{f})^\wedge = (\hat{f})^*$ and $(f^*)^\wedge = (\hat{f})^-$ for all $f \in \mathbf{L}^1(\mathscr{C})$, where the notation is taken from Volume 1, p. 31.

(2) If $f \in \mathbf{L}^1(\mathscr{C})$, $\zeta \in \mathscr{C}^\wedge$ and $\omega \in \mathscr{C}$, then

$$(T_\omega f)^\wedge(\zeta) = \zeta(\omega)\hat{f}(\zeta). \tag{14.1.12}$$

(3) If $\zeta, \eta \in \mathscr{C}^\wedge$, then

$$\hat{\zeta}(\eta) = \begin{cases} 1 & \text{if } \eta = \zeta \\ 0 & \text{if } \eta \neq \zeta. \end{cases} \tag{14.1.13}$$

When $f \in \mathbf{L}^1(\mathscr{C})$ we shall refer to the formal series

$$\sum_{\zeta \in \mathscr{C}^\wedge} \hat{f}(\zeta)\zeta$$

as the *Walsh-Fourier series* of f, and to $\hat{f}$ as the *Walsh-Fourier transform* of f.

14.1.9. **Trigonometric Polynomials on $\mathscr{C}$.** We write $\mathbf{T}(\mathscr{C})$ for the complex linear span of $\mathscr{C}^\wedge$, and call its elements *trigonometric polynomials on* $\mathscr{C}$. After the characters themselves, the most important examples are the functions P_N (for $N \in Z_+$), where P_N is called the *Nth Paley polynomial* and is defined by

$$P_N = \sum_{\zeta \in \mathscr{A}_N} \zeta. \tag{14.1.14}$$

The sequence $(P_N)_{N \in Z_+}$ is termed the *Paley kernel*.

It is clear that $\mathbf{T}(\mathscr{C})$ is a linear space stable under translation and under pointwise products.

For every $N \in Z_+$ and every $\zeta \in \mathscr{A}_N$, ζ assumes the value 1 at every point of $\mathscr{C}_N$. It follows that $\mathbf{T}(\mathscr{C}) \subset \mathbf{S}(\mathscr{C})$ (see (14.1.7)). The converse inclusion also holds, as we now show.

14.1.10. $\mathbf{T}(\mathscr{C}) = \mathbf{S}(\mathscr{C})$.

Proof. It remains to prove that $\mathbf{S}(\mathscr{C}) \subset \mathbf{T}(\mathscr{C})$. Assume $f \in \mathbf{S}(\mathscr{C})$ and choose $N \in Z_+$ such that f is constant on every coset of $\mathscr{C}_N$. Then f is a finite linear combination of characteristic functions of cosets of $\mathscr{C}_N$ and, by translation invariance, it suffices to prove that the characteristic function of $\mathscr{C}_N$ belongs to $\mathbf{T}(\mathscr{C})$. However, this characteristic function equals

$$\prod_{|n| \leq N} (1^{\mathscr{C}} + P_n) \tag{14.1.15}$$

which is clearly an element of $\mathbf{T}(\mathscr{C})$.

14.1.11. $\mathbf{T}(\mathscr{C})$ is uniformly dense in $\mathbf{C}(\mathscr{C})$, and $\|\cdot\|_p$-dense in $\mathbf{L}^p(\mathscr{C})$ for $1 \leqslant p < \infty$.

Proof. Immediate from 14.1.6, 14.1.10 and the fact that $\mathbf{C}(\mathscr{C})$ is $\|\cdot\|_p$-dense in $\mathbf{L}^p(\mathscr{C})$ whenever $1 \leqslant p < \infty$.

Define (cf. Volume 1, p. 29) $\mathbf{c}_0(\mathscr{C}^\wedge)$ to be the set of all complex-valued functions g on $\mathscr{C}^\wedge$ such that, for every $\varepsilon > 0$, the set $\{\zeta \in \mathscr{C}^\wedge : |g(\zeta)| > \varepsilon\}$ is finite. The next theorem is an exact analogue of the Riemann-Lebesgue Lemma 2.3.8.

14.1.12. If $f \in \mathbf{L}^1(\mathscr{C})$ then $\hat{f} \in \mathbf{c}_0(\mathscr{C}^\wedge)$.

Proof. This follows from 14.1.11, (14.1.13), 14.1.8, (1), and the fact that $\mathbf{C}(\mathscr{C})$ is dense in $\mathbf{L}^1(\mathscr{C})$.

Convergence of the Walsh-Fourier series of a function to that function is, in stark contrast to the situation for T, all that could be wished for; see 14.1.15 and Exercise 14.23. Over T, the blame for failure of $\mathbf{L}^1$-convergence rests with the Dirichlet kernels refusing to form an approximate identity (cf. Volume 1, p. 155 and 10.3.2). On the other hand for $\mathscr{C}$ we have the following result.

14.1.13. The Paley kernel forms an approximate identity in $\mathbf{L}^1(\mathscr{C})$ (see 3.2.1). More specifically,

$$P_N = 2^{2N+1}\chi_{\mathscr{C}_N} \qquad \text{for all } N \in Z_+; \tag{14.1.16}$$

$$\hat{P}_N = \chi_{\mathscr{A}_N} \qquad \text{for all } N \in Z_+; \tag{14.1.17}$$

$$(P_N)_{N=0}^{\infty} \text{ forms an approximate identity in } \mathbf{L}^1(\mathscr{C}). \tag{14.1.18}$$

Proof. First we observe that

$$P_N = \prod_{|n| \leq N} (1^{\mathscr{C}} + \rho_n). \tag{14.1.19}$$

From this, (14.1.16) is evident. Next, by multiplying out the product appearing in (14.1.19), (14.1.17) is a restatement of the orthogonality relations 14.1.7. Finally to demonstrate (14.1.18) we observe that by (14.1.16),

$$\int_{\mathscr{C}} |P_N| \, d\lambda = \int_{\mathscr{C}} P_N \, d\lambda = 1,$$

and

$$\lim_{N \to \infty} \int_{\mathscr{C} \setminus \mathscr{C}_M} |P_N| \, d\lambda = 0 \text{ for any fixed } M \in Z_+. \tag{14.1.20}$$

14.1.14. For every $N \in Z_+$ and every $f \in \mathbf{L}^1(\mathscr{C})$, we define

$$s_N f = \sum_{\zeta \in \mathscr{A}_N} \hat{f}(\zeta)\zeta. \tag{14.1.21}$$

This symmetric partial sum of the Walsh-Fourier series of f could, of course, have been defined using convolution on $\mathscr{C}$. Indeed, more extensive analysis over $\mathscr{C}$ would (cf. Chapter 3) necessitate the definition of convolution over $\mathscr{C}$:

$$\begin{aligned} f * g(\omega) &= \int_{\mathscr{C}} f(\omega\phi)g(\phi) \, d\lambda(\phi) \\ &= \int_{\mathscr{C}} f(\omega\phi^{-1})g(\phi) \, d\lambda(\phi) \end{aligned} \tag{14.1.22}$$

for $f, g \in \mathbf{L}^1(\mathscr{C})$. Then, $s_N f = P_N * f$ for every $N \in Z_+$ and every $f \in \mathbf{L}^1(\mathscr{C})$; see Exercise 14.19.

14.1.15. (1) If $f \in \mathbf{C}(\mathscr{C})$, then $\|s_N f - f\|_\infty \to 0$ as N tends to infinity.

(2) If $f \in \mathbf{L}^p(\mathscr{C})$ and $1 \leqslant p < \infty$, then $\|s_N f - f\|_p \to 0$ as N tends to infinity.

(3) If $f \in \mathbf{L}^1(\mathscr{C})$ and $\hat{f}(\zeta) = 0$ for all $\zeta \in \mathscr{C}^\wedge$ then $f(\omega) = 0$ for λ-almost all $\omega \in \mathscr{C}$.

The reader is invited to prove these assertions in Exercise 14.20 (cf. 3.2.1 and 3.2.2). Regarding (1), see Volume 1, p. 155.

There is an $\mathbf{L}^2$-theory of functions on $\mathscr{C}$ just as satisfying as that for T. The procedure is closely analogous to that in Chapter 8, so we relegate the proof of the following result to Exercise 14.2.1.

14.1.16. The Walsh-Fourier transform is a Hilbert-space isomorphism from $\mathbf{L}^2(\mathscr{C})$ onto $1^2(\mathscr{C}^\wedge)$, the latter denoting the space of square-summable sequences on $\mathscr{C}^\wedge$.

14.1.17. The reader may verify that Hölder's inequality holds for functions on $\mathscr{C}$ just as it does for functions on T: if $1 \leqslant p \leqslant \infty, f \in \mathbf{L}^p(\mathscr{C})$ and $g \in \mathbf{L}^{p'}(\mathscr{C})$, then the pointwise product fg belongs to $\mathbf{L}^1(\mathscr{C})$ and

$$\|fg\|_1 \leqslant \|f\|_p \|g\|_{p'}. \tag{14.1.23}$$

Directly from this we obtain, for all λ-measurable functions f on $\mathscr{C}$,

$$\|f\|_p \leqslant \|f\|_q \qquad \text{if } 0 < p < q; \tag{14.1.24}$$

$$\int_{\mathscr{C}} |f|^p \, d\lambda \leqslant \left\{\int_{\mathscr{C}} |f|^{p_1} \, d\lambda\right\}^{s_1} \left\{\int_{\mathscr{C}} |f|^{p_2} \, d\lambda\right\}^{s_2}, \tag{14.1.25}$$

whenever $p_1, p_2, s_1, s_2 > 0$, $p = p_1 s_1 + p_2 s_2$ and $s_1 + s_2 = 1$.

14.1.18. We have not mentioned the probabilistic notion of *independence*, nor stressed the importance of the Rademacher characters in providing a model for such independence. This point of view, which leads to the use of probabilistic methods in harmonic analysis (and is at the heart of the results mentioned in 14.3.6), is vital for current developments; we refer the reader to the delightful introduction [Kac] and then to Kahane's timely and substantial monograph [Kah_3].

14.1.19. To exemplify the analogy between Walsh-Fourier and (classical) Fourier series (noted in Exercise 14.17), we mention only the following result (see Yano [1], Theorem 7) which should be compared with 7.3.1: If $(a_n)_{n=0}^{\infty}$ is a quasi-convex sequence (see 7.1.2) and $a_n \to 0$ then $\sum_{n=0}^{\infty} a_n F^{-1}(n)$ is the Walsh-Fourier series of a function $f \in \mathbf{L}^1(\mathscr{C})$, to which the Walsh-Fourier series converges (the bijection F is defined in Exercise 14.17). Whilst in 7.3.1 the Fourier series is also nonnegative whenever $(a_n)_{n=0}^{\infty}$ is actually convex and monotonically decreasing to zero, this conclusion fails for the Cantor group result (see Coury [1], § 4).

In spite of the fact that harmonic analysis over $\mathscr{C}$ is usually simpler than that over T, there remain a multitude of topics yet to be studied. Three areas which have been studied concern:

(1) Questions of λ-almost everywhere convergence (see MR **30** # 2282; **34** # 8075; **35** # 4667; **36** # 599; **38** # 6296; **39** # 3222; **41** # 4113; **49** ## 983, 5691; **50** ## 2803, 7939, 14045, 14046; **51** ## 10990, 13578; **52** ## 1150, 1155, 3871, 11458, 14826; **53** ## 3598, 13991, 13996, 13997; **54** ## 3281, 5728, 5730; **55** ## 3670, 3671; **57** # 10349; **58** ## 6893, 6895, 6898, 6899, 6900, 23330, 23331, 23332);

(2) Walsh-Fourier series with well-behaved coefficients (see MR **22** # 863; **29** # 2597; **39** # 3227; **41** # 4112; **49** ## 11132, 7685, 5671; **50** # 2802; **51** ## 8727, 13576; **52** # 8805; **54** # 5727; **55** ## 3668, 8676; **58** ## 6894, 6897, 12165);

(3) Differentiability of Walsh-Fourier series (see MR **38** # 6298; **52** # 11457; **53** ## 1148, 8774, 8775, 13995; **54** # 13446; **55** ## 3669, 13149; **56** # 9176; **57** ## 7017, 7020; **58** ## 29796, 29799).

Other papers relevant to the study of Fourier series over $\mathscr{C}$ include: MR **36** # 1909; **37** # 3272; **43** # 5244; **48** # 11896; **49** # 9530; **50** ## 5369, 13578, 14046; **51** ## 3789, 6258; **52** ## 6309, 14828; **53** ## 1149, 3589, 13994; **54** # 3280; **55** # 3667; **57** # 7016; **58** # 12177.

14.2 Rademacher Series Convergent in $\mathbf{L}^2(\mathscr{C})$.

As announced at the beginning of this chapter, we study "most" series $\sum_{n \in Z} \pm \hat{f}(n) e_n$ by considering series $\sum_{\zeta \in \mathscr{R}} \phi(\zeta)\zeta$. If K is a subset of $\mathscr{C}^\wedge$, a function f in $\mathbf{L}^1(\mathscr{C})$ is called *K-spectral* if and only if $\hat{f}(\zeta) = 0$ for all $\zeta \in \mathscr{C}^\wedge \backslash K$. We denote by $\mathbf{L}_K^p(\mathscr{C})$, $\mathbf{C}_K(\mathscr{C})$ and $\mathbf{T}_K(\mathscr{C})$ the spaces of all K-spectral elements of $\mathbf{L}^p(\mathscr{C})$, $\mathbf{C}(\mathscr{C})$ and $\mathbf{T}(\mathscr{C})$ respectively (cf. 15.1.1).

The positive results that are essential for further developments in this book refer to some unexpected properties of functions in $\mathbf{L}_{\mathscr{R}}^2(\mathscr{C})$. These results are stated as 14.2.1 and 14.2.2; although in fact 14.2.2 implies 14.2.1 (see Exercise 14.4), the latter is used as a stepping-stone to the former and is best stated and proved separately.

14.2.1. **The Basic Inequalities.** To every $p > 0$ there correspond positive numbers A_p and B_p such that for all $f \in \mathbf{L}_{\mathscr{R}}^2(\mathscr{C})$,

$$A_p \| f \|_2 \leqslant \| f \|_p \leqslant B_p \| f \|_2; \tag{14.2.1}$$

moreover, B_p may be taken to be $(k!)^{1/p}$ where $2k$ is the least even integer not less than p; and, if $0 < p < 2$, A_p may be taken to be $2^{(p-2)/2p}$.

Proof. We begin by considering the right-hand inequality in (14.2.1). Suppose first that $p = 2k$ is an even integer and that $f \in \mathbf{T}_{\mathscr{R}}(\mathscr{C})$ (the latter restriction being easily removed by use of 14.1.11 and Fatou's Lemma).

Let $g = f^k$ be the kth pointwise power of f. Then $\hat{g}$ is the k-fold convolution of $\hat{f}$ and so, for all $\zeta \in \mathscr{C}^\wedge$,

$$\hat{g}(\zeta) = \sum \hat{f}(\zeta_1) \cdot \hat{f}(\zeta_2) \cdots \hat{f}(\zeta_k), \tag{14.2.2}$$

where the sum is over all k-tuples $(\zeta_1, \zeta_2, \cdots, \zeta_k)$ in $\mathscr{R}^k$ whose product $\zeta_1 \cdot \zeta_2 \cdots \zeta_k$ equals ζ. But by (14.1.3), two k-tuples of elements of $\mathscr{R}$ have the same product if and only if one is a permutation of the other. So there are exactly $k!$ such k-tuples $(\zeta_1, \cdots, \zeta_k)$ having a product equal to ζ. This, together with the Cauchy-Schwarz inequality, shows that

$$|\hat{g}(\zeta)|^2 \leqslant \left[\sum_{\zeta_1 \cdots \zeta_k = \zeta} 1^2\right]\left[\sum_{\zeta_1, \cdots, \zeta_k = \zeta} |\hat{f}(\zeta_1)^2 \cdots |\hat{f}(\zeta_k)|^2\right]$$

$$= k!\left[\sum_{\zeta_1 \cdots \zeta_k = \zeta} |\hat{f}(\zeta_1)|^2 \cdots |f(\zeta_k)|^2\right].$$

Hence

$$\|f\|_{2k}^{2k} = \|g\|_2^2 \leqslant k! \sum_{\zeta \in \mathscr{C}^\vee} \sum_{\zeta_1 \cdots \zeta_k = \zeta} |\hat{f}(\zeta_1)|^2 \cdots |\hat{f}(\zeta_k)|^2$$

$$= k! \sum_{(\zeta_1, \cdots, \zeta_k) \in \mathscr{R}^k} |\hat{f}(\zeta_1)|^2 \cdots |\hat{f}(\zeta_k)|^2$$

$$= k!\left[\sum_{\zeta \in \mathscr{R}} |\hat{f}(\zeta)|^2\right]^k$$

$$= k!\|f\|_2^{2k}, \tag{14.2.3}$$

where the penultimate step follows by the multinomial theorem, according to which

$$\left(\sum_{\zeta \in \mathscr{R}} a_\zeta\right)^k = \sum_{(\zeta_1, \cdots, \zeta_k) \in \mathscr{R}^k} a_{\zeta_1} \cdots a_{\zeta_k} \tag{14.2.4}$$

for every complex-valued function $\zeta \to a_\zeta$ with domain $\mathscr{R}$.

For other choices of p we need only refer to (14.1.24) to deduce that the right-hand half of (14.2.1) holds with $B_p = (k!)^{1/p}$, where $2k$ is the least even integer not less than p.

The left-hand inequality in (14.2.1) is trivial when $p \geqslant 2$ in view of 14.1.16 and (14.1.24). If $0 < p < 2$, we write $2 = s_1 p + s_2 4$, where $s_1 > 0$, $s_2 > 0$ and $s_1 + s_2 = 1$. Then (14.1.25) combines with the case $p = 4$ of the right-hand half of (14.2.1) and the estimate $B_4 \leqslant 2^{1/4}$ to yield

$$\sum_{\zeta \in \mathscr{R}} |\hat{f}(\zeta)|^2 = \int_{\mathscr{C}} |f|^2\, d\lambda$$

$$\leqslant \left[\int_{\mathscr{C}} |f|^p\, d\lambda\right]^{s_1}\left[\int_{\mathscr{C}} |f|^4\, d\lambda\right]^{s_2}$$

$$\leqslant \left[\int_{\mathscr{C}} |f|^p\, d\lambda\right]^{s_1}\left[2^{1/4}\left(\sum_{\zeta \in \mathscr{R}} |\hat{f}(\zeta)|^2\right)^{1/2}\right]^{4s_2}$$

which in turn leads to

$$\left[\int_{\mathscr{C}} |f|^p\, d\lambda\right]^{1/p} \geqslant 2^{(p-2)/2p}\left[\sum_{\zeta \in \mathscr{R}} |\hat{f}(\zeta)|^2\right]^{1/2}, \tag{14.2.5}$$

thereby completing the proof.

14.2.2. **Supplement to 14.2.1.** Suppose again that $f \in \mathbf{L}^2_{\mathscr{R}}(\mathscr{C})$. Then for every real number μ,

$$\int_{\mathscr{C}} \exp [\mu |f|^2]\, d\lambda < \infty. \tag{14.2.6}$$

In fact, if $|\mu| \, \|f\|_2^2 < 1$, then

$$\int_{\mathscr{C}} \exp [\mu |f|^2]\, d\lambda \leqslant (1 - |\mu| \, \|f\|_2^2)^{-1}. \tag{14.2.7}$$

Proof. The arguments are similar to those used in 13.9.2 to prove the inequalities (12.9.14) and (12.9.15). By (14.2.1) and the estimate $B_{2k} \leqslant (k!)^{1/2k}$,

$$\begin{aligned}\int_{\mathscr{C}} \exp [\mu |f|^2]\, d\lambda &= \sum_{k \in Z_+} \mu^k (k!)^{-1} \|f\|_{2k}^{2k} \\ &\leqslant \sum_{k \in Z_+} (|\mu| \, \|f\|_2^2)^k \\ &= (1 - |\mu| \, \|f\|_2^2)^{-1}\end{aligned}$$

provided that $|\mu| \, \|f\|_2^2 < 1$, and this establishes (14.2.7).

For (14.2.6) we need to free $|\mu|$ from dependence on $\|f\|_2$. Whenever $|\mu| > 0$ we do this by choosing N so large that $\sum_{n > N} |\hat{f}(\rho_n)|^2$ is small enough for (14.2.7) to apply with $f - s_N f$ in place of f. It then suffices to observe that

$$|f|^2 \leqslant 2[|f - s_N f|^2 + |s_N f|^2].$$

14.2.3. **Pointwise Convergence Almost Everywhere.** In Exercise 14.23 the reader is invited to prove that if $f \in \mathbf{L}^1(\mathscr{C})$ then the series $\sum_{\zeta \in \mathscr{C}^\wedge} \hat{f}(\zeta)\zeta(\omega)$ is convergent for λ-almost all $\omega \in \mathscr{C}$. Concerning $\mathscr{R}$-spectral functions it can be shown ([Z_1], p. 212; [Ba_1], pp. 230–233; [Kac], pp. 31–33) that if $\sum_{\zeta \in \mathscr{R}} |\hat{f}(\zeta)|^2 = \infty$, then $\sum_{\zeta \in \mathscr{C}^\wedge} \hat{f}(\zeta)\zeta(\omega)$ is λ-almost everywhere nonsummable by Cesàro means (or, indeed, by any of the usual linear summability methods used in analysis). Thus for $\mathscr{R}$-spectral Walsh-Fourier series, 14.2.1 provides the dichotomy: if $\sum_{\zeta \in \mathscr{R}} |\hat{f}(\zeta)|^2 < \infty$ then the Walsh-Fourier series of f converges λ-almost everywhere; if $\sum_{\zeta \in \mathscr{R}} |\hat{f}(\zeta)|^2 = \infty$, the same series diverges (and is nonsummable) λ-almost everywhere.

14.3. Applications to Fourier Series.

In this section we suppose that $(c_n)_{n \in Z}$ is a given two-way-infinite sequence such that

$$\sum_{n \in Z} |c_n|^2 < \infty, \tag{14.3.1}$$

and get quite a lot of mileage out of considering $(c_n)_{n \in Z}$ to be either the Fourier coefficients of an element $\sum_{n \in Z} c_n \rho_n$ of $\mathbf{L}^2_{\mathscr{R}}(\mathscr{C})$ or of an element $\sum_{n \in Z} c_n e_n$ of $\mathbf{L}^2 \equiv \mathbf{L}^2(T)$. For every $\omega \in \mathscr{C}$ and $x \in T$ we write

$$f_\omega(x) = \sum_{n \in Z} \rho_n(\omega) c_n e^{inx}, \tag{14.3.2}$$

the series being convergent in mean in $\mathbf{L}^2$; thus $f_\omega \in \mathbf{L}^2$ for every $\omega \in \mathscr{C}$. (The notation $f_\omega(x)$ is conventionally sloppy. What is intended is that, for every $\omega \in \mathscr{C}$, f_ω is an element of $\mathbf{L}^2$ such that

$$\hat{f}_\omega(n) = \rho_n(\omega) c_n \qquad \text{for all } n \in Z;$$

for a given $x \in T$ (or R), the series on the right of (14.3.2) is purely formal. The sloppiness can, of course, be removed; cf. [EG], pp. 20–23.) However, as we shall now proceed to show, much more than this is true for most values of ω.

14.3.1. Suppose that (14.3.1) holds and that f_ω is defined by (14.3.2). Then for λ-almost all $\omega \in \mathscr{C}$,

$$\frac{1}{2\pi} \int_T \exp\,[\mu |f_\omega(x)|^2]\, dx < \infty \tag{14.3.3}$$

for every real number μ.

Proof. We may obviously assume that $\mu > 0$. Moreover, it is clearly enough to show that, for every $\mu > 0$, (14.3.3) holds for λ-almost all $\omega \in \mathscr{C}$.

For every $\omega \in \mathscr{C}$, every real x and every $N \in Z_+$, define

$$s_N(x, \omega) = \sum_{|n| \leqslant N} \rho_n(\omega) c_n e^{inx}.$$

Then (14.3.1) and 14.2.2 show that, for some real number A independent of N,

$$\int_{\mathscr{C}} \exp\,[\mu |s_N|^2]\, d\lambda \leqslant A \qquad \text{for all } N \in Z_+.$$

Thus

$$\frac{1}{2\pi} \int_T \left\{ \int_{\mathscr{C}} \exp\,[\mu |s_N(x, \omega)|^2]\, d\lambda(\omega) \right\} dx \leqslant A.$$

By Fubini's theorem ([HS], (21.12)), this can be written

$$J_N \equiv \int_{\mathscr{C}} \left\{ \frac{1}{2\pi} \int_T \exp\,[\mu |s_N(x, \omega)|^2]\, dx \right\} d\lambda(\omega) \leqslant A. \tag{14.3.4}$$

Now

$$\frac{1}{2\pi}\int_T |f_\omega(x) - s_N(x, \omega)|^2\, dx = \sum_{|n|>N} |c_n \rho_n(\omega)|^2$$
$$= \sum_{|n|>N} |c_n|^2,$$

which tends to zero as $N \to \infty$, *uniformly* with respect to $\omega \in \mathscr{C}$. Because of this, a standard argument (see, for example, [W], Theorem 4.5a) shows that nonnegative integers $N_1 < N_2 < \cdots$ may be chosen independent of $\omega \in \mathscr{C}$, such that for every $\omega \in \mathscr{C}$ one has

$$\lim_{k\to\infty} s_{N_k}(x, \omega) = f_\omega(x)$$

for almost all $x \in T$. This being so, if we write

$$I(\omega) = \frac{1}{2\pi}\int_T \exp\,[\mu |f_\omega(x)|^2]\, dx$$

and

$$I_0(\omega) = \liminf_{k\to\infty} \frac{1}{2\pi}\int_T \exp\,[\mu |s_{N_k}(x, \omega)|^2]\, dx,$$

two applications of Fatou's lemma imply that

$$I(\omega) \leqslant I_0(\omega)$$

and that

$$\int_{\mathscr{C}} I_0(\omega)\, d\lambda(\omega) \leqslant \liminf_{k\to\infty} J_{N_k}.$$

Therefore, by (14.3.4),

$$\int_{\mathscr{C}} I_0(\omega)\, d\lambda(\omega) \leqslant A < \infty,$$

which entails that $I(\omega) \leqslant I_0(\omega) < \infty$ for λ-almost all $\omega \in \mathscr{C}$.

Remarks. We have commenced the proof by using the s_N in order that no trouble be experienced in appealing to Fubini's theorem. Had f_ω been used from the start, knowledge of the measurability of $f_\omega(x)$ in the pair (ω, x) would have been necessary, and this property is by no means evident.

Similarly, we have introduced I_0 as an intermediary simply to avoid the necessity of showing that I is measurable (which is not very difficult). Alternatively, of course, one could use versions of Fatou's lemma in which measurability is not assumed (see, for example, [HS], (9.39) or [E], Proposition 4.5.4).

14.3.2. If $f \in \mathbf{L}^2$, it is possible to choose the $\pm$ signs in such a way that

$$\sum_{n \in Z} \pm \hat{f}(n) e^{inx} \tag{14.3.5}$$

is the Fourier series of a function g such that

$$\frac{1}{2\pi} \int_T \exp\,[\mu |g(x)|^2]\, dx < \infty \tag{14.3.6}$$

for every real number μ (so that, in particular, $g \in \mathbf{L}^p$ for every $p < \infty$).

Proof. It suffices to take $c_n = \hat{f}(n)$ and $g = f_\omega$, $\omega \in \mathscr{C}$ being chosen so that (14.3.3) holds. Since the series (14.3.2) converges in mean in $\mathbf{L}^2(T)$,

$$\hat{g}(n) = \hat{f}_\omega(n) = \rho_n(\omega) c_n = \rho_n(\omega) \hat{f}(n) = \pm \hat{f}(n).$$

14.3.3. There exist a function g satisfying (14.3.6) for every real μ, and a sequence of $\pm$ signs, such that

$$\sum_{n \in Z} \pm \hat{g}(n) e^{inx}$$

is not the Fourier series of any function in $\bigcup_{p>2} \mathbf{L}^p$.

Proof. Apply 14.3.2, choosing $f \in \mathbf{L}^2$ so that $f \notin \bigcup_{p>2} \mathbf{L}^p$. For example, suppose that f is bounded except in the neighbourhood of the origin, where it equals $|x|^{-1/2}(\log 1/|x|)^{-\alpha}$, α being greater than 1/2.

14.3.4. ***Remarks.*** (1) Various strengthenings of 14.3.3 are derivable with more work. It can be shown (Edwards [7], Theorem (2.8)) that a continuous g may be chosen which satisfies the conditions stipulated in 14.3.3. Furthermore, g may be chosen so that $\sum_{n \neq 0} n^{-\alpha} |\hat{g}(n)|^\beta < \infty$ for all $\alpha > 0$ and $\beta > 0$ (see MR **53**, # 1168).

(2) It is easy to deduce from the proof of 14.3.1 that there exists an absolute constant B with the following property: to every $f \in \mathbf{L}^2$ and every $\alpha > 0$ corresponds a set $\mathscr{S} \subset \mathscr{C}$ satisfying $\lambda(\mathscr{S}) < B/\alpha$ and such that

$$\| f_\omega \|_p \leqslant \alpha p^{1/2} \| f \|_2$$

for all $\omega \in \mathscr{C} \backslash \mathscr{S}$ and all p such that $1 \leqslant p < \infty$; compare 15.3.1. Uchiyama [2] has discussed the reverse type of inequality and has shown in particular that, for every $f \in \mathbf{L}^2$ and every $\varepsilon > 0$, there exists a number $B_{\varepsilon, f} > 0$ such that

$$\| (s_N f)_\omega \|_1 \geqslant B_{\varepsilon, f} \| s_N f \|_2,$$

for all $N \in Z_+$ and all $\omega \in \mathscr{C}$ save perhaps those belonging to a set of λ-measure at most ε.

14.3.5. **Pointwise Convergence Almost Everywhere and Related Matters.** More detailed arguments based upon the results stated in 14.2.3 serve to show that, if (14.3.1) holds, then λ-almost all the series on the right of (14.3.2) are pointwise convergent almost everywhere. If, however, (14.3.1) is false, then λ-almost all these series are almost everywhere pointwise non-summable (compare the remarks in 14.2.3). A consequence of this is the following assertion, due to Littlewood:

(1) If, for every choice of the $\pm$ signs, the series $\sum_{n \in Z} \pm c_n e^{inx}$ is a Fourier-Lebesgue series, then $\sum_{n \in Z} |c_n|^2 < \infty$. For the details, see [$Z_1$], pp. 214–215 and/or [$Ba_1$], pp. 234–235.

From (1) it is possible (cf. Exercise 14.7) to deduce a little more, namely:

(2) If, for every choice of the $\pm$ signs, the series $\sum_{n \in Z} \pm c_n e^{inx}$ is a Fourier-Stieltjes series, then $\sum_{n \in Z} |c_n|^2 < \infty$.

It is possible (cf. Exercise 14.5) to show that (2) is equivalent to

(2′) If $\sum_{n \in Z} |c_n \hat{f}(n)| < \infty$ for every $f \in \mathbf{C}$, then $\sum_{n \in Z} |c_n|^2 < \infty$.

The theorem (2′) is due to Orlicz, Paley and Sidon independently and almost simultaneously. See also Section 16.9 below; Edwards [7]; Edwards, Hewitt and Ritter [1].

A different approach to (1), (2) and (2′) is due to Helgason [1], where multipliers (see Chapter 16 below) form part of the theme. Some of the features of particular cases of Helgason's arguments, many of which refer to a fairly general class of Banach algebras, are sketched in Exercises 14.11–14.14. See also Helgason [2], [3].

In a somewhat similar and simpler vein is the following statement:

(3) If, for every choice of the $\pm$ signs, the series $\sum_{n \in Z} \pm c_n e^{inx}$ is the Fourier series of a function in $\mathbf{L}^\infty$, then $\sum_{n \in Z} |c_n| < \infty$.

Indeed, by 10.5.2, the hypothesis entails that

$$\lim_{N \to \infty} \sum_{|n| \leq N} \left(1 - \frac{|n|}{N+1}\right) \pm c_n \hat{f}(n)$$

exists finitely for every $f \in \mathbf{L}^1$. The same is therefore true when $c_n \hat{f}(n)$ is replaced by its real and imaginary parts. Hence, by suitable choice of the $\pm$ signs, we conclude that

$$\lim_{N \to \infty} \sum_{|n| \leq N} \left(1 - \frac{|n|}{N+1}\right) |c_n \hat{f}(n)|$$

exists finitely for every $f \in \mathbf{L}^1$, so that (see 5.3.4)

$$\sum_{n \in Z} |c_n \hat{f}(n)| < \infty$$

for every $f \in \mathbf{L}^1$. Reference to Exercise 3.14 leads from this to the stated conclusion.

It is known that in each of the statements (1), (2) and (3) it suffices to impose the respective hypothesis, not for *all* choices of the $\pm$ signs, but merely for a suitable set of choices. One type of "suitable set" is a set of positive λ-measure in $\mathscr{C}$. A second type is a nonmeagre subset (in the product topology, of course) of $\mathscr{C}$. The technique for doing this is similar to that used in Edwards [7].

14.3.6. **Uniform Convergence and Related Matters.** In view of 14.3.1 and 14.3.2 it is natural to ask under what conditions on $(c_n)_{n \in Z}$ it is possible to choose the $\pm$ signs in such a way that $\sum_{n \in Z} \pm c_n e^{inx}$ is the Fourier series of a continuous function; or is uniformly convergent on T.

Let us write $\mathbf{C}_{\text{a.s.}}$ (to indicate almost sure continuity) for the set of all $f \in \mathbf{L}^2$ such that for λ-almost all $\omega \in \mathscr{C}$, the series $\sum_{n \in Z} \rho_n(\omega) \hat{f}(n) e^{inx}$ is the Fourier series of a continuous function on T. Though it is not obvious, $\mathbf{C}_{\text{a.s.}}$ becomes a Banach space when equipped with the norm

$$\|f\|_{\text{a.s.}} = \int_{\mathscr{C}} \left\| \sum_{n \in Z} \rho_n(\omega) \hat{f}(n) e^{inx} \right\|_\infty d\lambda(\omega).$$

(It has, for instance, to be proved that $\|f\|_{\text{a.s.}}$ thus defined is finite for every $f \in \mathbf{C}_{\text{a.s.}}$.)

We mention two characterisations of $\mathbf{C}_{\text{a.s.}}$. Firstly, $f \in \mathbf{C}_{\text{a.s.}}$ if and only if $f = \sum_{n=1}^\infty h_n * k_n$ with $h_n \in \mathbf{L}^2$, $k_n \in \mathbf{L}_\phi$ and $\sum_{n=1}^\infty \|h_n\|_2 \|k_n\|_\phi < \infty$, wherein $\phi(u) = u(1 + \log(1 + u))^{1/2}$ for nonnegative real u, and $\|g\|_\phi$ denotes the norm $\inf\{\mu > 0 : \int \phi(|g|/\mu) \leqslant 1\}$ in the Orlicz space $\mathbf{L}_\phi$ (see Pisier [2]). The appearance here of ϕ is not as mysterious as it might first seem; ϕ is, up to equivalence of Orlicz functions, dual to the function $\psi(u) = \exp(u^2) - 1$, and the Orlicz space $\mathbf{L}_\psi$ of functions for which there is some $\mu > 0$ with $\int \psi(|f|/\mu) < \infty$ has already been heralded in (14.3.3).

Furthermore, if $|||f|||$ denotes the infimum of $\sum_{n=1}^\infty \|h_n\|_2 \|k_n\|_\phi$ over all representations $f = \sum_{n=1}^\infty h_n * k_n$, then $|||\cdot|||$ and $\|\cdot\|_{\text{a.s.}}$ are equivalent norms on $\mathbf{C}_{\text{a.s.}}$.

The second characterisation we mention follows either from the last one or from more general probabilistic arguments. It states that $f \in \mathbf{C}_{\text{a.s.}}$ if and only if

$$\int_0^1 (\log F(r))^{1/2} \, dr < \infty,$$

where, by virtue of the hypothesis $f \in \mathbf{L}^2$, the definition $d(u, v) = \|T_u f - T_v f\|_2$ gives a translation-invariant pseudo-metric, and $F(r)$ is defined to be $\inf\{n$: there is an open cover $A_1, \cdots, A_n$ of T in which every A_j has d-length not larger than $2r\}$. Furthermore, the norm $\|\cdot\|_I$ on $\mathbf{C}_{\text{a.s.}}$, defined by

$$\|f\|_I = \|f\|_2 + \int_0^1 (\log F(r))^{1/2} \, dr,$$

is equivalent to $\|\cdot\|_{\text{a.s.}}$.

Again, the appearance of the square root of the log term is to some extent explained by being inverse to the exponential of the square which occurs in (14.3.3).

Though it has been suppressed in this summary, underpinning both these characterisations are results of Dudley and Fernique on characterising certain

Gaussian processes with continuous sample paths. For an exposition of these results, the reader is referred to Marcus and Pisier [1].

Returning to the question of uniform convergence, we write $\mathbf{U}_{\text{a.s.}}$ for the set of functions f in $\mathbf{L}^2$ for which $\sum_{n \in Z} \rho_n(\omega)\hat{f}(n)e^{inx}$ is a uniformly convergent Fourier series for λ-almost all $\omega \in \mathscr{C}$. A surprisingly tight solution to the question as to when $f \in \mathbf{U}_{\text{a.s.}}$ was given by Salem and Zygmund [1] using classical methods; see also [Ba_1], p. 331. Salem and Zygmund showed that, if the remainders

$$r_N = \left\{ \sum_{|n| > N} |\hat{f}(n)|^2 \right\}^{1/2}$$

satisfy the condition

$$\sum_{N=2}^{\infty} \frac{r_N}{N(\log N)^{1/2}} < \infty, \tag{14.3.7}$$

then $f \in \mathbf{U}_{\text{a.s.}}$. The conclusion is valid, in particular, when

$$\sum_{n \neq 0} |\hat{f}(n)|^2 \log^{1+\varepsilon} |n| < \infty$$

for some $\varepsilon > 0$, a case discussed earlier by Paley and Zygmund and deducible quite rapidly from 14.2.2. (see [Z_1], p. 219). The condition

$$\sum_{n \neq 0} |\hat{f}(n)|^2 \log |n| < \infty$$

is, however, not enough to ensure the desired result: this may be seen by applying 15.1.4 and 15.2.4 to the lacunary series

$$\sum_{k=2}^{\infty} \frac{e^{i2^k x}}{k \log k}.$$

Compare also the remarks in 10.4.5 and 10.4.6.

Ideas similar to those used to characterise $\mathbf{C}_{\text{a.s.}}$ have enabled Pisier to characterise $\mathbf{U}_{\text{a.s.}}$. Firstly, for $f \in \mathbf{L}^2$, we define a translation invariant pseudo-metric d on T by

$$d(x, y) = \sigma(x - y) = \left\{ \int_{\mathscr{C}} \left| \sum_{n \in Z} \rho_n(\omega)\hat{f}(n)e^{inx} - \sum_{n \in Z} \rho_n(\omega)\hat{f}(n)e^{iny} \right|^2 d\lambda(\omega) \right\}^{1/2}$$

and define the nondecreasing rearrangement $\bar{\sigma}$ of σ by

$$\bar{\sigma}(u) = \sup \{ y \in R : \{ x \in T : \sigma(x) < y \} \text{ has invariant measure (on } T) \text{ less than } u \}.$$

(This ensures that the probability distributions of $\bar{\sigma}$ and σ coincide.) Now we can state that $f \in \mathbf{U}_{\text{a.s.}}$ if and only if

$$\int_0^1 \frac{\bar{\sigma}(s)}{s(\log 4/s)^{1/2}} \, ds < \infty. \tag{14.3.8}$$

The reader may wonder at the appearance of $\bar{\sigma}$ here. To allay his doubts, we mention that when $\hat{f}$ is real and F denotes the nonincreasing rearrangement of $\hat{f}$, $2\bar{\sigma}(1/N) \geqslant \{\sum_{|n|>N} (F(n))^2\}^{1/2}$ for every integer $N \geqslant 1$; using this, we may compare (14.3.8) with (14.3.7). For an expansive treatment we again refer to Marcus and Pisier [1].

Incidentally, from the result mentioned above, Pisier also deduces that, if $f \in \mathbf{C}_{\text{a.s.}}$, then so does the function g such that $\hat{g}$ equals the nonincreasing rearrangement of $\hat{f}$.

For an earlier approach to these questions, see Billard [1], [Kah$_3$] and Fielder, Jurkat and Körner [1].

14.4 Comments on the Hausdorff-Young Theorem and Its Dual

It has been seen in 13.5.1(1) that

$$\sum_{n \in Z} |\hat{f}(n)|^{p'} < \infty$$

whenever $1 \leqslant p \leqslant 2$ and $f \in \mathbf{L}^p$. We can now verify that this assertion is no longer true when $p > 2$; compare 13.5.3(2).

Indeed, if $p > 2$, then $p' < 2$ and a sequence $(c_n)_{n \in Z}$ may be chosen so that

$$\sum_{n \in Z} |c_n|^2 < \infty, \qquad \sum_{n \in Z} |c_n|^{p'} = \infty.$$

By 14.3.2, if $p < \infty$, there exists $f \in \mathbf{L}^p$ such that $|\hat{f}(n)| = |c_n|$ $(n \in Z)$, so that

$$\sum_{n \in Z} |\hat{f}(n)|^{p'} = \sum_{n \in Z} |c_n|^{p'} = \infty.$$

If, on the other hand, $p = \infty$, the proposed extension signifies that $\sum |\hat{f}(n)| < \infty$ for each $f \in \mathbf{L}^\infty$. This is negatived by the results mentioned in 14.3.6, and is otherwise clear from many simple examples (see Exercise 1.5).

From 14.3.5(2) it can be shown that the dual result, 13.5.1(2), is also false when $p > 2$. In other words, if $p > 2$ there exist functions $\phi \in \ell^p(Z)$ such that the distribution $\hat{\phi}$ is not a Radon measure. A more explicit proof of this will appear in Section 15.4. Even stronger results are known for a general class of groups; see Gaudry [2], Theorem 2.5.

14.5 A Look at Some Dual Results and Generalizations

Inasmuch as functions on a set S and taking the values ± 1 are simply and obviously expressible in terms of characteristic functions χ_A of subsets A of S, it turns out that the arguments leading to 14.3.2 will in fact establish something slightly stronger and expressible as follows: if $\phi \in \ell^2(Z)$ then, for "almost all" subsets A of Z, $(\chi_A \phi)^\wedge$ belongs to $\mathbf{L}^p$ for every finite p. On the other hand, the results given in [Z$_1$], p. 214, imply somewhat more than 14.3.5(1) and 14.3.5(2), namely: if ϕ is a tempered function on Z that does not belong to $\ell^2(Z)$, then, for "almost all" subsets A of Z, the distribution $(\chi_A\phi)^\wedge$ is not a Radon measure. (The term "almost all" is to be interpreted by reversing the passage from $\pm$ 1-valued functions to characteristic functions and relating the former to $\mathscr{C}$, as has been done hitherto in this chapter.)

It is tempting to contemplate the duals of these results, that is, to consider what can be said about the transforms $(\chi_A f)^\wedge$, where f is a given integrable function on the group T and A denotes a variable measurable subset of that group. Curiously enough, little if anything appears in the classical literature concerning this matter. Related questions have recently been discussed for a category of general groups by Figà-Talamanca [1], [2] and Gaudry ([2], Theorems 2.6 and 2.7). Owing to the topological differences between the smooth compact group T and the discrete group Z, the dual results are in our case a little different from what one might perhaps expect on the basis of the results mentioned in the preceding paragraph.

By using techniques similar to those employed in Gaudry [2] and Edwards [8], it can be shown that, if $f \in \mathbf{L}^1$, and if to each A belonging to a nonmeager set of measurable subsets of T there corresponds an index $p < 2$ such that $(\chi_A f)^\wedge \in \ell^p$, then f is null. (On the other hand, of course, $(\chi_A f)^\wedge \in \ell^2$ for every measurable A whenever $f \in \mathbf{L}^2$.) In this statement the term "nonmeager" is to be interpreted in terms of the metric space whose elements are (equivalence classes modulo null sets of) measurable subsets A of T, the metric being defined by

$$d(A, B) = \frac{1}{2\pi}\int |\chi_A - \chi_B|\,dx.$$

(It happens quite frequently that the meager subsets of a complete metric space play a role somewhat similar to that filled by the null sets of a measure space, especially in situations where no natural countably additive measure is available. Compare also the closing remarks in 14.3.5.) An alternative formulation asserts that if ϕ is a function in $\mathbf{A}(Z)$, and if to each A belonging to a nonmeager set of measurable subsets of T there corresponds an index $p < 2$ and a function $\psi \in \ell^p$ such that $\chi_A\hat{\phi} = \hat{\psi}$, then $\phi = 0$.

A pointer to the existence of analogues of such results for the groups R^m (which analogues are in fact embraced in the results obtained by Figà-Talamanca and Gaudry mentioned above) seems first to have arisen in the work of Hörmander [1] on the multiplier problem for the spaces $\mathbf{L}^p(R^m)$. Corresponding problems for the circle group T are dealt with in some detail in Chapter 16.

For extensions to compact groups, see Figà-Talamanca and Rider [2].

EXERCISES

14.1. Verify (without reference to Tychonoff's theorem) that $\mathscr{C}$ is compact. Show that the group (pointwise) product is a continuous function from $\mathscr{C} \times \mathscr{C}$ to $\mathscr{C}$.

14.2. Suppose that $g \in \mathbf{L}^2(\mathscr{C})$ is such that

$$\int_{\mathscr{C}} \exp\,[\mu\,|g(\omega)|^2]\; d\lambda(\omega) = \infty$$

for some real number μ. Prove that $g \notin \mathbf{L}^2_{\mathscr{R}}(\mathscr{C})$, and deduce that

$$f = g - \sum_{\zeta \in \mathscr{R}} \hat{g}(\zeta)\zeta$$

belongs to $\mathbf{L}^2(\mathscr{C})$, is nonnull, and yet

$$\int_{\mathscr{C}} f(\omega)\rho_n(\omega)\; d\lambda(\omega) = 0 \qquad \text{for all } n \in Z.$$

(This establishes, in particular, that the set $\mathscr{R}$ of Rademacher characters of $\mathscr{C}$, is not a complete orthonormal set in $\mathbf{L}^2(\mathscr{C})$. As an alternative proof of this incompleteness, consider $f = \rho_n\,\rho_m$, where n and m are distinct integers.)

14.3. Verify the claims made about annihilators in 14.1.4.

14.4. Show that 14.2.2 implies 14.2.1, though perhaps with values of B_p and A_p differing from the given ones by a nonzero constant factor.

14.5. (1) Using 14.3.2, show that if $(c_n)_{n \in Z}$ has the property that

$$\sum_{n \in Z} |c_n \hat{f}(n)| < \infty$$

for each f such that

$$\frac{1}{2\pi} \int \exp\,[\mu|f(x)|^2]\, dx < \infty$$

for all real numbers μ, then $\sum_{n \in Z} |c_n|^2 < \infty$.

(2) Assuming the result stated in 14.3.5(2), show that the conclusion of (1) is valid when the hypothesis is merely that $\sum_{n \in Z} |c_n \hat{f}(n)| < \infty$ for each $f \in \mathbf{C}$.

Note: For numerous similar results, see Mahmudov [1] and MR **30** # 5113a, b.

14.6. (1) By using part (1) of the preceding exercise, show that if $(c_n)_{n \in Z}$ has the property that for some $q > 1$ the series $\sum_{n \in Z} \pm c_n e^{inx}$ is, for all choices of the $\pm$ signs, the Fourier series of a function in $\mathbf{L}^q$, then $\sum_{n \in Z} |c_n|^2 < \infty$. [Compare 14.3.5(1).]

(2) Use part (2) of the preceding exercise to prove statement 14.3.5(2).

Hints: If, for example, $\sum \pm c_n e^{inx}$ is always a Fourier-Stieltjes series, the series $\sum \pm c_n \hat{f}(n)$ is Cesàro-summable for each $f \in \mathbf{C}$ and each choice of $\pm$ signs.

14.7. Give a direct proof that 14.3.5(1) implies 14.3.5(2).

Hints: Assuming that $\sum \pm c_n e^{inx}$ is always a Fourier-Stieltjes series, show that $\sum \pm c_n \hat{f}(n) e^{inx}$ is always a Fourier-Lebesgue series for $f \in \mathbf{L}^1$. Apply 14.3.5(1) and the hints to Exercise 3.14.

14.8. Assuming the results due to Paley and Zygmund stated in 14.3.6, prove that if

$$\sum_{n \neq 0} |c_n|^2 \log^{1+\varepsilon} |n| < \infty$$

for some $\varepsilon > 0$, then there exists a continuous function f such that, for some choice of the $\pm$ signs, $\hat{f}(n) = \pm c_n$ $(n \in Z)$.

14.9. Assume that

$$f = \sum_{|n| \leqslant N} c_n \rho_n,$$

where $N \in Z_+$ and $c_{-N}, \cdots, c_N$ are complex numbers. Prove that

$$\sum_{\zeta \in \mathscr{C}^\wedge} |\hat{f}(\zeta)| \leqslant 2\|f\|_\infty \tag{1}$$

and conclude that

$$\sum_{\zeta \in \mathscr{C}^\wedge} |\hat{f}(\zeta)| \leqslant 4\|f\|_{2N+1}. \tag{2}$$

Remark. Denote by $\mathbf{A}(\mathscr{C})$ the set of all continuous complex-valued functions f on $\mathscr{C}$ such that $\sum_{\zeta \in \mathscr{C}^\wedge} |\hat{f}(\zeta)| < \infty$; cf. 10.6.1. It follows from (1) that $\mathbf{C}_{\mathscr{R}}(\mathscr{C}) \subset \mathbf{A}(\mathscr{C})$. In the terminology introduced in 15.1.1 below, $\mathscr{R}$ is thus a Sidon subset of $\mathscr{C}^\wedge$; cf. 15.1.4(c).

Hints: Assume first that $\hat{f}$ is real-valued and show that there exists $\omega_0 \in \mathscr{C}$ such that $\operatorname{sgn} c_n = \rho_n(\omega_0)$ for all $n \in Z$ such that $|n| \leqslant N$.

For (2) observe that f is constant on the cosets of $\mathscr{C}_N$ and use (14.1.24).

14.10. Let ϕ be a complex-valued function on Z, and let $1 \leqslant p \leqslant \infty$. Consider the following statement:

$$(c_p) \qquad \sum_{n \in Z} |\phi(n)\hat{f}(n)| < \infty \qquad \text{for every } f \in \mathbf{L}^p.$$

Prove that

(1) (c_1) is true if and only if $\phi \in \ell^1(Z)$;

(2) if (c_p) is true, then $\phi \in \ell^2(Z)$;

(3) if $2 \leqslant p \leqslant \infty$, then (c_p) is true if and only if $\phi \in \ell^2(Z)$;

(4) if $1 < p < 2$ and $\phi \in \ell^p(Z)$, then (c_p) is true;

(5) if $1 < p < \infty$ and (c_p) is true, then

$$\sum_{n \in Z} a_{|n|} |\phi(n)| < \infty$$

for every sequence $(a_n)_{n=0}^\infty \downarrow 0$ such that

$$\sum_{n=1}^\infty n^{p-2} a_n{}^p < \infty.$$

Hints: Recall Exercise 3.14, Subsections 7.3.5, 8.2.1, 13.5.1(1), and Exercise 14.5(2).

14.11. Denote by $\mathbf{F}$ the set of $f \in \mathbf{L}^1$ such that

$$\|f * g\|_1 \leqslant B_f \|\hat{g}\|_\infty \qquad \text{for all } g \in \mathbf{L}^1, \tag{1}$$

and let $\|f\|_*$ denote the smallest admissible value of B_f in (1). Verify that

(a) $\|f\|_1 \leqslant \|f\|_*$ for $f \in \mathbf{F}$, and that $\mathbf{L}^2 \subset \mathbf{F}$ and $\|f\|_* \leqslant \|f\|_2$ for $f \in \mathbf{L}^2$;

(b) $\mathbf{F}$ is a Banach space under the norm $\| \cdot \|_*$;

(c) $\mathbf{F}$ consists precisely of those $f \in \mathbf{L}^1$ such that $\hat{f}\phi \in \mathbf{A}(Z)$ for each $\phi \in \mathbf{c}_0(Z)$.

These results are due to Helgason [1].

Hints: For (c), observe that if $f \in \mathbf{F}$, f defines a linear map $F: \hat{g} \to f * g$ of $\mathbf{A}(Z)$ into $\mathbf{L}^1$; show that it is possible to continuously extend F from $\mathbf{A}(Z)$ to $\mathbf{c}_0(Z)$. For the converse, if f is as in (c), consider the linear map $S: \phi \to (\hat{f}\phi)^\wedge$ from $\mathbf{c}_0(Z)$ into $\mathbf{L}^1$ and apply the closed graph theorem.

14.12. The notation is as in Exercise 14.11. Prove that $\mathbf{F} = \mathbf{L}^2$ as linear spaces. (The result is due to Helgason [1]; actually, as is easily derived from the proof sketched below, the norms $\| \cdot \|_*$ and $\| \cdot \|_2$ are equivalent.)

Hints: In view of Exercise 14.11(a), it suffices to prove that $\mathbf{F} \subset \mathbf{L}^2$; and to do this it suffices to show that

$$\|f\|_2 \leqslant \text{const } \|f\|_* \qquad \text{for all } f \in \mathbf{T}. \tag{1}$$

Let $f = \sum_{|n| \leqslant N} \hat{f}(n) e_n$. Denote by G the product group T^{2N+1} and let $t = (t_{-N}, \cdots, t_N)$ denote a generic element of G. By considering

$$g = \sum_{|n| \leqslant N} \exp(it_n) e_n,$$

show that $\|f\|_* \geqslant \|f * g\|_1$ and hence that

$$\|f\|_* \geqslant \sup_{t \in G} \frac{1}{2\pi} \int \Big| \sum_{|n| \leqslant N} \hat{f}(n) \exp(it_n) e^{inx} \Big| \, dx$$

$$\geqslant I\Big\{\frac{1}{2\pi} \int \Big|\sum \cdots\Big| \, dx\Big\},$$

where I denotes the normalized invariant integral on G. Using translation-invariance, deduce that

$$\|f\|_* \geqslant I(|F|), \tag{2}$$

where $F(t) = \sum_{|n| \leqslant N} \hat{f}(n) \exp(it_n)$. Verify that

$$I(|F|^4) = 2\{\sum |\hat{f}(n)|^2\}^2 - \sum |\hat{f}(n)|^4 \geqslant 2\|f\|_2^4, \tag{3}$$

and that

$$I(|F|^2) \leqslant \{I(|F|)\}^{2/3} \{I(|F|^4)\}^{1/3}, \tag{4}$$

and so derive (1) from (2), (3) and (4).

14.13. Let ϕ be a complex-valued function on Z such that $\phi\psi \in \mathbf{A}(Z)$ for each $\psi \in \mathbf{c}_0(Z)$. Prove that $\phi \in \ell^2(Z)$.

Hints: Let $\mathbf{S}$ denote the set of ϕ with the stated property. By the preceding exercise, $\mathbf{A}(Z) \cap \mathbf{S} \subset \ell^2(Z)$. Use the closed graph theorem to show that

$$\|(\phi\psi)^\wedge\|_1 \leqslant \text{const } \|\psi\|_\infty$$

whenever $\psi \in \mathbf{c}_0(Z)$ and $\phi \in \mathbf{S}$, and so deduce that $\mathbf{S} \subset \mathscr{F}\mathbf{M}$ (recall 12.3.9). Observe that as a consequence, if $\phi \in \mathbf{S}$ and $h \in \mathbf{L}^1$, then $\hat{h}\phi \in \mathbf{A}(Z) \cap \mathbf{S} \subset \ell^2(Z)$, and apply the closed graph theorem again to deduce that $\|\hat{h}\phi\|_2 \leqslant \text{const } \|h\|_1$ for $h \in \mathbf{L}^1$.

14.14. Let ϕ be a complex-valued function on Z such that $\phi\omega \in \mathscr{F}\mathbf{M}$ for each ± 1-valued function ω on Z. Prove that $\phi \in \ell^2(Z)$. [This statement is equivalent to 14.3.5(2).]

Hints: The hypothesis signifies that $\phi\chi \in \mathscr{F}\mathbf{M}$ for every $\chi \in \mathbf{K}$, where $\mathbf{K}$ denotes the set of all characteristic functions of subsets of Z. By using the category theorem (see I, A.3 and Edwards [7]), show that there exists a number B such that

$$\|(\phi\chi)^\wedge\|_1 \leqslant B \qquad \text{for all } \chi \in \mathbf{K}. \tag{1}$$

Aim at deducing that

$$\phi\psi \in \mathscr{F}\mathbf{M} \quad \text{for all } \psi \in \ell^\infty(Z), \tag{2}$$

and then use Exercise 14.13 and the hints to Exercise 14.7.

To prove (2), consider any real-valued $\psi \in \ell^\infty(Z)$ satisfying $0 \leqslant \psi < 1$. For $k, r = 1, 2, \cdots$, define $A_{k,r} = \{n \in Z : (r-1)/k \leqslant \psi(n) < r/k\}$, $B_{k,r} = A_{k,1} \cup \cdots \cup A_{k,r}$, $B_{k,0} = \varnothing$, and introduce the function

$$\psi_k = \sum_{r=1}^{k} \frac{r}{k} \chi_{A_{k,r}},$$

observing that $\phi\psi_k \to \phi\psi$ as $k \to \infty$. Since $\psi_k = \chi_k - (1/k) \sum_{r=1}^{k-1} \chi_r$, where $\chi_r = \chi_{B_{k,r}}$, (1) shows that

$$\|(\phi\psi_k)^\wedge\|_1 \leqslant 2B.$$

Use 12.3.9 to infer that $\phi\psi \in \mathscr{F}\mathbf{M}$.

14.15. Show that the dual group of $\mathscr{C}^\wedge$ is isomorphic to $\mathscr{C}$ (see Volume 1, p. 20). Identify the annihilator of $\mathscr{A}_N$, that is, identify $\{x \in (\mathscr{C}^\wedge)^\wedge : x(\zeta) = 1 \text{ for all } \zeta \in \mathscr{A}_N\}$.

14.16. (1) Prove that the function $\Phi: \mathscr{C} \to [0, 1]$ given by

$$\Phi(\omega) = \frac{1 + \omega(0)}{3} + \frac{1 + \omega(1)}{3^2} + \frac{1 + \omega(-1)}{3^3} + \frac{1 + \omega(2)}{3^4} + \frac{1 + \omega(-2)}{3^5} + \cdots$$

is a homeomorphism from $\mathscr{C}$ onto the Cantor ternary set (with induced topology), but that Φ is not a group homomorphism (with addition modulo 1 taken as the group operation on $[0, 1]$) nor an isomorphism of measure spaces. (Recall that two measure spaces $(X, \mathscr{M}, \mu)$ and $(Y, \mathscr{N}, \nu)$ are isomorphic if there is a μ-negligible subset E of X, a ν-negligible subset F of Y and a bijection $f: X\backslash E \to Y\backslash F$, such that for all $N \in \mathscr{N}$, $f^{-1}(N\backslash F) \in \mathscr{M}$ and $\mu(f^{-1}(N\backslash F)) = \nu(N)$.)

(2) Prove that, in contrast, the function $\Psi: \mathscr{C} \to [0, 1]$ such that

$$\Psi(\omega) = \frac{1-\omega(0)}{2} + \frac{1-\omega(1)}{2^2} + \frac{1-\omega(-1)}{2^3} + \frac{1-\omega(2)}{2^4} + \frac{1-\omega(-2)}{2^5} + \cdots$$

for all $\omega \in \mathscr{C}$, fails to be a bijection, because for every rational $r \in [0, 1]$ of the form $r = k2^{-j}$ where $k, j \in Z_+$ are such that $k \leqslant 2^j$ and $j \geqslant 1$, there exist $\omega, \phi \in \mathscr{C}$ such that $\omega \neq \phi$ and $\Psi(\omega) = \Psi(\phi) = r$. Let us denote by Ψ_1 the restriction of Ψ to $\{\omega \in \mathscr{C}: \omega \text{ is not eventually } 1\}$. Prove that Ψ_1 is a group homomorphism (with addition modulo 1 in $[0, 1]$), that Ψ_1 is not a homeomorphism (with the usual topology on $[0, 1]$), but that Ψ is an isomorphism of measure spaces (when $[0, 1]$ is endowed with the usual Lebesgue structure). What is the image of $\mathscr{C}_N$ under Ψ_1?

(3) From (2) it follows that two compact Abelian groups which are isomorphic as groups and measure spaces, need not be homeomorphic. Suppose two infinite compact Abelian groups are homeomorphic and isomorphic as measure spaces. Need they be isomorphic as groups?

(4) The (classical) Rademacher functions are defined by

$$r_n(t) = \text{sign}\,(\sin\,(2^{n+1}\pi t))$$

for every $n \in Z_+$ and every $t \in [0, 1]$; see 14.1.1 in the first edition of this book. Prove that, for every $n \in Z$,

$$\rho_n = r_{j(n)} \circ \Psi_1,$$

where $j(n)$ equals $-n/2$ if n is even and equals $(1+n)/2$ if n is odd. Can you deduce what the classical definition of the Walsh functions is (see 14.1.3)?

14.17. Define $F: \mathscr{C}^\wedge \to Z_+$ by

$$F(\rho_n) = \begin{cases} 2^{2n} & \text{if } n \geqslant 0 \\ 2^{-2n-1} & \text{if } n < 0 \end{cases};$$

$$F(1^{\mathscr{C}}) = 0;$$

$$F(\textstyle\prod_{n\in\Phi} \rho_n) = \sum_{n\in\Phi} F(\rho_n) \text{ for every nonvoid finite subset } \Phi \text{ of } Z.$$

Prove that F is a bijection, which takes $\mathscr{A}_N$ onto $\{0, 1, \cdots, 2^{2N+1}-1\}$. The ordering defined on $\mathscr{C}^\wedge$ to make F increasing is called the *Paley order* on $\mathscr{C}^\wedge$. (Using this notation we can write a Walsh-Fourier series as $\sum_{n=0}^{\infty} c_n F^{-1}(n)$, which highlights the similarity with Fourier series over T and suggests analogous results; see 14.1.19.

14.18. Even though the (unweighted) partial sums of the Walsh-Fourier series of a continuous function on $\mathscr{C}$ converge uniformly to f, $\mathbf{C}(\mathscr{C})$ does not collapse too dramatically; in particular $\mathbf{C}(\mathscr{C})\backslash\mathbf{A}(\mathscr{C})$ is nonvoid (in spite of the result in Exercise 14.9). (Recall that from the results of Chapter 7 follow examples of continuous functions on T whose Fourier series do not converge absolutely.) In the present exercise the reader is asked to prove in two different ways that $\mathbf{C}(\mathscr{C})\backslash\mathbf{A}(\mathscr{C})$ is nonvoid.

(1) Let $(c_N)_{N=1}^{\infty}$ denote a decreasing sequence of nonnegative real numbers such that $\sum_{N=1}^{\infty} c_N = \infty$. Show that the Walsh-Fourier series

$$\sum_{N=1}^{\infty} (-1)^N 2^{-(2N-1)} c_N (P_N - P_{N-1}) \tag{1}$$

converges uniformly on $\mathscr{C}$ to a continuous function f, say, for which

$$\sum_{\zeta \in \mathscr{C}^{\wedge}} |\hat{f}(\zeta)| = 3 \sum_{N=1}^{\infty} c_N = \infty. \tag{2}$$

Hints: For (1), use (14.1.16) to identify $P_N(\omega) - P_{N-1}(\omega)$ for $\omega \in \mathscr{C}_N$ and for $\omega \in \mathscr{C}_{N-1}\backslash\mathscr{C}_N$. For (2), use (14.1.17).

(2) Show that $\mathbf{A}(\mathscr{C})$ is a Banach space when endowed with the norm

$$\| f \|_{\mathbf{A}} = \sum_{\zeta \in \mathscr{C}^{\wedge}} |\hat{f}(\zeta)|.$$

Hence show that $\mathbf{A}(\mathscr{C})$ is a proper subset of $\mathbf{C}(\mathscr{C})$.

Hints: The inclusion $\mathbf{C}(\mathscr{C}) \subset \mathbf{A}(\mathscr{C})$ together with the closed graph theorem B.3.3 shows that for some $K \in Z_+$,

$$\| f \|_{\mathbf{A}} \leqslant K \| f \|_{\infty} \qquad \text{for all } f \in \mathbf{C}(\mathscr{C}).$$

Then (look ahead to the proof of 15.3.1), there exists $L \in Z_+$ such that

$$\| g \|_2 \leqslant L \| g \|_1 \qquad \text{for all } g \in \mathbf{T}(\mathscr{C}),$$

which inequality can be readily contradicted.

14.19. Pursue, as far as you are able and interested, the study of convolution of functions on $\mathscr{C}$ (see 14.1.14). For instance if $\mathbf{M}(\mathscr{C})$ is defined to be the Banach space of continuous linear functionals on $\mathbf{C}(\mathscr{C})$, can you extend convolution of functions to convolution of measures, and prove the properties corresponding to those over T (see §12.7)?

14.20. Prove the claims made in 14.1.15 (1), (2) and (3).

14.21. Prove, using the method of Chapter 8, that the Walsh-Fourier transform is an isomorphism between the Hilbert spaces $\mathbf{L}^2(\mathscr{C})$ and $l^2(\mathscr{C}^{\wedge})$; that is, $f \to \hat{f}$ is bijective and linear, and $\int_{\mathscr{C}} f\bar{g}\, d\lambda = \sum_{\mathscr{C}^{\wedge}} \hat{f}\bar{\hat{g}}$ for all $f, g \in \mathbf{L}^2(\mathscr{C})$.

14.22. (1) Prove Hölder's inequality (14.1.23) over $\mathscr{C}$, and its corollaries (14.1.24) and (14.1.25).

(2) Prove that the invariant integral I on $\mathscr{C}$ is invariant under the left shift (see 14.1.5).

14.23. Prove that, if $f \in \mathbf{L}^1(\mathscr{C})$, then for λ-almost all $\omega \in \mathscr{C}$,

$$\lim_{N \to \infty} s_N f(\omega) = f(\omega).$$

This result is in strong contrast with the analogue for T; see Kolmogorov's example mentioned in 10.3.4 and 10.3.5. In view of 14.1.14 and (14.1.17), the result can be recast as: if $f \in \mathbf{L}^1(\mathscr{C})$ then, for λ-almost all $\omega \in \mathscr{C}$,

$$\lim_{N \to \infty} 2^{2N+1} \int_{\omega \mathscr{C}_N} f \, d\lambda = f(\omega),$$

which is a fundamental theorem of calculus for $\mathbf{L}^1(\mathscr{C})$. (As usual, $\omega \mathscr{C}_N$ denotes the coset of $\mathscr{C}_N$ containing ω).

Hint: For $f \in \mathbf{L}^1(\mathscr{C})$, define the maximal function Mf by

$$Mf(\omega) = \sup \left\{2^{2N+1} \int_{\omega \mathscr{C}_N} |f| \, d\lambda : N \in Z_+\right\} \qquad \text{for all } \omega \in \mathscr{C}.$$

Aim first to show that $f \to Mf$ is of weak type (1, 1) (see (13.7.5)). Do this by showing that, if $\|f\|_1 < t$, then $\{\omega \in \mathscr{C} : |Mf(\omega)| > t\}$ can be partitioned into a countable number of sets of the form $E_{\omega, N} \equiv \omega \mathscr{C}_N$ such that

$$t < \lambda(E_{\omega, N})^{-1} \int_{E_{\omega, N}} |f| \, d\lambda$$

and use this fact to show that $\lambda(|Mf| > t) \leqslant t^{-1} \int_{\mathscr{C}} |f| \, d\lambda$. The desired result now follows by choosing a sequence $(g_k)_{k \in Z_+}$ of continuous functions such that $\|f - g_k\|_1 \to 0$ as $k \to \infty$ and $\lim_{k \to \infty} g_k(\omega) = f(\omega)$ for λ-almost all $\omega \in \mathscr{C}$, and observing that

$$\begin{aligned} |s_N f - f| &\leqslant |s_N f - s_N g_k| + |s_N g_k - g_k| + |g_k - f| \\ &\leqslant M(f - g_k) + |s_N g_k - g_k| + |g_k - f|. \end{aligned}$$

For $(g_k)_{k \in Z_+}$ one may (by 14.1.15 (1)) take a suitable subsequence of $(s_N f)_{N \in Z_+}$. (For an elaboration of the partitioning argument, see the statement and proof of Lemma 2.2.1 in [EG].)

14.24. (1) Prove that, for λ-almost every $\omega \in \mathscr{C}$,

$$\lim_{N \to \infty} (2N + 1)^{-1} \sum_{|j| \leqslant N} \omega(j) = 0$$

(and so infer that most members of $\mathscr{C}$ have, asymptotically, an equal number of $\pm$ signs).

(2) Prove that for almost all real numbers x between 0 and 1, if x_n denotes the nth digit in the binary expansion of x, then

$$\lim_{n\to\infty} n^{-1} \sum_{j=1}^{n} x_j = \tfrac{1}{2},$$

which may be interpreted as asserting that most numbers have equal quantities of 0's and 1's in their binary expansion. Observe that x_n is well defined only as long as x is not a dyadic rational (see Exercise 14.16 (2)), and that the set of dyadic rationals is null in [0, 1].

Hint: For (1) define f_N on $\mathscr{C}$ by

$$f_N(\omega) = (2N+1)^{-4}\left(\sum_{|j|\leqslant N} \rho_j(\omega)\right)^4$$

and show that $\sum_{N=1}^{\infty} \|f_N\|_1 < \infty$. For (2), use Exercise 14.16 (2) to replace $\mathscr{C}$ by [0, 1].

14.25. Suppose that f is a function on $\mathscr{C}$ of the form

$$f(\omega) = F(\omega(-M), \cdots, \omega(M)) \qquad \text{for all } \omega \in \mathscr{C},$$

where $M \in Z_+$ and F is a complex-valued function on $K \equiv \prod_{j=-M}^{M} K_j$ and $K_j = \{-1, +1\}$ for all $j \in \{-M, \cdots, M\}$.

Prove that

$$I(f) = \lim_{M\to\infty} 2^{-2M-1} \sum F, \tag{1}$$

the sum extending over K.

Deduce that, if in addition there are functions $G_{-M}, \cdots, G_M$ on $\{-1, +1\}$ such that

$$F(\varepsilon_{-M}, \cdots, \varepsilon_M) = \prod_{j=-M}^{M} G_j(\varepsilon_j)$$

for all $(\varepsilon_{-M}, \cdots, \varepsilon_M) \in K$, then

$$I(f) = 2^{-2M-1} \prod_{j=-M}^{M} (G_j(-1) + G_j(1)). \tag{2}$$

More particularly still, if $\alpha_{-M}, \cdots, \alpha_M \in Z$, then

$$I(\rho_{-M}^{\alpha_{-M}}, \cdots, \rho_M^{\alpha_M}) = 1 \quad \text{or} \quad 0 \tag{3}$$

according as α_j is even for all $j \in \{-M, \cdots, M\}$ or not. (Cf. equation (14.1.5) in the first edition of this book.)

Remark. The equation (3) can be deduced from 14.1.7 and the independence of $\mathscr{R}$ (see (14.1.3)).

CHAPTER 15

Lacunary Fourier Series

As the name suggests, a *lacunary trigonometric series* is, roughly speaking, a trigonometric series $\sum_{n \in Z} c_n e^{inx}$ in which $c_n = 0$ for all integers n save perhaps those belonging to a relatively sparse subset E of Z. Examples of such series have appeared momentarily in Exercises 5.6 and 6.13. Indeed for the Cantor group $\mathscr{C}$, the good behaviour of a lacunary Walsh-Fourier series

$$\sum_{\zeta \in \mathscr{R}} c_\zeta \zeta$$

(whose coefficients vanish outside the subset $\mathscr{R}$ of $\mathscr{C}^\wedge$) has already been noted: by Exercise 14.9, if the lacunary series belongs to $\mathbf{C}(\mathscr{C})$ then it belongs to $\mathbf{A}(\mathscr{C})$; and, by 14.2.1, if it belongs to $\mathbf{L}^p(\mathscr{C})$ for some $p > 0$, then it also belongs to $\mathbf{L}^q(\mathscr{C})$ for $q \in [p, \infty)$. In this chapter we shall be mainly concerned with lacunary Fourier series on the circle group and will deal more systematically with some (though by no means all) aspects of their curious behaviour.

The classical theory concentrated to a large extent on the case of series that exhibit *Hadamard gaps*, that is, series for which the corresponding set E is of the form $\{\pm n_k : k = 1, 2, \cdots\}$, where the n_k are positive integers forming a Hadamard sequence:

$$\inf_k \frac{n_{k+1}}{n_k} > 1;$$

see the exercises just cited and Section 8.6; [Z_1], pp. 203–212, 215, 247; [Z_2], pp. 131–132; [Ba_1], pp. 178–181; [Ba_2], Chapter XI. Less extreme forms of lacunarity have also been examined (see [Z_1], pp. 222 ff.; [KS], Chapitre XII; Moeller and Frederickson [1]; [I], pp. 86 ff.; Izumi and Kahane [1], Izumi [1]); but we shall concentrate mainly on the phenomena that are typified by Hadamard gap series.

It has become possible to disentangle from any explicit assumptions of lacunarity some of the characteristic properties of such series by means of inequalities and functional analytic statements referring to the so-called

"Sidon subsets" of Z. (The name, coined by Kahane, is explained by the fact that Sidon established some of these properties for series with Hadamard gaps, others being due to Banach.) We shall in the main follow the account given by Rudin ([R], Chapter 5), which applies to compact Abelian groups in general. Extensions to non-Abelian compact groups were initiated by Hewitt and Zuckerman [1], and, for an exposition, the reader is referred to [HR], §37; see also Rudin [6], Figà-Talamanca [1], [DR] Chapter 5, [Kz] Chapter V, and the monograph devoted to Sidon sets, [LR].

The approach via functional analytic inequalities and theorems is fairly new, the first such approach appearing in print being that due to Hewitt and Zuckerman [1]. The functional analytic background is explained briefly in general terms in [E], Section 8.8. Many other concepts of lacunarity have since been examined in this way and for general compact groups; see Section 15.5 and 15.8 and the references cited immediately above.

Lacunarity for general orthogonal expansions on subintervals of R are discussed in [KSt], Kap VII.

The dual aspects of the topics discussed in the main portion of this chapter will be mentioned briefly in Section 15.7.

The harmonic analysis and synthesis of general continuous functions on R^n, a topic mentioned in passing in 11.2.3(4), leads to the introduction of "complex" Sidon sets; for this we must refer the reader to Gilbert [1] and the references cited there.

The reader may find it helpful to examine the brief survey article Kahane [4], the historical notes in [HR], Volume II, pp. 445–449, and [LR], which is a more recent and systematic account covering almost all the principal themes dealt with in this chapter.

15.1 Introduction of Sidon Sets

15.1.1. **Some Definitions and Notations.** Let E be a subset of Z.

A distribution F will be said to be *E-spectral* if and only if $\hat{F}(n) = 0$ for all $n \in Z \backslash E$; cf. 14.2 above.

It will be convenient when $p \geqslant 1$ to denote by $\mathbf{L}_E{}^p$ the set of all E-spectral functions in $\mathbf{L}^p$; by $\mathbf{M}_E$ the set of all E-spectral measures; by $\mathbf{C}_E$ the set of E-spectral functions in $\mathbf{C}$; by $\mathbf{T}$ the set of all trigonometric polynomials; and by $\mathbf{T}_E$ the set of all E-spectral trigonometric polynomials. The reader is left to verify that $\mathbf{L}_E{}^p$ is a closed linear subspace of $\mathbf{L}^p$ (if $p \geqslant 1$); that $\mathbf{C}_E$ is a closed linear subspace of $\mathbf{C}$; that $\mathbf{M}_E$ is a closed linear subspace of $\mathbf{M}$; and that $\mathbf{T}_E$ is a linear subspace of $\mathbf{T}$.

We also denote by $\ell^p(E)$ $(1 \leqslant p \leqslant \infty)$ the set of complex-valued functions ϕ on E such that

$$\|\phi\|_p = \begin{cases} [\sum_{n\in E} |\phi(n)|^p]^{1/p} & \text{if } p < \infty \\ \sup_{n\in E} |\phi(n)| & \text{if } p = \infty, \end{cases}$$

is finite; $\mathbf{c}_0(E)$ is the set of $\phi \in \ell^\infty(E)$ such that

$$\lim_{n\in E, |n|\to\infty} |\phi(n)| = 0,$$

and we regard $\mathbf{c}_0(E)$ as a normed subspace of $\ell^\infty(E)$.

We write ℓ^p and $\mathbf{c}_0$ for $\ell^p(Z)$ and $\mathbf{c}_0(Z)$, respectively. The reader will note that there is a natural injection of $\ell^p(E)$ into ℓ^p and of $\mathbf{c}_0(E)$ into $\mathbf{c}_0$, obtained by extending each function on E so as to be zero on $Z\backslash E$.

If $\mathbf{S}$ is any set of functions on Z, $\mathbf{S}|E$ will denote the set of restrictions $\phi|E$ of functions $\phi \in \mathbf{S}$.

If $\mathbf{H}$ is a set of distributions, $\mathscr{F}\mathbf{H}$ will denote the set of functions on Z that are Fourier transforms of elements of $\mathbf{H}$. Notice that $\mathscr{F}\mathbf{L}^1 = \mathbf{A}(Z)$, in the notation introduced in 2.3.9. One may thus write (for example) $\mathscr{F}\mathbf{L}^1 \subset \mathbf{c}_0$ (by 2.3.8), $\mathscr{F}\mathbf{L}^2 = \ell^2$ (by 8.3.1), and $\mathscr{F}\mathbf{M} \subset \ell^\infty$ (by 12.2.9 and 12.5.3(1)).

The following result, which appears as Exercise 2.19, will be used later.

15.1.2. Given any finite subset F of Z and any $\varepsilon > 0$, there exists a trigonometric polynomial t such that

$$\|t\|_1 \leqslant 1 + \varepsilon, \quad 0 \leqslant \hat{t}(n) \leqslant 1 \quad \text{for all } n \in Z, \qquad \hat{t}(n) = 1 \quad \text{for all } n \in F.$$

Remarks. If $\{0\}$ is a proper subset of F, we cannot in general arrange that $t \geqslant 0$. For suppose that $\{0, n_0\} \subset F$, where $n_0 \neq 0$, and that $t \geqslant 0$. Then

$$(2\pi)^{-1} \int_0^{2\pi} t(x)(1 - e^{-in_0x})\, dx = 0,$$

hence

$$(2\pi)^{-1} \int_0^{2\pi} t(x)(1 - \cos n_0 x)\, dx = 0,$$

and so, since $t \geqslant 0$ and $n_0 \neq 0$, $t(x) = 0$ a.e. But then $\hat{t}(0) = 0$, a contradiction since $0 \in F$.

However, if $0 \notin F$, it can be arranged that $t \geqslant 0$. In fact, define $P = F \cup (-F)$ and

$$t_N = F_N + \sum_{n\in P} (1 - \hat{F}_N(n))e_n + \sum_{n\in P} (1 - \hat{F}_N(n)).$$

Then $t_N \in T$, $t_N \geqslant 0$,

$$\hat{t}_N(n) = \hat{F}_N(n) + (1 - \hat{F}_N(n)) + 0 = 1 \qquad \text{for all } n \in P$$

and

$$\| t_N \|_1 = \hat{t}_N(0) = \hat{F}_N(0) + 0 + \sum_{n \in P} (1 - \hat{F}_N(n))$$
$$= 1 + \sum_{n \in P} (1 - \hat{F}(n))$$

which is at most $1 + \varepsilon$, if N is chosen sufficiently large (depending upon P and ε).

15.1.3. **Sidon Sets.** A subset E of Z is termed a *Sidon set* if and only if there exists a nonnegative number B, possibly depending on E, such that

$$\|\hat{f}\|_1 \leqslant B \cdot \|f\|_\infty \tag{15.1.1}$$

for every $f \in \mathbf{T}_E$. The smallest such B, namely

$$\sup \{\|\hat{f}\|_1/\|f\|_\infty : f \in \mathbf{T}_E \quad \text{and} \quad \|f\|_\infty \neq 0\}$$

is called the *Sidon constant* of E, usually denoted hereinafter by B_E.

The above terminology is suggested by a theorem of Sidon ([Z_1], p. 247; [Ba_2], p. 246) for series with Hadamard gaps, which corresponds closely to 15.1.4(b) and (c) below. Parts (d) and (e) of 15.1.4 show, on the other hand, that all Sidon sets share certain properties which were established by Banach for series with Hadamard gaps (see [Z_2], p. 131).

15.1.4. **Fundamental Criteria.** If E is a subset of Z, the following five statements about E are equivalent:

(a) E is a Sidon set;
(b) $\|\hat{f}\|_1 < \infty$ for each $f \in \mathbf{L}_E{}^\infty$;
(c) $\|\hat{f}\|_1 < \infty$ for each $f \in \mathbf{C}_E$;
(d) $\mathscr{F}\mathbf{M}|E = \ell^\infty(E)$;
(e) $\mathscr{F}\mathbf{L}^1|E = \mathbf{c}_0(E)$.

Proof. This is conveniently broken into parts.

(1) Let us show that (a) implies (b), beginning with the remark that, if $f \in \mathbf{L}_E{}^\infty$, then $\sigma_N f = F_N * f \in \mathbf{T}_E$. So, by (a) and (15.1.1), we have

$$\|(\sigma_N f)^\wedge\|_1 \leqslant B \cdot \|\sigma_N f\|_\infty = B \cdot \|F_N * f\|_\infty \leqslant B \cdot \|F_N\|_1 \cdot \|f\|_\infty$$
$$= B \cdot \|f\|_\infty,$$

the penultimate step following from 3.1.6. Now

$$\|(\sigma_N f)^\wedge\|_1 = \sum_{|n| \leqslant N} \left(1 - \frac{|n|}{N+1}\right) |\hat{f}(n)|$$
$$\geqslant \tfrac{1}{2} \cdot \sum_{|n| \leqslant \frac{1}{2}N} |\hat{f}(n)|,$$

whence it appears that

$$\sum_{|n| \leqslant \frac{1}{2}N} |\hat{f}(n)| \leqslant 2B \cdot \|f\|_\infty .$$

From this, (b) follows easily on letting $N \to \infty$.

(2) The implication (b) $\Rightarrow$ (c) is a trivial consequence of the inclusion $\mathbf{C} \subset \mathbf{L}^\infty$.

(3) It will next be shown that (c) implies (a). If (c) holds, the map $f \to \hat{f}$ is a 1–1 linear map of $\mathbf{C}_E$ onto $\ell^1(E)$. The inverse map is evidently continuous, in view of the simple inequality $\|f\|_\infty \leqslant \|\hat{f}\|_1$. The open mapping theorem (I, B.3.2) entails that this mapping is bicontinuous, which requires that an inequality of the form (15.1.1) be valid for E-spectral trigonometric polynomials f, that is, that (a) is true.

At this stage we have established the equivalence of (a), (b), and (c).

(4) Next on the list comes a proof that (c) implies (d). Suppose that (c) holds and that $\phi \in \ell^\infty(E)$. The mapping

$$f \to \sum_{n \in E} \hat{f}(n)\phi(n)$$

is then a continuous linear functional on $\mathbf{C}_E$. By the Hahn-Banach theorem (I, B.5.1) and 12.2.3, there exists a measure $\mu \in \mathbf{M}$ such that

$$\sum_{n \in E} \hat{f}(n)\phi(n) = \mu(f)$$

for $f \in \mathbf{C}_E$. If herein we take $f = e_n$, where $n \in E$, and write λ for $\check{\mu}$ (defined in 12.6.8), then $\lambda \in \mathbf{M}$ and it appears that $\phi(n) = \hat{\lambda}(n)$. Thus $\phi \in \mathscr{F}\mathbf{M}|E$, showing that (d) holds.

(5) To deduce (e) from (d), we remark that the latter combines with the open mapping theorem (Volume 1, Appendix B.3.2) to entail the existence of a number B' such that to each $\phi \in \ell^\infty(E)$ corresponds a measure $\mu \in \mathbf{M}$ such that

$$\hat{\mu}|E = \phi, \qquad \|\mu\|_1 \leqslant B' \cdot \|\phi\|_\infty . \tag{15.1.2}$$

The reader will observe that an appeal to the open mapping theorem is justified by 12.7.1. Suppose now that $\psi \in \mathbf{c}_0(E)$ and $\|\psi\|_\infty \leqslant 1$. Let

$$E_k = \{n \in E : 2^{-k} < |\psi(n)| \leqslant 2^{-k+1}\},$$

where $k = 1, 2, \cdots$, so that E_k is a finite subset of E. Let ψ_k be defined to be equal to ψ on E_k and to be zero elsewhere on E. According to (15.1.2), we may choose for each k a measure $\mu_k \in \mathbf{M}$ such that

$$\hat{\mu}_k|E = \psi_k, \qquad \|\mu_k\|_1 \leqslant B' \cdot 2^{1-k}. \tag{15.1.3}$$

Moreover, by 15.1.2, we may choose trigonometric polynomials t_k such that

$$\hat{t}_k = 1 \text{ on } E_k, \qquad \|t_k\|_1 \leqslant 2. \tag{15.1.4}$$

Put

$$f = \sum_{k=1}^{\infty} t_k * \mu_k.$$

Since $t_k * \mu_k$ is a trigonometric polynomial and since, by (15.1.3), (15.1.4), and 12.7.4

$$\|t_k * \mu_k\|_1 \leqslant B' \cdot 2^{2-k},$$

it follows that $f \in \mathbf{L}^1$ and

$$\hat{f}(n) = \sum_{k=1}^{\infty} \hat{t}_k(n)\hat{\mu}_k(n)$$

for all $n \in Z$. In particular, if $n \in E_k$, (15.1.3) and (15.1.4) yield

$$\hat{f}(n) = \hat{\mu}_k(n) = \psi_k(n) = \psi(n).$$

Since also $\hat{f}(n) = 0$ at all points $n \in E$ at which $\psi(n) = 0$ (that is, at all points of E not belonging to some E_k), it is seen that $\hat{f}|E = \psi$. Thus $\mathscr{F}\mathbf{L}^1|E = \mathbf{c}_0$ and (e) is derived.

(6) Finally, let us show that (e) implies (a), thus completing the circle. If (e) holds, the open mapping theorem (Volume 1, Appendix B.3.2) comes into play once more and shows that there exists a number B'' such that to each $\psi \in \mathbf{c}_0(E)$ corresponds a function $f \in \mathbf{L}^1$ such that

$$\hat{f}|E = \psi, \qquad \|f\|_1 \leqslant B'' \cdot \|\psi\|_\infty. \tag{15.1.5}$$

Let g be any E-spectral trigonometric polynomial, and define $\psi(n)$ to be $|\hat{g}(n)|/\hat{g}(n)$ or 0 according as $\hat{g}(n) \neq 0$ or $\hat{g}(n) = 0$. Then $\psi \in \mathbf{c}_0(E)$ and $\|\psi\|_\infty \leqslant 1$. Choose f as in (15.1.5). Then, since $\|f\|_1 \leqslant B''$, 3.1.6 yields

$$\sum_{n \in Z} |\hat{g}(n)| = \sum_{n \in Z} \hat{f}(n)\hat{g}(n) = f * g(0)$$
$$\leqslant \|f\|_1 \cdot \|g\|_\infty \leqslant B''\|g\|_\infty,$$

showing that (a) holds (with B'' in place of B).

Remarks. (1) Part (3) of the preceding proof used the open mapping theorem to show that, if (c) holds, then there exists a number B such that

$$\|\hat{f}\|_1 \leqslant B\|f\|_\infty \tag{15.1.6}$$

for every $f \in \mathbf{C}_E$. The reader will find it instructive to construct a proof using the uniform boundedness principle (Volume 1, Appendix B.2.1); see Exercise 15.1. A similar conclusion, with $\mathbf{L}_E^{\infty}$ in place of $\mathbf{C}_E$, follows from (b).

(2) If E is a Hadamard set, somewhat sharper versions of 15.1.4(b) and (c) are valid; see 15.2.6.

(3) As we have seen, an inequality of the form (15.1.6) holds for $f \in \mathbf{L}_E^{\infty}$ whenever E is a Sidon subset of Z. From 15.1.4(e) it is clear that no such inequality is valid for general trigonometric polynomials f. More specifically, an example of D. J. Newman [2] shows that there is an absolute constant c such that to every positive integer N corresponds a trigonometric polynomial f of degree at most N for which $\|f\|_{\infty} = 1$ and $\|\hat{f}\|_1 \geqslant \frac{1}{2}N^{1/2} - c$.

(4) Let E be a Sidon set. It follows from 15.1.4(d) that the operator $T: \mu \to \hat{\mu}|E$ effects a homomorphism of the convolution algebra $\mathbf{M}$ onto the algebra $\ell^{\infty}(E)$ with pointwise multiplication; and from 15.1.4(e) that $T|\mathbf{L}^1$ effects a homomorphism of the convolution algebra $\mathbf{L}^1$ onto the algebra $\mathbf{c}_0(E)$ (again with pointwise multiplication). On the other hand, it can be shown (Edwards [12], Theorem 1) that, if E is any infinite subset of Z, there exists no *isomorphism* of any subalgebra of $\mathbf{M}$ onto either of $\ell^{\infty}(E)$ or $\mathbf{c}_0(E)$, or indeed onto any Banach algebra $\mathbf{B}$ of the type specified in Exercise 11.24. See also Remark (11) below.

This fact indicates that, if E is an infinite Sidon set (see Section 15.2 for examples), although $\hat{\mu}|E$ can be specified as freely as one can expect, one remains largely in the dark as to how freely one can simultaneously specify $\hat{\mu}|Z\backslash E$. One certainly cannot simultaneously demand that $\hat{\mu}|Z\backslash E$ shall coincide with an arbitrarily given element of $\ell^{\infty}(Z\backslash E)$ or of $\mathbf{c}_0(Z\backslash E)$. Can one, however, demand that simultaneously $\hat{\mu}|Z\backslash E$ shall belong to $\mathbf{c}_0(Z\backslash E)$? Perhaps the most important single result in the positive direction is due to Drury [3]. It may be stated as follows; cf. [LR], Chapter 3. Suppose that E is a Sidon set with Sidon constant B, that $0 < \varepsilon \leqslant 1$, and that $\phi \in \ell^{\infty}(E)$; then there exists $\mu \in \mathbf{M}$ such that $\hat{\mu}|E = \phi$, $|\hat{\mu}(n)| \leqslant \varepsilon$ for all $n \in Z\backslash E$, and

$$\|\mu\| \leqslant 512B^4\varepsilon^{-1}.$$

This seemingly innocuous extension of 15.1.4(d) incorporates a most important step in the study of lacunarity and will be referred to again in the sequel.

Fournier has shown (see [LR], Theorem 2.20) that Drury's result is best possible in the sense that, if E is an infinite Sidon set and $\phi \in \ell^{\infty}(E)$, then there is a measure $\mu \in \mathbf{M}$ such that $\hat{\mu}|E = \phi$ and $\hat{\mu}|Z\backslash E \in \mathbf{c}_0(Z\backslash E)$ if and only if $\phi \in \mathbf{c}_0(E)$.

Using Drury's result, Hartman and Wells have independently shown that, if E is a Sidon set, 15.1.4(d) can be strengthened in another direction. Roughly speaking, a measure is called *continuous* if it assigns zero measure to every point of T. (This is one instance where it is easier to regard a measure as assigning values to certain subsets of T; see 12.2.3 above and $[E_1]$, 1.7.) The Hartman-Wells Theorem (see [LR], Chapter 4) states: if E is a Sidon set and $\phi \in \ell^\infty(E)$, then there is a continuous measure μ such that $\hat{\mu} | E = \phi$.

(5) The proof of 15.1.4 could be considerably shortened by making use of general duality theory for Banach spaces as set out in Chapter 8 of [E] (for example).

For instance, the equivalence of (a) and (e) of 15.1.4 follows from a general theorem ([E], Corollary 8.6.15 or [KS], p. 141) which asserts that if $\mathbf{E}$ and $\mathbf{F}$ are Banach spaces and U a continuous linear operator from $\mathbf{E}$ into $\mathbf{F}$, then U maps $\mathbf{E}$ onto $\mathbf{F}$ if and only if the adjoint operator U' is a topological isomorphism of $\mathbf{F}'$ into $\mathbf{E}'$ (see I, B.1.7). On applying this result to the case in which $\mathbf{E} = \mathbf{L}^1$, $\mathbf{F} = c_0(E)$, and $Uf = \hat{f} | E$, the equivalence of (a) and (e) appears almost immediately.

(6) The proof of 15.1.4 actually shows that the following numbers coincide:

(a) the Sidon constant of E

(b) $\sup \{\|\hat{f}\|_1 / \|f\|_\infty : f \in \mathbf{L}_E^\infty \text{ and } \|f\|_\infty \neq 0\}$

(c) $\sup \{\|\hat{f}\|_1 / \|f\|_\infty : f \in \mathbf{C}_E \text{ and } \|f\|_\infty \neq 0\}$

(d) $\inf \{\sup \{\|\mu\|_1 / \|\phi\|_\infty : \mu \in \mathbf{M} \text{ and } \hat{\mu} | E = \phi\} : \phi \in \mathbf{l}^\infty(E) \text{ and } \|\phi\|_\infty \neq 0\}$

(e) $\inf \{\sup \{\|f\|_1 / \|\phi\|_\infty : f \in \mathbf{L}^1 \text{ and } \hat{f} | E = \phi\} : \phi \in \mathbf{c}_0(E) \text{ and } \|\phi\|_\infty \neq 0\}$.

Sets whose Sidon constant takes on certain extreme values (particularly in the non-abelian setting) have been investigated in Cartwright, Howlett and McMullen [1].

(7) The space $\mathbf{C}_E$ in 15.1.4(c) may be replaced by a range of smaller spaces (see Edwards, Hewitt and Ross [1]). Indeed E is Sidon whenever one of the following spaces is contained in $\mathbf{A}$:

(a) $\mathbf{A}^p \equiv \{f \in \mathbf{C} : \hat{f} \in \ell^p(Z)\}$, where $1 < p < \infty$

(b) $\mathbf{A}^{1+} \equiv \bigcap_{1 < p < \infty} \mathbf{A}^p$

(c) $\mathbf{U} \equiv \{f \in \mathbf{C} : s_n f \to f \text{ uniformly as } n \to \infty\}$

(d) $\mathbf{A}_\omega \equiv \{f \in \mathbf{C} : \omega \hat{f} \in \ell^1(Z)\}$, where ω is a given element of $\mathbf{c}_0(Z)$.

(8) From the equivalence of (a), (b) and (c) in 15.1.4, it follows that, for all subsets E of Z, if $\mathbf{C}_E \subset \mathbf{A}$ then $\mathbf{L}_E^\infty \subset \mathbf{C}$. The converse is false, however; see Rosenthal [3], [HR], (37.25.g) and 15.8.3 below.

(9) Regarding 15.1.4(d), a symmetric subset E of Z is Sidon, if and only if every $\psi \in \mathbf{l}^\infty(E)$ satisfying $\psi(-n) = \overline{\psi(n)}$ for all $n \in E$, belongs to $\mathscr{F}\mathbf{M} | E$. This results from the fact that every $\phi \in \ell^\infty(E)$ can be written $\phi = \psi_1 + i\psi_2$, where $\psi_1(n) = 2^{-1}(\phi(n) + \overline{\phi(-n)})$ and $\psi_2(n) = (2i)^{-1}(\phi(n) - \overline{\phi(-n)})$ for all $n \in E$.

Similarly, regarding 15.1.4(b) and 15.1.4(c), if $E \subset Z$ is symmetric, then E is Sidon if and only if

$$\|\hat{f}\|_1 \leqslant \text{const. } \|f\|_\infty$$

for every real-valued $f \in \mathbf{T}_E$. This is so since, if $g \in \mathbf{T}_E$, then $f_1 = \operatorname{Re} g$ and $f_2 = \operatorname{Im} g$ both belong to $\mathbf{T}_E$, are real-valued, and satisfy

$$\begin{aligned}\|\hat{g}\|_1 &\leqslant \|\hat{f}_1\|_1 + \|\hat{f}_2\|_1 \leqslant \text{const. } \|f_1\|_\infty + \text{const. } \|f_2\|_\infty \\ &\leqslant 2 \text{ const. } \|g\|_\infty.\end{aligned}$$

See also 15.1.6 below.

(10) If $E \subset Z$ is Sidon, so too is $E \cup (-E)$. To see this, appeal to the first part of (9): if $\psi \in \ell^\infty(E \cup (-E)$ satifies $\psi(-n) = \overline{\psi(n)}$ for all $n \in E \cup (-E)$, choose $\mu \in \mathbf{M}$ such that $\hat{\mu} | E = \psi | E$ and define $\nu = 2^{-1}(\mu + \bar{\mu})$; then

$$\hat{\nu}(n) = 2^{-1}(\hat{\mu}(n) + \overline{\hat{\mu}(-n)}) \qquad \text{for all } n \in Z$$

and it is easy to verify that

$$\hat{\nu} | (E \cup (-E)) = \psi.$$

(11) The content of 15.1.4(e) is that every function belonging to $\mathbf{c}_0(E)$ has an extension to Z of the form $\hat{f}$ for some $f \in \mathbf{L}^1$. On the other hand, Dunkl and Ramirez [1] have shown that, if E is infinite, then (contrary to what one might suppose) there exists no continuous linear map V of $\mathbf{c}_0(E)$ into $\mathbf{L}^1$ such that

$$(V\phi)^\wedge(n) = \phi(n) \qquad \text{for all } \phi \in \mathbf{c}_0(E) \text{ and all } n \in E.$$

Concerning this, see Exercise 15.23 below.

We turn next to a refinement of criterion 15.1.4(d).

15.1.5. **Supplement to 15.1.4.** Let E be a subset of Z. In order that E be a Sidon set, it is sufficient (and necessary, by 15.1.4) that for every function ϕ on E taking only the values 1 and -1, there exists a measure $\mu \subset \mathbf{M}$ such that

$$\sup \{|\phi(n) - \hat{\mu}(n)| : n \in E\} < 1.$$

Proof. We will show that 15.1.4(c) is satisfied.

Let $f \in \mathbf{C}_E$ and suppose first that $\hat{f}$ is real-valued. Define the ± 1-valued function ϕ on E so that $\phi \cdot \hat{f} = |\hat{f}|$. By hypothesis, there exists a measure $\mu \in \mathbf{M}$ and a positive number δ satisfying

$$\sup \{|\phi(n) - \hat{\mu}(n)| : n \in E\} \leqslant 1 - \delta. \tag{15.1.7}$$

If $\lambda = \frac{1}{2}(\mu + \mu^*)$, it is easily seen that λ also satisfies (15.1.7), $\hat{\lambda}$ being just the real part of $\hat{\mu}$. Moreover,

$$|\hat{f} \cdot \hat{\lambda} - |\hat{f}|| = |\hat{f}| \cdot |\hat{\lambda} - \phi| \leqslant (1 - \delta) \cdot |\hat{f}| \text{ on } E.$$

Putting $g = \lambda * f$, it follows that $g \in \mathbf{C}_E$ and

$$\hat{g} = \hat{\lambda} \cdot \hat{f} \geqslant \delta \cdot |\hat{f}| \text{ on } E. \tag{15.1.8}$$

Let F be any finite subset of E and choose a trigonometric polynomial t as in 15.1.2, taking $\varepsilon = 1$. Then (15.1.8) yields

$$\begin{aligned}\delta \cdot \sum_{n \in F} |\hat{f}(n)| &\leqslant \sum_{n \in F} \hat{t}(n)\hat{g}(n) \leqslant \sum_{n \in Z} \hat{t}(n)\hat{g}(n) \\ &= t * g(0) \leqslant \|t\|_1 \cdot \|g\|_\infty \leqslant 2\|\mu\|_1 \cdot \|f\|_\infty .\end{aligned}$$

The last-written term being independent of F, it follows that $\hat{f} \in \ell^1$.

If $\hat{f}$ is not real-valued, we write $f = f_1 + if_2$, where $f_1 = \frac{1}{2}(f + f^*)$, $f_2 = -\frac{1}{2}i(f - f^*)$, so that $\hat{f}_1$ and $\hat{f}_2$ are real-valued and f_1 and f_2 belong to $\mathbf{C}_E$. By what we have established, $\hat{f}_1$ and $\hat{f}_2$ belong to ℓ^1, so that the same is true of $\hat{f} = \hat{f}_1 + i\hat{f}_2$. This completes the proof.

Remark. A slightly shorter proof runs as follows. Having proved (15.1.8), it follows that g is continuous and positive definite. Hence by 9.2.8, $\hat{g} \in \mathbf{l}^1$ and so, since (15.1.8) implies that $|\hat{f}| \leqslant \delta^{-1}\hat{g}$, $\hat{f} \in \mathbf{l}^1$ and the proof is complete.

15.1.6. If E is assumed to be symmetric (that is, $E = -E$), then we may in 15.1.5 assume that each ϕ referred to is either even or odd.

For, assuming again that $f \in \mathbf{C}_E$ and that f is real-valued, we may write $f = f_e + f_o$, where $f_e = \frac{1}{2}(f + \check{f}) \in \mathbf{C}_E$ is even, and $f_o = \frac{1}{2}(f - \check{f}) \in \mathbf{C}_E$ is odd. The two components f_e and f_o may be treated separately and call for even and odd functions ϕ, respectively.

15.1.7. It is evident from 15.1.4(b) or (c) that any subset of a Sidon set is a Sidon set; that $E \cup F$ is a Sidon set whenever E is a Sidon set and F is a finite subset of Z that if E is a Sidon set and $n \in Z$ then the translate $n + E$ is a Sidon set; and that $-E$ is a Sidon set if (and only if) E is a Sidon set.

15.2 Construction and Examples of Sidon Sets

In this section we give some structural properties of a subset E of Z which ensure that E is a Sidon set, and which permit us to show in particular that each Hadamard set is a Sidon set. We have already met this situation in Chapter 14 for the Cantor group. Reference back to 14.2.1 will convince the reader that one crucial point in proving that $\sum_{\zeta \in \mathscr{R}} c_\zeta \zeta$ is the Walsh-Fourier series of a function in $\mathbf{L}^p(\mathscr{C})$ (for each $p < \infty$) whenever it is the Walsh-Fourier series of a function in $\mathbf{L}^1(\mathscr{C})$ was that, apart from order, each element of $\mathscr{C}^\wedge$ can be written uniquely as a product of elements of $\mathscr{R}$. In trying the same technique in Z, it suffices to bound the number of representations of an integer as a sum of elements of E.

15.2.1. **Notation.** We call a subset F of Z *asymmetric* if $n \in F\backslash\{0\}$ implies $-n \notin F$. For every subset E of Z, every positive integer s, and every $n \in Z$, denote by $R_s(n, E)$ the number of asymmetric subsets F of $E \cup (-E)$ having exactly s elements and satisfying

$$n = \sum_{m \in F} m. \tag{15.2.1}$$

A subset E of Z is called a *Rider set* if there is a number B such that, for all $n \in Z$,

$$R_s(n, E) \leqslant B^s \qquad \text{for all positive integers } s; \tag{15.2.2}$$

the smallest such B is termed the *Rider constant* of E.

A finite union of Rider sets is termed a *Stechkin set*; note that if E is a Stechkin set then so too is $E \cup (-E)$.

15.2.2. **An Arithmetical Criterion.** Every symmetric Stechkin set E is a Sidon set.

Proof. We may assume without loss of generality that 0 does not belong to E. By assumption, E is the union of finitely many Rider sets $E_1, \cdots, E_t$ which, again without loss of generality, we suppose are symmetric and pairwise disjoint. Let B denote the maximum of the Rider constants of $E_1, \cdots, E_t$. Since $B \geqslant 1$, setting $b = (3tB^2)^{-1}$ yields $b \leqslant \frac{1}{3}$ and $Bb \leqslant \frac{1}{3}$. Let ϕ be an arbitrary function on E taking only the values $\pm b$. Our aim is to apply 15.1.5 and 15.1.6, in doing which it will suffice to deal with the case in which ϕ is assumed to be even.

Fix j and assume that E_j is enumerated as $\pm n_1, \pm n_2, \cdots$, where $0 < n_1 < n_2 < \cdots$. Define

$$\left.\begin{aligned} f_k(x) &= 1 + \phi(n_k)e^{in_kx} + \phi(-n_k)e^{-in_kx} \\ &= 1 + 2\phi(n_k)\cos n_kx \geqslant 0. \end{aligned}\right\} \tag{15.2.3}$$

(If ϕ were odd, we should argue with

$$\begin{aligned} f_k(x) &= 1 - i\phi(n_k)e^{in_kx} - i\phi(-n_k)e^{-in_kx} \\ &= 1 + 2\phi(n_k)\sin n_kx \geqslant 0.) \end{aligned}$$

Write also

$$t_N(x) = \prod_{k=1}^{N} f_k(x), \tag{15.2.4}$$

which is a nonnegative trigonometric polynomial. Evidently,

$$t_N(x) = \prod_{r=1}^{N} \sum_{u \in B} F(r, u)$$

wherein $B = \{-1, 0, 1\}$ and

$$F(r, u) = \psi(n_r)^{|u|}(\exp(in_r x))^u.$$

At this point we make use of the formula (true for all complex-valued functions F on $\{1, \cdots, N\} \times B$)

$$\prod_{r=1}^{N} \sum_{u \in B} F(r, u) = \sum_{w \in W} \prod_{r=1}^{N} F(r, w(r)),$$

wherein W is the set of all functions w from $\{1, 2, \cdots, N\}$ into B. Thus

$$t_N = \sum_{w \in W} \prod_{r=1}^{N} \phi(n_r)^{|w(r)|} \exp(in_r w(r)x)$$

$$= \sum_{w \in W} \prod_{r=1}^{N} D(w, r) \tag{15.2.5}$$

say. Defining $W(s)$ to be the set of all $w \in W$ such that

$$\sum_{r=1}^{N} |w(r)| = s,$$

W is the disjoint union of $W(0)$, $W(1)$, $\cdots$, $W(N)$. Furthermore, $W(0)$ comprises only the zero function in W, and

$$\sum_{w \in W(0)} \prod_{r=1}^{N} D(w, r) = 1;$$

and $W(1)$ comprises only these functions in W with singleton supports, and

$$\sum_{w \in W(1)} \prod_{r=1}^{N} D_r = \sum_{k=1}^{N} \phi(n_k)e^{in_k x} + \sum_{k=1}^{N} \phi(n_k)e^{-in_k x}.$$

Thus, by (15.2.5),

$$t_N(x) = 1 + \sum_{k=1}^{N} \phi(n_k)e^{in_k x} + \sum_{k=1}^{N} \phi(n_k)e^{-in_k x}$$

$$+ \sum_{s=2}^{N} \sum_{w \in W(s)} \prod_{r=1}^{N} D(w, r). \tag{15.2.6}$$

Expressing $W(s)$ as the disjoint union of the $W(s, n)$, where $W(s, n)$ is the set of all $w \in W(s)$ such that

$$\sum_{r=1}^{N} w(r)n_r = n,$$

and noting that $W(s, n)$ is nonvoid for only finitely many n, it follows that the fourth term on the right of (15.2.6) is

$$\sum_{n \in Z} c_N(n)e^{inx},$$

where

$$c_N(n) = \sum_{s=2}^{N} \sum_{w \in W(s,\, n)} \prod_{r=1}^{N} \phi(n_r)^{|w(r)|}.$$

Accordingly, since ϕ assumes only the values $\pm b$,

$$\begin{aligned} |c_N(n)| &\leqslant \sum_{s=2}^{N} \sum_{w \in W(s,\, n)} b^s \\ &= \sum_{s=2}^{N} b^s \cdot \#(W(s, n)), \end{aligned}$$

where, for every finite set A, $\#(A)$ denotes the cardinal of A. For every $w \in W(s, n)$ define $\theta(w)$ to be the set

$$\{w(r)n_r \colon r \in \{1, 2, \cdots, N\} \text{ and } w(r) \neq 0\}.$$

It is simple (if a little tedious) to check that, for every $s \in \{1, 2, \cdots, N\}$, θ is an injective map of $W(s, n)$ into the set of all asymmetric subsets F of E_j having exactly s elements and satisfying

$$\sum_{m \in F} m = n.$$

Hence, by (15.2.2),

$$\#(W(s, n)) \leqslant R_s(n, E_j) \leqslant B^s$$

for all $n \in Z$ and all $s \in \{1, 2, \cdots, N\}$. Thus, for all $n \in Z$,

$$|c_N(n)| \leqslant \sum_{s=n}^{N} b^s B^s \leqslant B^2 b^2 (1 - Bb)^{-1} \leqslant (6t^2 B^2)^{-1}. \tag{15.2.7}$$

This shows that, in particular,

$$\|t_N\|_1 = 1 + c_N(0) \leqslant 1 + (6t^2 B^2)^{-1}.$$

Applying 12.3.9, it is seen that a subsequence (t_{N_p}) of $(t_N)_{N=1}^{\infty}$ converges weakly to a measure $\mu_j \in \mathbf{M}$. This entails that $\lim_{p \to \infty} \hat{t}_{N_p}(n) = \hat{\mu}_j(n)$, so that

(15.2.5) and (15.2.7) then show that

$$\begin{aligned} |\hat{\mu}_j(n) - \phi(n)| &\leqslant (6t^2B^2)^{-1} \qquad (n \in E_j), \\ |\hat{\mu}_j(n)| &\leqslant (6t^2B^2)^{-1} \qquad (n \in E \backslash E_j). \end{aligned} \tag{15.2.8}$$

If we put $\mu = \mu_1 + \cdots + \mu_t$, (15.2.8) shows that for $n \in E$

$$|\hat{\mu}(n) - \phi(n)| \leqslant t \cdot (6t^2B^2)^{-1} = \tfrac{1}{2}b.$$

It now remains merely to appeal to 15.1.5 and 15.1.6.

15.2.3. **Another Arithmetical Criterion.** Every asymmetric Stechkin set E is a Sidon set.

Proof. This is very similar to that of 15.2.2: the set E_j is this time enumerated as $n_1, n_2, \cdots$ and f_k is defined as

$$\begin{aligned} f_k(x) &= 1 + \phi(n_k)e^{in_kx} + \phi(n_k)e^{-in_kx} \\ &= 1 + 2\phi(n_k)\cos n_kx, \end{aligned}$$

from which point the argument proceeds as before.

Remarks. (1) The proofs of 15.2.2 and 15.2.3 witness the introduction of trigonometric polynomials of the form

$$\prod_{k=1}^{N} (1 + \alpha_k \cos n_k x)$$

and corresponding infinite products

$$\prod_{k=1}^{\infty} (1 + \alpha_k \cos n_k x).$$

Such products are termed *Riesz products*; they appear in connection with various problems in harmonic analysis. For more about them, see [Z_1], pp. 208–212; [Ba_2], pp. 246–249; [LR], Chapter [2]; Keogh [2]; Brown [4]; and [MG], Chapter 7.

(2) For more criteria like 15.2.3, see Rider [4]. In particular, the conclusions of both 15.2.2 and 15.2.3 still hold when the assumption that E be Stechkin is weakened to E being a finite union $E = \bigcup_{j=1}^{t} E_j$ of sets E_j, each of which satisfies (15.2.2) merely when $n = 0$. The reader is asked for a proof of this fact in Exercise 15.24.

15.2.4. **Hadamard Sets Are Sidon Sets.** Let E be a *Hadamard set*, that is, $E = \{\pm n_1, \pm n_2, \cdots\}$, where $0 < n_1 < n_2 < \cdots$ and

$$q \equiv \inf \frac{n_{k+1}}{n_k} > 1.$$

It is then very simple to see that E can be partitioned into a finite number of Hadamard sets E_j, for each of which the corresponding value of q exceeds 3. But then, for any $n \in Z$,

$$R_s(n, E_j) \leqslant 1. \tag{15.2.9}$$

Indeed, if an integer n admitted two (or more) representations (15.2.1), we should have a relation of the form

$$\alpha_1 n_{r_1} + \alpha_2 n_{r_2} + \cdots = 0,$$

wherein $r_1 > r_2 > \cdots$ and α_i is ± 1 or ± 2. Then, however, one would deduce that

$$\begin{aligned} n_{r_1} &\leqslant 2(n_{r_2} + n_{r_3} + \cdots) \leqslant 2n_{r_1}(q^{-1} + q^{-2} + \cdots) \\ &\leqslant 2(q-1)^{-1} n_{r_1} < n_{r_1}, \end{aligned}$$

a contradiction.

By (15.2.9) and 15.2.2, E is both a Stechkin and a Sidon set.

Indeed, the same argument shows that any finite union of Hadamard sets is a Sidon set.

However, there exist Sidon sets that are not finite unions of Hadamard sets; see Exercise 15.3.

15.2.5. Any infinite subset A of Z contains an infinite Sidon set E.

Proof. Choose freely any nonzero $n_1 \in A$. Suppose that $n_1, \cdots, n_k$ have already been selected from $A \backslash \{0\}$. The set S_k of integers of the form

$$\alpha_1 n_{r_1} + \alpha_2 n_{r_2} + \cdots + \alpha_s n_{r_s},$$

where $1 \leqslant r_1 < r_2 < \cdots < r_s \leqslant k$ and where each α_i is $\pm \frac{1}{2}$, ± 1, or ± 2, and at most one is of the latter form, is finite. So one may select a nonzero integer

$$n_{k+1} \in A \backslash S_k.$$

If $E = \{n_1, n_2, \cdots\}$, it is clear that no $n \in E \cup \{0\}$ admits a representation in the form

$$n = \pm n_{k_1} \pm n_{k_2} \pm \cdots \pm n_{k_s}$$

with distinct k_i and $s > 1$, so that 15.2.3 shows that E is a Sidon set. Evidently, E is an infinite subset of A.

The proofs of 15.2.2 and 15.2.3 lead to a refinement of 15.1.4(b) valid for certain Sidon sets.

15.2.6. Suppose that E is a Stechkin subset of Z (in particular, suppose that E is a finite union of Hadamard sets; see the proof of 15.2.4.). Then

$\hat{f} \in \ell^1$ for any E-spectral $f \in \mathbf{L}^1$ such that each of $\operatorname{Re} f$ and $\operatorname{Im} f$ is essentially bounded above or essentially bounded below.

Proof. The proof is based on the observation that the measure μ constructed in the proof of 15.2.2 (or of 15.2.3) is positive, being the weak limit in $\mathbf{M}$ of nonnegative trigonometric polynomials. We now modify the proof of 15.1.5 so as to make use of this additional information about μ.

By a combination of changes from f to $-f$ and from f to $\bar{f}$, it is seen to be enough to deal with the case in which $\operatorname{Re} f \leqslant m$ a.e. and $\operatorname{Im} f \leqslant m$ a.e. (Notice that $-E$ satisfies the stated conditions whenever E does so, and that $\bar{f}$ is $(-E)$-spectral whenever f is E-spectral.) In this case, we can, as in the proof of 15.1.5, decompose f into a sum $f_1 + if_2$, where f_1 and f_2 are E-spectral, $\hat{f}_1$ and $\hat{f}_2$ are real-valued, and moreover $\operatorname{Re} f_1 \leqslant m$ a.e. and $\operatorname{Re} f_2 \leqslant m$ a.e. So it will suffice to deal with the case in which $\operatorname{Re} f \leqslant m$ a.e. and $\hat{f}$ is real-valued. Since alteration of f on a null set leaves $\hat{f}$ unchanged, we may as well assume that $\operatorname{Re} f \leqslant m$ everywhere.

These preliminary reductions having been made, choose the ± 1-valued function ϕ on E such that $\phi \cdot \hat{f} = |\hat{f}|$, setting $\phi(n) = 1$ whenever $n \in E$ and $\hat{f}(n) = 0$. Choose a positive measure μ so that

$$|\phi(n) - \hat{\mu}(n)| \leqslant 1 - \delta \qquad \text{for all } n \in E.$$

Multiplying through by $|\hat{f}(n)|$ and using the definition of ϕ, it appears that

$$\operatorname{Re} \hat{\mu}(n)\hat{f}(n) \geqslant \delta \cdot |\hat{f}(n)| \qquad \text{for all } n \in E. \tag{15.2.10}$$

Consider now the function $g = \mu * f$. Since μ is positive,

$$\operatorname{Re} g = \mu * \operatorname{Re} f \leqslant \mu(m \cdot 1) \equiv m' < \infty .$$

Therefore [see (5.1.6), (5.1.8), and (5.1.9)]

$$\operatorname{Re} (\sigma_N g) = \operatorname{Re} (F_N * g) = F_N * \operatorname{Re} g \leqslant m' .$$

In particular, evaluating at the origin and remembering that $\hat{g} = \hat{\mu} \cdot \hat{f}$,

$$\operatorname{Re} \sum_{|n| \leqslant N} \left(1 - \frac{|n|}{N+1}\right) \hat{\mu}(n)\hat{f}(n) \leqslant m' .$$

By (15.2.10) and the assumption that f is E-spectral, this implies that

$$\sum_{|n| \leqslant N} \left(1 - \frac{|n|}{N+1}\right) |\hat{f}(n)| \leqslant m'\delta^{-1}.$$

Since the right-hand term here is independent of N, it follows that $\hat{f} \in \ell^1$.

At this point see also Exercise 15.12. For a still deeper analogous result, due to Zygmund, see [Ba_2], pp. 249–257.

15.2.7. **Finite unions of Sidon sets.** Drury [3] has proved that the union of two Sidon sets is again a Sidon set. The reader will find it an easy exercise to show that this is a consequence of the result due to Drury mentioned in 15.1.4, Remark (4). We refer to [LR], Chapter 3, for an exposition of Drury's result.

15.2.8. **An Arithmetical Property of Sidon Sets.** Given a subset E of Z and a positive integer N, we denote by $\alpha_E(N)$ the largest integer α such that some arithmetic progression of N terms contains α elements of E. In other words, if we define $\alpha_E(N, a, b)$ for $a, b \in Z$ and $b \neq 0$ to be the number of terms of the arithmetic progression $a + b, a + 2b, \cdots, a + Nb$ which fall in E, then $\alpha_E(N)$ is the supremum of $\alpha_E(N, a, b)$ as a and b vary.

It can be shown (see Exercise 15.8) that if E is a Sidon set, then

$$\alpha_E(N) \leqslant B_E \log N \qquad \text{for all } N \geqslant 3; \tag{15.2.11}$$

and that on the other hand, if E is the Sidon set $\{1, 2, 2^2, 2^3, \cdots\}$, then

$$\alpha_E(N) > \frac{\log N}{\log 2}$$

for $N = 2^k$.

In particular, if E is a Sidon set, the number of elements n of E satisfying $|n| \leqslant N$ is $O(\log N)$ for large positive integers N. So, for example, the range of a nonconstant polynomial function on Z into Z is never a Sidon set.

It can also be shown (see Exercise 15.22) that if E is a Sidon set, then E does not contain the sum $\{n + m\colon n \in F,\ m \in G\}$ of two infinite sets F and G. For a further discussion of the arithmetical properties of Sidon sets, see [LR], Chapter 6.

There are two refinements of the estimate (15.2.11) (see [LR], Chapter 6). Each establishes that, for a certain set $\mathscr{M}$ of well-behaved subsets of Z,

$$\#(E \cap M) \leqslant B_E \cdot \log(\#(M)) \tag{15.2.12}$$

for all $M \in \mathscr{M}$ where, for every finite set F, $\#(F)$ denotes the cardinal of F. In both of these refinements the set $\mathscr{M}$ is large enough to cover the situation dealt with in Exercise 15.22 and to arrange for (15.2.12) to yield (15.2.11).

There seems to be little hope of an arithmetic characterisation of Sidon subsets of Z. However for the Cantor group $\mathscr{C}$, the situation is far better. There, a Sidon set is defined (naturally enough) to be a subset E of $\mathscr{C}^\wedge$ for which there exists a number $B \geqslant 0$ such that

$$\|\hat{f}\|_1 \leqslant B\|f\|_\infty$$

for all $f \in \mathbf{T}_E(\mathscr{C})$. Thus Exercise 14.9 amounts to showing that the set $\mathscr{R}$ of Rademacher characters is a Sidon set. Now for subsets of $\mathscr{C}^\wedge$ an arithmetic characterisation is possible (though not easy): a subset of

$\mathscr{C}^\wedge$ is a Sidon set if and only if it is a finite union of independent subsets of $\mathscr{C}^\wedge$ (see (14.1.3)). This theorem is due to Malliavin-Brameret and Malliavin [1]; the reader may care to read Pisier's proof (Pisier [2]) which characterises Sidon sets using, instead of an arithmetic condition, the Orlicz spaces mentioned in 14.3.6.

15.3 Further Inequalities Involving Sidon Sets

We now use the properties of Rademacher characters discussed in Chapter 14 to derive some fundamental functional analytic inequalities stemming from the definition of Sidon sets.

15.3.1. Let E be a Sidon set, with Sidon constant B. If $\mu \in \mathbf{M}_E$ is an E-spectral measure, then $\mu \in \mathbf{L}^p$ for all $p < \infty$, and

$$\|\mu\|_p \leqslant Bp^{1/2}\|\mu\|_2 \qquad \text{for all } p \text{ such that } 2 < p < \infty, \tag{15.3.1}$$

$$\|\mu\|_2 \leqslant B2^{1/2}\|\mu\|_1. \tag{15.3.2}$$

Proof. We start by deducing the analogous inequalities for Rademacher-spectral trigonometric polynomials on the Cantor group $\mathscr{C}$. From 14.2.1 it follows, since $m! \leqslant m^m$ whenever $m \in Z^+$, that for all $m \in Z^+$ and for all $g \in \mathbf{T}_{\mathscr{R}}(\mathscr{C})$,

$$\|g\|_{2m} \leqslant m^{1/2}\|g\|_2, \tag{15.3.3}$$

$$\|g\|_2 \leqslant 2^{1/2}\|g\|_1. \tag{15.3.4}$$

Now whenever $2 < p < \infty$, we may write $2m - 2 \leqslant p \leqslant m$ for some integer m satisfying $1 < m \leqslant p$. Then, by (15.3.3),

$$\|g\|_p \leqslant \|g\|_{2m} \leqslant m^{1/2}\|g\|_2 \leqslant p^{1/2}\|g\|_2. \tag{15.3.5}$$

In (15.3.5) and (15.3.4) we have the desired analogues of (15.3.1) and (15.3.2).

Suppose now that $f \in \mathbf{T}_E$ is an E-spectral trigonometric polynomial over the circle group T, and define on $T \times \mathscr{C}$ the function

$$g(x, \omega) = g_\omega(x) = g_x(\omega) = \sum_{n \in Z} \hat{f}(n)\rho_n(\omega)e_n(x) \tag{15.3.6}$$

where ρ_n is the nth Rademacher character (see 14.1.3). Since E is a Sidon set, reference to 15.1.4(5) confirms that to each $\omega \in \mathscr{C}$ corresponds a measure $\mu_\omega \in \mathbf{M}$ such that $\|\mu_\omega\|_1 \leqslant B$ and

$$\hat{\mu}_\omega(n) = \rho_n(\omega) \qquad \text{for all } n \in E. \tag{15.3.7}$$

Since f and g_ω are E-spectral trigonometric polynomials on the circle, this shows that $f = \mu_\omega * g_\omega$ and therefore (see 12.7.3) that

$$\|f\|_p \leqslant \|\mu_\omega\|_1 \|g_\omega\|_p \leqslant B\left[(2\pi)^{-1} \int_T |g(x, \omega)|^p \, dx\right]^{1/p} \tag{15.3.8}$$

for all $\omega \in \mathscr{C}$. Integrating this inequality over $\mathscr{C}$ with respect to ω and using (15.3.5),

$$\|f\|_p \leqslant Bp^{1/2}\left[\sum_{n \in Z} |\hat{f}(n) e_n(x)|^2\right]^{1/2} = Bp^{1/2}\|f\|_2. \tag{15.3.9}$$

On the other hand, we also have $g_\omega = \mu_\omega * f$ for all $\omega \in \mathscr{C}$ and therefore

$$(2\pi)^{-1} \int_T |g(x, \omega)| \, dx \leqslant \|\mu_\omega\|_1 \|f\|_1 \leqslant B\|f\|_1. \tag{15.3.10}$$

But applying (15.3.4) to g_x and integrating over the circle with respect to x gives

$$2\pi\left[\sum_{n \in Z} |\hat{f}(n)|^2\right]^{1/2} \leqslant 2^{1/2} \int_T \left[\int_{\mathscr{C}} |g(x, \omega)| \, d\lambda(\omega)\right] dx.$$

This, the Fubini-Tonelli theorem, and (15.3.10) combine to show that

$$\|f\|_2 \leqslant 2^{1/2}B\|f\|_1. \tag{15.3.11}$$

At this stage we have, therefore, established (15.3.1) and (15.3.2) for E-spectral trigonometric polynomials f. For the rest, we apply these special cases to the functions $f_N = F_N * \mu \in \mathbf{T}_E$. Thus suppose first that μ is known to belong to $\mathbf{L}^1$. Then, by 6.1.1., $f_N \to \mu$ in $\mathbf{L}^1$, on applying (15.3.11) and (15.3.9) to the differences $f_{N'} - f_N$, it is seen that $(f_N)_{N=1}^{\infty}$ is Cauchy in $\mathbf{L}^p$ for every $p < \infty$, hence is convergent in $\mathbf{L}^p$ for such p. The limit can only be μ, so that $\mu \in \mathbf{L}^p$ for every $p < \infty$. Moreover on applying (15.3.11) and (15.3.9) to the f_N and using Fatou's Lemma, (15.3.2) and (15.3.1) follow.

Finally, if we assume merely that $\mu \in \mathbf{M}$, (15.3.11) may be applied to the f_N to yield

$$\|f_N\|_2 \leqslant 2^{1/2}B\|F_N * \mu\|_1 \leqslant 2^{1/2}B\|\mu\|_1. \tag{15.3.12}$$

At this point 12.3.10(2) comes into play and asserts that some subsequence of $(f_N)_{N=1}^{\infty}$ is weakly convergent in $\mathbf{L}^2$. Since $f_N = F_N * \mu$ converges weakly in $\mathbf{M}$ to μ (see Exercise 12.17), it follows that $\mu \in \mathbf{L}^2 \subset \mathbf{L}^1$. The arguments of the preceding paragraph are now applicable and complete the proof.

Remarks. (1) The dependence on p of the constant appearing on the right-hand side of (15.3.2) renders it possible to show that indeed $\exp(c|f|^2)$ is integrable for any number c and any $f \in \mathbf{L}_E^1$, E being a Sidon set; see Exercise 15.4 and cf. 14.2.2.

(2) The estimate (15.3.2) is the best possible in the sense that, given any infinite subset E of Z and any integer $k \geq 1$, there exists an $f \in \mathbf{T}_E$ such that

$$\|f\|_k \geq 2^{-5/2}k^{1/2}\|f\|_2;$$

see Exercises 15.14 to 15.16. This result is due to Rudin ([6], Theorem 3.4).

(3) Prompted by 15.3.1, a subset E of Z is called a Λ-*set* (cf.15.5.3) if there is a constant B (independent of p) such that, for every $p \in (2, \infty)$,

$$\|f\|_p \leqslant Bp^{1/2}\|f\|_2,$$

for all $f \in \mathbf{T}_E$. Thus 15.3.1 demonstrates that every Sidon set is a Λ-set. Pisier [1] has shown that, conversely, every Λ-set is a Sidon set; the techniques used by Pisier are related to those mentioned in 14.3.6.

The next result shows that all Sidon sets share another property similar to 15.1.4(d) and (e).

15.3.2. If E is a Sidon set, then $\mathscr{F}\mathbf{C}|E = \ell^2(E)$.

Proof. This will be carried out in two stages, in the first of which we show in particular that $\mathscr{F}\mathbf{L}^\infty|E = \ell^2(E)$.

(1) Let $\phi \in \ell^2(E)$. Regarding $\mathbf{L}_E^1$ as a normed linear subspace of $\mathbf{L}^1$, (15.3.2) shows that the mapping

$$g \to \sum_{n \in Z} \hat{g}(n)\phi(n)$$

is a continuous linear functional on $\mathbf{L}_E^1$ of norm at most $2^{1/2}B\|\phi\|_2$. The Hahn-Banach theorem (I, B.5.1) and I, C.1 combine to show that as a consequence there exists a function $f \in \mathbf{L}^\infty$ such that $\|f\|_\infty \leqslant 2^{1/2}B\|\phi\|_2$ and

$$\sum_{n \in Z} \hat{g}(n)\phi(n) = \frac{1}{2\pi}\int g(x)f(-x)\,dx$$

for each $g \in \mathbf{L}_E^1$. On taking $g = e_n$, where $n \in E$, this shows that $\hat{f}(n) = \phi(n)$ for $n \in E$.

(2) Next suppose that ϕ is as before and partition E into finite sets $E_1, E_2, \cdots$ such that

$$\Big[\sum_{n \in E_k} |\phi(n)|^2\Big]^{1/2} \leqslant ck^{-2}$$

where c is independent of k. Define ϕ_k to be equal to ϕ on E_k and to be zero on $E \backslash E_k$. According to (1), there exists for each k a function $f_k \in \mathbf{L}^\infty$ such that $\hat{f}_k|E = \phi_k$ and $\|f_k\|_\infty \leqslant 2^{1/2}\,Bck^{-2}$. Referring to 15.1.2, choose trigonometric polynomials t_k so that

$$\|t_k\|_1 \leqslant 2, \qquad \hat{t}_k = 1 \text{ on } E_k.$$

Then $h_k = t_k * f_k \in \mathbf{C}$ and $\|h_k\|_\infty \leqslant \|t_k\|_1 \cdot \|f_k\|_\infty \leqslant 2^{3/2}Bck^{-2}$. Therefore

$$h = \sum_{k=1}^{\infty} h_k$$

belongs to $\mathbf{C}$, the series being uniformly convergent. By the latter token, $\hat{h} = \sum_{k=1}^{\infty} \hat{h}_k$. So, for $n \in E$, one has

$$\begin{aligned}\hat{h}(n) &= \sum_{k=1}^{\infty} \hat{h}_k(n) = \sum_{k=1}^{\infty} \hat{t}_k(n) \cdot \hat{f}_k(n) = \sum_{k=1}^{\infty} 1 \cdot \phi_k(n) \\ &= \phi(n).\end{aligned}$$

Thus $\phi = \hat{h}|E$ and the proof is complete.

15.3.3. Let E be a Sidon set and suppose that $1 < p < 2$. Then

$$\left[\sum_{n \in E} |\hat{f}(n)|^2\right]^{1/2} \leqslant Bp'^{1/2} \cdot \|f\|_p$$

holds for each $f \in \mathbf{L}^p$.

Proof. Let g be any E-spectral trigonometric polynomial:

$$g(x) = \sum_{n \in E} c_n e^{inx},$$

c_n being zero for all but a finite set of $n \in E$. Applying (15.3.1) with $p'(< \infty)$ in place of p, we have by 3.1.4

$$\begin{aligned}\Big|\sum_{n \in E} \hat{f}(n) c_n\Big| &= \Big|\frac{1}{2\pi}\int f(x) g(-x)\,dx\Big| \leqslant \|f\|_p \cdot \|g\|_{p'} \\ &\leqslant \|f\|_p \cdot Bp'^{1/2} \cdot \left[\sum_{n \in E} |c_n|^2\right]^{1/2}.\end{aligned}$$

This entails that

$$\left[\sum_{n \in E} |\hat{f}(n)|^2\right]^{1/2} \leqslant Bp'^{1/2} \cdot \|f\|_p,$$

as one sees on choosing c_n to coincide with $\overline{\hat{f}(n)}$ on larger and larger finite subsets of E.

Remarks. (1) If $2 \leqslant p \leqslant \infty$, the inequality in 15.3.3 is valid for arbitrary subsets E of Z, the factor $Bp'^{1/2}$ being replaced by unity. (This is due to the Parseval formula and the inequality $\|f\|_2 \leqslant \|f\|_p$ for $2 \leqslant p \leqslant \infty$.)

(2) The hypothesis $p > 1$ in 15.3.3 cannot be removed. More precisely, if $E \subset Z$ and if $\hat{f}|E \in \bigcup_{q<\infty} \ell^q(E)$ for every $f \in \ell^1$, then E is finite.

Indeed the hypothesis implies (by Appendix B.2.1 in Volume 1) that there exists $q < \infty$ such that

$$\left(\sum_{n \in E} |\hat{f}(n)|^q\right)^{1/q} \leqslant \text{const.} \|f\|_1 \qquad \text{for every } f \in \mathbf{L}^1. \qquad (15.3.13)$$

From this and Appendix C in Volume 1 it follows that every $\psi \in \ell^{q'}$ which vanishes on $Z \backslash E$ has the form $\hat{g}$ for some $g \in \mathbf{L}^\infty$. If E were infinite, there would exist (by 15.2.5) an infinite Sidon set $S \subset E$. By 15.1.4(b), therefore, every $\psi \in \ell^{q'}$ which vanishes on $Z \backslash S$ satisfies $\psi \in \ell^1$, in particular

$$\ell^{q'}(S) \subset \ell^1(S). \qquad (15.3.14)$$

Since $q' > 1$ and S is infinite, the inclusion (15.3.14) is false and a contradiction emerges.

15.3.4. **Factorization in $\mathbf{L}^p$ $(p > 1)$.** Throughout this subsection it is assumed that $1 < p \leqslant \infty$.

It has been proved in 7.5.1 that $\mathbf{L}^1 * \mathbf{L}^1 = \mathbf{L}^1$; in Section 8.4 that $\mathbf{L}^2 * \mathbf{L}^2 \subset \mathbf{A}$, which is evidently a proper subset of $\mathbf{L}^2$; and in Exercise 13.20 that $\mathbf{L}^p * \mathbf{L}^p$ is always a proper subset of $\mathbf{L}^p$. Indeed, Exercise 13.20 shows that any $f \in \mathbf{L}^p$ satisfying

$$\sum_{n \in Z} |\hat{f}(n)|^{p'/2} = \infty$$

is a prime element of $\mathbf{L}^p$ (that is, does not belong to $\mathbf{L}^p * \mathbf{L}^p$); that such functions f exist, follows from 13.5.3(1), since $p'/2 < p'$.

To this we may now add, as a corollary of 15.3.3, that any $f \in \mathbf{L}^p$ satisfying

$$\sum_{n \in E} |\hat{f}(n)| = \infty$$

for some Sidon set $E \subset Z$ is a prime element of $\mathbf{L}^p$.

For some further results about the impossibility of factorizing a general element of $\mathbf{L}^p$, see Edwards [6].

15.3.5. **Comment on 15.3.1.** It has been seen in 15.3.1 that $\mathbf{M}_E \subset \mathbf{L}^p$ whenever E is a Sidon set and $p < \infty$. One cannot here take $p = \infty$. Indeed, it is not difficult to see that there exist no infinite subsets E of Z having the property that $\bigcap_{p<\infty} \mathbf{L}_E{}^p \subset \mathbf{L}^\infty$; see Exercise 15.15, which leads to an extension of the results mentioned in Remark (2) following 15.3.1.

15.3.6. **Application to Homomorphisms.** Let α be a mapping of Z into $Z \cup \{\infty\}$ and let us agree to set $\hat{f}(\infty) = 0$ whenever $f \in \mathbf{L}^1$. Consider the homomorphism T_α of the convolution algebra $\mathbf{L}^1$ into the convolution algebra $\mathbf{D}$ defined by the formula

$$(T_\alpha f)^\frown = \hat{f} \circ \alpha. \tag{15.3.15}$$

It is then natural to ask for necessary and sufficient conditions on α in order that T_α shall map $\mathbf{L}^p$ into $\mathbf{L}^q$ for given values of p and q, in which case we shall for brevity say that α is of *class* (p, q).

The solution for maps of class $(1, 1)$ is stated in Subsection 4.2.6, and from this it is clear that such maps α cannot be specified with appreciable freedom on any infinite subsets of Z. The same is true of maps of class $(1, q)$ when $1 < q \leqslant \infty$, as appears from Exercise 12.49.

In this subsection we will apply what has been learned about Sidon sets to show that, on the contrary, if $1 < p \leqslant \infty$ and $1 < q < \infty$, maps α of class (p, q) can be prescribed with a fair amount of freedom on certain infinite subsets of Z. (Compare the results in Subsection 16.4.3(1) applying to multipliers.)

More precisely, suppose that $1 < p \leqslant \infty$, $1 < q < \infty$, that α is of class (p, q), that E is a subset of Z, that $E \cap \alpha^{-1}(Z) \equiv S$ is a Sidon subset of Z, and that α' is a map of Z into $Z \cup \{\infty\}$ such that

(1) α' agrees on $Z \backslash E$ with α;

(2) $\alpha'(E) \cap Z$ is a Sidon set;

(3) there exists a number B such that, for $m \in \alpha'(E) \cap Z$, $E \cap \alpha'^{-1}(\{m\})$ has at most B elements.

We claim that then α' is of class (p, q).

Proof. Let $f \in \mathbf{L}^p$. By (2), (3) and 15.3.3,

$$\sum_{n \in E} |\hat{f} \circ \alpha'(n)|^2 \leqslant B \cdot \sum_{m \in \alpha'(E) \cap Z} |\hat{f}(m)|^2 < \infty.$$

Also, since S is a Sidon set and α is of class (p, q), 15.3.3 gives

$$\sum_{n \in E} |\hat{f} \circ \alpha(n)|^2 = \sum_{n \in S} |(T_\alpha f)^\frown(n)|^2 < \infty.$$

Hence

$$\sum_{n \in E} |\hat{f} \circ \alpha'(n) - \hat{f} \circ \alpha(n)|^2 < \infty,$$

so that there exists a function $h \in \mathbf{L}^2$ such that $\hat{h} = (\hat{f} \circ \alpha' - \hat{f} \circ \alpha)\chi_E$. But then, by 15.3.1, $h \in \mathbf{L}^q$. It is furthermore clear from (1) that

$$T_{\alpha'} f = T_\alpha f + h \in \mathbf{L}^q,$$

and hence that α' is of class (p, q).

Remarks. (a) The other hypotheses being granted, (2) and (3) are satisfied whenever $\alpha' | E$ is a permutation of E.

(b) On taking $\alpha \equiv \infty$ (so that $T_\alpha = 0$) and $E = Z$, one infers directly that a map β of Z into $Z \cup \{\infty\}$ is of class (p, q), where $1 < p \leqslant \infty$ and $1 < q < \infty$, whenever $Z \cap \beta(Z)$ is a Sidon set and there exists a number B such that, for each $m \in Z$, $\beta^{-1}(\{m\})$ has at most B elements.

15.4 Counterexamples Concerning the Parseval Formula and Hausdorff-Young Inequalities

15.4.1. In 13.5.1 it has been shown that $\mathscr{F}\mathbf{L}^p \subset \ell^{p'}$ for $1 \leqslant p \leqslant 2$; by 8.3.1, this inclusion relation can be replaced by equality when $p = 2$. Now we can verify that the inclusion is proper whenever $1 \leqslant p < 2$. This is indeed trivial if $p = 1$. If $1 < p < 2$, choose any infinite Sidon set E; since $p' > 2$ one may choose ϕ in $\ell^{p'}(E)$ not in $\ell^2(E)$; extend ϕ to Z so that $\phi(Z \backslash E) = \{0\}$. Then, by 15.3.3, ϕ fails to belong to $\mathscr{F}\mathbf{L}^p$. It can also be shown (compare 2.3.9) that, if $1 \leqslant p < 2$, then $\mathscr{F}\mathbf{L}^p$ is a meagre subset of $\ell^{p'}$.

In case $p = 1$ we have, of course, $\mathbf{A}(Z) \equiv \mathscr{F}\mathbf{L}^1 \subset \mathbf{c}_0$. Here again the inclusion is proper. This (and more) has been proved in 2.3.9. Also, 10.1.6 shows that $(\{\log(2 + |n|)\}^{-1})_{n \in Z} \in \mathbf{c}_0$ does not belong to $\mathscr{F}\mathbf{L}^1$. Again, as follows from the proof of 15.1.4, the relation $\mathscr{F}\mathbf{L}^1 = \mathbf{c}_0$ would entail that $\mathscr{F}\mathbf{M} = \ell^\infty$, which is false by 12.7.8. The same relation would also entail, via 10.5.2, that $\hat{f} \in \ell^1$ whenever $f \in \mathbf{C}$; this too is false (see 7.2.2 and 8.3.2).

15.4.2. As was heralded in 13.5.3(2), and proved in one way in Section 14.4, 13.5.1(2) is false for $p > 2$. To see this in another way, let E be any infinite Sidon set and choose a complex-valued function ϕ on Z which vanishes on $Z \backslash E$ and which is such that $\phi \in \ell^p$ and $\phi \notin \ell^2(Z)$; this is possible precisely because $p > 2$. Then 15.3.1 entails that ϕ does not belong to $\mathscr{F}\mathbf{M}$, a fortiori ϕ does not belong to $\mathscr{F}\mathbf{L}^{p'}$. The series

$$\sum_{n \in Z} \phi(n) e^{inx},$$

although convergent in $\mathbf{D}$, is not even weakly convergent in $\mathbf{M}$ and, a fortiori again, not convergent in $\mathbf{L}^{p'}$. A specific example is the series

$$\sum_{k=1}^{\infty} k^{-1/2} \cos 2^k x.$$

That 13.5.1(1) is also false when $p > 2$ can be proved in a similar fashion, thus: suppose again that E is an infinite Sidon set and let $r \to n_r$ be an injection of $\{1, 2, \cdots\}$ into E. Define the function ϕ on Z by setting $\phi(n) = \{r^{1/2} \log(1 + r)\}^{-1}$ if $n = n_r$ $(r = 1, 2, \cdots)$ and $\phi(n) = 0$ otherwise. Then $\phi \in \ell^2(Z)$ and $\phi(Z \backslash E) = 0$. By 15.3.1, therefore, there exists a function f belonging to $\mathbf{L}^p$ for every $p < \infty$, such that $\hat{f} = \phi$. However, it is evident that $\hat{f} = \phi$ fails to belong to $\ell^{p'}(Z)$ whenever $p > 2$ (in which case $p' < 2$).

15.5 Sets of Type (p, q) and of Type $\Lambda(p)$

In this section we shall touch briefly upon a milder type of lacunarity than that exhibited by E-spectral functions with E a Sidon set. The new concepts were introduced and studied by Rudin [6] upon which this account is based. Figà-Talamanca and Rider [1] and Rider [4] have extended some of Rudin's considerations to compact groups in general, but we have no space for an account of these developments; a good account appears in [LR], Chapter 5.

In 15.3.1 we have seen that, if E is a Sidon set, then to each exponent p satisfying $0 < p < \infty$ corresponds a number $B_p = B_p(E) \geqslant 0$ such that

$$\|f\|_p \leqslant B_p \cdot \|f\|_1$$

for each E-spectral trigonometric polynomial f. The sets E we are about to consider are characterized by similar, but weaker, inequalities.

15.5.1. Sets of Type (p, q). In what follows we suppose that all exponents p, q, r, and s lie in the real interval $(0, \infty)$.

If $p < q$, a subset E of Z is said to be of *type* (p, q) if and only if there exists a number $B = B(p, q, E) \geqslant 0$ such that

$$\|f\|_q \leqslant B \cdot \|f\|_p \tag{15.5.1}$$

for all E-spectral trigonometric polynomials f. We observe once and for all that if $p \geqslant 1$ and $q \geqslant 1$, and if an inequality of this sort holds for all $f \in \mathbf{T}_E$, then each E-spectral function in $\mathbf{L}^p$ belongs to $\mathbf{L}^q$ and the same inequality continues to hold for such functions. (This is an almost immediate consequence of the fact that $\mathbf{T}_E$ is everywhere dense, relative to the topology defined by $\|\cdot\|_p$, in $\mathbf{L}_E{}^p$, provided $1 \leqslant p < \infty$; compare 2.4.4, 6.1.1, and 6.2.1. The reader is advised to provide a proof of this assertion.)

By way of example, 15.3.1 affirms that any Sidon set E is of type $(1, q)$ whenever $q > 1$.

If E is of type (p, q), so too is any subset of E.

By using the open mapping theorem (I, B.3.2), it may be shown that E is of type (p, q), where $1 \leqslant p < q$, if and only if $\mathbf{L}_E{}^p = \mathbf{L}_E{}^q$; see Exercise 15.5.

Inasmuch as $\|f\|_r \leqslant \|f\|_s$ whenever $r < s$, it is evident that if $p_1 < p_2 < q_1 < q_2$, and if E is of type (p_1, q_2), then E is of type (p_2, q_1).

The next assertion is less trivial.

15.5.2. If $0 < p < q < r < \infty$, then E is of type (p, r) if and only if it is of type (q, r).

Proof. By what has been said above, if E is of type (p, r) then it is of type (q, r).

Suppose conversely that E is of type (q, r), so that

$$\|f\|_r \leqslant B \cdot \|f\|_q \tag{15.5.2}$$

for some $B \geqslant 0$ and all $f \in \mathbf{T}_E$. Now an application of Hölder's inequality shows that

$$\|f\|_q^{q(r-p)} \leqslant \|f\|_p^{p(r-q)} \cdot \|f\|_r^{r(q-p)}. \tag{15.5.3}$$

On combining these inequalities, it is found that

$$\|f\|_q^{p(r-q)} \leqslant B^{r(q-p)} \cdot \|f\|_p^{p(r-q)},$$

or

$$\|f\|_q \leqslant B' \cdot \|f\|_p,$$

and substitution from this into (15.5.2) leads to

$$\|f\|_r \leqslant BB' \cdot \|f\|_p$$

for all $f \in \mathbf{T}_E$, showing that E is of type (p, r).

15.5.3. **Sets of Type $\Lambda(q)$.** Taking the cue suggested by 15.5.2, a subset E of Z will be said to be of *type* $\Lambda(q)$, where $0 < q < \infty$, if and only if there exists an exponent p satisfying $0 < p < q$ such that E is of type (p, q). Frequently we shall write $E \in \Lambda(q)$ to signify that E is of type $\Lambda(q)$.

By 15.3.1, the remarks in 15.5.1, and 15.5.2, a Sidon set E is of type $\Lambda(q)$ whenever $0 < q < \infty$. However, there exist subsets E of Z which are of type $\Lambda(q)$ for all q satisfying $0 < q < \infty$ and which are *not* Sidon sets; see Rudin [6], Theorem 4.11, and Exercise 15.7.

It is evident that $\Lambda(q_2) \subset \Lambda(q_1)$ whenever $q_1 < q_2$. In the other direction, Bachelis and Ebenstein [1] have shown that if $E \in \Lambda(q_1)$ for some q_1 such that $1 \leqslant q_1 < 2$, then $E \in \Lambda(q_2)$ for some $q_2 > q_1$; the proof involves reflexivity of the Banach space $\mathbf{L}_E^1$.

Any subset of a set of type $\Lambda(q)$ is again of type $\Lambda(q)$.

By partial analogy with each of 15.1.4, 15.3.1, 15.3.2, and 15.3.3, one has the following criterion (Rudin [6], Theorems 5.1 and 5.4).

15.5.4. **Criterion for Sets of Type $\Lambda(p)$.** If E is a subset of Z, and if $1 < p < \infty$, then the following five statements are equivalent:

(a) $E \in \Lambda(p)$;
(b) $\mathbf{M}_E \subset \mathbf{L}^p$;
(c) $\mathbf{L}_E^1 \subset \mathbf{L}^p$;
(d) $\mathscr{F}\mathbf{L}^\infty | E = \mathscr{F}\mathbf{L}^{p'} | E$;
(e) $\mathscr{F}\mathbf{C} | E = \mathscr{F}\mathbf{L}^{p'} | E$.

If furthermore $p > 2$, these are equivalent to

(f) $\mathscr{F}\mathbf{L}^{p'} | E \subset \ell^2(E)$.

(As usual, p' is defined by $1/p + 1/p' = 1$.)

Proof. This will be broken into several parts.

(1) Let us begin by showing that (a) implies (b). Assuming (a), take any $\mu \in \mathbf{M}_E$ and apply to the trigonometric polynomials $F_N * \mu \in \mathbf{T}_E$ the inequality

$$\|f\|_p \leqslant B \cdot \|f\|_1.$$

Since $\|F_N * \mu\|_1 \leqslant \|\mu\|_1$ (by 12.7.3), it follows that the numbers $\|F_N * \mu\|_1$ are bounded with respect to N. A repetition of the final phase of the proof of 15.3.1 leads thence to the conclusion that $\mu \in \mathbf{L}^p$.

(2) The implication (b) $\Rightarrow$ (c) is trivial.

(3) To show that (c) implies (a), it suffices to appeal to Exercise 15.5.

It is now certain that (a), (b), and (c) are equivalent.

(4) Now we show that (a) implies (d). Assuming (a), take any $g \in \mathbf{L}^{p'}$ and consider the linear functional

$$f \to \frac{1}{2\pi} \int \check{f} \cdot g \, dx \tag{15.5.4}$$

on $\mathbf{T}_E$, which we regard as a subspace of $\mathbf{L}^1$. According to (a) and the Hahn-Banach theorem (I, B.5.1), this linear functional has a continuous extension to $\mathbf{L}^1$. By I, C.1, this extension must be of the form

$$f \to \frac{1}{2\pi} \int \check{f} \cdot h \, dx, \tag{15.5.5}$$

where $h \in \mathbf{L}^\infty$. Comparing (15.5.4) and (15.5.5) in the case where $f = e_n$ with $n \in E$, it is seen that $\hat{g}|E = \hat{h}|E$, and (d) is thereby established.

(5) To show that (d) implies (e), we again take $g \in \mathbf{L}^{p'}$. By 7.5.1, $g = f * g_1$ for suitably chosen $f \in \mathbf{L}^1$ and $g_1 \in \mathbf{L}^{p'}$. Assuming (d), there exists $h_1 \in \mathbf{L}^\infty$ such that $\hat{h}_1|E = \hat{g}_1|E$. Put $h = f * h_1$. Then $h \in \mathbf{C}$ and, for $n \in E$,

$$\hat{h}(n) = \hat{f}(n)\hat{h}_1(n) = \hat{f}(n)\hat{g}_1(n) = \hat{g}(n),$$

so that (e) is verified.

(6) The proof that (e) implies (a) is a little more complicated. Assuming (e), to each $f \in \mathbf{L}^{p'}$ corresponds at least one (usually many) $g \in \mathbf{C}$ such that $\hat{g}|E = \hat{f}|E$. Since g is not generally uniquely determined by f, we must use a quotient space (see I, B.1.8). Knowledge of f in fact determines g up to addition of elements of $\mathbf{C}_F$, where $F = Z\backslash E$, and thus determines uniquely the element $\dot{g}$ of the quotient linear space $\mathbf{C}/\mathbf{C}_F$ (that is, the coset $g + \mathbf{C}_F$ modulo $\mathbf{C}_F$ corresponding to g). Now $\mathbf{C}$ is a Banach space, and $\mathbf{C}_F$ is a closed linear subspace of $\mathbf{C}$. Consequently (compare the substance of 11.4.7) the quotient linear space $\mathbf{C}/\mathbf{C}_F$ is turned into a Banach space by defining on it the norm

$$\|\dot{g}\| = \inf\{\|g + h\| : h \in \mathbf{C}_F\}. \tag{15.5.6}$$

We now have a linear operator T from $\mathbf{L}^{p'}$ into $\mathbf{C}/\mathbf{C}_F$ defined by $Tf = \dot{g}$, where g is any element of $\mathbf{C}$ such that $\hat{g}|E = \hat{f}|E$. It is simple to show (see

Exercise 15.6) that T has a closed graph. Hence (I, B.3.3) T is continuous from $\mathbf{L}^{p'}$ into $\mathbf{C}/\mathbf{C}_F$. This signifies that there exists a number $B \geqslant 0$ such that to each $f \in \mathbf{L}^{p'}$ corresponds at least one $g \in \mathbf{C}$ such $\hat{g}|E = \hat{f}|E$ and

$$\|g\|_\infty \leqslant B \cdot \|f\|_{p'}.$$

This being so, if $u \in \mathbf{T}_E$ one has

$$\left|\frac{1}{2\pi}\int \check{u}f\,dx\right| = \left|\frac{1}{2\pi}\int \check{u}g\,dx\right| \leqslant \|u\|_1 \cdot \|g\|_\infty$$

$$\leqslant \|u\|_1 \cdot B\|f\|_{p'},$$

and this for each $f \in \mathbf{L}^{p'}$. Appeal to Exercise 3.6 permits the inference that $\|u\|_p \leqslant B \cdot \|u\|_1$ for each $u \in \mathbf{T}_E$, showing that (a) holds.

This proves that (a) to (e) are equivalent.

(7) Finally, it is evident from the Parseval formula that (e) implies (f) without the additional restriction $p > 2$. On the other hand, to show that (f) implies (a) when $p > 2$, we note that (f) and 8.3.1 combine to show that in this case $\mathscr{F}\mathbf{L}^{p'}|E = \mathscr{F}\mathbf{L}^2|E$, and an argument exactly like that utilized in (6) shows that E is of type $(2, p)$ and so that $E \in \Lambda(p)$.

Remarks. (1) Stage 7 of the above proof may alternatively be accomplished as follows. Assume 15.5.4(f). Then (by the closed graph theorem; see Volume 1, Appendix B.3.3)

$$\left[\sum_{n \in E} |\hat{f}(n)|^2\right]^{1/2} \leqslant \text{const.}\|f\|_{p'} \qquad \text{for all } f \in \mathbf{L}^{p'}.$$

Hence, for all $g \in \mathbf{T}_E$,

$$\left|(2\pi)^{-1}\int_0^{2\pi} f(x)g(-x)\,dx\right| = \sum_{n \in E} |\hat{f}(n)\hat{g}(n)|$$

$$\leqslant \left(\sum_{n \in E} |\hat{f}(n)|^2\right)^{1/2} \|g\|_2$$

$$\leqslant \text{const.}\|f\|_{p'}\|g\|_2.$$

So, by the converse of Hölder's inequality (Exercise 3.6),

$$\|g\|_{p'} \leqslant \text{const.}\|g\|_2 \qquad \text{for all } g \in \mathbf{T}_E.$$

Thus E is of type $(2, p)$. If $p > 2$, this entails that E is $\Lambda(p)$.

(2) It has been proved (Bachelis, MR **42** # 6523) that, if $k \in \{1, 2, 3, \ldots\}$, then $E \subset Z$ is of type $\Lambda(2k)$, if and only if $|\hat{f}| \in \mathscr{F}\mathbf{L}^p$ for every $f \in \mathbf{L}_E{}^p$.

See also Exercise 16.30.

15.5.5. **Properties of Sets of Type $\Lambda(p)$.** Rudin ([6], pp. 216–223) establishes a number of structural properties of sets of type $\Lambda(p)$, somewhat analogous to the results in Section 15.2; see Exercises 15.7 to 15.11. Besides the results mentioned in these exercises, we quote one more such conclusion involving the function $r_s(n, E)$, defined to be the number of representations of n in the form $n = n_1 + \cdots + n_s$ with $n_1, \cdots, n_s \in E$ (compare the definition of $R_s(n, E)$ in 15.2.1), namely: if E is a subset of $Z \cap [0, \infty)$ and if $s > 1$ is an integer, then

(a) if E is of type $\Lambda(2s)$, we have

$$\limsup_{N \to \infty} N^{-1} \sum_{n=0}^{N} r_s(n, E)^2 < \infty;$$

(b) if E is a finite union of sets $E_1, \cdots, E_t$ such that

$$\sup_n r_s(n, E_j) < \infty \qquad \text{for all } j \text{ such that } 1 \leqslant j \leqslant t,$$

then E is of type $\Lambda(2s)$.

See Rudin [6], Theorem 4.5, Exercise 15.11, and [Ba_2], p. 258.

It is only fair to add that there is a considerable gap between the necessary condition (a) and the sufficient condition (b) in order that E be of type $\Lambda(2s)$. Also, as Rudin points out, (b) cannot be weakened to the demand that

$$\limsup_{N \to \infty} N^{-1} \sum_{n=0}^{N} r_s(n, E) < \infty. \tag{15.5.7}$$

For, if E consists of the perfect squares and $s = 2$, condition (15.5.7) is satisfied; but (a) is not satisfied and E is therefore not of type $\Lambda(4)$. It is moreover apparently unknown whether the set of all perfect squares is of type $\Lambda(p)$ for *any* p whatsoever; see Rudin [6], p. 219. More generally, if k is a positive integer and $p > 2k$, then the set E of all kth powers of positive integers is not of type $\Lambda(p)$. In fact, by 7.3.5 (ii), if $0 < \alpha < 1$ the series

$$\sum_{n=1}^{\infty} n^{-\alpha} \cos nx$$

is the Fourier series of a function $f \in \mathbf{L}^{p'}$ whenever $\alpha p > 1$. Since $p > 2k$, α may be chosen so that $0 < \alpha < 1$, $\alpha p > 1$ and $2\alpha k \leqslant 1$, the last clause implying that

$$\sum_{n \in E} n^{-2\alpha} = \sum_{m=1}^{\infty} m^{-2\alpha k} = \infty.$$

Thus, 15.5.4(f) fails and E is not of type $\Lambda(p)$.

15.6 Pointwise Convergence and Related Matters

As usual, we have neglected the study of pointwise convergence, except insofar as 15.1.4(b) and (c) trivially entail the absolute and uniform pointwise convergence of Fourier series of functions in $\mathbf{L}_E^\infty$ and $\mathbf{C}_E$, respectively, whenever E is a Sidon set. A further basic result of this sort, applying to Fourier series with Hadamard gaps, is contained in Exercise 6.13.

Parallel to what comes about for Rademacher series (see 14.2.3) and as might be suggested in some measure by 15.3.1, there is a rather startling dichotomy concerning the pointwise convergence of a lacunary trigonometric series

$$\sum_{n \in Z} c_n e^{inx} \tag{15.6.1}$$

in which $c_n = 0$, except when $n = \pm n_k$ for some Hadamard sequence (n_k). It turns out that if

$$\sum_{n \in Z} |c_n|^2 < \infty,$$

then the series (15.6.1) is pointwise convergent almost everywhere; whereas if

$$\sum_{n \in Z} |c_n|^2 = \infty,$$

then the series (15.6.1) is pointwise divergent almost everywhere, and indeed fails at almost all points to be summable by any one of the usual summability methods. (The first assertion follows from Exercise 6.13 and 8.3.1.)

Usually treated in connection with pointwise convergence is the circumstance that the behavior on small subsets of the sum function of a lacunary trigonometric series largely determines its global behavior. A variant of one such result appears in Exercise 15.17.

For further details concerning these and other fascinating topics, see [Z_1], Chapter V; [Ba_2], Chapter XI; [M], Chapitre VIII; Moeller [1]; Moeller and Frederickson [1]; Emel'janov [1].

15.7 Dual Aspects: Helson Sets

The dual aspects of the problems discussed hitherto in this chapter arise when the groups T and Z are interchanged. They are of more recent origin, being in fact the outcome of work of Helson [4] on analogous problems for the group R of real numbers. The Sidon subsets discussed earlier are peculiar to discrete groups (such as Z): their analogues for nondiscrete groups (such as R and T) are termed Helson sets. Rudin's general treatment ([R], Chapter 5) is designed to cover both concepts from a common point of view insofar as this is possible; see also [KS], Chapitre XI. (There are differences of detail that demand separate treatment at certain points, however.) Here we have space merely to indicate some of the analogies and unsolved problems. For other developments in the case of R, see Helson and Kahane [1].

15.7.1. **Restatement of 15.1.3.** The definition of Sidon subsets of Z given in 15.1.3 can be reformulated as the statement that a subset E of Z is a Sidon set if and only if there exists a number $B = B_E \geqslant 0$ such that

$$\|\phi\|_1 \leqslant B \,.\, \|\hat{\phi}\|_\infty$$

for all $\phi \in \ell^1(Z)$ satisfying $\operatorname{supp} \phi \subset E$.

As has been said in the Remarks following 12.13.3, $\ell^1(Z)$ can be identified with the set of (bounded Radon) measures on Z. This observation paves the way for the appropriate definition of a Helson subset of T, which is as follows.

15.7.2. **Helson Sets Defined.** A subset E of T is termed a *Helson set* if and only if there exists a number $B = B_E \geqslant 0$ such that

$$\|\mu\|_1 \leqslant B \,.\, \|\hat{\mu}\|_\infty \tag{15.7.1}$$

for all measures $\mu \in \mathbf{M}$ such that $\operatorname{supp} \mu \subset E$. (Concerning supports of measures, see 12.11.4.)

For reasons stemming from the nondiscrete character of T, it is customary (as in [R]) to restrict attention to those Helson sets that are closed in T; we have avoided imposing this restriction from the outset, solely in order to heighten the analogy with 15.7.1.

Somewhat surprisingly, it is the case that E is a Helson set if and only if there exists a number $B = B_E \geqslant 0$ such that

$$\|\mu\|_1 \leqslant B \,.\, \limsup_{|n| \to \infty} |\hat{\mu}(n)| \tag{15.7.2}$$

for all measures μ such that $\operatorname{supp} \mu \subset E$; see [KS], p. 143, and compare McGehee [2].

15.7.3. **Analogues of 15.1.4 and 15.1.5.** There is a valid analogue of 15.1.4 for Helson sets, the most significant portion of which asserts that a closed subset E of T is a Helson set if and only if each continuous complex-valued function on E is the restriction to E of the transform $\hat{\phi}$ of a suitably chosen $\phi \in \ell^1(Z)$, that is, the restriction to E of an element of $\mathbf{A}$. In other words, in the notation of 10.6.2(8), a closed set E is a Helson set if and only if $\mathbf{A}(E) = \mathbf{C}(E)$. (This property is used by Rudin to *define* closed Helson sets; see also Kahane [6].) The proof differs in no essential respect from that of 15.1.4.

There is also a valid analogue of 15.1.5.

It should be remarked at this point that general functional analytic principles lead to yet other equivalent formulations of the definition of Helson sets. We cite a few examples and mention a number of corollaries that relate Helson sets to other categories of sets already encountered in Subsection 12.11.5. (As the reader will perceive, harmonic analysis on a group G gives rise to a somewhat bewildering variety of classes of subsets of G. Among the major unsolved problems of the subject are to be found those of determining reasonably direct and verifiable criteria for membership of any one such class, as well as that of determining relationships between the various classes. Except in the trivial case in which G is finite, not a single one of these problems has yet received a satisfactory solution.)

If we reintroduce the notation established in 12.11.5(3), an easy application of the closed graph theorem (I, B.3.3) combined with 12.3.9 will show that a closed set E is a Helson set if and only if $\mathbf{M}(E)$ is closed in $\mathbf{P}(E)$, or (what is equivalent) is closed in $\mathbf{P}$. (Moreover, by a result from general duality theory ([E], Theorem 8.10.5) and 12.3.9 once more, E is a Helson set if and only if $\mathbf{M}(E)$ is weakly closed in $\mathbf{P}$.)

Again, it is not difficult to show that a closed set E is a Helson set if and only if $\mathbf{M}(E) = \mathbf{P}^0(E)$; see [KS], p. 142 and Edwards [9]. It is always the case that $\mathbf{M}(E) \subset \mathbf{P}^0(E)$, $\mathbf{P}^0(E)$ being defined as in 12.11.5(3) above.

It follows [compare 12.11.5(3)] that a closed set E which supports no true pseudomeasures [so that $\mathbf{M}(E) = \mathbf{P}(E)$] is a Helson set, and that a closed set E which is simultaneously a Helson set [so that $\mathbf{M}(E) = \mathbf{P}^0(E)$] and a spectral synthesis set [so that $\mathbf{P}^0(E) = \mathbf{P}(E)$] supports no true pseudomeasures.

15.7.4. **Examples of Helson Sets: Kronecker Sets.** Although there is no difficulty in exhibiting finite Helson sets, the production of infinite Helson sets is a good deal more complicated. This task has been discharged through the intervention of the so-called Kronecker sets.

The nomenclature, which is by now pretty firmly rooted, is due to Rudin; however, such sets were first introduced and constructed by Hewitt and Kakutani [1] and might well have been named accordingly.

A subset K of T is termed a *Kronecker set* if and only if each continuous complex-valued function f on K such that $|f(\dot{x})| = 1$ $(\dot{x} \in K)$ is the limit, uniformly on K, of characters e_n $(n \in Z)$. (Here and again below we represent points of T in the form $\dot{x}$, where x is a real number and $\dot{x}$ the coset modulo $2\pi Z$ containing x.)

It is known ([R], Theorem 5.2.2) that there exist Kronecker sets K in T which are homeomorphic with Cantor's ternary set on the line (see Exercise 12.44 and [HS], pp. 70–71); any such set K is perfect and uncountable.

On the other hand ([R], Theorem 5.6.6), every closed Kronecker set is a Helson set (in T).

15.7.5. **Further Examples: Independent Sets.** Another category of infinite Helson sets in T arises in the following way.

A subset E of T is termed *independent* if, whenever $\dot{x}_1, \cdots, \dot{x}_k$ are *distinct* elements of E, the relations $n_1, \cdots, n_k \in Z$, $n_1\dot{x}_1 + \cdots + n_k\dot{x}_k = \dot{0}$ entail that $n_j\dot{x}_j = \dot{0}$ for $j = 1, 2, \cdots, k$. (cf. (14.1.3).)

It is almost evident that any Kronecker set is independent, and that any element $\dot{x}$ of a Kronecker set is of infinite order (that is, $n\dot{x} \neq \dot{0}$ for any integer $n \neq 0$). It is also true ([R], Theorem 5.1.3) that any *finite* independent set, each of whose elements is of infinite order, is a Kronecker set.

The major point to be made is that furthermore any countable closed independent set is a Helson set ([R], Theorem 5.6.7; [KS], p. 148); compare Exercise 15.21.

From this it follows incidentally that there exist closed independent Helson sets which contain elements of finite order and which are therefore not

Kronecker sets: the sets $\{\dot{1}, \dot{\pi}\}$ and $\{\dot{1}, (2^{1/2})^{\cdot}, \dot{\pi}\}$ are trivial examples of such sets.

It is also known ([KS], p. 148) that there exist perfect independent sets which are not Helson sets.

For more examples and counter examples, see Körner [1].

15.7.6. **Inclusion Relations.** If we denote by $\mathscr{H}$, $\mathscr{I}$, and $\mathscr{K}$ the classes of closed Helson, closed independent, and closed Kronecker sets, respectively, we can summarize the known relations between these classes in the following scheme:

$$\mathscr{K} \subset \mathscr{I}, \qquad \mathscr{K} \subset \mathscr{H}$$

$$\mathscr{I} \not\subset \mathscr{H}, \qquad \mathscr{H} \not\subset \mathscr{K}, \qquad \mathscr{I} \not\subset \mathscr{K},$$

$$\{countable\} \cap \mathscr{I} \subset \mathscr{H},$$

where $\{countable\}$ denotes the class of countable subsets of T. In particular, $\mathscr{H}$, $\mathscr{I}$, and $\mathscr{K}$ are all different.

The relation $\mathscr{H} \not\subset \mathscr{K}$ has been amplified in the work of Wik [2] and Kaufman [1]; Wik produces examples of sets E which satisfy (15.7.1) with $B = 1$, for all $\mu \in \mathbf{M}$ such that supp $\mu \in E$, and which are not Kronecker sets. For more details of this and other results, see [LP].

15.7.7. **Another Characterization of Helson Sets.** Helson sets can be characterized in terms of approximation of functions. Thus it is known (Edwards [6]) that a closed set E in T is a Helson set if and only if the following statement is true: to each continuous complex-valued function f on E corresponds a number $c = c(f)$ such that f is the limit, uniformly on E, of a sequence $(f_r)_{r=1}^{\infty}$ of functions of the form

$$f_r = \sum_{j=1}^{N(r)} \alpha_{rj} \cdot g_{rj} * h_{rj},$$

where the α_{rj} are complex numbers satisfying

$$\sum_{j=1}^{N(r)} |\alpha_{rj}| \leqslant 1 \qquad \text{for all } r \in \{1, 2, \cdots\}$$

and the g_{rj} and h_{rj} are functions in $\mathbf{L}^2$ satisfying

$$\|g_{rj}\|_2 \leqslant c, \qquad \|f_{rj}\|_2 \leqslant c \qquad \text{for all } r \in \{1, 2, \cdots\} \qquad \text{and all } j \in \{1, 2, \cdots, N(r)\}.$$

15.7.8. **Perfect Helson Sets.** It has been mentioned in 12.11.5(3) that one can construct nonvoid perfect sets E which support no true pseudomeasures. Any such set E, satisfying as it does $\mathbf{P}(E) \subset \mathbf{M}$, has the property that $\mathbf{P}^0(E) \subset \mathbf{M}$ and is therefore a Helson set (see 15.7.2).

This remark, taken together with the substance of Subsections 15.7.4, 15.7.5, 15.7.6, and Exercise 15.21, makes it abundantly clear that a characterization of Helson sets in group-theoretic and topological terms represents an extremely formidable undertaking. A solution does not appear to be in sight.

15.7.9. **Measures Supported by Helson Sets.** What Helson originally proved in [4] for the case of the group R is also true for T namely: if E is a Helson subset of T, there exists no nonzero measure $\mu \in \mathbf{M}$ such that $\operatorname{supp} \mu \in E$ and $\hat{\mu} \in c_0(Z)$; in particular, $c_{\alpha *} \cdot (E) = 0$ for $0 \leqslant \alpha \leqslant 1$ (see 12.12.3 and 12.12.7), and E is a set of uniqueness in the wide sense (as defined in 12.12.8).

This result, which follows immediately from the characterization of Helson sets contained in (15.7.2), provides some indication of the necessary sparseness of Helson sets. It appears as Theorem 5.6.10 in [R].

15.7.10. **Characterization Problems.** Closed Helson sets are defined in functional analytic terms (see 15.7.2). Other characterizations of this type have been mentioned in 15.7.3 and 15.7.7; see also Rosenthal [2]. It would plainly be of the greatest interest to characterize closed Helson sets in group-theoretical (or arithmetical) and topological terms, but no such characterization appears to be even remotely attainable at present. (Much the same is true, as we have seen, of the spectral synthesis sets and the Sidon sets.)

The most that seems to be known in this direction is that Helson sets certainly do possess specifiable and rather specialized arithmetical properties; see Exercise 15.21 for indications of some such aspects.

15.7.11. **Carleson Sets.** It has been recorded in 15.7.3 that a closed subset E of T is a Helson set if and only if each continuous complex-valued function f on E is the restriction to E of a function

$$\sum_{n \in Z} c_n e^{inx}, \tag{15.7.3}$$

wherein the c_n are complex numbers satisfying

$$\sum_{n \in Z} |c_n| < \infty . \tag{15.7.4}$$

In 1952 Carleson [2] was led to introduce the class of closed subsets E of T having a similar property, the sole difference being that the sequence (c_n) is to satisfy the additional condition

$$c_n = 0 \qquad \text{for all } n \in Z \text{ satisfying } n < 0. \tag{15.7.5}$$

These sets came to be termed *Carleson sets.*

It is trivial that any Carleson set is a Helson set. In 1960 Wik [1] established the entirely unexpected result that the converse is also true: Carleson sets and Helson sets are the same things. For an account of these matters, see [KS], Chapitre XI.

15.7.12. **Relations with Dirichlet Series. Bohr Sets.** Over 50 years ago, Harald Bohr proved a result about Dirichlet series that can be formulated in the following way: if P denotes the set $\{2, 3, 5, \cdots\}$ of prime positive integers, there exists a number $B \geqslant 0$ such that

$$\sum_{n \in P} |\phi(n)| \leqslant B \cdot \sup_{t \in R} \Big| \sum_{n \in Z'} \phi(n) n^{-it} \Big| \tag{15.7.6}$$

for all complex-valued functions ϕ with compact supports defined on the set Z' of positive integers.

Rider [3] has attached the label *Bohr set* to each subset E of Z' such that (15.7.6) holds with E in place of P and a suitable number $B = B_E \geqslant 0$. He exhibits an interesting connection between such Bohr sets and certain Sidon subsets of the character group of the product of denumerably many copies of T, and uses this to produce further examples of Bohr sets, and an example of an infinite subset of Z' containing no infinite Bohr set (compare 15.2.5).

In connection with the definition of Bohr sets, compare Exercise 15.20.

15.7.13 Finite unions of Helson sets. The substance of 15.2.7 above suggests the problem of proving that the union of two Helson sets is a Helson set. This problem has a relatively lengthy history; see MR **39** # 6020; **40** # 3815; **43** # # 7866, 7867. The first complete proof is due to Varopoulos [5]. Other proofs soon followed; see Saeki [2] and Herz [4]. See also McGehee's fine review (MR **46** # 5939) of Herz [4].

15.8 Other Species of Lacunarity.

There is an abundance of species of lacunarity now on the market (see, for instance, [LR], Chapters 7 to 10); here we mention only a few.

Some variants of the spaces involved in 15.1.4 have been seen to lead back to Sidon sets (see 15.1.4, Remark (7)); naturally enough, other variants lead to new types of lacunarity.

15.8.1. ***p*-Sidon sets.** Suppose $p \in [1, 2]$. A subset E of Z is termed a *p-Sidon set* if $f \in \mathbf{C}_E$ implies $\hat{f} \in \ell^p(Z)$ (cf. 15.1.4(c)). Evidently every subset of Z is 2-Sidon and a subset of Z is 1-Sidon if and only if it is Sidon. It is far from obvious that there are p-Sidon sets (with $1 < p < 2$) which are non-Sidon; however they do exist and were first constructed in Edwards and Ross [1], to which the reader is referred for the basic results concerning p-Sidon sets. See also the references in [LR], 10.6.

15.8.2. ***W*-Sidon Sets.** If E is a subset of Z and W is a complex-valued function on E then E is called a *W-Sidon set* if $f \in \mathbf{C}_E$ implies $\sum_{n \in E} |W(n)\hat{f}(n)| < \infty$ (cf. 15.1.4(c) again).

This weighted version of Sidonicity is, in its theory, intermediate between the areas of lacunarity and multiplier theory (see Chapter 16). If $W \in \ell^2(E)$, every subset E is W-Sidon; and it is not obvious that there are W-Sidon sets (with $W \notin \ell^2(E)$) which are not Sidon. For such examples, for the basic results, and for the connection with p-Sidon sets, see Sanders [1].

15.8.3. **Rosenthal Sets.** From 15.1.4 it follows that if E is a Sidon set then $\mathbf{L}_E^\infty = \mathbf{C}_E$. A subset E of Z is called a *Rosenthal set* if $\mathbf{L}_E^\infty = \mathbf{C}_E$. Rosenthal [3] has constructed Rosenthal sets which are not Sidon; in fact Blei has proved that every non-Sidon set contains a Rosenthal set; see [LR], 10.4.

15.8.4. **Interpolation Sets.** Recall that a measure $\mu \in \mathbf{M}$ is termed *discrete* (or *atomic*; see Exercise 12.51) if it is expressible in the form

$$\mu = \sum_{k=1}^{\infty} c_k \varepsilon_{a_k}$$

for some choice of a sequence $(a_k)_{k=1}^\infty$ of distinct elements of T, and of a sequence $(c_k)_{k=1}^\infty$ of complex numbers satisfying $\sum_{k=1}^\infty |c_k| < \infty$. ($\varepsilon_x$ here denotes the Dirac measure at x; see 12.2.3).

A subset E of Z is called an *interpolation set* if, for every $\phi \in \ell^\infty(E)$, there is a discrete measure $\mu \in \mathbf{M}$ satisfying $\hat{\mu} | E = \phi$. Although every interpolation set is Sidon, the converse is false; we refer to the references in [LR], 10.10.

15.8.5. **Riesz Sets.** From 15.3.1 it follows that if E is a Sidon set then $\mathbf{M}_E = \mathbf{L}_E^1$. There is an old theorem due to F. and M. Riesz (see [R], 8.2.1) which guarantees that the set $E = Z_+$ also has this property. Meyer [2] has studied such sets and named them *Riesz sets*. Once again, see [LR], 10.5 for further references.

15.8.6. **Fatou-Zygmund Sets.** A complex-valued function ϕ on a subset E of Z is called *hermitian* if, whenever n and $-n$ both belong to E, $\phi(-n) = \overline{\phi(n)}$.

A subset E of Z is called a *Fatou-Zygmund set* if there is a constant B such that, for all hermitian elements ϕ of $\ell^\infty(E)$, there exists a positive measure $\mu \in \mathbf{M}$ such that $\hat{\mu} | E = \phi$ and $\|\mu\|_1 \leqslant B\|\phi\|_\infty$. (Recall that a measure μ is positive if $\mu(f) \geqslant 0$ for every nonnegative $f \in \mathbf{C}$.) It is easy to see that every Fatou-Zygmund set is Sidon; the rather surprising converse was proved by Drury (see [LR], 3.6) using a modification of the techniques used to prove his result quoted in Remark (4) following 15.1.4. For further results concerning Fatou-Zygmund sets, we refer to [LR], Chapter 7.

15.8.7. **Associated Sets.** Suppose that K is a nonempty compact subset of T and that E is a subset of Z. We say that E and K are *strictly associated* if there is a constant B such that

$$\|\hat{f}\|_1 \leqslant B \|f\chi_K\|_\infty$$

for all $f \in \mathbf{T}_E$ (here χ_K denotes the characteristic function of the set K). By definition, if E is a Sidon set then E and T are strictly associated. However Déchamps-Gondim has proved that if E is a Sidon set then E is strictly associated with every compact subset K of T having nonempty interior. See [LR], Chapters 8 and 9 for an exposition of this and related topics.

EXERCISES

15.1. Use the uniform boundedness principle (I, B.2.1) to prove the following statement: if $\mathbf{V}$ is a closed linear subspace of $\mathbf{C}$, and if p is an exponent satisfying $1 \leqslant p \leqslant \infty$ and such that $\hat{f} \in \ell^p$ for each $f \in \mathbf{V}$, then there exists a number $B \geqslant 0$ such that $\|\hat{f}\|_p \leqslant B \cdot \|f\|_\infty$ for each $f \in \mathbf{V}$.

15.2. Show that if E is a finite union of Hadamard sets, and if for $\xi > 0$ we denote by $N(\xi)$ the number of elements n of E satisfying $\xi < |n| < 2\xi$, then $N(\xi)$ is bounded with respect to ξ.

15.3. Let E consist of all numbers of the form

$$3^{2^m+2} + 3^{2^m+j},$$

where $j \in \{0, 1, 2, \cdots, 2^{m-1}\}$ and $m \in \{0, 1, 2, \cdots\}$. Prove that E is a Sidon set but is not a finite union of Hadamard sets (Hewitt and Zuckerman [1]).

Hints: Use 15.2.3 and the preceding exercise.

15.4. Show that if E is a Sidon set and $f \in \mathbf{L}_E{}^1$, then

$$\frac{1}{2\pi}\int \exp\,(c|f|^2)\,dx < \infty$$

for any real number c.

Hint: Use 15.3.1 and look again at the proof of 14.2.2.

15.5. Prove that a subset E of Z is of type (p, q), where $1 \leqslant p < q < \infty$, if and only if $\mathbf{L}_E{}^p = \mathbf{L}_E{}^q$.

Hints: For the "if" assertion, use the open mapping (or the closed graph) theorem. For the "only if" statement, use the remarks in 15.5.1.

15.6. The notations being as in the proof of 15.5.4, construct a detailed proof of the statement that the operator T has a graph closed in $\mathbf{L}^{p'} \times (\mathbf{C}/\mathbf{C}_F)$.

15.7. Suppose that E is of type $\Lambda(q)$, where $q > 2$, so that $\|f\|_q \leqslant B_q\|f\|_2$ for $f \in \mathbf{T}_E$. Let $\alpha_E(N)$ be the largest integer α such that some arithmetic progression of N terms contains α elements of E; see 15.2.8. Prove that

$$\alpha_E(N) \leqslant CB_q{}^2N^{2/q}$$

for all N, C being an absolute constant. (See Rudin [6], Theorem 3.5.)

Remarks. It is shown by Rudin [6], Theorem 4.11) that if

$$\lim_{N \to \infty} N^{-\varepsilon}\alpha(N) = 0 \qquad \text{for each } \varepsilon > 0,$$

then there exists a set E of type $\Lambda(q)$ for every $q < \infty$ for which $\alpha_E(N) > \alpha(N)$ for infinitely many N. In view of the next exercise, if α be chosen so that $\limsup_{N \to \infty} \alpha(N)/\log N = \infty$, any such set E fails to be a Sidon set.

Hints: Suppose that some N-termed arithmetic progression

$$\{a + b, a + 2b, \cdots, a + Nb\}$$

contains numbers $n_1, \cdots, n_\alpha$ each lying in E. Write $Q(x) = e^{imbx + iax} F_N(bx)$, where $m = \frac{1}{2}N$ or $\frac{1}{2}(N + 1)$ according as N is even or odd, and

$$f(x) = \sum_{k=1}^{\alpha} e^{in_k x},$$

and observe that

$$\tfrac{1}{2}\alpha \leqslant \sum_{k=1}^{\alpha} \hat{Q}(n_k) = \frac{1}{2\pi} \int \check{f} \cdot Q \, dx.$$

Now use Hölder's inequality and a suitable majorant for $\|F_N\|_{q'}$.

15.8. Show that if E is a Sidon set, then

$$\alpha_E(N) \leqslant B_E \cdot \log N \qquad (N \geqslant 3).$$

(See Rudin [6], Theorem 3.6.)

Remark. If E is the set $\{2^k : k = 0, 1, 2, \cdots\}$, then E is a Sidon set and yet

$$\alpha_E(N) > \frac{\log N}{\log 2}$$

for $N \leqslant 2^k$.

Hint: Use 15.3.1 and the preceding exercise.

15.9. Show that if E is of type $\Lambda(1)$, then E does not contain arbitrarily long arithmetic progressions, and indeed that $\alpha_E(N) < N$ for all sufficiently large N. (See Rudin [6], Theorem 4.1.)

Hints: Take p, $0 < p < 1$, so that $\|f\|_1 \leqslant B \cdot \|f\|_p$ for $f \in \mathbf{T}_E$. Assuming E to contain $a + b, a + 2b, \cdots, a + Nb$, where $b \neq 0$, consider

$$f(x) = e^{iax} \sum_{n=1}^{N} e^{inbx}.$$

15.10. Suppose that E_1 and E_2 are of type $\Lambda(p)$ and that $E = E_1 \cup E_2$. Prove that

(1) if $p > 2$, then E is of type $\Lambda(p)$;

(2) if $p > 1$, and if $E_1 \subset [0, \infty)$, $E_2 \subset (-\infty, 0)$, then again E is of type $\Lambda(p)$.

See Rudin [6], Theorem 4.4.

Hints: We may assume that E_1 and E_2 are disjoint. Suppose in case (1) that $\|f_i\|_p \leqslant B_i \cdot \|f_i\|_2$ for $f_i \in \mathbf{T}_{E_i}$, $i = 1, 2$. Decompose any $f \in \mathbf{T}_E$ into $f_1 + f_2$ with $f_i \in \mathbf{T}_{E_i}$. For (2), take s satisfying $1 < s < p$, so that now $\|f_i\|_p \leqslant B_i \cdot \|f_i\|_s$. Proceed as for (1), using 12.10.3.

15.11. Prove the results appearing as (a) and (b) in 15.5.5.

Hints: Assume $E = \{n_k\}$, where $0 \leqslant n_1 < n_2 < \cdots$. Consider

$$f = \sum_{j=1}^{k} e_{n_j} \in \mathbf{T}_E .$$

Putting $r(n) = r_s(n, E)$, check that f^s is of the form

$$r(0) + r(1)e^{ix} + r(2)e^{2ix} + \cdots + r(n_k)e^{in_k x} + \cdots ;$$

the succeeding coefficients are immaterial. Deduce that

$$\sum_{m=0}^{n_k} r^2(m) \leqslant \|f\|_{2s}^{2s} \leqslant B^{2s} \cdot \|f\|_2^{2s} = B^{2s} \cdot k^s .$$

Now use the result of Exercise 15.7 and the obvious fact that $\alpha_E(n_k + 1) \geqslant k$. This leads to (a).

As for (b), by Exercise 15.10 it may be assumed that $t = 1$. Consider any $f = \sum_k a(k)e_{n_k} \in \mathbf{T}_E$. Then

$$f^s(x) = \sum_m b_m e_m ,$$

where $b_m = \sum a(k_1) \cdots a(k_s)$, summed over all representations

$$m = n_{k_1} + \cdots + n_{k_s} .$$

Among these representations for a given m, choose one with indices $k_1(m)$, say, which maximizes $|a(k_1) \cdots a(k_s)|$. Then, if $r_s(n, E) \leqslant B$, we have $|b_m|^2 \leqslant B^2 |a(k_1(m)) \cdots a(k_s(m))|^2$. So, obviously,

$$\sum_m |b_m|^2 \leqslant B^2 [\sum_k |a(k)|^2]^s .$$

Now use the Parseval formula for f^s and for f.

15.12. Extend 15.2.6 to E-spectral Radon measures μ such that

$$\operatorname{Re} \mu(u) \leqslant m \cdot \|u\|_1 , \qquad \operatorname{Im} \mu(u) \leqslant m \cdot \|u\|_1$$

for some $m \geqslant 0$ independent of u and each nonnegative $u \in \mathbf{C}$.

Hint: Consider the functions $\mu * f \in \mathbf{L}_E^1$, where f is a nonnegative integrable function, apply 15.2.6, and finally appeal to Exercise 3.14.

15.13. Using 15.3.2, show that there exists a continuous function f and a sequence $\varepsilon_n \downarrow 0$ such that

$$\sum_{n>0} |\hat{f}(n)|^{2-\varepsilon_n} = \infty;$$

see the remarks in 8.3.2.

15.14. Assume that k is a positive integer and that $n_1, \cdots, n_k$ are distinct integers. Prove that there exists a choice of $\pm$ signs such that the trigonometric polynomial

$$f(x) = \sum_{m=1}^{k} \pm e^{in_m x}$$

satisfies

$$\|f\|_k \geqslant 2^{-5/2} k^{1/2} \|f\|_2.$$

Hints: Assume first that $k = 2N + 1$ is odd and relabel $n_1, \cdots, n_k$, $n_{-N}, \cdots, n_N$. Apply the conclusion of Exercise 14.9 to

$$g(\omega, x) = \sum_{j=-N}^{N} \rho_j(\omega) e^{in_j x}$$

to derive

$$\int_{\mathscr{C}} \left((2\pi)^{-1} \int_0^{2\pi} |g(\omega, x)|^k \, dx \right) d\lambda(\omega) \geqslant 4^{-k} k^k.$$

Then define $f(x) = g(\omega_0, x)$ for a suitably chosen $\omega_0 \in \mathscr{C}$.

If $k = 2N + 2$ is even, observe that, if f is chosen as above, then $\|f\|_k \geqslant \|f\|_{2N+1}$.

15.15. Let $\mathbf{V}$ be an infinite dimensional closed invariant subspace of $\mathbf{L}^a$, where $1 \leqslant a < \infty$. Show that the relation

$$\bigcap_{p<\infty} (\mathbf{V} \cap \mathbf{L}^p) \subset \mathbf{L}^\infty \tag{1}$$

is false.

Hints: Make $\mathbf{H} = \bigcap_{p<\infty} (\mathbf{V} \cap \mathbf{L}^p)$ into a Fréchet space (I, B.1.3). Assuming (1) to hold, use the closed graph theorem (I, B.3.3) to deduce that there exists $B \geqslant 0$ such that $\|f\|_\infty \leqslant B\|f\|_p$ for some $p < \infty$ and all $f \in \mathbf{H}$. By Remark (ii) following 11.2.1, $e_n \in \mathbf{H}$ for infinitely many $n \in Z$. Derive a contradiction from 15.1.4(d) and 15.2.5.

15.16. Suppose that $\mathbf{V}$ and a are as in the preceding exercise. Define

$$B_p = \sup\{\|f\|_p : f \in \mathbf{V} \cap \mathbf{T}, \|f\|_a \leqslant 1\},$$

where $\mathbf{T}$ denotes (as usual) the set of all trigonometric polynomials. Prove that

$$\lim_{p\to\infty} B_p = \infty.$$

15.17. Let E be a Hadamard set, as in 15.2.4. It can be shown (see, for example, [Z_1], pp. 203–204) that, if S is a measurable subset of T having measure $m(S) > 0$, and if $\lambda > 1$, there exists an integer $\nu = \nu(S, \lambda, q) > 0$ such that

$$\|t\|_2^2 \leqslant \lambda \cdot m(S)^{-1} \cdot \int_S |t(x)|^2 \, dx \tag{1}$$

for all E-spectral trigonometric polynomials t satisfying $\hat{t}(n) = 0$ for $|n| < \nu$. Assuming this, prove the following result: if F is an E-spectral distribution, Ω a nonvoid open subset of T, and if F coincides on Ω with a function f which is analytic on Ω, then F is equal globally to an analytic function. (See [Z_1], p. 206 for a more refined "pointwise" analogue applying to E-spectral functions. Compare also Exercise 8.15.)

Hints: First extend (1) to functions more general than E-spectral trigonometric polynomials. Then apply this extension of (1) to the functions $F * v_N$, where v_N is as in Exercise 12.5. Finally, use Exercise 2.8.

15.18. Let E and Ω be as in the preceding exercise. What can be said about the global nature of any E-spectral distribution F which is such that F coincides on Ω with a function in $\mathbf{L}^2$?

What if $\mathbf{L}^2$ is here replaced by $\mathbf{T}$?

15.19. Suppose that $\alpha > 0$, that $b \in Z$ and $b > 1$, and that $(c_k)_{k=1}^{\infty}$ is a bounded sequence. Define $f_\alpha \in \mathbf{C}$ by

$$f_\alpha(x) = \sum_{k=1}^{\infty} c_k b^{-\alpha k} \cos b^k x.$$

Prove that

(1) $\Omega_\infty f_\alpha(a) = O(|a|^\alpha)$ as $a \to 0$, if $0 < \alpha < 1$;

(2) if

$$\sum_{k=1}^{\infty} |c_k|^2 b^{(1-\alpha)2k} = \infty,$$

then f_α is not of bounded variation.

Remarks. Taking $c_k = 1$, Weierstrass showed that f_α is nowhere differentiable in the pointwise sense whenever α is sufficiently small; that the same is true whenever $0 < \alpha < 1$ was established by Hardy. It is interesting to note that, although (1) is false if $\alpha = 1$ and $c_k = 1$, yet in this case it is true that

$$\sup_x |f_1(x + a) + f_1(x - a) - 2f_1(x)| = O(|a|) \qquad \text{as } a \to 0;$$

see [Z_1], p. 47.

Hints: For (1), see [Z_1], p. 47. For (2), use 12.5.10 and 15.3.1.

15.20. Do there exist any infinite subsets E of Z corresponding to which a number $B \geqslant 0$ exists such that

$$\sum_{n \in E} |\hat{f}(n)| \leqslant B \|f\|_\infty$$

for each $f \in \mathbf{T}$? Justify your answer.

15.21. Let E be a subset of $[0, 2\pi)$ which is closed when regarded as a subset of T and which has the following property: there exist numbers $a_k > 0$ and positive integers N_k $(k = 1, 2, \cdots)$ such that $\lim_{k\to\infty} N_k = \infty$ and $\{a_k, 2a_k, \cdots, N_k a_k\} \subset E$.

Prove that E (viewed as a subset of T) is not a Helson set (and therefore, by 15.7.3, supports true pseudomeasures).

Remarks. Any set E of the specified sort is, of course, highly nonindependent when viewed as a subset of T; see 15.7.5.

As an example, one may take $E = \{0, 1, \frac{1}{2}, \frac{1}{3}, \cdots\}$; this provides an instance of a countable closed set that supports true pseudomeasures; see 12.11.5(3).

There are stronger results, analogous to that cited in 15.2.8 for Sidon sets; see [KS], p. 146, Théorème VIII. Compare also with Exercises 15.7 to 15.9.

Hints: By hypothesis, for any positive integer N, the set S_N of numbers $a > 0$ such that $\{a, 2a, \cdots, Na\} \subset E$ is nonvoid. Define $\delta_N = \inf_{a\in S_N} \sup \|\mu\|_1$, the supremum being taken over all measures μ of the form $\mu = \sum_{j=1}^{N} c_j \varepsilon_{ja}$ for which $\|\mu\|_{\mathbf{P}} \leqslant 1$. Show that $\delta_N \to \infty$ as $N \to \infty$.

15.22. Prove that, if E is a Sidon set, then

$$\sup \{\min \{\#(F), \#(G)\} : F + G \subset E\} < \infty \tag{1}$$

where, for every finite set F, $\#(F)$ denotes the cardinal of F, and where $F + G = \{n + m : n \in F \text{ and } m \in G\}$. In particular, a Sidon set does not contain the sum of two infinite sets.

Hints: Assume (1) fails, let $n \in Z$ satisfy $n > 1$, and choose sets $F, G \subset Z$ such that $F + G \subset E$ and $\min \{\#(F), \#(G)\} \geqslant n^3$. Letting $F_1 \subset F$ have exactly n elements, first choose $m_1 \in G \backslash F_1$, and then show that it is possible to select, for each $k \in Z$ satisfying $2 \leqslant k \leqslant n$, an element m_k from G so that

$$m_k \notin (F_1 + (-F_1) + \{m_1, \cdots, m_{k-1}\}) \cup F_1.$$

Now, with $F_1 = \{l_1, \cdots, l_n\}$ and $G_1 = \{m_1, \cdots, m_n\}$, estimate both the $\mathbf{A}$ and $\mathbf{C}$ norms of the trigonometric polynomial

$$t_n = \sum_{j,k=1}^{n} u_{jk} e_{l_j} e_{m_k}$$

where (u_{jk}) is an $n \times n$ unitary matrix satisfying $|u_{jk}| = n^{-1/2}$ for all j, k.

15.23. (i) Assume that E is an infinite subset of Z. Prove that there exists no continuous linear map V of $\mathbf{c}_0(E)$ into $\mathbf{L}^1$ such that

$$(V\phi)\hat{}\,(n) = \phi(n) \qquad \text{for all } \phi \in \mathbf{c}_0(E) \text{ and all } n \in E;$$

cf. Remark (11) following 15.1.4.

(ii) Prove that, if K is an infinite closed subset of T, then there exists no continuous linear map U of $\mathbf{C}(K)$ into $\mathbf{A}$ such that $(Uf)|K = f$ for all $f \in \mathbf{C}(K)$.

Hints: For (i), assume the existence of V with the stated properties. Obviously, V is injective. Also, since $\|V\phi\|_\infty \geqslant \|(V\phi)^\wedge\|_\infty = \|\phi\|_\infty$ for all $\phi \in \mathbf{c}_0(E)$, V^{-1} is continuous. Deduce that V is a linear homeomorphism of $\mathbf{c}_0(E)$ onto a closed linear subspace of $\mathbf{L}^1$. Using Appendix C.2 in Volume 1, conclude that $\mathbf{c}_0(E)$ is weakly sequentially complete. Infer that the constant function 1 with domain E belongs to $\mathbf{c}_0(E)$. Since E is infinite, this is a contradiction.

For (ii), argue similarly, to the point where it is deducible that, for every $a \in K$, the characteristic function $\chi_{\{a\}, K}$ of $\{a\}$ relative to K, belongs to $\mathbf{C}(K)$. Obtain a contradiction by choosing $a \in K$ which belongs to the closure in K of $K \backslash \{a\}$.

A different style of proof is given by Graham [1].

15.24. Verify the claim made in 15.2.3, Remark (2), namely that if E is a symmetric subset of Z satisfying

$$R_s(0, E) \leqslant B^s \qquad \text{for all positive integers } s,$$

then E is a Rider set.

Hint: Apply the technique, used to prove (15.2.7), to the function

$$t_N = \prod_{k=1}^{N} u_k,$$

wherein

$$u_k = 1 + (2B)^{-1}(e_{-n_k} + e_{n_k}).$$

CHAPTER 16

Multipliers

A little less than sixty years ago, Fekete [1] discussed some generalities and some particular questions pertaining to what has since come to be known as the problem of Fourier "multiplier (or factor, or conversion) sequences (or functions)"; see [Z_1], pp. 175–179, 378, where references will be found to other special results due to many authors, and also the remarks in 16.3.8, 16.3.9, and 16.7.6.

In this chapter we aim to describe a general approach to such problems that places emphasis on the so-called "multiplier operators" associated with such functions, an approach that appears to have been vitalized by Wendel [1], [2] in connection with the isomorphism problem for group algebras (mentioned in 4.2.7). In all the most important cases, these multiplier operators belong to a category of operators that are very simply characterizable in algebraic and topological terms involving convolutions; see Section 16.2.

The *leit-motiv* of this chapter is accordingly the association with each multiplier function ϕ of a corresponding multiplier operator U_ϕ, followed by the representation of U_ϕ as convolution with some distribution A (from which ϕ is easily recaptured as the Fourier transform of A), and a struggle to tie down the nature of A as closely as possible in a manner depending on the range and image spaces of U_ϕ. It is in this characterization of A that the real sting lies. This task, along with the discussion of some important special cases, occupies Section 16.3. The discussion of other important special cases continues throughout Sections 16.4 to 16.6.

The number of particular cases of the multiplier problem that have been effectively solved is limited, and it will become clear to the reader of this chapter that there is no lack of enticing unsolved problems. A few such problems, as well as some extensions of the multiplier concept, are mentioned in Section 16.7.

In Section 16.8 we relate multiplier problems with questions concerning direct-sum decompositions of standard function spaces in terms of their closed invariant subspaces (compare 2.2.1). Concerning the study of multipliers in the general setting of Banach (and topological) algebras, see [La].

16.1 Preliminaries

The so-called *multiplier problem* can be formulated in quite general terms as follows.

16.1.1. **Multiplier Functions and Operators.** Suppose two sets, **F** and **G**, of distributions are given. It is required to determine necessary and sufficient conditions on the complex-valued function ϕ on Z in order that the implication

$$f \in \mathbf{F} \Rightarrow \phi \cdot \hat{f} \in \mathscr{F}\mathbf{G} \tag{16.1.1}$$

shall be valid.

In (16.1.1), as elsewhere in this book, $\mathscr{F}\mathbf{G}$ denotes the set of transforms $\hat{g}$ of elements g of **G**.

A function ϕ satisfying (16.1.1) is termed a (*Fourier*) *multiplier of type* (**F**, **G**), and we shall then write $\phi \in (\mathbf{F}, \mathbf{G})$. If $\mathbf{F} = \mathbf{L}^p$ and $\mathbf{G} = \mathbf{L}^q$, where, *as will be assumed throughout this chapter*, $1 \leqslant p \leqslant \infty$ and $1 \leqslant q \leqslant \infty$, it is customary to write simply (p, q) in place of $(\mathbf{L}^p, \mathbf{L}^q)$.

The most interesting cases are those in which both **F** and **G** are chosen from the arsenal of standard function- or measure-spaces, such as $\mathbf{C}^k$ $(0 \leqslant k \leqslant \infty)$, $\mathbf{L}^p$ $(1 \leqslant p \leqslant \infty)$, or **M**. However, complete solutions are not yet available even for all these cases. Where necessary and sufficient conditions *are* known, they are frequently expressed by membership of ϕ to $\mathscr{F}\mathbf{H}$ for some set **H** of distributions. As will by now be clear, the verification of such a relation is more often than not a most formidable task.

The reader will by now have guessed that analogous multiplier problems present themselves in cases where the underlying group G is something more general than T. Although the methods employed in this chapter are frequently specializations of those applicable in the more general situation, we have space only to make a general reference to Brainerd and Edwards [1] and the research papers referred to therein, together with a few more specific references and comments at appropriate places. The case in which $G = Z$, the group dual to T, is in some respects simpler (inasmuch as measures and "distributions" on Z are just functions on Z) and in other respects more complicated, and here too there remain numerous unsolved problems; we have insufficient space to deal with this dual problem, but see Exercise 16.29 and the references cited in 16.4.7 and 16.4.9(3). In this connection it may be noted that the multipliers of $\ell^2(Z)$ correspond exactly with the so-called doubly infinite *Toeplitz matrices*; these entities and their close relatives occupy a small niche in analysis, for an introduction to which the reader is referred to [Z_1], p. 168 and Widom [1]. See also Hirschman [2], Widom [2].

Without missing any cases of current interest, much repetition will be avoided by making the following *standing hypotheses* concerning **F** and **G**:

With the sole exception that **G** *is sometimes chosen to be* **D**, *it will be assumed that each of* **F** *and* **G** *is a linear space containing* $\mathbf{C}^\infty$ *and contained in* **D**; *each is stable under the translation operators* T_a; *each is a Banach or a Fréchet space* (see Volume 1, Appendix B) *and convergence of a sequence in* **F** *or in* **G** *implies its distributional convergence* (see Section 12.3); *and each operator of translation, or of convolution with a trigonometric polynomial, is a continuous endomorphism of* **F** *and of* **G**. (**D** itself fails to satisfy these conditions only insofar as it is neither a Banach space nor a Fréchet space.)

We shall make frequent use of the device that associates with a multiplier ϕ of type (**F**, **G**) the corresponding *multiplier operator* U_ϕ with domain **F** and range in **G**, defined in the following way. The uniqueness theorem 2.4.1 and 12.5.4(1) shows that if ϕ is a multiplier of type (**F**, **G**), then to each $f \in \mathbf{F}$ corresponds precisely one $g \in \mathbf{G}$ such that $\hat{g} = \phi \cdot \hat{f}$; we then define $U_\phi f = g$.

The notation $U_\phi f$ will continue to be used, whenever $f \in \mathbf{D}$ and ϕ is a complex-valued function on Z such that $\phi \cdot \hat{f}$ is tempered, to denote that distribution g such that $\hat{g} = \phi \cdot \hat{f}$ [see 12.5.3(2)].

The examples that follow illustrate a number of features common to most multiplier operators which will be considered formally and in some detail in subsequent sections.

16.1.2. **Some Examples.** (1) As a general comment, observe that if $\mathbf{F}_1 \subset \mathbf{F}$ and $\mathbf{G}_1 \supset \mathbf{G}$, then $(\mathbf{F}_1, \mathbf{G}_1) \supset (\mathbf{F}, \mathbf{G})$. This is evident from the defining property (16.1.1).

(2) The inclusions $\ell^1 \subset (\mathbf{M}, \mathbf{C})$ and $\ell^2 \subset (\mathbf{M}, \mathbf{L}^q)$ for $q \leqslant 2$ are evident, in view of the case $m = 0$ of 12.5.3(1) and 8.3.1.

(3) If ϕ has the form $\phi(n) = a_{|n|}$, where $(a_N)_{N=0}^\infty$ is decreasing to zero and convex, then ϕ belongs to each of $(\mathbf{M}, \mathbf{L}^1)$, $(\mathbf{L}^\infty, \mathbf{C})$, and $(\mathbf{L}^p, \mathbf{L}^p)$; moreover, U_ϕ is expressible as convolution with an integrable function; see Exercise 16.1.

(4) It is relatively simple (see Exercise 16.2) to show that $(\mathbf{L}^2, \mathbf{L}^2) = \ell^\infty$. In this connection the reader should observe that if $\phi \in \ell^\infty$, then there exists a unique pseudomeasure A such that $\hat{A} = \phi$ (see Section 12.11), and the associated multiplier operator U_ϕ is defined by convolution with A:

$$U_\phi f = A * f \tag{16.1.2}$$

for $f \in \mathbf{L}^2$.

Observe also that each multiplier operator of type $(\mathbf{L}^2, \mathbf{L}^2)$ is continuous and commutes with the translation operators T_a; this follows from (16.1.2),

(8.2.2), and 12.6.2. As will be seen in 16.2.1, this too is a characteristic feature.

(5) From 12.7.2 and 12.7.3 one may (see Exercise 16.3) derive the following inclusions

$$\mathscr{F}\mathbf{M} \subset (\mathbf{C}, \mathbf{C}), \qquad \mathscr{F}\mathbf{M} \subset (\mathbf{L}^p, \mathbf{L}^p), \qquad \mathscr{F}\mathbf{M} \subset (\mathbf{M}, \mathbf{M}).$$

It will appear in due course (see 16.3.2, 16.3.3, and Remark (2) following 16.3.5) that indeed

$$\mathscr{F}\mathbf{M} = (\mathbf{C}, \mathbf{C}) = (\mathbf{L}^\infty, \mathbf{L}^\infty) = (\mathbf{L}^1, \mathbf{L}^1) = (\mathbf{M}, \mathbf{M}).$$

However, as will be seen in 16.4.3(1), $\mathscr{F}\mathbf{M}$ does not exhaust $(\mathbf{L}^p, \mathbf{L}^p)$ if $1 < p < \infty$.

Here again all the corresponding multiplier operators U_ϕ are linear, continuous, commute with translations, and admit a representation as convolution with a suitable distribution.

16.1.3. **The Determination of** $(\mathbf{C}^\infty, \mathbf{D})$. We aim to show that $(\mathbf{C}^\infty, \mathbf{D})$ comprises just the tempered functions ϕ on Z, that is, the functions that satisfy a majorization of the form

$$\phi(n) = O(|n|^k) \qquad \text{as } |n| \to \infty \tag{16.1.3}$$

for some (ϕ-dependent) integer k. It will appear that $(\mathbf{C}^\infty, \mathbf{C}^\infty)$ also comprises exactly the tempered functions on Z.

It is on the one hand clear from 12.1.1 that any tempered sequence ϕ belongs to $(\mathbf{C}^\infty, \mathbf{C}^\infty)$, and a fortiori to $(\mathbf{C}^\infty, \mathbf{D})$.

Suppose on the other hand that ϕ belongs to $(\mathbf{C}^\infty, \mathbf{D})$. Then $\phi\hat{f}$ is tempered whenever $f \in \mathbf{C}^\infty$. We wish to conclude from this that ϕ is tempered. However, were this not the case, there would exist a sequence (n_k) of integers such that $0 < |n_k| \to \infty$ and

$$|\phi(n_k)| > |n_k|^k.$$

But then, by 12.1.1 again,

$$f(x) = \sum |n_k|^{-\frac{1}{2}k} \exp{(in_k x)}$$

would belong to $\mathbf{C}^\infty$, and yet

$$|\phi(n_k)\hat{f}(n_k)| > |n_k|^{\frac{1}{2}k},$$

contrary to the hypothesis that $\phi\hat{f}$ is tempered.

The reader will observe that the associated multiplier operator U_ϕ is given by

$$U_\phi f = A * f, \tag{16.1.4}$$

where A is the distribution

$$A = \sum_{n \in Z} \phi(n) e^{inx}; \tag{16.1.5}$$

see 12.5.3(2). This operator U_ϕ is linear, commutes with translations and with convolutions, and is easily seen to be continuous from $\mathbf{C}^\infty$ into any one of the "natural" spaces lying between $\mathbf{C}^\infty$ and $\mathbf{D}$ (such as $\mathbf{C}$, $\mathbf{L}^p$, and $\mathbf{M}$).

As has been indicated, we shall see in Section 16.3 that virtually all multiplier operators are expressible in the form (16.1.4), A being a suitably chosen distribution.

16.1.4. **Properties of Multiplier Operators.** In all the examples considered in 16.1.2 and 16.1.3 the standing hypotheses on $\mathbf{F}$ and $\mathbf{G}$ (laid down in 16.1.1) are fulfilled; and in all these examples it has appeared that the multiplier operators U_ϕ concerned have the following properties:

(1) U_ϕ is linear, continuous, commutes with the translation operators T_a and with convolution with trigonometric polynomials (the last condition means that $U_\phi(t * f) = U_\phi t * f = t * U_\phi f$ for each $t \in \mathbf{T}$ and each $f \in \mathbf{F}$);

(2) U_ϕ admits a convolution representation

$$U_\phi f = A * f,$$

where A is some distribution depending upon ϕ.

In the course of the next two sections we shall see that properties (1) and (2) persist in all the cases of practical interest, and that indeed these properties come close to characterizing the multiplier operators. In particular, in Section 16.3 we shall establish a convolution representation formula of the type (16.1.4) for multipliers of various types $(\mathbf{F}, \mathbf{G})$. The approach to be adopted there will cover the case of $(\mathbf{C}^\infty, \mathbf{D})$, but it also adapts to certain other choices of $(\mathbf{F}, \mathbf{G})$ in such a way as to be more directly productive of the best results.

Although the case of $(\mathbf{C}^\infty, \mathbf{D})$ provides a sort of "universal covering theorem," inasmuch as virtually all multiplier operators of interest have restrictions to $\mathbf{C}^\infty$ which are multiplier operators of type $(\mathbf{C}^\infty, \mathbf{D})$, far too much information is thrown away by this process of restriction for the outcome to have any lasting interest in the discussion of multiplier operators of type $(\mathbf{F}, \mathbf{G})$ for numerous other natural choices of $\mathbf{F}$ and $\mathbf{G}$.

16.2 Operators Commuting with Translations and Convolutions; $\mathfrak{m}$-operators.

We begin this section by summarizing in formal terms the essential properties of multiplier operators already encountered.

16.2.1. **Characteristic Properties of Multiplier Operators.** Let ϕ be a multiplier of type $(\mathbf{F}, \mathbf{G})$ and U_ϕ the associated multiplier operator mapping $\mathbf{F}$ into $\mathbf{G}$ (see 16.1.1). Then

(1) U_ϕ is linear;

(2) U_ϕ commutes with translations; that is,

$$U_\phi T_a = T_a U_\phi$$

for each group element a;

(3) U_ϕ commutes with convolution by trigonometric polynomials; that is,

$$U_\phi(t * f) = U_\phi t * f = t * U_\phi f \tag{16.2.1}$$

for every $f \in \mathbf{F}$ and every trigonometric polynomial t; and, more generally, (16.2.1) holds whenever $t \in \mathbf{F}, f \in \mathbf{F}$, and $t * f \in \mathbf{F}$;

(4) U_ϕ is continuous from $\mathbf{F}$ into $\mathbf{G}$.

Note: The reader is reminded of the standing hypotheses concerning $\mathbf{F}$ and $\mathbf{G}$ imposed throughout this chapter. It is well to remark, however, that quite often not all of these hypotheses are essential for the truth of any one assertion made: this will usually be quite clear from a close examination of the proofs.

Proof. The first three statements are almost immediate consequences of the definition of U_ϕ (namely, $U_\phi f = g$ signifies exactly that $f \in \mathbf{F}$, $g \in \mathbf{G}$, and $\hat{g} = \phi \cdot \hat{f}$) combined with the properties of convolution vis-à-vis the Fourier transformation (see especially 12.5.5 and 12.6.5) and the uniqueness theorem for Fourier transforms (see 2.4.1 in the case of functions, and 12.5.4 in the case of distributions). The reader is urged to write out in full detail at least one of the proofs.

The proof of (4) is a little less immediate, being based upon the closed graph theorem (see I, B.3.3). According to this theorem, to show that U_ϕ is continuous, it suffices to prove that: if a sequence (f_n) extracted from $\mathbf{F}$ converges in $\mathbf{F}$ to the limit f, and if at the same time the sequence $(g_n) = (U_\phi f_n)$ converges in $\mathbf{G}$ to the limit g, then necessarily $g = U_\phi f$. Now, with the aid of the standing hypotheses, from $f_n \to f$ in $\mathbf{F}$ and $g_n \to g$ in $\mathbf{G}$ it follows that $\hat{f}_n \to \hat{f}$ and $\hat{g}_n \to \hat{g}$ pointwise on Z (see the definition of distributional convergence in 12.3.1). But, by the definition of U_ϕ, we have $\hat{g}_n = \phi \cdot \hat{f}_n$ for each n. Passage to the limit as $n \to \infty$ shows that therefore $\hat{g} = \phi \cdot \hat{f}$, which signifies exactly that $g = U_\phi f$ and thus completes the proof.

Remark. Although $\mathbf{D}$ is not a Fréchet space, 16.2.1 still holds for multipliers of type $(\mathbf{F}, \mathbf{D})$. The only modification needed is in the proof of (4), the form of the closed graph theorem stated in I, B.3.3 being no longer applicable (though the closed graph theorem can be shown to be valid for the case in hand; see, for example, [E], Chapter 6 and Exercise 8.43). Often, a simple direct proof of the continuity of U_ϕ is possible; see Exercise 16.4.

Before proceeding to a converse of 16.2.1, we shall deal with a simple corollary thereof.

16.2.2. **An Application of 16.2.1.** It has been noted in 16.1.2(4) that $(\mathbf{L}^2, \mathbf{L}^2) = \ell^\infty$. A good deal more true is true, namely, $(\mathbf{A}, \mathbf{P}) = (\mathbf{A}, \mathbf{A}) = \ell^\infty$.

Indeed, since $(\mathbf{A}, \mathbf{A}) \subset (\mathbf{A}, \mathbf{P})$, and since it is quite evident that $\ell^\infty \subset (\mathbf{A}, \mathbf{A})$, it suffices to show that $(\mathbf{A}, \mathbf{P}) \subset \ell^\infty$.

Now, if $\phi \in (\mathbf{A}, \mathbf{P})$, 16.2.1(4) affirms that the associated multiplier operator U_ϕ is continuous from $\mathbf{A}$ into $\mathbf{P}$. This signifies the existence of a number $c \geqslant 0$ such that

$$\| U_\phi f \|_P \leqslant c \| f \|_{\mathbf{A}}$$

for each $f \in \mathbf{A}$. Taking $f = e_n$, and noting that $U_\phi e_n = \phi(n) e_n$ and $\|e_n\|_{\mathbf{A}} = \|e_n\|_{\mathbf{P}} = 1$, it follows that $|\phi(n)| \leqslant c$ for all $n \in Z$, and so that $\phi \in \ell^\infty$.

We now return to the general development and establish a converse of 16.2.1.

16.2.3. **A Converse of 16.2.1.** (1) If U is a linear operator mapping $\mathbf{F}$ into $\mathbf{D}$ such that

$$U(t * f) = t * Uf \tag{16.2.2}$$

for each trigonometric polynomial t and each $f \in \mathbf{F}$, then there exists a function $\phi \in (\mathbf{F}, \mathbf{D})$ such that

$$Uf = U_\phi f \qquad (f \in \mathbf{F}). \tag{16.2.3}$$

(2) If U is a linear operator mapping $\mathbf{F}$ into $\mathbf{D}$ which commutes with translations, then there exists a function ϕ such that

$$Ut = U_\phi t \tag{16.2.4}$$

for each trigonometric polynomial t.

Proof. (1) In (16.2.2) we take $t = e_n = f$ and so deduce that $Ue_n = \phi(n)e_n$, where $\phi(n) = (Ue_n)\hat{}(n)$. Then linearity of U shows that $(Ut)\hat{} = \phi \cdot \hat{t}$ for each trigonometric polynomial t. From (16.2.2) again,

$$[U(t * f)]\hat{} = \hat{t} \cdot (Uf)\hat{};$$

and, from what we have just established, we have also

$$[U(t * f)]\hat{} = \phi \cdot (t * f)\hat{} = \phi \cdot \hat{t} \cdot \hat{f}.$$

By comparison, therefore, $\hat{t} \cdot (Uf)\hat{} = \phi \cdot \hat{t} \cdot \hat{f}$ for all trigonometric polynomials t, which implies (16.2.3).

(2) Let $u_n = Ue_n$. Since U commutes with translations,

$$\begin{aligned} T_a u_n = T_a U e_n = U T_a e_n &= U(e^{-ina} e_n) \\ &= e^{-ina} u_n, \end{aligned}$$

by linearity of U. It is easily seen (for example, by taking the Fourier transform of this relation) to follow that $u_n = \phi(n) e_n$ for some complex-valued function ϕ on Z, and linearity of U now leads to (16.2.4).

Remark. In case (2) it is not evident that the function ϕ belongs to $(\mathbf{F}, \mathbf{D})$, there being no assurance (without further hypotheses upon U) that (16.2.4) continues to hold when t is replaced by an arbitrary element of $\mathbf{F}$. However, we do have the following corollary.

16.2.4. **Another Converse of 16.2.1.** Suppose that the trigonometric polynomials are everywhere dense in $\mathbf{F}$. Suppose too that U is a linear operator mapping $\mathbf{F}$ into $\mathbf{G}$ such that *either*

(1) U satisfies (16.2.2) for each trigonometric polynomial t and each $f \in \mathbf{F}$; *or*

(2) U is continuous and commutes with translations.

Then $U = U_\phi$ for some $\phi \in (\mathbf{F}, \mathbf{G})$.

Proof. Case (1) is already disposed of by 16.2.3(1).

In case (2), we know that (16.2.4) holds for each trigonometric polynomial t. Moreover, given $f \in \mathbf{F}$, there is a sequence $(t_k)_{k=1}^{\infty}$ of trigonometric polynomials converging in $\mathbf{F}$ to f. Then, by the assumed continuity of U, the sequence $(Ut_k) = (U_\phi t_k)$ converges in $\mathbf{G}$ to Uf. From this it follows that $(U_\phi t_k)^\wedge = \phi \cdot \hat{t}_k$ converges pointwise to $\phi \cdot \hat{f}$, so that the limit of $(U_\phi t_k)_{k=1}^{\infty}$ in $\mathbf{G}$, say g, must be such that $\hat{g} = \phi \cdot \hat{f}$. It follows thence that $\phi \in (\mathbf{F}, \mathbf{G})$ and that $U = U_\phi$.

16.2.5. **The Cases $\mathbf{F} = \mathbf{L}^\infty$ and $\mathbf{F} = \mathbf{M}$.** Although 16.2.4 shows that in the majority of interesting cases there is identity between the multiplier operators and those continuous linear operators which commute either with translations or with convolutions, there are one or two interesting cases where the identification is as yet in doubt. In these cases the doubt persists either because one of the spaces $\mathbf{F}$ and $\mathbf{G}$ involved does not satisfy the standing hypotheses laid out in 16.1.1, or because the trigonometric polynomials are not everywhere dense in $\mathbf{F}$.

Two especially significant such cases are those in which $\mathbf{F} = \mathbf{L}^\infty$ and $\mathbf{F} = \mathbf{M}$, respectively. Even here, however, it is still true that any linear operator from $\mathbf{F}$ into $\mathbf{G}$, which commutes with translations or with convolutions with trigonometric polynomials, satisfies (16.2.4) for some function ϕ and all trigonometric polynomials t. If furthermore U is continuous, for the *weak* topology on $\mathbf{L}^\infty$ (or $\mathbf{M}$), in the sense that $Uf_k \to Uf$ in $\mathbf{G}$ whenever $f_k \to f$ weakly in $\mathbf{L}^\infty$ (or in $\mathbf{M}$), then it would again follow that $\phi \in (\mathbf{F}, \mathbf{G})$ and $U = U_\phi$. In this connection it should be recalled (from 12.3.9 and 12.3.10) that $f_k \to f$ weakly in $\mathbf{L}^\infty$ (or in $\mathbf{M}$) if and only if

$$\frac{1}{2\pi}\int f_k g \, dx \to \frac{1}{2\pi}\int fg \, dx \qquad (g \in \mathbf{L}^1)$$

$$[\text{or } f_k(g) \to f(g) \qquad (g \in \mathbf{C})].$$

The reader should also bear in mind that if $f \in \mathbf{L}^\infty$ (or $f \in \mathbf{M}$), then $\sigma_N f \to f$ weakly in $\mathbf{L}^\infty$ (or weakly in $\mathbf{M}$); see Exercise 16.5.

16.2.6. **Commutativity with Translations and with Convolutions.** In the preceding results we have had occasion to consider both continuous linear operators that commute with translations, and those that commute with convolutions. In many important instances one can show a priori that these two categories of operators are identical. For instance, any continuous linear operator U from $\mathbf{C}^k$ or $\mathbf{L}^p$ $(1 \leqslant p < \infty)$ into $\mathbf{C}^h$, or $\mathbf{L}^q$, or $\mathbf{M}$, or $\mathbf{D}$ commutes with translations if and only if it commutes with convolutions; see Exercise 16.6. As one might expect from 16.2.5, however, this equivalence is in doubt if the domain of U is $\mathbf{L}^\infty$ or $\mathbf{M}$, unless we demand continuity with respect to *weakly* convergent sequences in $\mathbf{L}^\infty$ or in $\mathbf{M}$.

16.2.7. **$\mathfrak{m}$-operators.** Much repetition will be saved if, guided by the foregoing results and discussion, we henceforth adopt the following definition. By an *$\mathfrak{m}$-operator of type* $(\mathbf{F}, \mathbf{G})$ is meant a continuous linear operator from $\mathbf{F}$ into $\mathbf{G}$ which (a) commutes with translations and (b) satisfies (16.2.2) for each trigonometric polynomial t and each $f \in \mathbf{F}$. In addition, we shall denote by $\mathfrak{m}(\mathbf{F}, \mathbf{G})$ the set of $\mathfrak{m}$-operators of type $(\mathbf{F}, \mathbf{G})$.

It is to be observed that 16.2.6 shows that, in case $\mathbf{F}$ is $\mathbf{C}^k$ or $\mathbf{L}^p$ $(p \neq \infty)$ and $\mathbf{G}$ is $\mathbf{C}^h$, $\mathbf{L}^q$, $\mathbf{M}$, or $\mathbf{D}$, a continuous linear operator from $\mathbf{F}$ into $\mathbf{G}$ belongs to $\mathfrak{m}(\mathbf{F}, \mathbf{G})$ provided it satisfies *either* (a) *or* (b) above.

From 16.2.1 and 16.2.3(1) it appears that $U \in \mathfrak{m}(\mathbf{F}, \mathbf{G})$ if and only if $U = U_\phi$ for some $\phi \in (\mathbf{F}, \mathbf{G})$.

There is in general no special difficulty in verifying that, if $A \in \mathbf{D}$, the convolution operator U defined by

$$Uf = A * f$$

belongs to $\mathfrak{m}(\mathbf{F}, \mathbf{G})$, provided only that it does indeed map $\mathbf{F}$ into $\mathbf{G}$; in particular, U always belongs to $\mathfrak{m}(\mathbf{C}^\infty, \mathbf{C}^\infty)$. The converse will be examined in 16.3.1.

It is a consequence of the preceding remark that almost all linear operators which arise naturally in harmonic analysis are $\mathfrak{m}$-operators (even though they may not crop up directly from multiplier problems). Moreover, any linear differential operator of order k with constant coefficients is an $\mathfrak{m}$-operator of type $(\mathbf{C}^k, \mathbf{C})$. Reference to Section 6.6 will confirm that almost all summability processes used in connection with Fourier series are definable in terms of sequences of $\mathfrak{m}$-operators of type $(\mathbf{L}^1, \mathbf{L}^1)$.

In 16.3.11 we shall discuss some relationships between $\mathfrak{m}$-operators and the translation operators T_a (which are themselves especially simple $\mathfrak{m}$-operators).

The pointwise theory of those $\mathfrak{m}$-operators defined by convolutions with distributions that are not measures (compare 12.8.2 and the beginning of Section 13.9) is, in those cases where pointwise existence theorems are valid at all, a part of a highly elaborate theory of singular integrals. Concerning this aspect of the subject, the reader is recommended to examine the survey article Calderón [2] and Cordes [1].

It may be added as an aside that, if **F** and **G** satisfy the standing hypotheses stipulated in 16.1.1, if **T** is everywhere dense in **F**, if U is a continuous linear map from **F** into **G**, and if U satisfies (a) (that is, if U commutes with translations), then U satisfies (b). The proof is left as an exercise for the reader.

16.2.8. **A Theorem of Stein about Sequences of $\mathfrak{m}$-operators.** For use in Section 16.5 in connection with multipliers we propose to state here without proof a deep and important theorem due to E. M. Stein ([1], Theorem 1 and Corollary 1) about sequences of $\mathfrak{m}$-operators of type $(\mathbf{L}^p, \mathbf{L}^p)$, where $1 \leqslant p \leqslant 2$. There is no reason at all why any reader who has persevered to the present stage should not tackle the original paper.

Stein's theorem reads as follows. Suppose that $1 \leqslant p \leqslant 2$ and that U_k $(k = 1, 2, \cdots)$ is a sequence of $\mathfrak{m}$-operators of type $(\mathbf{L}^p, \mathbf{L}^p)$. Assume that to each $f \in \mathbf{L}^p$ corresponds a (possibly f-dependent) set E of positive measure such that

$$\limsup_{k\to\infty} |U_k f(x)| < \infty \qquad \text{for all } x \in E. \tag{16.2.5}$$

Define

$$U^*f(x) = \sup_{k \geqslant 1} |U_k f(x)| \qquad (\leqslant \infty), \tag{16.2.6}$$

The conclusion is that the operator U^* is of weak type (p, p) that is (see 13.7.5), there exists a number c such that for each number $\kappa > 0$ and each $f \in \mathbf{L}^p$

$$m(\{x \in [0, 2\pi) : U^*f(x) > \kappa\}) \leqslant c\kappa^{-p}\|f\|_p^p, \tag{16.2.7}$$

where m denotes Lebesgue measure.

In the paper cited, Stein uses this theorem to great effect in the discussion of diverse problems; he also shows that various extensions are possible. Unfortunately, the theorem is definitely false for $p > 2$ (Stein [1], p. 157).

Some at least of the roots of Stein's theorem are due to Calderón and appear in the discussion on pp. 165–166 of [Z_2]. A proof of Stein's Lemma 1 ([1], p. 146), which makes no explicit reference to probability theory, is also to be found on p. 166 of [Z_2]. For further developments, see Coifman [1], Stein and Zygmund [1], Sawyer [1] and Gilbert [1].

16.3 Representation Theorems for $\mathfrak{m}$-operators

In this section we begin the study of the representation of $\mathfrak{m}$-operators in terms of convolution. In particular, and first of all, we shall recover the result established in 16.1.3 for multiplier operators of type $(\mathbf{C}^\infty, \mathbf{D})$. However, as has been heralded in 16.1.4, a different approach will be used. This method makes little explicit use of the Fourier transformation. As developed in this section, it will lead to complete solutions of the representation problem for most of those cases in which a fully effective solution is known.

Both the statements and the proofs of our first theorem are variants (for periodic functions and distributions) of those given first by Schwartz ([S$_2$], pp. 53–54; [E], pp. 332–335) for functions and distributions on R^m.

16.3.1. **Multipliers from $\mathbf{C}^\infty$ to D.** (1) To each $U \in \mathfrak{m}(\mathbf{C}^\infty, \mathbf{D})$ corresponds a distribution $A \in \mathbf{D}$ such that

$$Uf = A * f \tag{16.3.1}$$

for $f \in \mathbf{C}^\infty$; and, conversely, if $A \in \mathbf{D}$, the equation (16.3.1) defines U as a member of $\mathfrak{m}(\mathbf{C}^\infty, \mathbf{D})$.

(2) $\mathfrak{m}(\mathbf{C}^\infty, \mathbf{D}) = \mathfrak{m}(\mathbf{C}^\infty, \mathbf{C}^\infty)$.

(3) $(\mathbf{C}^\infty, \mathbf{D}) = (\mathbf{C}^\infty, \mathbf{C}^\infty)$ comprises exactly all tempered functions on Z.

(4) If $U \in \mathfrak{m}(\mathbf{F}, \mathbf{G})$, there exists a distribution $A \in \mathbf{D}$ such that (16.3.1) holds for each $f \in \mathbf{F}$.

Proof. (1) The converse portion stems from 12.6.2 and 12.6.3. The direct assertion can be proved by writing $U = U_\phi$, where $\phi \in (\mathbf{C}^\infty, \mathbf{D})$ (see 16.2.3), using 16.1.3, and taking for A that distribution for which $\hat{A} = \phi$. The proof to be given here proceeds along different lines which are of interest in other connections, ignoring as it does any relationship between U and multipliers of type $(\mathbf{C}^\infty, \mathbf{D})$.

Choose an approximate identity $(k_i)_{i=1}^\infty$, each k_i being a member of $\mathbf{C}^\infty$. If $f \in \mathbf{C}^\infty$, $k_i * f \to f$ in $\mathbf{C}^\infty$ so that continuity of U entails

$$Uf = \mathbf{D}\text{-}\lim_{i\to\infty} U(k_i * f).$$

From 16.2.1 and 16.2.7 it appears that this may be written in the form

$$Uf = \mathbf{D} - \lim_{i\to\infty} A_i * f, \tag{16.3.2}$$

where

$$A_i = Uk_i \in \mathbf{D}. \tag{16.3.3}$$

Now (16.3.2) signifies that

$$A_i * f * \check{g}(0) = A_i * f(g)$$

converges, as $i \to \infty$, to $Uf(g)$, and hence is bounded with respect to i, for each $f, g \in \mathbf{C}^\infty$. On the other hand, as the reader will verify easily on using 12.1.1, each $h \in \mathbf{C}^\infty$ is expressible as $f * g$ with f and g suitably selected from $\mathbf{C}^\infty$. Thus the numbers $A_i(h) = A_i * \check{h}(0)$ are bounded with respect to i for each $h \in \mathbf{C}^\infty$.

At this point 12.3.7 shows that a subsequence of (A_i) may be chosen which converges in $\mathbf{D}$ to a limit A. By dropping terms, we may therefore assume that $A_i \to A$ in $\mathbf{D}$. This being so, 12.6.3(2) shows that

$$A_i * f \to A * f \quad \text{in } \mathbf{C}^\infty \tag{16.3.4}$$

whenever $f \in \mathbf{C}^\infty$. Comparing (16.3.4) and (16.3.2), (16.3.1) is seen to follow.

(2) The inclusion $\mathfrak{m}(\mathbf{C}^\infty, \mathbf{C}^\infty) \subset \mathfrak{m}(\mathbf{C}^\infty, \mathbf{D})$ being evident, it suffices to prove the reverse inclusion.

But, if $U \in \mathfrak{m}(\mathbf{C}^\infty, \mathbf{D})$, (16.3.1) and 12.6.3(1) combine to show that u maps $\mathbf{C}^\infty$ continuously into $\mathbf{C}^\infty$ (and not merely into $\mathbf{D}$).

(3) This has been established in 16.1.3. (It also follows from 16.2.1 and the Remark following that theorem, (1) of the present theorem, 12.5.3, and 12.6.5.)

(4) If $U \in \mathfrak{m}(\mathbf{F}, \mathbf{G})$ the standing hypotheses on $\mathbf{F}$ and $\mathbf{G}$ ensure[1] that the restriction of U to $\mathbf{C}^\infty$ belongs to $\mathfrak{m}(\mathbf{C}^\infty, \mathbf{D})$. Hence (1) shows that (16.3.1) holds for some $A \in \mathbf{D}$ and each $f \in \mathbf{C}^\infty$. If $f \in \mathbf{F}$ is given and t is any trigonometric polynomial, $t * f$ belongs to $\mathbf{C}^\infty$ and so

$$U(t * f) = A * t * f.$$

On the other hand, by 16.2.7(b),

$$U(t * f) = t * Uf.$$

Comparison of the last two equations, valid for any trigonometric polynomial t, shows that $Uf = A * f$. (We are here using the associativity and commutativity of convolution.)

Remarks. (1) There are restricted analogues of 16.3.1 for the case in which T is replaced by a more general group; see Brainerd and Edwards [1], Edwards [10], [11], and Gaudry [1], [2]. See also Taibleson [2].

(2) The result 16.3.1, although satisfying, represents the beginning rather than the end of multiplier problems. The interesting questions, many of them still without effective solutions, are of the following nature: given the pair $(\mathbf{F}, \mathbf{G})$ and an $\mathfrak{m}$-operator (or a multiplier operator) U of type $(\mathbf{F}, \mathbf{G})$, 16.3.1 shows that there is a distribution A such that (16.3.1) holds at any rate for each $f \in$; $\mathbf{C}^\infty$ the remaining problem is to effectively determine conditions on the distribution A in order that $A * f$ shall belong to $\mathbf{G}$ whenever $f \in \mathbf{F}$. If $\mathbf{C}^\infty$ is dense in $\mathbf{F}$ and if $\mathbf{F}$ and $\mathbf{G}$ are Banach spaces, this is so if and only if

$$\|A * f\|_{\mathbf{G}} \leqslant \text{const } \|f\|_{\mathbf{F}}$$

for $f \in \mathbf{C}^\infty$. Among the remaining results of this section appear instances in which an effective answer is known. In Sections 16.4 and 16.5, however, we handle some cases where no complete and effective answer is forthcoming.

The preceding representation theorem will now be used as a stepping stone to others of a similar nature.

[1] Since the closed graph theorem shows that the injection of $\mathbf{C}^\infty$ into $\mathbf{F}$ is continuous.

16.3.2. Multipliers from C to C or L^1 to M. If $U \in \mathfrak{m}(\mathbf{C}, \mathbf{C})$ (or $\mathfrak{m}(\mathbf{L}^1, \mathbf{M})$) there exists a measure $\mu \in \mathbf{M}$ such that

$$Uf = \mu * f \tag{16.3.5}$$

for $f \in \mathbf{C}$ (or for $f \in \mathbf{L}^1$); and conversely. In particular, $(\mathbf{C}, \mathbf{C}) = (\mathbf{L}^1, \mathbf{M}) = \mathscr{F}\mathbf{M}$.

Proof. If U is defined by (16.3.5), then 12.7.2 (or 12.7.3) and 12.6.5 show that $U \in \mathfrak{m}(\mathbf{C}, \mathbf{C})$ (or $U \in \mathfrak{m}(\mathbf{L}^1, \mathbf{M})$). Conversely, if U belongs to either of these two categories, 16.3.1(4) entails that $Uf = A * f$ for some $A \in \mathbf{D}$ and all $f \in \mathbf{C}$ (or $f \in \mathbf{L}^1$); and then 12.8.4 shows that A must be a measure.

In the case of $\mathfrak{m}(\mathbf{L}^1, \mathbf{M})$ the conclusion $A \in \mathbf{M}$ may otherwise be reached by noting that the A_i appearing in the proof of 16.3.1(1) form a norm-bounded sequence of measures and so, by 12.3.9, a subsequence $(A_{i_k})_{k=1}^{\infty}$ converges weakly in $\mathbf{M}$ to a measure μ. Since it is known that $A_i * f \to A * f$ for each $f \in \mathbf{C}^{\infty}$, A and μ must coincide.

The final assertion follows from what is already established, if appeal be made to 16.2.1 and 12.6.5.

Remarks. (1) Other proofs are possible; for the case of $\mathfrak{m}(\mathbf{C}, \mathbf{C})$, for example, one might argue along the lines adopted in 16.3.5.

(2) It can be shown (see Exercise 16.20) that $\mathfrak{m}(\mathbf{C}, \mathbf{L}^{\infty}) = \mathfrak{m}(\mathbf{C}, \mathbf{C})$.

16.3.3. L^1-Multipliers. If $U \in \mathfrak{m}(\mathbf{L}^1, \mathbf{L}^1)$ there exists a measure $\mu \in \mathbf{M}$ such that (16.3.5) holds for $f \in \mathbf{L}^1$; and conversely. In particular,

$$(\mathbf{L}^1, \mathbf{L}^1) = \mathscr{F}\mathbf{M}.$$

Proof. Once again the converse assertion ensues from 12.7.3 and 12.6.5. The direct assertion stems from 16.3.2, since $\mathbf{L}^1$ is a subspace of $\mathbf{M}$ (see 12.3.8). As before, the statement about multiplier functions is contained in what precedes, using 16.2.1.

Remarks. It can also be shown that $(\mathbf{M}, \mathbf{M}) = (\mathbf{C}, \mathbf{L}^{\infty}) = (\mathbf{L}^{\infty}, \mathbf{L}^{\infty}) = \mathscr{F}\mathbf{M}$; see Exercises 16.7 and 16.8.

16.3.4. Multipliers from L^1 to L^p ($p > 1$). If $U \in \mathfrak{m}(\mathbf{L}^1, \mathbf{L}^p)$, where $1 < p \leqslant \infty$, then there exists a function $k \in \mathbf{L}^p$ such that

$$Uf = k * f \tag{16.3.6}$$

for each $f \in \mathbf{L}^1$; and conversely. In particular, $(\mathbf{L}^1, \mathbf{L}^p) = \mathscr{F}\mathbf{L}^p$ for $1 < p \leqslant \infty$.

Proof. The converse statement is a consequence of 3.1.2, 3.1.6, and the associativity of convolution.

For the direct assertion, one begins by observing that appeal to 16.3.1(4) shows that there exists a distribution A such that $Uf = A * f$ for $f \in \mathbf{L}^1$. The continuity of U then entails the existence of a number $b \geqslant 0$ such that

$$\|A * f\|_p \leqslant b \cdot \|f\|_1 \tag{16.3.7}$$

for $f \in \mathbf{L}^1$. Suppose now that f is allowed to vary along an approximate identity (k_i), and put $A_i = A * k_i \in \mathbf{L}^p$, as in the proof of 16.3.1(1). Then (16.3.7) shows that the numbers $\|A_i\|_p$ are bounded with respect to i, and 12.3.10 affirms that some subsequence of (A_i) converges weakly in $\mathbf{L}^p$ to a limit, say k. On the other hand, since (k_i) is an approximate identity, the sequence (A_i) converges in $\mathbf{D}$ to A, as follows from 12.3.2(3) and 12.6.6. It follows that $A = k$, so that (16.3.6) holds for $f \in \mathbf{L}^1$.

Finally, the last assertion stems from what is already established via 16.2.1 and 12.6.5.

Remarks. Without altering the essence of the preceding proof, explicit reference to 16.3.1 could be avoided by simply mimicking the proof of the latter in the present context: in other words, explicit reference to A is avoided, the function k being obtained directly as a weak limiting point in $\mathbf{L}^p$ of the sequence $(Uk_i) = (A_i)$ by appealing to 12.3.10 as before. The reader is urged to construct such a proof in detail.

The next two results introduce a different technique, which is frequently successful when discussing $\mathfrak{m}$-operators of type $(\mathbf{F}, \mathbf{C})$.

16.3.5. **Multipliers from $\mathbf{L}^p$ to $\mathbf{C}$.** If $U \in \mathfrak{m}(\mathbf{L}^p, \mathbf{C})$, where $1 \leqslant p \leqslant \infty$, then there exists a function $h \in \mathbf{L}^{p'}$ such that

$$Uf = h * f$$

holds for $f \in \mathbf{L}^p$; and conversely. In particular, $(\mathbf{L}^p, \mathbf{C}) = \mathscr{F}\mathbf{L}^{p'}$ for $1 \leqslant p \leqslant \infty$.

Proof. The converse assertion follows from 3.1.2, 3.1.4, and the associativity of convolution. To prove the direct assertion, we suppose first that $p < \infty$ and consider the linear functional $f \to Uf(0)$ defined on $\mathbf{L}^p$. The assumed continuity of U ensures that this functional is continuous. Hence (I, C.1) there exists a function $h \in \mathbf{L}^{p'}$ such that

$$Uf(0) = h * f(0).$$

If we apply this formula with $T_{-x}f$ in place of f, using 3.1.2 and the fact that U commutes with translations, we find that

$$\begin{aligned} Uf(x) &= T_{-x}Uf(0) = UT_{-x}f(0) = h * T_{-x}f(0) \\ &= T_{-x}(h * f)(0) = h * f(x), \end{aligned}$$

which is the desired result.

The preceding proof breaks down when $p = \infty$ (where?) and we proceed as follows. Take an approximate identity $(f_i)_{i=1}^{\infty}$ composed of trigonometric polynomials, and let h_i be the trigonometric polynomial Uf_i. If $f \in \mathbf{L}^\infty$, Uf is continuous (by hypothesis), and so $\lim_i f_i * Uf(0)$ exists finitely, being in fact equal to $Uf(0)$. By (16.2.2) this signifies that $\lim_i h_i * f(0)$ exists finitely for each $f \in \mathbf{L}^\infty$. Consequently (see I, C.2), there exists a function $h \in \mathbf{L}^1$ such that

$$\lim_i f_i * Uf(0) = h * f(0)$$

for each $f \in \mathbf{L}^\infty$, that is,

$$Uf(0) = h * f(0)$$

for each $f \in \mathbf{L}^\infty$. From this point on, the argument proceeds exactly as before.

The final statement is deduced in the by-now-customary fashion, using 16.2.1 and 12.6.5.

Remarks. (1) In a sense, 16.3.5 contains a sort of analogue of the converse of Hölder's inequality (given in Exercise 3.6); see Exercise 16.9.

(2) By 16.4.1, 16.3.3, and 16.3.4, it ensues that $\mathfrak{m}(\mathbf{L}^p, \mathbf{L}^\infty) = \mathfrak{m}(\mathbf{L}^1, \mathbf{L}^{p'})$ coincides with the set of operators by convolution with elements of $\mathbf{L}^{p'}$ if $1 \leqslant p < \infty$, or with elements of $\mathbf{M}$ if $p = \infty$. For $1 \leqslant p < \infty$, this conclusion may be deduced from 16.3.5 in the manner indicated in Exercise 16.19.

16.3.6. **Multipliers from M to C.** If $U \in \mathfrak{m}(\mathbf{M}, \mathbf{C})$, then there exists a function $h \in \mathbf{C}$ such that

$$U\mu = h * \mu \tag{16.3.8}$$

for $\mu \in \mathbf{M}$; and conversely. In particular, $(\mathbf{M}, \mathbf{C}) = \mathscr{F}\mathbf{C}$.

Proof. We give a proof of the first assertion. The rest follows in the usual fashion and will be left for the reader to verify.

If $U \in \mathfrak{m}(\mathbf{M}, \mathbf{C})$ then, since $\mathbf{L}^1$ is a subspace of $\mathbf{M}$, 16.3.5 shows that there exists a function $h \in \mathbf{L}^\infty$ such that

$$Uf = h * f \tag{16.3.9}$$

for $f \in \mathbf{L}^1$. Now, if $\mu \in \mathbf{M}$, then $f = t * \mu \in \mathbf{L}^1$ for each trigonometric polynomial t. Hence, using (16.3.9), (16.2.2) and the commutativity of convolution, we obtain

$$h * (t * \mu) = U(t * \mu) = t * U\mu,$$

and thence $U\mu = h * \mu$. Finally, taking $\mu = \varepsilon$ (the Dirac measure at the origin; see 12.2.3), it appears at once that $h = U\varepsilon \in \mathbf{C}$.

16.3.7. **Bounded and Uniform Convergence Multipliers.** In Section 10.3 it has been seen that

$$\sup_N \|s_N f\|_\infty = \infty$$

for suitable continuous functions f. With this in mind, we pose the following question. Which complex-valued functions ϕ on Z have the property that, for each continuous function f, the series

$$\sum_{n \in Z} \phi(n)\hat{f}(n)e^{inx} \tag{16.3.10}$$

has uniformly bounded partial sums? An answer to this multiplier problem will be given in 16.3.8, for which we proceed to lay some foundations.

By taking

$$f(x) = \sum_{n \in Z} (1 + n^2)^{-1} e^{inx},$$

it becomes evident that any function ϕ possessing the stated property is tempered, so that (16.3.10) is certainly the Fourier series of some distribution g (see 12.5.3(2)). Moreover, if ϕ has the stated property, it is the case that

$$\sup_N \|s_N g\|_\infty < \infty. \tag{16.3.11}$$

According to Exercise 12.22, therefore, $g \in \mathbf{L}^\infty$.

At this stage let us introduce the set $\mathbf{L}_b^\infty$ of functions $g \in \mathbf{L}^\infty$ for which (16.3.11) holds. It follows from what precedes that our problem is precisely that of determining which functions ϕ belong to $(\mathbf{C}, \mathbf{L}_b^\infty)$. As usual, we shall approach this problem in terms of the associated multiplier operator U_ϕ from $\mathbf{C}$ into $\mathbf{L}_b^\infty$.

We shall first of all wish to be sure that $U_\phi \in \mathfrak{m}(\mathbf{C}, \mathbf{L}_b^\infty)$. In seeking confirmation of this by appeal to 16.2.1 and 16.2.7, it is sufficient to check that $\mathbf{L}_b^\infty$ can be made into a Banach space in such a way as to satisfy the standing hypotheses laid out in 16.1.1. This can be achieved by taking as the norm on $\mathbf{L}_b^\infty$ the function

$$\|g\| = \sup_N \|s_N g\|_\infty, \tag{16.3.12}$$

and recalling from Exercise 12.22 the inequality

$$\|g\|_\infty \leqslant \|g\|. \tag{16.3.13}$$

Armed with this, the verification is simple (see Exercise 16.10). Recall that $s_N g = D_N * g$ for every distribution g.

Now we are ready to state and prove the main result.

16.3.8. (1) The operators $U \in \mathfrak{m}(\mathbf{C}, \mathbf{L}_b^\infty)$ are precisely those of the form

$$Uf = \mu * f, \tag{16.3.14}$$

where $\mu \in \mathbf{M}$ is such that

$$m \equiv \sup_N \|D_N * \mu\|_1 < \infty. \tag{16.3.15}$$

(2) A function ϕ has the property that, for each continuous function f, the series (16.3.10) has uniformly bounded partial sums, if and only if $\phi = \hat{\mu}$ for some $\mu \in \mathbf{M}$ satisfying (16.3.15), and in that case

(3) the series (16.3.10) is uniformly convergent for each continuous function f, and

(4) the series (16.3.10) is convergent in norm in $\mathbf{L}^p$ whenever $f \in \mathbf{L}^p$ and $1 \leqslant p < \infty$.

Proof. (1) If (16.3.14) and (16.3.15) hold, we have for $f \in \mathbf{C}$:

$$\begin{aligned}\sup_N \|s_N Uf\|_\infty &= \sup_N \|D_N * \mu * f\|_\infty \\ &\leqslant \sup_N \|D_N * \mu\|_1 \cdot \|f\|_\infty \\ &\leqslant m \cdot \|f\|_\infty < \infty,\end{aligned}$$

as follows by use of 3.1.6. This shows that U maps $\mathbf{C}$ continuously into $\mathbf{L}_b^\infty$. The remaining properties of U called for in 16.2.7 are almost evidently attained, so that $U \in \mathfrak{m}(\mathbf{C}, \mathbf{L}_b^\infty)$.

Suppose conversely that $U \in \mathfrak{m}(\mathbf{C}, \mathbf{L}_b^\infty)$. Then 16.3.1(4) is applicable and shows that there exists a distribution A such that

$$Uf = A * f \tag{16.3.16}$$

holds for $f \in \mathbf{C}$. Thanks to the continuity of U, there exists a number m for which, on account of (16.3.16), we have

$$\sup_N \|s_N(A * f)\|_\infty \leqslant m \cdot \|f\|_\infty,$$

that is,

$$\sup_N \|D_N * A * f\|_\infty \leqslant m \cdot \|f\|_\infty$$

for $f \in \mathbf{C}$. Noticing that $D_N * A$ is a trigonometric polynomial, Exercise 3.6 yields

$$\sup_N \|D_N * A\|_1 \leqslant m.$$

From this and 12.3.9 it is easily inferred that A is in fact a measure $\mu \in \mathbf{M}$. This completes the proof of (1).

(2) This follows in the familiar manner from (1) applied to $U = U_\Phi$, provided the remarks in 16.3.7 are borne in mind.

(3) Let $f \in \mathbf{C}$ and $\varepsilon > 0$ be given. Choose a trigonometric polynomial t satisfying

$$\|f - t\|_\infty \leqslant (2m)^{-1}\varepsilon;$$

this is possible by 2.4.4. For all N we have by 3.1.6 and (16.3.15)

$$\|s_N(\mu * f) - s_N(\mu * t)\|_\infty \leqslant m \cdot \|f - t\|_\infty \leqslant \tfrac{1}{2}\varepsilon. \tag{16.3.17}$$

Since $\mu * t$ is a trigonometric polynomial, $s_N(\mu * t) = \mu * t$ for $N > N_0$, where N_0 depends upon t. Hence, by (16.3.17), we have

$$\|s_N(\mu * f) - s_{N'}(\mu * f)\|_\infty \leqslant \varepsilon$$

provided N and N' exceed N_0, which shows that the series (16.3.10) is uniformly convergent.

(4) The proof follows the same lines as that of (3); note that (by 3.1.6 again)

$$\|s_N(\mu * f) - s_N(\mu * t)\|_p \leqslant m \cdot \|f - t\|_p,$$

and that (by 2.4.4 again) t may be chosen to satisfy $\|f - t\|_p \leqslant (2m)^{-1}\varepsilon$. (If $1 < p < \infty$, the assertion is also a consequence of 12.7.3 and 12.10.1.)

Remarks. (1) The condition (16.3.15) is not generally satisfied when $\mu \in \mathbf{L}^1$ (see Exercise 10.2); it is satisfied, however, if $\mu \in \mathbf{L}^p$ for some $p > 1$, or if $\mu \in \mathbf{L}^1$ and $\mu \cdot \log^+ |\mu| \in \mathbf{L}^1$ (see 12.10.1 and 12.10.2).

(2) The analogous problem of determining $(\mathbf{F}, \mathbf{C}_u)$, where $\mathbf{C}_u$ is the space of continuous functions with uniformly convergent Fourier series, has been solved in a number of cases by diverse authors, including Tomíc, Karamata, Katayama, and Goes; see *Mathematical Reviews* 20 #7184 and 21 #7392. The conditions obtained sometimes vary from one author to the next, and it may well be a nontrivial task to prove that the apparently different conditions are indeed equivalent. For example, Karamata shows that in case $\phi(n) = \phi(-n)$, the conditions on

$$\mu = 2 \sum_{n=1}^{\infty} \phi(n) \cos nx$$

(convergent in $\mathbf{D}$) required in 16.3.8 can be expressed in the form

$$\sum_{n=1}^{\infty} n^{-1} \sum_{\nu \geqslant n} |\phi(\nu) - \phi(\nu + 1)| < \infty .$$

In the form given here, 16.3.8 was proved in a somewhat different way by Karamata [1] for the case in which $\phi(n) = \phi(-n)$. He also remarked that in this case a sufficient condition is that

$$\phi(n) = O\left(\frac{1}{\log n}\right) \quad (n \to \infty), \qquad \sum_{n=0}^{\infty} (n + 1)|\Delta^2\phi(n)| < \infty ;$$

that this implies (16.3.15) follows from partial summation [compare formula (7.3.5)].

(3) For some other results bearing directly or otherwise on multipliers, see Goes [1], [2].

16.3.9. **Lipschitz Multipliers.** The discussion and arguments appearing in 16.3.7 and 16.3.8 can be modified quite easily so as to characterize those functions ϕ such that, for each continuous function f, the series (16.3.10) is the Fourier series of a continuous function g satisfying a Lipschitz condition

$$\|T_a g - g\|_\infty = O(\omega(a)), \qquad (16.3.18)$$

where ω is a given nonnegative function such that $\omega(a)/a$ is bounded away from zero for small $a \neq 0$, and such that $\omega(a) \to 0$ as $a \to 0$. The O-constant in (16.3.18) may, of course, depend upon g.

The set $\mathbf{\Lambda}_\omega$ of continuous functions g satisfying a Lipschitz condition (16.3.18) may be formed into a Banach space by introducing the norm

$$\|g\|_\omega = \|g\|_\infty + \sup_{a \neq 0} \omega(a)^{-1} \|T_a g - g\|_\infty , \qquad (16.3.19)$$

and $\mathbf{\Lambda}_\omega$ then satisfies the standing hypotheses laid down in 16.1.1.

The problem posed at the outset of this subsection is that of determining $(\mathbf{C}, \mathbf{\Lambda}_\omega)$ and may be approached by studying the operators U of type $\mathfrak{m}(\mathbf{C}, \mathbf{\Lambda}_\omega)$. The conclusion proves to be that these operators are precisely those of the form

$$Uf = k * f, \tag{16.3.20}$$

wherein $k \in \mathbf{L}^1$ is such that

$$\sup_{a \neq 0} \omega(a)^{-1} \|T_a k - k\|_1 < \infty; \tag{16.3.21}$$

the functions ϕ of type $(\mathbf{C}, \mathbf{\Lambda}_\omega)$ are accordingly precisely those of the form $\phi = \hat{k}$, where k is as just described.

In Exercise 16.11, the reader is invited to construct a detailed proof of these statements.

Remarks. Zygmund [1] determines, among other things, the functions of type $(\mathbf{\Lambda}_\omega, \mathbf{\Lambda}_\omega)$ for $\omega(\delta) = \delta^\alpha$, where $0 < \alpha \leqslant 1$. He shows that a necessary and sufficient condition for ϕ to be of this type is that

$$\Phi(x) = \sum_{n \neq 0} (in)^{-1} \phi(n) e^{inx}$$

(convergent in $\mathbf{D}$) be a function of class Λ^1_*, that is (see [Z₁], p. 45), that

$$\sup_{a > 0} a^{-1} \|T_a \Phi + T_{-a}\Phi - 2\Phi\|_1 < \infty.$$

The function Φ is, apart from an insignificant term, the indefinite integral of the function k which would appear in the corresponding representation of the form (16.3.20) for $U = U_\phi$, and it is interesting to note that the second difference $T_a\Phi + T_{-a}\Phi - 2\Phi$ now takes the place of the first difference $T_a k - k$.

See also MR **44** # 7208.

16.3.10. **Positive $\mathfrak{m}$-operators.** An operator $U \in \mathfrak{m}(\mathbf{F}, \mathbf{G})$ will be said to be *positive*, if $Uf \geqslant 0$ whenever $f \in \mathbf{F}$ and $f \geqslant 0$. (In this context a relation $h \geqslant 0$ is to be understood in the distributional sense, as described in Exercise 12.7: if h is a continuous function, the relation signifies that $h(x) \geqslant 0$ for all x; if h is a general integrable function, it signifies that $h(x) \geqslant 0$ for almost all x.)

From 16.3.1(4) and Exercise 12.7 it follows that any positive $U \in \mathfrak{m}(\mathbf{F}, \mathbf{G})$ is representable in the form

$$Uf = \mu * f \qquad (f \in \mathbf{F}) \tag{16.3.22}$$

where μ is a positive measure.

If $\mathbf{F} = \mathbf{G} = \mathbf{C}$, $\mathbf{L}^p$, or $\mathbf{M}$, 3.1.6 and 12.7.3 show that any $\mu \in \mathbf{M}$ (positive or not) yields, via (16.3.22), an operator $U \in \mathfrak{m}(\mathbf{F}, \mathbf{F})$; evidently, this U is positive if and only if μ is positive.

On the other hand, if $p < q$, not every measure μ yields, via (16.3.22), an element U of $\mathfrak{m}(p, q) = \mathfrak{m}(\mathbf{L}^p, \mathbf{L}^q)$. Effective necessary and sufficient conditions on μ in order that U shall belong to $\mathfrak{m}(p, q)$ are apparently unknown (even assuming that μ is positive); but see the partial results in 16.4.4 and 16.4.6.

It is perhaps worth remarking that (16.3.22) can be established in most cases of interest without any appeal to distributional notions. By thus working within the domain of functions and Radon measures, the arguments become at once extendible to more general groups (see Brainerd and Edwards [1], Part I, Section 3).

In this connection we may mention in passing the following facts and implied problems. Suppose that G is a (Hausdorff locally compact) group, and that k is a measurable function on G such that $k * f$ exists in some sense for each $f \in \mathbf{C}_c(G)$ (the space of continuous functions on G with compact supports); and suppose further that

$$\|k * f\|_q \leqslant \text{const} \, \|f\|_q$$

for a given $q \in [1, \infty]$ and each $f \in \mathbf{C}_c(G)$. What can be deduced about k?

If G is the circle group, the case of the Hilbert distribution (see Sections 12.8 and 12.9) shows that $k * f$ may exist as a Cauchy principal value and satisfy the stated conditions for each $q \in (1, \infty)$, even though k is nonintegrable. Much the same is true of the Hilbert transform on R or R^n.

Let us consider further the case in which G is assumed to be noncompact and k is given to be integrable over each compact subset of G. This ensures that $k * f$ exists pointwise as an absolutely convergent integral whenever $f \in \mathbf{C}_c(G)$. The stated inequality is easily seen to imply that $k \in \mathbf{L}^1(G)$, if q is 1 or ∞; but this implication is generally not valid if $q \in (1, \infty)$. The additional hypothesis that k be nonnegative permits the deduction that $k \in \mathbf{L}^1(G)$ for certain classes of groups G; see Brainerd and Edwards [1], Part I, Section 3. Rather surprisingly, not even this much is true for general G and $q \in (1, \infty)$. In fact, Theorem 9 of Kunze and Stein [1] shows that, if G_0 is the group of real 2×2 unimodular matrices, then

$$\|k * f\|_2 \leqslant \text{const} \, \|f\|_2$$

whenever $k \in \mathbf{L}^p(G_0)$ for some p satisfying $1 \leqslant p < 2$ and $f \in \mathbf{L}^2(G_0)$; if one takes $1 < p < 2$, the noncompactness of G_0 ensures that one can choose nonnegative functions k belonging to $\mathbf{L}^p(G_0)$ but not to $\mathbf{L}^1(G_0)$.

For further developments, see Gilbert [3]; Cowling [1], [2]; MR **35** # 3008; **38** # # 269, 5997; **42** # 6522; **51** # 13594.

16.3.11. $\mathfrak{m}$-operators and Translations. Reverting for a moment to generalities, we observe that from 16.3.1 and use of the Hahn-Banach theorem (much as in 11.2.2; compare also part (2) of the proof of 11.1.2) it follows that any $U \in \mathfrak{m}(\mathbf{F}, \mathbf{F})$ leaves stable each closed (translation-) invariant subspace of $\mathbf{F}$; compare Subsections 3.1.8 to 3.1.10. [When $U = U_\phi$ and $\mathbf{F}$ is $\mathbf{C}$

or $\mathbf{L}^p$ ($1 \leqslant p < \infty$), this may also be deduced from the specification of the closed invariant subspaces of $\mathbf{F}$ obtained in 11.1.1 and 11.2.1.] An equivalent formulation of this property of U is as follows (compare 3.1.9): for each $f \in \mathbf{F}$, Uf is the limit in $\mathbf{F}$ of linear combinations of translates of f.

It is natural to ask whether, conversely, any endomorphism U of $\mathbf{F}$ (U may or may not be assumed a priori to be continuous) that leaves stable each closed invariant subspace of $\mathbf{F}$ is necessarily a member of $\mathfrak{m}(\mathbf{F}, \mathbf{F})$. An affirmative answer is given in some (but by no means all) cases in Edwards [11], and further results have since been given by Johnson [4].

On combining the results obtained in Section 1 of Edwards [11] with 16.3.1 one obtains a type of abstract characterization of convolution operators on $\mathbf{C}$ and on $\mathbf{L}^p$ ($1 \leqslant p < \infty$), namely: any endomorphism U of $\mathbf{C}$ or of $\mathbf{L}^p$ (continuity is not assumed) that leaves stable each closed invariant subspace of $\mathbf{C}$ or of $\mathbf{L}^p$ is of the form $Uf = A * f$ for some $A \in \mathbf{D}$; U is therefore necessarily continuous and, in the case of $\mathbf{C}$, A is necessarily a measure (by 16.3.2). Some similar results appear in Edwards [10].

16.3.12. **Convolution as a Bilinear Operator.** We turn aside briefly in order to justify some remarks made in Subsection 3.1.10.

Let B denote a continuous bilinear mapping from $\mathbf{C}^\infty \times \mathbf{C}^\infty$ into $\mathbf{D}$ with the property that

$$B(T_a f, g) = T_a B(f, g) = B(f, T_a g) \tag{16.3.23}$$

for $f, g \in \mathbf{C}^\infty$ and $a \in R/2\pi Z$. Observe that continuity follows from bilinearity whenever B is positive in the sense that $B(f, g) \geqslant 0$ for all nonnegative functions f and g in $\mathbf{C}^\infty$; compare the hints to Exercise 12.9.

It will be shown that

(1) there exists a distribution A such that

$$B(f, g) = A * f * g \tag{16.3.24}$$

for $f, g \in \mathbf{C}^\infty$;

(2) if B is positive, then A is a positive measure;

(3) if B is positive, and if furthermore

$$\operatorname{supp} B(f, g) \subset \operatorname{supp} f + \operatorname{supp} g, \tag{16.3.25}$$

whenever $f, g \in \mathbf{C}^\infty$, then

$$B(f, g) = \text{const}\, f * g \tag{16.3.26}$$

for $f, g \in \mathbf{C}^\infty$.

Proof. Fix $g \in \mathbf{C}^\infty$ and consider the mapping $U_g: f \to B(f, g)$ from $\mathbf{C}^\infty$ into $\mathbf{D}$. It is evident that U_g is linear, continuous, and [by (16.3.23)] commutes with translations. By 16.3.1, therefore, there exists $A_g \in \mathbf{D}$, which is evidently uniquely determined by g, such that

$$B(f, g) = A_g * f \qquad (f, g \in \mathbf{C}^\infty). \tag{16.3.27}$$

It is also evident that, if B is positive, then A_g is positive and hence (Exercise 12.7) is a positive measure.

The map $U: g \to A_g$ is clearly linear from $\mathbf{C}^\infty$ into $\mathbf{D}$, and, by (16.3.23) once more, U commutes with translations. A little thought will show that U is also continuous from $\mathbf{C}^\infty$ into $\mathbf{D}$ (remember that $\mathbf{C}^\infty * \mathbf{C}^\infty$ fills out $\mathbf{C}^\infty$). A second appeal to 16.3.1 shows that there exists $A \in \mathbf{D}$ such that

$$A_g = A * g \qquad (g \in \mathbf{C}^\infty). \tag{16.3.28}$$

Once again, if B is positive, A must be a positive measure (Exercise 12.7 once more). In any case, (16.3.24) follows on combining (16.3.27) and (16.3.28).

Finally, on assuming that (16.3.25) obtains, it follows that $\operatorname{supp} A \subset \{\dot{0}\}$ (take $f = g = v_n$, as in Exercise 12.5, and consider what happens as $n \to \infty$). Then, by Exercise 12.33, positivity of B entails that $A = \text{const } \varepsilon$, and (16.3.26) follows from (16.3.24).

Remark. If the bilinear operator B maps $\mathbf{C} \times \mathbf{C}$ continuously into $\mathbf{P}$ and satisfies (16.3.23) and (16.3.25) for $f, g \in \mathbf{C}^\infty$, then (16.3.26) holds for $f, g \in \mathbf{C}$. (In this case, there is no need to assume positivity.)

In fact, (16.3.24) holds for $f, g \in \mathbf{C}^\infty$. The continuity of B on $\mathbf{C} \times \mathbf{C}$ then shows that A must be such that $A * f * g$ belongs to $\mathbf{P}$ and coincides with $B(f, g)$ whenever $f, g \in \mathbf{C}$; in particular, continuity shows that A itself belongs to $\mathbf{P}$. In addition, (16.3.25) shows that $\operatorname{supp} A \subset \{\dot{0}\}$, and the stated conclusion follows once again from Exercise 12.33.

16.4 Multipliers of Type $(\mathbf{L}^p, \mathbf{L}^q)$

This and the two following sections will be devoted to a brief study of the problem of multipliers of type $(\mathbf{L}^p, \mathbf{L}^q)$ and of the associated multiplier operators. For pairs (p, q) of the special forms $(2, 2)$ and $(1, q)$, solutions have already been obtained; see 16.1.2, 16.3.3, and 16.3.4. As will appear in 16.4.1, the solution for a pair (q', ∞) can be derived from that for $(1, q)$. Apart from these special cases, there is as yet no complete and effective solution, and we are able here to give only a few conditions, some sufficient and others necessary, in order that a given complex-valued function ϕ on Z shall be a multiplier of type $(\mathbf{L}^p, \mathbf{L}^q)$.

Except where other ranges of the parameters p and q are specified, it is hereafter supposed that $1 \leqslant p, q \leqslant \infty$; the conjugate exponents p' and q' then satisfy the same inequalities.

For brevity we shall write (p, q) in place of $(\mathbf{L}^p, \mathbf{L}^q)$ and $\mathfrak{m}(p, q)$ in place of $\mathfrak{m}(\mathbf{L}^p, \mathbf{L}^q)$.

From 16.3.1 it appears that $\mathfrak{m}(p, q)$ is in a one-to-one correspondence with a certain set of distributions in such a way that $U \in \mathfrak{m}(p, q)$ is associated with that distribution A for which

$$Uf = A * f \tag{16.4.1}$$

for $f \in \mathbf{L}^p$; the distribution A must be such that

$$\|A * f\|_q \leqslant \text{const } \|f\|_p \tag{16.4.2}$$

for $f \in \mathbf{L}^p$. On the other hand, if $1 \leqslant p < \infty$, a simple continuity argument shows that if (16.4.2) holds for $f \in \mathbf{C}^\infty$ then it and (16.4.1) hold for each $f \in \mathbf{L}^p$, $A * f$ being then of necessity (the distribution generated by) an element of $\mathbf{L}^q$ whenever $f \in \mathbf{L}^p$. We shall frequently discuss multiplier operators U in terms of the associated distributions A.

The smallest permissible value of the constant appearing on the right-hand side of (16.4.2) will be denoted by $\|A\|_{p,q}$ or by $\|U\|_{p,q}$. The latter is just the (p, q)-norm of U, as defined in 13.2.1. On occasions we shall also write $\|\phi\|_{p,q} = \|U_\phi\|_{p,q}$ for $\phi \in (p, q)$ and speak loosely of U_ϕ as a multiplier.

In view of the alias just explained, it is natural and convenient to use the symbol $\mathfrak{m}(p, q)$ to denote also the set of distributions A which satisfy an inequality of the form (16.4.2). With the usual conventions, together with those explained in 13.1.2 and 13.2.1, one may say that $\mathfrak{m}(p, q)$ comprises those distributions A for which $\|A\|_{p,q} < \infty$. In view of 16.2.7, one may accordingly write $(p, q) = \mathscr{F}\,\mathfrak{m}(p, q)$, the set of sequences (functions on Z) of the form $\hat{A}$ obtained when A ranges over $\mathfrak{m}(p, q)$.

From 12.7.3(1), 12.11.1, and the arguments used in 16.2.2 it follows that

$$\mathbf{M} \subset \mathfrak{m}(p, p) \subset \mathbf{P}. \tag{16.4.3}$$

Further, and trivially,

$$\mathfrak{m}(p_1, q_1) \supset \mathfrak{m}(p_2, q_2) \qquad \text{whenever } p_1 \geqslant p_2,\ q_1 \leqslant q_2. \tag{16.4.4}$$

As we shall see forthwith, the dependence of $\mathfrak{m}(p, q)$ and (p, q) on the parameters p and q can be further clarified by the use of simple general arguments from duality theory in conjunction with the Riesz-Thorin convexity theorem 13.4.1. [A general discussion of duality theory will be found in Chapter 8 of [E]; see also 16.7.5(2).]

16.4.1. **The relation $\mathfrak{m}(p, q) = \mathfrak{m}(q', p')$.** Since $(p')' = p$ and $(q')' = q$, the stated equality will follow as soon as it is established that

$$\mathfrak{m}(p, q) \subset \mathfrak{m}(q', p').$$

The proof will proceed in three stages.

(1) Suppose first that $p \neq \infty$ and that $A \in \mathfrak{m}(p, q)$, so that (16.4.2) holds (with const $= \|A\|_{p,q}$) for $f \in \mathbf{L}^p$. Hölder's inequality then shows that

$$\begin{aligned} |A * f * g(0)| &= \left|\frac{1}{2\pi}\int (A * f)\cdot \check{g}\, dx\right| \\ &\leqslant \|A\|_{p,q}\|f\|_p\|g\|_{q'} \end{aligned}$$

for any $g \in \mathbf{L}^{q'}$. So, by the Hahn-Banach theorem (see I, B.5.1) and the results in I, C.1, to any such g corresponds a unique $g' \in \mathbf{L}^{p'}$ such that

$$A * f * g(0) = f * g'(0) \tag{16.4.5}$$

for all $f \in \mathbf{L}^p$ and

$$\|g'\|_{p'} \leqslant \|A\|_{p,q}\|g\|_{q'}.$$

From (16.4.5) it follows that $g' = A * g$, and the last-written inequality thus signifies that $A \in \mathfrak{m}(q', p')$ and that $\|A\|_{q',p'} \leqslant \|A\|_{p,q}$. This completes the proof in case $p \neq \infty$.

(2) Next consider the case in which $p = \infty$ and $q \neq 1$. The same type of argument as was used in (1), but making appeal to 12.2.9 in place of I, C.1, shows that any $A \in \mathfrak{m}(\infty, q)$ belongs to $\mathfrak{m}(\mathbf{L}^{q'}, \mathbf{M})$. However, since $q' \neq \infty$, Exercise 12.23 can be applied to show that A maps $\mathbf{L}^{q'}$ into $\mathbf{L}^1$ (and not merely into $\mathbf{M}$). Thus $\mathfrak{m}(\infty, q) \subset \mathfrak{m}(q', 1)$, and again $\|A\|_{q',p'} \leqslant \|A\|_{p,q}$.

(3) Finally, if $p = \infty$ and $q = 1$, there is nothing to prove, because $q' = p = \infty$ and $p' = q = 1$. Thus all cases are covered and the proof is complete.

It is worth noting that the inequality $\|A\|_{q',p'} \leqslant \|A\|_{p,q}$, known to be true in all cases, yields also

$$\|A\|_{p,q} \equiv \|A\|_{(p')',(q')'} \leqslant \|A\|_{q',p'},$$

and therefore

$$\|A\|_{q',p'} = \|A\|_{p,q}. \tag{16.4.6}$$

The reader will also observe that, as a corollary, one obtains the relation $(p, q) = (q', p')$ and the equality resulting from (16.4.6) after replacement of A by a function $\phi \in (p, q)$.

16.4.2. A Convexity Theorem. Suppose that $A \in \mathfrak{m}(p_j, q_j)$ for $j = 0, 1$ and that

$$\frac{1}{p} = \frac{1-t}{p_0} + \frac{t}{p_1}, \qquad \frac{1}{q} = \frac{1-t}{q_0} + \frac{t}{q_1}$$

for some t satisfying $0 \leqslant t \leqslant 1$. Then $A \in \mathfrak{m}(p, q)$ and

$$\|A\|_{p,q} \leqslant \|A\|_{p_0,q_0}^{1-t}\|A\|_{p_1,q_1}^{t}.$$

(The reader is left to formulate the analogous statement concerning multiplier functions ϕ.)

Proof. The result is an immediate consequence of applying 13.4.1 to the operator T with domain $\mathbf{L}^\infty$ defined by $Tf = A * f$. To say that $A \in \mathfrak{m}(p, q)$ means exactly that T is of type (p, q) in the sense in which the term "type" is defined in 13.2.1. The details may be left to the reader's care; see Exercise 16.12.

Remark. We see that, in particular, if ϕ is a given complex-valued function on Z, the set of exponents p for which $\phi \in (p, p)$ is always either an open interval (a, a') or a closed interval $[a, a']$, where $1 \leqslant a \leqslant 2$.

16.4.3. **Some Proper Inclusion Relations.** We revert to the relations (16.4.3) and show that, except in those cases in which we know already [see 16.3.3 and Remark (2) following 16.3.5] that equality obtains, the inclusions are strict. Indeed, we shall prove somewhat more than this.

(1) If $1 < p \leqslant \infty$ and $1 \leqslant q < \infty$, then $\mathfrak{m}(p, q)$ contains distributions that are not measures; in particular, if $1 < p < \infty$,

$$\mathbf{M} \subsetneqq \mathfrak{m}(p, p), \qquad \mathscr{F}\mathbf{M} \subsetneqq (p, p) \tag{16.4.7}$$

Proof. Take any infinite Sidon subset E of Z (see 15.2.4 or 15.2.5) and choose any bounded complex-valued function ϕ on Z that vanishes on $Z \backslash E$ and which does not belong to $\ell^2(Z)$. Since $p > 1$, 15.3.3 shows that

$$\sum_{n \in E} |\hat{f}(n)|^2 < \infty$$

for each $f \in \mathbf{L}^p$. So, since ϕ is bounded and vanishes on $Z \backslash E$, 8.3.1 entails that

$$g = \sum_{n \in Z} \phi(n)\hat{f}(n)e_n \in \mathbf{L}^2.$$

Evidently, $\hat{g}$ vanishes on $Z \backslash E$, so that 15.3.1 ensures that $g \in \mathbf{L}^q$ for any finite q. It thus appears that $\phi \in (p, q)$ and $U_\phi \in \mathfrak{m}(p, q)$. Yet, since $\phi \notin \ell^2(Z)$, 15.3.1 shows that $\phi \notin \mathscr{F}\mathbf{M}$.

For the case $q \leqslant p$, an alternative proof is provided by the substance of 16.4.8.

(2) If $1 \leqslant p \leqslant \infty$ and $p \neq 2$, then

$$\mathfrak{m}(p, p) \subsetneqq \mathbf{P}, \qquad (p, p) \subsetneqq \ell^\infty(Z). \tag{16.4.8}$$

Proof. The relations (16.4.8) are true if $p = 1$ or $p = \infty$ (see 16.3.3, Exercise 16.7, and either Example 12.7.8 or Exercise 12.37). So, by 16.4.1, we may assume that $2 < p < \infty$. In this case, take $g \in \mathbf{L}^2$, $g \notin \mathbf{L}^p$. By 14.3.2, there exists a ± 1-valued function ϕ on Z such that $\phi \cdot \hat{g}$ is the Fourier transform of some $f \in \mathbf{L}^p$. Since then $\hat{g} = \phi \cdot \hat{f}$ and $g \notin \mathbf{L}^p$, ϕ cannot belong to (p, p) in spite of the fact that $\phi \in \ell^\infty(Z) = \mathscr{F}\mathbf{P}$.

Remarks. (1) If we use Remark 14.3.4(2) in place of 14.3.2, the proof of 16.4.3(2) will yield the following assertion. If α is a mapping of Z into $Z_+ \equiv \{0, 1, 2, \cdots\}$ such that $\sup_{m \in Z_+} \# \alpha^{-1}(\{m\}) < \infty$ (see 11.3.4 for the notation), and if $1 \leqslant p \leqslant \infty$, $p \neq 2$, there is a ± 1-valued function ω on Z_+ such that $\phi = \omega \circ \alpha \notin (p, p)$.

(2) On making use of 14.3.5(1), it can be seen that (16.4.8) remains valid if therein $\mathfrak{m}(p, p)$ and (p, p) are replaced by $\mathfrak{m}(p, q)$ and (p, q), respectively, provided the relations $q \leqslant 2 \leqslant p$ are *not* fulfilled. (If $q \leqslant 2 \leqslant p$, it is evident that $\mathfrak{m}(p, q) = \mathbf{P}$ and $(p, q) = \ell^\infty(Z)$.)

The next three subsections are concerned with some rather crude conditions, some necessary and others sufficient, in order that a given complex-valued function ϕ on Z shall belong to (p, q), or that a given distribution A shall belong to $\mathfrak{m}(p, q)$.

16.4.4. Some Necessary Conditions. Suppose that $\phi \in (p, q)$. Then

(1) if $1 \leqslant p \leqslant \infty$, $1 < q \leqslant 2$,

$$\sum_{n \in Z} (1 + |n|)^{-q'/p' - \varepsilon} |\phi(n)|^{q'} < \infty \tag{16.4.9}$$

for each $\varepsilon > 0$;

(2) if $2 \leqslant p < \infty$, $1 \leqslant q \leqslant \infty$,

$$\sum_{n \in Z} (1 + |n|)^{-p/q - \varepsilon} |\phi(n)|^{p} < \infty \tag{16.4.10}$$

for each $\varepsilon > 0$.

Proof. Since (2) derives from (1) on the basis of 16.4.1, it suffices to prove (1).

By Exercise 7.8, if $\delta > 0$, there exists $f \in \mathbf{L}^p$ such that $\hat{f}(n) = |n|^{-1/p' - \delta}$ for $n \neq 0$. Since $1 < q \leqslant 2$, and since $\phi \hat{f} \in \mathscr{F}\mathbf{L}^q$ by hypothesis, 13.5.1(1) shows that $\phi \hat{f} \in \ell^{q'}$. This conclusion is evidently equivalent to (16.4.9).

16.4.5. Remarks. (1) If $p = 1$ a stronger result is implied by 16.3.4 and 13.5.1(1). If $q \leqslant p$ (that is, $q' \geqslant p'$) the conclusions are obviously trivial (since ϕ is necessarily bounded); thus the only interesting conclusions are those in which either $1 < p < q \leqslant 2$ or $2 \leqslant p < q < \infty$.

(2) If $p < 2 < q$, the results in 16.4.4 do not apply directly. However, if $\phi \in (p, q)$, then $\phi \in (p, 2)$ and $\phi \in (2, q)$; to each of these two relations 16.4.4 can be applied. It thus results that, if $r = \max(p', q)$, then

$$\sum_{n \in Z} (1 + |n|)^{-2/r - \varepsilon} |\phi(n)|^2 < \infty$$

for each $\varepsilon > 0$.

(3) The results stated in 16.4.4 can be slightly strengthened by appealing to 7.3.4 and 7.3.5 to show that if $p \geqslant 1$ and $(a_n)_{n=1}^{\infty}$ is such that $a_n \downarrow 0$ and

$$\sum_{n=1}^{\infty} n^{p-2} a_n{}^p < \infty,$$

then there exists an $f \in \mathbf{L}^p$ such that $\hat{f}(n) = a_{|n|}$ or $\operatorname{sgn} n \cdot a_{|n|}$ for $n \in Z$, $n \neq 0$. Taking a_n to be of the form $n^{-1/p'}(\log n)^{-c}$ for $n \geqslant 2$, where $c > 1/p$, it will be seen that the factor $(1 + |n|)^{-\varepsilon}$ could be replaced by

$$\{\log(2 + |n|)\}^{-q'/p - \varepsilon}$$

in (16.4.9), and by $\{\log(2 + |n|)\}^{-p/q' - \varepsilon}$ in (16.4.10).

Naturally, further refinements of the same nature can be made, but there is no indication of anything definitive being forthcoming in this fashion.

16.4.6. **Some Sufficient Conditions.** Suppose that $1 \leqslant p \leqslant q \leqslant \infty$ and that positive numbers r and s are defined by the relations

$$1 - \frac{1}{r} = \frac{1}{p} - \frac{1}{q}, \qquad \frac{1}{s} = \frac{1}{p} - \frac{1}{q}.$$

Then (1) $\mathbf{L}^r \subset \mathfrak{m}(p, q)$ and $\|A\|_{p,q} \leqslant \|A\|_r$ for $A \in \mathbf{L}^r$;

(2) if $s \leqslant 2$, then $\ell^s(Z) \subset (p, q)$ and $\|\phi\|_{p,q} \leqslant \|\phi\|_s$ for $\phi \in \ell^s(Z)$;

(3) if $1 < p \leqslant 2 \leqslant q < \infty$, and if ϕ is a complex-valued function on Z such that

$$M \equiv \sup_{n \in Z} (1 + |n|)^{1/s} |\phi(n)| < \infty,$$

then $\phi \in (p, q)$ and $\|\phi\|_{p,q} \leqslant A_p A_{q'} M$, where A_t is defined for $1 < t \leqslant 2$ as in Exercise 13.9(1). (Cf. Hörmander [1], Theorem 1.11, p. 106.)

Proof. Statement (1) is an immediate deduction from 13.6.1. Statement (2) follows from (1) combined with the Hausdorff-Young theorem 13.5.1(2), according to which any $\phi \in \ell^s(Z)$ can be written in the form $\hat{A}$ for some $A \in \mathbf{L}^{s'}$ satisfying $\|A\|_{s'} \leqslant \|\phi\|_s$.

As for (3), for any $f \in \mathbf{L}^p$ the case $r = 2$ of Exercise 13.9(1) shows that

$$\{\sum_{n \in Z} (1 + |n|)^{-2\alpha} |\hat{f}(n)|^2\}^{1/2} \leqslant A_p \|f\|_p,$$

where $\alpha = 1/p - \frac{1}{2}$. Writing $\beta = 1/q - \frac{1}{2}$, we have therefore

$$\begin{aligned}\sum_{n \in Z} |(1 + |n|)^{-\beta} \phi(n) \hat{f}(n)|^2 &\leqslant \sum_{n \in Z} M^2 (1 + |n|)^{-2\beta - 2/s} |\hat{f}(n)|^2 \\ &= M^2 \sum_{n \in Z} (1 + |n|)^{-2\alpha} |\hat{f}(n)|^2 \\ &\leqslant M^2 A_p^{\,2} \|f\|_p^2.\end{aligned}$$

The case $s = 2$ of Exercise 13.9(2) now shows that $\phi \hat{f} = \hat{g}$ for some $g \in \mathbf{L}^q$ satisfying

$$\|g\|_q \leqslant A_{q'} \cdot M A_p \|f\|_p.$$

Thus $\phi \in (p, q)$ and $\|\phi\|_{p,q} \leqslant A_p A_{q'} M$, as alleged.

16.4.7. **Some Results of Hirschman.** We propose to consider two interesting results concerning (p, p) given by Hirschman ([1], pp. 231–236). Throughout this subsection ϕ denotes a complex-valued function on Z. For $\beta > 0$ Hirschman defines (compare 8.7.2).

$$V_\beta(\phi) = \sup \{\sum_{k=1}^{r-1} |\phi(n_{k+1}) - \phi(n_k)|^\beta\}^{1/\beta},$$

the supremum (which may be ∞) being taken over all strictly increasing sequences $(n_k)_{k=0}^r$ of integers. The results are as follows:

(1) If

$$\phi(n) = O(|n|^{-\delta}) \qquad \text{as } |n| \to \infty$$

for some $\delta > 0$ and

$$V_\beta(\phi) < \infty$$

for some $\beta > 2$, then $\phi \in (p, p)$ whenever

$$\frac{2\beta}{\beta + 2} < p < \frac{2\beta}{\beta - 2}.$$

(If the condition $\beta > 2$ is here replaced by $1 \leqslant \beta < 2$, 8.7.3 and 3.1.6 show at once that $\phi \in (p, p)$ for all p satisfying $1 \leqslant p \leqslant \infty$.)

(2) If

$$\phi(n) = O(|n|^{-\alpha}) \qquad \text{as } |n| \to \infty,$$

where $0 < \alpha \leqslant \frac{1}{2}$, then $\phi \in (p, p)$ whenever

$$\frac{2}{1 + 2\alpha} < p < \frac{2}{1 - 2\alpha}.$$

Of these, (2) follows from (1), because the hypotheses of (2) entail that

$$V_\beta(\phi) < \infty$$

for $\beta > 1/\alpha$. Before giving the fairly elaborate proof of (1), it is worth noting that both Hirschman (loc. cit., p. 235) and Zygmund ([Z_1], pp. 200–202) give results bearing upon the membership to (p, p) for various values of p of certain functions ϕ of the form

$$\phi(n) = \begin{cases} n^{-b} \exp(icn^a) & (n > 0) \\ 0 & (n \leqslant 0), \end{cases}$$

where a, b, and c are positive real numbers. (Zygmund's results, which are in a sense complementary to those of Hirschman, take the form of estimates of the behavior of the sum function

$$\hat{\phi}(x) = \sum_{n=1}^{\infty} n^{-b} \exp(icn^a) e^{inx}$$

near its only possible singularity at the origin.)

Hirschman also obtains somewhat similar results for the case in which the underlying group is Z or R^m. In this connection see also 16.4.9(3) and de Leeuw [1].

Proof of (1). In view of 16.4.1, it suffices to deal with the case in which

$$\frac{2\beta}{\beta + 2} < p \leqslant 2. \tag{16.4.11}$$

Furthermore, we may suppose that ϕ is real-valued, that $\phi(0) = 0$, and that $\delta < 1$. We therefore write

$$\alpha = \frac{2}{\delta} > 2,$$

so that

$$|\phi(n)| \leqslant K(1 + |n|)^{-2/\alpha}. \tag{16.4.12}$$

The idea of the proof is to approximate ϕ by functions ϕ_m ($m = 0, 1, 2, \cdots$), in the sense that

$$\phi = \lim_{m \to \infty} \phi_m \tag{16.4.13}$$

pointwise on Z, where the ϕ_m are to be chosen rather carefully. The proof is presented in several stages.

(a) Construction of the ϕ_m. We begin by interpolating ϕ linearly between the integers, thus obtaining a real-valued function defined on R; this new function will be denoted by ϕ again.

Let m be any nonnegative integer.

If $|\phi(x)| < 2^{-m}$ for all $x > 0$, define $x_1 = 0$; otherwise, let x_1 be the largest $x > 0$ such that $\phi(x) = \pm 2^{-m}$. If there exist numbers x satisfying $0 < x < x_1$ and $\phi(x) = \phi(x_1) \pm 2^{-m}$, let x_2 be the largest; otherwise, let $x_2 = 0$. This procedure will terminate after a finite number of steps with a number x_N, so that $0 = x_N < x_{N-1} < \cdots < x_1$.

The numbers $x_{-1} < x_{-2} < \cdots < x_{-M} = 0$ are defined in an exactly analogous fashion.

Define ϕ_m by the relations

$$\phi_m(n) = \begin{cases} 0 & \text{if } n = 0, \text{ or } n > x_1, \text{ or } n < x_{-1} \\ \phi(x_k) & \text{if } x_{k+1} < n \leqslant x_k,\ 1 \leqslant k \leqslant N - 1, \\ \phi(x_{-k}) & \text{if } x_{-k} \leqslant n < x_{-k-1},\ 1 \leqslant k \leqslant M - 1. \end{cases}$$

It is almost evident that (16.4.13) holds; in fact, $|\phi_m(n) - \phi(n)| \leqslant 2^{-m}$, so that (16.4.13) holds uniformly. Also, (16.4.12) and a careful study of the definition of ϕ_m will show that

$$|\phi_m(n)| \leqslant K_1(1 + |n|)^{-2/\alpha}; \tag{16.4.14}$$

the reader may find some use for a rough figure at this point.

(b) The next step is to verify that, if $0 < \varepsilon < 1$,

$$V_{2-\varepsilon}(\phi_m)^{2-\varepsilon} \leqslant 2^{(\beta - 2 + \varepsilon)m} V_\beta(\phi)^\beta. \tag{16.4.15}$$

To begin with, an argument like that appearing in the last paragraph on p. 227 of Hirschman [1] will show that any sum

$$\sum_{j=0}^{s-1} |\phi_m(n_{j+1}) - \phi_m(n_j)|^{2-\varepsilon},$$

in which $n_0 < n_1 < \cdots < n_s$ are integers, is majorized by

$$\sum_h |\phi(x_{h+1}) - \phi(x_h)|^{2-\varepsilon}.$$

Next, since $|\phi(x_{h+1}) - \phi(x_h)|$ is either 0 or a positive integer multiple of 2^{-m}, this last-written sum is majorized by

$$2^{(\beta - 2 + \varepsilon)m} \sum_h |\phi(x_{h+1}) - \phi(x_h)|^\beta.$$

Finally, since ϕ is linear between successive integers, the remaining sum does not exceed $V_\beta(\phi)^\beta$, whence (16.4.15).

(c) We now define $\psi_0 = \phi_0$ and $\psi_m = \phi_m - \phi_{m-1}$ for $m = 1, 2, \cdots$, so that

$$\phi_m = \sum_{u=0}^{m} \psi_u$$

and therefore

$$\|\phi_m\|_{p,p} \leqslant \sum_{m=0}^{\infty} \|\psi_m\|_{p,p}. \tag{16.4.16}$$

From (16.4.14) and (16.4.15) it follows immediately that

$$|\psi_m(n)| \leqslant K_2(1 + |n|)^{-2/\alpha} \tag{16.4.17}$$

and

$$V_{2-\varepsilon}(\psi_m)^{2-\varepsilon} \leqslant 4 \cdot 2^{(\beta - 2 + \varepsilon)m} V_\beta(\phi)^\beta_\varepsilon. \tag{16.4.18}$$

At this point we apply 8.7.3 to ψ_m and use (16.4.17) and (16.4.18) to derive the inequality

$$\|\psi_m\|_{1,1} \leqslant A_{\alpha,\beta,\varepsilon} 2^{\{(\beta - 2 + \varepsilon)(\alpha - 2)/2(\alpha - 2 + \varepsilon)\}m}. \tag{16.4.19}$$

In addition, since $|\phi - \phi_m| \leqslant 2^{-m}$, we have

$$\|\psi_m\|_{2,2} = \|\psi_m\|_\infty \leqslant 2.2^{-m} \tag{16.4.20}$$

By (16.4.11),

$$\frac{1}{p} = \frac{1-t}{1} + \frac{t}{2}$$

where

$$t > \frac{\beta - 2}{\beta}. \tag{16.4.21}$$

So, using 16.4.2, (16.4.19), and (16.4.20), we see that

$$\sum_{m=0}^{\infty} \|\psi_m\|_{p,p} < \infty, \tag{16.4.22}$$

provided

$$\frac{(\beta - 2 + \varepsilon)(\alpha - 2)(1 - t)}{2(\alpha - 2 + \varepsilon)} - t < 0,$$

which, in view of (16.4.21), is true for sufficiently small $\varepsilon > 0$.

(d) From (16.4.16) and (16.4.22) it appears that

$$\sup_m \|\phi_m\|_{p,p} < \infty.$$

This relation, together with (16.4.13), entails that $\phi \in (p, p)$; see Exercise 16.21. The proof is therefore complete.

16.4.8. **The Hilbert Distribution as a Multiplier.** Let H denote the Hilbert distribution, as defined in Section 12.8. The distribution H is a pseudomeasure but (as was seen in 12.8.1) not a measure. By Marcel Riesz's theorem 12.9.1, $H \in \mathfrak{m}(p, p)$ whenever $1 < p < \infty$. It follows immediately that $H * \mu \in \mathfrak{m}(p, p)$ whenever $1 < p < \infty$ and $\mu \in \mathbf{M}$. On the other hand, by 12.8.4(1) and the fact that H is not a measure, $H * \mu \in \mathbf{M}$ is true for but a meager set of $\mu \in \mathbf{M}$ (compare 16.4.3(1)). See also [EG], Section 6.7.

16.4.9. **Further Results.** (1) Hörmander [1] makes a detailed study of multipliers for the case in which the group T is replaced by a group R^m. He obtains analogues of 16.4.6 in the shape of his Theorems 1.11 and 2.4 and his Corollary 1.2. Hörmander's work also includes similar results ([1], Theorem 2.5) which are more complicated and which, as far as the author is aware, have no published analogues for the group T; further results of the same nature have been obtained by Littman [1]. Still for the case $G = R^m$, Schwartz [3] discussed multipliers that are direct analogues of the conjugate function operator studied in Sections 12.8, 12.9, and 13.9; Krée [1] has continued work on this theme. See also Calderón [3].

Calderón [2] provides a survey of these and many other related questions. (We may point out here that both Hörmander and Calderón speak of "translation invariant operators" when the operators in question in fact commute with translations.) See also Cordes [1] and Peetre [1], [2], [3].

Concerning multipliers on more general groups, see in addition Gaudry [1], [2], [3] and Figà-Talamanca [3].

(2) For a discussion of some of the properties of functions $\alpha : Z \to Z$ such that $\exp \circ (i\alpha) \in (p, p)$, see Edwards [13].

(3) Hahn [1] gives some interesting extensions of the results discussed in 16.4.7, together with a proof of the inclusion

$$\ell^p(Z) * \ell^{p'}(Z) \subset (r, r) \quad \text{if} \quad 1 \leqslant p \leqslant 2, r_p \leqslant r \leqslant r_p', \tag{16.4.23}$$

where $r_p = 2p/(3p - 2)$. Hahn's proof of (16.4.23) is fairly simple and ties up with the general developments mentioned in 13.4.2(2). It may be briefly described in the following fashion.

Denote by ϕ and ψ complex-valued functions on Z having finite supports and by f and g functions in $\mathbf{L}^\infty$. Define

$$Q(\phi, \psi, f, g) = \sum_{n \in Z} \phi * \psi(n)\hat{f}(n)\hat{g}(n).$$

By 3.1.6, the Cauchy-Schwarz inequality, and (8.2.2),

$$\begin{aligned} |Q(\phi, \psi, f, g)| &\leqslant \|\phi * \psi\|_\infty \|\widehat{fg}\|_1 \leqslant \|\phi\|_1 \|\psi\|_\infty \|\hat{f}\|_2 \|\hat{g}\|_2 \\ &= \|\phi\|_1 \|\psi\|_\infty \|f\|_2 \|g\|_2. \end{aligned} \tag{16.4.24}$$

Similarly,

$$\begin{aligned}|Q(\phi, \psi, f, g)| &= |(\phi * \psi)^{\wedge} * f * g(0)| = |(\hat{\phi}\hat{\psi}) * f * g(0)| \\ &\leq \|\hat{\phi}\hat{\psi}\|_1 \|f * g\|_\infty \leq \|\hat{\phi}\|_2 \|\hat{\psi}\|_2 \|f\|_1 \|g\|_\infty \\ &= \|\phi\|_2 \|\psi\|_2 \|f\|_1 \|g\|_\infty.\end{aligned} \tag{16.4.25}$$

Since Q is quadrilinear (that is, linear in each of its four arguments when the other three remain fixed), an appropriate convexity theorem ([Z_2], p. 106) leads from (16.4.24) and (16.4.25) to the inequality

$$|Q(\phi, \psi, f, q)| \leq \|\phi\|_p \|\psi\|_q \|f\|_r \|q\|_s$$

whenever

$$\left(\frac{1}{p}, \frac{1}{r}, \frac{1}{q}, \frac{1}{s}\right) = t\left(\frac{1}{1}, \frac{1}{\infty}, \frac{1}{2}, \frac{1}{2}\right) + (1 - t)\left(\frac{1}{2}, \frac{1}{2}, \frac{1}{1}, \frac{1}{\infty}\right)$$

for some $t \in [0, 1]$. In other words,

$$|Q(\phi, \psi, f, g)| \leq \|\phi\|_p \|\psi\|_{p'} \|f\|_r \|g\|_{r'}$$

provided $1 \leq p \leq 2$ and $r = r_p = 2p/(3p - 2)$. In view of the definition of Q and the converse of Hölder's inequality, it follows from this that

$$\|\phi * \psi\|_{r,r} \leq \|\phi\|_p \|\psi\|_{p'} \tag{16.4.26}$$

for $1 \leq p \leq 2$ and $r = r_p$. Appeal to Exercise 16.21 will now show that (16.4.26) holds whenever $\phi \in \ell^p(Z)$, $\psi \in \ell^{p'}(Z)$ and $r = r_p$. By 16.4.2, (16.4.26) holds also for the same ϕ and ψ, provided r lies in the interval $[r_p, r_p']$. This establishes (16.4.23).

(4) The area of harmonic analysis known as Littlewood-Paley theory has close connections with multiplier theory. For an account of some aspects of this, see [EG].

(At this point it is perhaps fair to comment on the bizarre so-called review of [EG] appearing in *Bull. Amer. Math. Soc.* **84** (1978), pp. 242–250. This so-called review is a splendid survey of the complex variable aspects of Littlewood-Paley theory, which aspects are merely mentioned and deliberately left aside in [EG]; see the fourth complete paragraph on p. 3 of [EG]. Only a tiny fraction of the said review is devoted to [EG] itself. It would therefore be a good plan to consult other reviews; for example, that appearing in *Austral. Math. Soc. Gazette* **5**(3) (1978), 100–108 and/or that to appear in *Mathematical Reviews*.

(5) For other aspects see Okikiolu [1]; Pigno [1], [2]; Price [1], [2]; Figà-Talamanca and Gaudry [2], [3], [4]; Gaudry [1], [2], [6]; Saeki [1]; Figà-Talamanca and Price [1]; Rivière and Sagher [1]; Cowling [3]; Goes [4]; Littman, McCarthy and Rivière [1], [2]; Doss [1]; Lanconelli [1]; Krée [1]; Edwards and Price [1].

16.5 A Theorem of Kaczmarz–Stein

The theorem of Kaczmarz [1] and Stein [1], Theorem 9 gives a condition that comes close to being necessary and sufficient in order that a bounded complex-valued function ϕ on Z shall belong to (p, p); see also

Goes [4]. A slightly modified account of the theorem will be given in this section. Its proof depends on the Marcinkiewicz interpolation theorem 13.8.1 and the theorem of Stein quoted in 16.2.8.

The first two subsections deal with some preliminaries.

16.5.1. **The Function Space $\mathbf{V}_p$.** If $1 \leqslant p \leqslant \infty$, $\mathbf{V}_p$ will denote the set of functions $\Phi \in \mathbf{L}^p$ such that

$$\sup \left\| \sum_{k=1}^{r} (T_{b_k}\Phi - T_{a_k}\Phi) \right\|_p < \infty, \tag{16.5.1}$$

the supremum being taken with respect to all finite sequences $([a_k, b_k])_{k=1}^r$ of nonoverlapping subintervals of $[0, 2\pi]$ (or of any other interval of length 2π).

Although the case $p = \infty$ will be of little direct concern to us in this section, it is interesting in connection with 16.3.3 and 16.3.5 to note that $\mathbf{V}_\infty$ comprises exactly those functions $\Phi \in \mathbf{L}^\infty$ that are equal almost everywhere to functions of bounded variation (that is, for which $D\Phi \in \mathbf{M}$; see 12.5.10). The proof of this is not completely trivial; see Exercise 16.15.

It is quite evident that $\mathbf{V}_p$ is a linear subspace of $\mathbf{L}^p$, and that $\mathbf{V}_q \subset \mathbf{V}_p$ if $p < q$.

16.5.2. From this point onward in this section, ϕ will denote a bounded complex-valued function on Z. We introduce the function Φ defined by

$$\Phi(x) = \sum_{n \neq 0} (in)^{-1}\phi(n)e^{inx}, \tag{16.5.2}$$

the series converging in $\mathbf{L}^2$ (for example).

It is to be observed that, if $\phi \in (p, p)$, and if we denote by β the function

$$\beta(x) = \sum_{n \neq 0} (in)^{-1}e^{inx}, \tag{16.5.3}$$

then Exercise 1.5 shows that $\beta \in \mathbf{L}^\infty \subset \mathbf{L}^{p'}$ and 16.4.1 then entails that $\Phi = U_\phi\beta \in \mathbf{L}^{p'}$. (It is true, of course, that in this case $\Phi \in \mathbf{L}^p$, too.) Moreover, in this case

$$U_\phi f = \{\phi(0) + D\Phi\} * f, \tag{16.5.4}$$

which shows by way of interest that U_ϕ corresponds to the distribution $A = \phi(0) + D\Phi$.

We can now state and prove the Kunze–Stein theorem.

16.5.3. (Kaczmarz–Stein) Suppose that $1 < p \leqslant 2$ and that the notation is as in 16.5.1 and 16.5.2. Then:

(1) if $\phi \in (p, p)$, then $\Phi \in \mathbf{V}_{p'}$;
(2) if $\Phi \in \mathbf{V}_{p'}$, then $\phi \in (r, r)$ for all exponents r satisfying $p < r < p'$.

Proof. We may and will assume throughout that $\phi(0) = 0$.

(1) We know already from 16.5.2 that $\Phi \in \mathbf{L}^{p'}$, and it thus remains to show that Φ satisfies (16.5.1) with p' in place of p.

Let $f \in \mathbf{L}^p$ and consider the function g defined by

$$g(x) = \int_0^x U_\phi f(y)\,dy.$$

Using the notation of 16.5.1, we then have

$$\begin{aligned}\Big|\sum_{k=1}^{r} \{g(-b_k) - g(-a_k)\}\Big| &\leqslant \sum_{k=1}^{r} \int_{-b_k}^{-a_k} |U_\phi f(y)|\,dy \\ &\leqslant 2\pi\|U_\phi f\|_1 \leqslant 2\pi\|U_\phi f\|_p \qquad (16.5.5)\\ &\leqslant 2\pi\|\phi\|_{p,p}\|f\|_p.\end{aligned}$$

On the other hand, by 6.2.8,

$$g(-b) - g(-a) = \sum_{n\neq 0} (in)^{-1}\phi(n)\hat{f}(n)(e^{-inb} - e^{-ina}),$$

which, by 10.5.4 and the fact that $\Phi \in \mathbf{L}^{p'}$ (see 16.5.2), is none other than

$$\Phi * f(-b) - \Phi * f(-a) = (T_b\Phi - T_a\Phi) * f(0),$$

the last step relying on 3.1.2. Thus (16.5.5) entails that

$$\Big|\Big\{\sum_{k=1}^{r} (T_{b_k}\Phi - T_{a_k}\Phi)\Big\} * f(0)\Big| \leqslant 2\pi\|\phi\|_{p,p}\|f\|_p,$$

which combines with the converse of Hölder's inequality (Exercise 3.6) to show that

$$\Big\|\sum_{k=1}^{r} (T_{b_k}\Phi - T_{a_k}\Phi)\Big\|_{p'} \leqslant 2\pi\|\phi\|_{p,p}.$$

Thus $\Phi \in \mathbf{V}_{p'}$ and this portion of the proof is complete.

(2) Let $f \in \mathbf{L}^p$ and write $\psi = \Phi * f$ and, for $m = 1, 2, \cdots$,

$$\psi_m(x) = m\Big\{\psi\Big(x + \frac{1}{m}\Big) - \psi(x)\Big\} = m\{T_{(-1/m)}\Phi - \Phi\} * f(x).$$

Since by hypothesis $\Phi \in \mathbf{L}^{p'} \subset \mathbf{L}^1$,

$$U^m : f \to \psi_m$$

is an $\mathfrak{m}$-operator of type $(\mathbf{L}^p, \mathbf{L}^p)$ (see 3.1.6 and 16.2.7).

The function ψ is of bounded variation since, by Hölder's inequality,

$$\begin{aligned}\Big|\sum_{k=1}^{r} \{\psi(b_k) - \psi(a_k)\}\Big| &\leqslant \Big\|\sum_{k=1}^{r} \{\Phi(b_k - x) - \Phi(a_k - x)\}\Big\|_{p'} \cdot \|f\|_p \\ &\leqslant \Big\|\sum_{k=1}^{r} (T_{-b_k}\Phi - T_{-a_k}\Phi)\Big\|_{p'} \cdot \|f\|_p\end{aligned}$$

for all finite sequences $([a_k, b_k])_{k=1}^{r}$ of nonoverlapping subintervals of $[0, 2\pi]$, and since $\Phi \in \mathbf{V}_{p'}$ by hypothesis. (This yields an independent proof that U^m is an $\mathfrak{m}$-operator of type $(\mathbf{L}^p, \mathbf{BV})$.)

By standard results ([W], Theorem 5.2e and Lemma 6.4b),

$$\lim_{m \to \infty} U^m f(x) = \psi'(x)$$

exists finitely for almost all x. (Note that we write ψ' and *not* $D\psi$: it is the pointwise, not the distributional, derivative of ψ that is here involved.)

Stein's theorem 16.2.8 now applies to show that the operator $f \to \psi'$ is of weak type (p, p).

Also, if f is absolutely continuous, the same is true of ψ (by 3.1.5), and we may then identify ψ' and $D\psi$. In this case, (16.5.4) shows that $\psi' = U_\Phi f$ (recall the assumption that $\phi(0) = 0$). Accordingly, reference to 13.7.2 and 13.7.5 shows that there exists a number A such that

$$D_{U_\Phi f}(t) \leqslant \left(\frac{A \|f\|_p}{t}\right)^p \tag{16.5.6}$$

for all absolutely continuous f and all $t > 0$.

The fact that U_Φ is of type (2, 2) can now be used to show that (16.5.6) continues to hold for any $f \in \mathbf{L}^2$, any such function f being approximable in $\mathbf{L}^2$ (and therefore in $\mathbf{L}^p$) by a sequence of absolutely continuous functions f_n; see Exercise 13.18. Knowing this, we may affirm that U_Φ, with domain $\mathbf{L}^2$, is of weak type (p, p) and of type (2, 2). This, together with the Marcinkiewicz interpolation theorem 13.8.1, shows that U_Φ is of type (r, r) for each r satisfying $p < r \leqslant 2$. Thus $\phi \in (r, r)$ for any such r. That the same is true for $2 \leqslant r < p'$ follows at once from 16.4.1, and the proof is complete. As was pointed out to me by Professor G. Goes, Kaczmarz [1] actually proves that, for every p such that $1 < p \leqslant \infty$, $\phi \in (p, 1)$ if and only if $\Phi \in V_{p'}$.

16.6 Banach Algebras Applied to Multipliers

It is very simple to verify that (p, p) is a complex commutative Banach algebra with identity (the constant function 1) relative to pointwise algebraic operations and the norm $\|\cdot\|_{p,p}$. For all p one has

$$(1, 1) = \mathscr{F}\mathbf{M} \subset (p, p) \subset (2, 2) = \ell^\infty(Z). \tag{16.6.1}$$

For $\phi \in (p, p)$,

$$\|\phi\|_\infty \leqslant \|\phi\|_{p,p}; \tag{16.6.2}$$

and for $\mu \in \mathbf{M}$,

$$\|\hat{\mu}\|_{p,p} \leqslant \|\mu\|_1. \tag{16.6.3}$$

The determination of all the continuous complex homomorphisms of (p, p) meets with difficulties that are presumably of the same order of magnitude as are encountered in the case of the algebra $\mathbf{M}$ of measures (see 12.7.4). To each $n \in Z$ corresponds the homomorphism $n \to \phi(n)$, but there are others. (The reader should provide a proof of the last statement.)

In view of this, we look at a subalgebra of (p, p) which is more tractable and still of some interest, namely, the closure $\mathfrak{m}_p$ in (p, p) of $\mathscr{F}\mathbf{L}^1$. From

(16.6.2) and 2.3.8 it appears that $\mathfrak{m}_p \subset \mathbf{c}_0(Z)$; as a consequence, $\mathfrak{m}_p$ has no identity element.

The introduction of $\mathfrak{m}_p$ is suggested by the step taken by Hörmander ([1], pp. 111–113) in his study of multipliers for groups R^m.

Reference to 4.1.3 leads to the conclusion that the evaluation maps $\phi \to \phi(n)$ $(n \in Z)$ exhaust the nontrivial continuous complex homomorphisms of the algebra $\mathfrak{m}_p$.

(Actually, 4.1.3 is to be applied to $\mathbf{E} = \mathbf{L}^1$ with the norm induced by that of $\mathfrak{m}(p, p)$. Conditions (a)–(d) in 4.1.3 are satisfied. Hence one concludes that every nontrivial continuous complex homomorphism of $\mathbf{E}$ is of the form $f \to \hat{f}(n)$ for some $n \in Z$. The same is therefore true of the nontrivial continuous complex homomorphisms of the closure in $\mathfrak{m}(p, p)$ of $\mathbf{E}$, as alleged.)

Knowing this, one can apply 11.4.15 to the algebra obtained by adjoining a formal identity to $\mathfrak{m}_p$ (see 11.4.1 and an analogous procedure in 11.4.16). The outcome is the following assertion.

16.6.1. **Functions of Multipliers.** Suppose that $\phi \in \mathfrak{m}_p$ and that the function Φ is defined and analytic on some open subset of the complex plane containing $\phi(Z) \cup \{0\}$ and satisfies $\Phi(0) = 0$. Then $\Phi \circ \phi \in \mathfrak{m}_p$.

This is the analogue, for the group T, of Hörmander's Theorem 1.18. Still following Hörmander ([1], pp. 111–113, Theorem 1.16), we can obtain a useful inclusion relation.

16.6.2. If $1 < p, q < \infty$, and if

$$\left|\frac{1}{q} - \frac{1}{2}\right| < \left|\frac{1}{p} - \frac{1}{2}\right| \tag{16.6.4}$$

then

$$(p, p) \cap \mathbf{c}_0(Z) \subset \mathfrak{m}_q.$$

Proof. In view of 16.4.1 and (16.6.4), we may assume that

$$\frac{1}{q} = \frac{t}{p} + \frac{1-t}{2}$$

for some t satisfying $0 \leqslant t < 1$.

Let $\phi \in (p, p) \cap \mathbf{c}_0(Z)$ and define $\phi_N = \phi \hat{F}_N$. Then it is trivial to assert that $\phi_N \in \mathscr{F}\mathbf{L}^1$. Moreover (see Exercise 16.16),

$$\|\phi_N\|_{p,p} \leqslant \|F_N\|_1 \|\phi\|_{p,p} \leqslant \|\phi\|_{p,p},$$

so that

$$\|\phi - \phi_N\|_{p,p} \leqslant 2\|\phi\|_{p,p} \tag{16.6.5}$$

Since $\phi \in \mathbf{c}_0(Z)$,

$$\|\phi - \phi_N\|_{2,2} = \|\phi - \phi_N\|_\infty \to 0 \qquad \text{as } N \to \infty. \tag{16.6.6}$$

Now we appeal to 16.4.2, which shows that

$$\|\phi - \phi_N\|_{q,q} \leqslant \|\phi - \phi_N\|_{p,p}^{t}\|\phi - \phi_N\|_{2,2}^{1-t}.$$

By (16.6.5) and (16.6.6), the right-hand member of this inequality tends to zero as $N \to \infty$ (remember that $0 \leqslant t < 1$). Thus $\phi \in \mathfrak{m}_q$ and the proof is complete.

16.6.3. **Remarks.** It is natural to ask whether the relation

$$(p, p) \cap \mathbf{c}_0(Z) \subset \mathfrak{m}_p \tag{16.6.7}$$

is true. The answer is affirmative when $p = 2$. The answer is negative when $p = 1$ or ∞: this is due to the existence of measures $\mu \notin \mathbf{L}^1$ for which $\hat{\mu} \in \mathbf{c}_0(Z)$. The existence of such measures in turn hinges on the existence of sets of multiplicity in the strict sense and having zero measure (see 12.12.8); although the question is a fairly delicate one, it is known that such sets exist in abundance (see $[Z_1]$, pp. 348–349; $[Z_2]$, pp. 147–152; [KS], Chapitres V and VI).

For $1 < p < 2$, (16.6.7) has been shown to be false by Gaudry and Figà-Talamanca [2].

It follows from 12.7.4 that the characters of T generate only a scant subset of *all* continuous complex homomorphisms of the Banach algebra $(1, 1) = \mathscr{F}M$; in particular the so-called *Wiener-Pitt phenomenon* asserts the existence of a measure $\mu \in \mathbf{M}$ such that $\hat{\mu}$ is real-valued on $Z \equiv T^{\wedge}$ but is not real-valued on the whole Gelfand space $\Gamma(\mathbf{M})$ (the notation is as in 11.4.9). Zafran [1] has shown that this phenomenon persists for, amongst others, the multiplier algebras $(p, p) \cap \mathbf{c}_0(Z)$ when $1 < p < 2$. See also Brown [3].

On page 178 of Volume 1, the functions which operate on $\mathbf{A}(T)$ and $\mathbf{A}(Z)$ are identified. Igari [2], [3] has shown that only entire functions operate on the Banach algebra (p, p) when $1 < p < 2$; that is, if $F:[-1, 1] \to C$ and $F \circ \varphi \in (p, p)$ for every $\varphi \in (p, p)$ such that $\varphi\colon \Gamma((p, p)) \to [-1, 1]$, then there exists $\varepsilon > 0$ such that F agrees on $(-\varepsilon, \varepsilon)$ with some entire function. For $p = 1$ Varopoulos [4] and for $1 < p < 2$ Zafran [2], [3] have shown that the same conclusion holds with (p, p) replaced throughout by the smaller algebra $(p, p) \cap \mathbf{c}_0(Z)$.

16.7 Further Developments

This section consists of brief comments on further results associated with multipliers and with extensions of the concept.[1] References are given to assist those readers who may wish to pursue the matters mentioned.

[1] For the majority of such developments, multipliers appear more naturally in their guise as operators (rather than as functions on some dual object); compare the opening paragraphs of Section 16.4

16.7.1. **Multipliers and Isomorphism Problems.** We have in 4.2.7 referred to the interest that attaches itself to the study of multipliers on account of their relationship with isomorphism problems. The reader should by now be in a better position to appreciate the details and the difficulties.

The question proves to be largely one of seeking to characterize in a sufficiently effective manner the multipliers $U \in \mathfrak{m}(\mathbf{F}, \mathbf{F})$, where $\mathbf{F}$ is (say) $\mathbf{L}^p$, $\mathbf{C}$, or $\mathbf{M}$, which map $\mathbf{F}$ in a one-to-one manner onto itself and which have certain additional properties (such as being bipositive or isometric).

It may be helpful to begin by looking at the problem in reverse, so to speak. Suppose that $n \in Z$, and that ζ is an automorphism of the (topological) group T. It is then easy to verify that the operator $S: f \to e_n(f \circ \zeta)$ is an automorphism of each of the convolution algebras $\mathbf{L}^p$ and $\mathbf{C}$ which is

(1) isometric in any case,
(2) bipositive (that is, is such that $Sf \geqslant 0$ if and only if $f \geqslant 0$) provided $n = 0$.

Moreover, if $a \in T$, the corresponding multiplier $U_a = S^{-1}T_aS$ (see 4.2.7) is given by $U_a f = e_n(-a) \cdot T_{\zeta(a)} f$, and is thus a scalar multiple of a translation operator. The knowledge of U_a determines $\zeta(a)$ uniquely.

More generally, if S is any automorphism of the convolution algebra $\mathbf{L}^p$ or $\mathbf{C}$ which satisfies (1) [or (2)], it is still visibly the case that U_a is an isometric (or bipositive) multiplier of $\mathbf{L}^p$ or $\mathbf{C}$. The question is: does this imply that U_a is a multiple of a translation operator?

One is thus led to pose this same question in respect of isometric or bipositive multipliers U of various other group algebras $\mathbf{F}$. An answer to this question provides an answer to the corresponding isomorphism problem for $\mathbf{F}$.

For $\mathbf{F} = \mathbf{L}^1$ the problem was solved affirmatively by Kawada [1] in the bipositive case and by Wendel [1], [2] for the isometric case; for $\mathbf{F} = \mathbf{M}$ and the isometric case by Johnson [1]; for $\mathbf{F} = \mathbf{L}^\infty$ and the isometric case by Gaudry ([3], Theorem 5.2.1); and for $\mathbf{F} = \mathbf{L}^p$ ($1 \leqslant p \leqslant \infty$, $p \neq 2$) and the isometric case by Strickartz [1] and Parrott [1]. Some other cases are discussed in Edwards [10] and Gaudry [4]. For a connected account of some of these results, see Gaudry [3], Chapter 5. See also Greenleaf [1], [3].

The results mentioned apply to groups more general than T. On the other hand, the problem of isomorphisms that are neither isometric nor bipositive would appear to remain largely open.

Related problems are discussed by Forelli [1], Rudin [7] and Rigelhof [1].

16.7.2. **Compact and Weakly Compact Multipliers.** It is customary to apply the adjective *compact* to a linear operator U from a Banach space $\mathbf{E}$ into a Banach space $\mathbf{F}$, if U transforms the closed unit ball of $\mathbf{E}$ into a relatively compact subset of $\mathbf{F}$; see, for example, [E], Section 9.2.

Gaudry ([3], Theorems 4.2.2 and 4.2.3) has given characterizations of multipliers $U \in \mathfrak{m}(p, q)$ that are compact in the above sense, the characterizations being in terms of the possibility of approximating U in a certain way by multipliers defined by convolution with functions. He gives also ([3], Theorem 4.2.5) a characterization of those multipliers $U \in \mathfrak{m}(1, 1)$ that are *weakly compact*, that is (see [E], Section 9.2), which transform the closed unit ball of $\mathbf{L}^1$ into a weakly relatively compact subset of $\mathbf{L}^1$: while (as we know) every multiplier of type $\mathfrak{m}(1, 1)$ is represented by convolution with a measure, the weakly compact multipliers of type $\mathfrak{m}(1, 1)$ are precisely those representable as convolution with a function in $\mathbf{L}^1$. These results apply to groups more general than T. See Exercise 16.25 and Akeman [1].

16.7.3. **Approximation of Multipliers.** Having decided (see 16.4.3(1)) that, if p is different from 1 and ∞, not every multiplier $U \in \mathfrak{m}(p, p)$ is representable as convolution with a measure, it is natural to ask whether and in what sense any such multiplier U can be approximated by multipliers U_μ of the form

$$U_\mu f = \mu * f,$$

where $\mu \in \mathbf{M}$. This problem has been tackled with success for general groups by Figà-Talamanca [1], [2] and by Figà-Talamanca and Gaudry [1]. For the case of the group T, some results of this sort are very easily obtainable; see Exercise 16.18.

This discussion has at the same time led to a characterization of $\mathfrak{m}(p, p)$ in terms of the dual of a suitable space of continuous functions on the underlying group. See also Rieffel [1].

16.7.4. **Generalized Multipliers.** If $\mathbf{B}$ denotes a Banach, or even a more general type of topological, algebra, it is natural to define a *left* (respectively, *right*) *multiplier* or *centralizer* of $\mathbf{B}$ to be a continuous linear space endomorphism U of $\mathbf{B}$ such that

$$U(xy) = (Ux)y \qquad (\text{respectively, } U(xy) = x(Uy))$$

for $x, y \in \mathbf{B}$.

We have in this book no space to spare for this extended concept; the interested reader should consult Helgason [1]; Wang [1]; Birtel [1], [2], [3]; Johnson [2], [3]; Máté [1], [2], [3]; [La].

16.7.5. **Two Variants of the Multiplier Problem.** We mention briefly two interesting and largely unsolved variants of the multiplier problems considered earlier in this chapter.

(1) *Multipliers of quotient spaces.* Suppose that $\mathbf{F}$ and $\mathbf{G}$ are as in 16.1.1, that S is a subset of Z, and that ϕ is a complex-valued function on S. The

problem is to determine necessary and sufficient conditions on ϕ in order that to each $f \in \mathbf{F}$ shall correspond at least one $g \in \mathbf{G}$ such that

$$\hat{g}(n) = \phi(n) \cdot \hat{f}(n) \qquad \text{for all } n \in S. \tag{16.7.1}$$

A function ϕ possessing this property may conveniently be said to be of type $(\mathbf{F}, \mathbf{G}; S)$.

Plainly, the original multiplier problem corresponds exactly to the case in which $S = Z$.

Equally plainly, if ϕ is the restriction to S of a multiplier of type $(\mathbf{F}, \mathbf{G})$, then ϕ is of type $(\mathbf{F}, \mathbf{G}; S)$. The nontrivial part of the problem thus remains in deciding whether, conversely, each ϕ of type $(\mathbf{F}, \mathbf{G}; S)$ is the restriction to S of some multiplier of type $(\mathbf{F}, \mathbf{G})$. The answer may, a priori at any rate, depend upon the choice of $\mathbf{F}$, of $\mathbf{G}$, and of S.

An affirmative answer has been established for the case $\mathbf{F} = \mathbf{G} = \mathbf{L}^1$ by Wells [1] and, independently, by Brainerd and Edwards [1], Part II, Theorem 3.3. In other words (see 16.3.3), a function ϕ is of type $(\mathbf{L}^1, \mathbf{L}^1; S)$ if and only if there exists a measure $\mu \in \mathbf{M}$ such that $\phi(n) = \hat{\mu}(n)$ for all $n \in S$. The arguments used in the latter reference could be adopted to deal with $(\mathbf{L}^1, \mathbf{M}; S)$ and $(\mathbf{L}^1, \mathbf{L}^p; S)$.

The answer is also affirmative when $\mathbf{F} = \mathbf{G} = \mathbf{L}^2$, as follows readily from the results of Chapter 8.

As far as the writer is aware, the problem is unsolved for all other choices of the pair $(\mathbf{F}, \mathbf{G})$ of the form $(\mathbf{L}^p, \mathbf{L}^q)$ and for the choice $\mathbf{F} = \mathbf{G} = \mathbf{C}$.

It is a simple matter to formulate the problem in terms of associated multiplier operators commuting with translations. Indeed, if we denote by $\mathbf{I} = \mathbf{I}(\mathbf{G}, S)$ the set of $g \in \mathbf{G}$ such that $\hat{g}(S) = 0$, the problem is virtually the same as that which asks for a representation theorem for those linear operators U from $\mathbf{F}$ into the quotient space $\mathbf{G}/\mathbf{I}$ that commute with translations. (Translation has a natural meaning in $\mathbf{G}/\mathbf{I}$, $T_a(g + \mathbf{I})$ being by definition the same as $T_a g + \mathbf{I}$.) In most cases of interest, $\mathbf{F}$ and $\mathbf{G}/\mathbf{I}$ will be such that the relevant operators U are continuous (compare with 16.2.1). The operator U_ϕ associated with a function ϕ of type $(\mathbf{F}, \mathbf{G}; S)$ will be defined so that $U_\phi f$ is the unique element $g + \mathbf{I}$ of $\mathbf{G}/\mathbf{I}$ defined by any $g \in \mathbf{G}$ satisfying (16.7.1).

(2) *Multipliers of invariant subspaces.* Viewed from the standpoint of multiplier operators in general, the problems discussed in (1) are closely related to those of multipliers of closed (translation-) invariant subspaces. The basis of the connection between the two types of problem lies in duality theory for topological linear spaces and especially the concept of adjoint operator (see [E], Chapter 8). We illustrate in the case of closed invariant subspaces (that is, closed ideals) $\mathbf{I}$ in $\mathbf{L}^p$, assuming that $1 < p < \infty$.

Let U be a multiplier operator of $\mathbf{I}$ that is, a continuous endomorphism of $\mathbf{I}$ which commutes with translations. We introduce the closed ideal $\mathbf{J}$ in $\mathbf{L}^{p'}$ defined by

$$\mathbf{J} = \{g \in \mathbf{L}^{p'} : f * g = 0 \text{ for all } f \in \mathbf{I}\}$$

and form the quotient space $\mathbf{L}^{p'}/\mathbf{J}$. The *adjoint operator* U' is the endomorphism of $\mathbf{L}^{p'}/\mathbf{J}$ defined in the following fashion: if $g \in \mathbf{L}^{p'}$, the linear functional on $\mathbf{I}$ defined by

$$f \to Uf * g(0)$$

is continuous and depends only on the coset $g + \mathbf{J}$. From I, C.1 it follows that there exists a unique coset $g' + \mathbf{J} = U'(g + \mathbf{J})$ such that

$$Uf * g(0) = f * g'(0) \qquad (f \in \mathbf{I}). \tag{16.7.2}$$

There is no difficulty in verifying that U' is a multiplier of $\mathbf{L}^{p'}/\mathbf{J}$, continuity of U' being interpreted relative to the quotient norm on $\mathbf{L}^{p'}/\mathbf{J}$ (see I, B.1.8).

The recovery of U from U' presents no trouble. If, for example, U' has been shown to be expressible in the form

$$U'(g + \mathbf{J}) = A * g + \mathbf{J} \qquad (g \in \mathbf{L}^{p'})$$

for some $A \in \mathbf{D}$, then (16.7.2) shows that

$$Uf = A * f \qquad (f \in \mathbf{I}).$$

A somewhat similar procedure allows one to pass from problems of type (2) back to problems of type (1).

Although the connections thus elicited between multipliers of invariant subspaces and multipliers of quotient spaces are theoretically satisfying and potentially useful, no *solved* cases of (1) are useful in studying type (2) problems.

For other aspects of this type of problem, see Wells [2]; Wada [1]; Meyer [1]; Glicksberg and Wik [1].

16.7.6. **Transformations of Fourier Coefficients.** It has become apparent that multiplier problems can be regarded as a study of subsets of the set of all continuous linear operators from $\mathbf{C}^\infty$ into $\mathbf{D}$ which commute with translations. One may ask how the picture changes if the condition of commutativity with translations is dropped.

It is quite simple to show that the continuous linear operators U from $\mathbf{C}^\infty$ into $\mathbf{D}$ are precisely those definable by a system of equations

$$(Uf)^\wedge(n) = \sum_{m \in Z} \phi(n, m)\hat{f}(m) \qquad \text{for all } f \in \mathbf{C}^\infty \text{ and all } n \in Z, \tag{16.7.3}$$

where ϕ is a complex-valued function on $Z \times Z$ determined by, and determining, U and subject to a majorization

$$|\phi(n, m)| \leqslant c(1 + |n|)^k(1 + |m|)^k \qquad \text{for all } n, m \in Z, \tag{16.7.4}$$

c and k being U-dependent numbers. In other words, such operators U may be regarded as linear transformations of Fourier coefficients defined by infinite matrices ϕ.

In this latter guise, such operators were discussed long ago by Hardy in special cases; more recent and more general studies are due to Bellman [1], Young [1], and Konyushkov [1]. In all cases, the principal aim has been to determine conditions on ϕ in order that U shall map $\mathbf{L}^p$ into itself.

16.8 Direct Sum Decompositions and Idempotent Multipliers

As an adjunct to this chapter, we return to the question (first broached in 2.2.1) of direct sum decompositions of the standard function spaces $\mathbf{C}$ and $\mathbf{L}^p$ in terms of their closed (translation-) invariant subspaces. Here we shall relate such decompositions to certain families of idempotent multipliers of $\mathbf{C}$ or $\mathbf{L}^p$, as the case may be; compare the remarks in 3.1.1.

The first two subsections are occupied by precise definitions of the decompositions and families involved.

Failing any special indication to the contrary $\mathbf{E}$ will, throughout this section, denote one of $\mathbf{C}$ or $\mathbf{L}^p$ $(1 \leqslant p < \infty)$.

16.8.1. **Direct Sum Decompositions.** By a *direct sum decomposition* of $\mathbf{E}$ we shall mean a family $(\mathbf{V}_i)_{i\in I}$ of closed (translation-) invariant subspaces $\mathbf{V}_i$ of $\mathbf{E}$ satisfying the following conditions:

(1) $\mathbf{V}_i \cap \mathbf{V}_j = \{0\}$ for all $i, j \in I$ such that $i \neq j$;

(2) the index set I is expressed in a definite way as the union of an increasing sequence $(I_r)_{r=1}^{\infty}$ of finite subsets I_r of I;

(3) to each $f \in \mathbf{E}$ corresponds at least one family $(f_i)_{i\in I}$ such that

$$f_i \in \mathbf{V}_i \qquad \text{for all } i \in I, \tag{16.8.1}$$

$$f = \lim_{r\to\infty} \sum_{i\in I_r} f_i \quad \text{in } \mathbf{E}. \tag{16.8.2}$$

Although it is tempting to write in place of (16.8.2) the relation

$$f = \sum_{i\in I} f_i,$$

some caution is required whenever I is infinite because there is no assurance whatsoever that the series will be unconditionally convergent; the specification of the sequence (I_r), which governs the grouping of terms, is one way of taking precautionary measures. Despite this, we shall frequently speak loosely of "a decomposition $(\mathbf{V}_i)$."

It will appear in the course of the proof of 16.8.3 that the family (f_i) referred to in (3) is uniquely determined by f, $(\mathbf{V}_i)$ and (I_r).

16.8.2. **Idempotent Decompositions of ε.** Our aim will be to relate any direct sum decomposition of **E** of the type described in 16.8.1 with a family $(\sigma_i)_{i \in I}$ of idempotent pseudomeasures (see Section 12.11) such that

$$\sigma_i \in \mathfrak{m}(\mathbf{E}, \mathbf{E}), \quad \sigma_i * \sigma_j = 0 \quad \text{for all } i, j \in I \text{ such that } i \neq j, \tag{16.8.3}$$

$$\sup_r \Big\| \sum_{i \in I_r} \sigma_i \Big\|_{\mathbf{E},\mathbf{E}} < \infty, \tag{16.8.4}$$

and

$$\lim_{r \to \infty} \sum_{i \in I_r} \sigma_i = \varepsilon. \tag{16.8.5}$$

The convergence referred to in (16.8.5) will be understood a priori to be weak convergence in **P** (the dual of **A**: see Section 12.11 and I, B.1.7); however, as will appear in stage (g) of the proof of 16.8.3, (16.8.4) ensures that (16.8.5) remains true with a stronger sense of convergence.

Idempotence of σ_i means, of course, that $\sigma_i * \sigma_i = \sigma_i$ or, what is equivalent, that $\hat{\sigma}_i$ is the characteristic function of some subset of Z.

The first clause of (16.8.3) signifies that $\sigma_i \in \mathbf{M}$ if $\mathbf{E} = \mathbf{C}$ or $\mathbf{L}^1$ (see 16.3.2) and $\sigma_i \in \mathfrak{m}(p, p)$ if $\mathbf{E} = \mathbf{L}^p$ (see the beginning of the present section). Moreover, $\|\cdot\|_{\mathbf{E},\mathbf{E}}$ means $\|\cdot\|_1$ or $\|\cdot\|_{p,p}$ according as **E** is **C** or $\mathbf{L}^p$, respectively. The second clause of (16.8.3) signifies that $S_i \cap S_j = \varnothing$ whenever $i, j \in I$ and $i \neq j$, where S_i denotes the support of $\hat{\sigma}_i$.

We shall speak of the family $(\sigma_i)_{i \in I}$ as an *idempotent decomposition of ε in* $\mathfrak{m}(\mathbf{E}, \mathbf{E})$.

We can now state and prove the principal result.

16.8.3. **Decomposition Theorem.** Let $(\mathbf{V}_i)_{i \in I}$ form a direct sum decomposition of **E**, as described in 16.8.1. Then there exists an idempotent decomposition $(\sigma_i)_{i \in I}$ of ε in $\mathfrak{m}(\mathbf{E}, \mathbf{E})$, as described in 16.8.2, such that:

(1) $\mathbf{V}_i = \{f \in \mathbf{E} : \sigma_i * f = f\} = \{f \in \mathbf{E} : \hat{f}(Z \backslash S_i) \subset \{0\}\}$ where S_i is the support of $\hat{\sigma}_i$;

(2) the decomposition of $f \in \mathbf{E}$ specified in 16.8.1(3) is unique and is given by

$$f_i = \sigma_i * f \qquad \text{for all } i \in I. \tag{16.8.6}$$

Conversely, given an idempotent decomposition $(\sigma_i)_{i \in I}$ of ε in $\mathfrak{m}(\mathbf{E}, \mathbf{E})$, let S_i be the support of $\hat{\sigma}_i$. Then the $\mathbf{V}_i$ defined as in (1) immediately above form a direct sum decomposition of **E** in which (16.8.6) holds for each $f \in \mathbf{E}$.

Proof. We begin with the direct assertion, the proof of which proceeds in a number of easy stages.

(a) By 11.1.2 and 11.2.1, to each $i \in I$ corresponds a subset S_i of Z such that $\mathbf{V}_i$ is the closed linear subspace of **E** generated by the e_n with $n \in S_i$ ($S_i = \varnothing$ if and only if $\mathbf{V}_i = \{0\}$); equivalently, $\mathbf{V}_i$ is the set of $f \in \mathbf{E}$ such that $\hat{f}(Z\backslash S_i) \subset \{0\}$. Conditions (1) and (3) of 16.8.1 entail that

$$S_i \cap S_j = \varnothing \qquad \text{for all } i, j \in I \text{ such that } i \neq j \tag{16.8.7}$$

and

$$Z = \bigcup_{i \in I} S_i. \tag{16.8.8}$$

(b) Given $f \in \mathbf{E}$, the decomposition referred to in 16.8.1(3) is unique.

To see this it is enough to show that the only family (f_i) such that $f_i \in \mathbf{V}_i$ for each i and

$$\lim_{r \to \infty} \sum_{i \in I_r} f_i = 0 \tag{16.8.9}$$

is that for which $f_i = 0$ for each i. But choose and fix any $j \in I$. If $\mathbf{V}_j = \{0\}$, then $f_j = 0$. If $\mathbf{V}_j \neq \{0\}$, then $S_j \neq \varnothing$. If $n \in S_j$, (a) shows that $\hat{f}_i(n) = 0$ whenever $i \neq j$. Accordingly, (16.8.9) entails that

$$\lim_{r \to \infty} \sum_{i \in I_r} \hat{f}_i(n) = 0,$$

which reduces to $f_j(n) = 0$. Thus $\hat{f}_j(S_j) \subset \{0\}$ and $\hat{f}_j(Z\backslash S_j) \subset \{0\}$, and 2.4.1 shows that $f_j = 0$.

(c) From (b) it follows that to each i corresponds a map $P_i: f \to f_i$ of $\mathbf{E}$ into $\mathbf{V}_i$; that P_i is linear; that $P_i^2 = P_i$ (so that P_i actually maps $\mathbf{E}$ onto $\mathbf{V}_i$); and that P_i commutes with translations (since each $\mathbf{V}_i$ is translation-invariant).

(d) Let us next show that each P_i is a continuous endomorphism of $\mathbf{E}$, in doing which we shall use the closed graph theorem (I, B.3.3).

Take any sequence $(f^k)_{k=1}^{\infty}$ extracted from $\mathbf{E}$ such that $f^k \to 0$ in $\mathbf{E}$ and $P_i f^k \to g$ in $\mathbf{E}$: we must show that $g = 0$.

We may, by (16.8.2) and (a), write

$$f^k = P_i f^k + h_k,$$

where $\hat{h}_k$ vanishes at all points of S_i, so that

$$\hat{f}^k = (P_i f^k)\hat{} \qquad \text{on } S_i.$$

On letting $k \to \infty$, it follows that $\hat{g} = \lim_{k \to \infty} \hat{f}^k = 0$ on S_i. Since $\mathbf{V}_i$ is closed in $\mathbf{E}$, $g \in \mathbf{V}_i$; since it has a Fourier transform vanishing on S_i, it follows that g_i must be 0, as required.

(e) By (c) and (d), P_i is a multiplier of $\mathbf{E}$, so that there exists a pseudo-measure $\sigma_i \in \mathfrak{m}(\mathbf{E}, \mathbf{E})$ such that

$$P_i f = \sigma_i * f \tag{16.8.10}$$

for each $f \in \mathbf{E}$. Since P_i is idempotent (see (c)), the same must be true of σ_i. Therefore $\hat{\sigma}_i$ is the characteristic function of a subset of Z which, since P_i maps $\mathbf{E}$ onto $\mathbf{V}_i$, must be S_i:

$$\hat{\sigma}_i = \chi_{S_i}. \tag{16.8.11}$$

(f) The relation (16.8.2) can now be written

$$\lim_{r \to \infty} \{\sum_{i \in I_r} \sigma_i\} * f = f \qquad \text{in } \mathbf{E} \tag{16.8.12}$$

for each $f \in \mathbf{E}$. From this (16.8.4) follows by appeal to the uniform boundedness principle (I, B.2.2), while (16.8.5) is an immediate consequence of (16.8.12), (16.8.4) and the remark that a sequence $(\lambda_r)_{r=1}^{\infty}$ of pseudomeasures converges weakly in $\mathbf{P}$ to a pseudomeasure λ if and only if $\lim_{r\to\infty} \hat{\lambda}_r = \hat{\lambda}$ boundedly on Z.

On collecting together the results (a) to (f), it is easily seen that the direct assertion is now completely proved.

(g) As for the converse, all that requires proof is the assertion that (16.8.4) and (16.8.5) together entail (16.8.12). Define $\lambda_r \in \mathfrak{m}(\mathbf{E}, \mathbf{E})$ to be

$$\sum_{i \in I_r} \sigma_i,$$

so that (16.8.4) signifies that the λ_r are equicontinuous endomorphisms of $\mathbf{E}$.

Now (16.8.5) ensures that (16.8.12) is true for each trigonometric polynomial f. Since the trigonometric polynomials are dense in $\mathbf{E}$ (see 2.4.4), equicontinuity of the λ_r shows that (16.8.12) does indeed persist for a general $f \in \mathbf{E}$.

To verify this final point (which is a perfectly general principle), observe that the equicontinuity of the λ_r means that

$$m \equiv \sup_r \|\lambda_r\|_{\mathbf{E},\mathbf{E}} < \infty,$$

so that

$$\|\lambda_r * f - f\|_{\mathbf{E}} \leqslant (m + 1)\|f\|_{\mathbf{E}} \tag{16.8.13}$$

for every $f \in \mathbf{E}$. Given $\delta > 0$ and $f \in \mathbf{E}$, choose a trigonometric polynomial f_0 such that

$$\|f - f_0\|_{\mathbf{E}} \leqslant \frac{\delta}{(m + 1)}.$$

An application of (16.8.13), with $f - f_0$ written in place of f, shows that

$$\|\lambda_r * f - f\|_{E} \leqslant \|\lambda_r * f_0 - f_0\|_{\mathbf{E}} + \delta.$$

So, since $\lim_{r\to\infty} \lambda_r * f_0 = f_0$ in $\mathbf{E}$,

$$\|\lambda_r * f - f\|_{\mathbf{E}} \leqslant 2\delta$$

if $r \geqslant r_0(\delta)$, and the convergence of $(\lambda_r * f)_{r=1}^{\infty}$ follows.

16.8.4. **Remarks and Special Cases.** (1) The summand f_i in (16.8.2) has now been identified with $\sigma_i * f$, that is, with

$$\sum_{n \in S_i} \hat{f}(n) e_n;$$

if S_i is infinite, this infinite sum is interpretable (since σ_i is known to define a continuous endomorphism of $\mathbf{E}$) as the limit in $\mathbf{E}$, as $N \to \infty$, of

$$\sum_{n \in S_i, |n| \leqslant N} \left(1 - \frac{|n|}{N + 1}\right) \hat{f}(n) e_n.$$

When $\mathbf{E} = \mathbf{L}^p$, where $1 < p < \infty$, the summation factors may be omitted.

Accordingly, 16.8.3 verifies in detail what might have been anticipated on heuristic grounds, namely, that a direct sum decomposition of $\mathbf{E}$ corresponds to a mode of bracketing of the terms of the Fourier series of a general $f \in \mathbf{E}$ in such a way that the resulting series is convergent in $\mathbf{E}$. (A bracket may contain infinitely many terms of the original series.)

(2) *The Cases* $\mathbf{E} = \mathbf{C}$ *or* $\mathbf{L}^1$. When $\mathbf{E} = \mathbf{C}$ or $\mathbf{L}^1$, each σ_i is a measure; and, by Helson's theorem [cited in 12.7.4(3)], S_i differs by a finite set from a periodic subset of Z. It is quite simple to verify that this means that σ_i is of the form

$$\sigma_i = u_i \cdot \mu_{a_i} + v_i, \tag{16.8.14}$$

where $a_i \in Z$, $a_i > 0$;

$$\mu_{a_i} = a_i^{-1} \sum_{k=0}^{a_i - 1} \varepsilon_{(2\pi k/a_i)},$$

which is the invariant measure of the cyclic subgroup of T of order a_i; u_i is a trigonometric polynomial of the form

$$u_i = \sum_{k=0}^{a_i - 1} c_{i,k} e_k,$$

each $c_{i,k}$ being 0 or 1; and v_i is a trigonometric polynomial.

The measures

$$\lambda_r = \sum_{i \in I_r} \sigma_i,$$

which are likewise idempotent, must have the same general form (16.8.14).

From the preceding remarks, it is easily seen how to construct direct sum decompositions of $\mathbf{E}$ of the type specified in 16.8.1 in which I is finite and each S_i is infinite.

On the other hand, there exist no decompositions in which each S_i is finite (that is, in which each $\mathbf{V}_i$ is of finite dimension). Indeed, in any such decomposition the equicontinuity of the endomorphisms λ_r (see part (g) of the proof of 16.8.3) would in this case mean that

$$\sup_r \|\lambda_r\|_1 < \infty. \tag{16.8.15}$$

However,

$$\lambda_r = \sum_{n \in T_r} e_n,$$

where

$$T_r = \bigcup_{i \in I_r} S_i$$

and the result cited in 12.7.4(4)(ii) shows that

$$\| \lambda_r \|_1 > \text{const} \log k_r$$

where k_r is the number of elements of T_r. Since $k_r \to \infty$ with r, this would contradict (16.8.15).

(3) *The Case* $\mathbf{E} = \mathbf{L}^p$ $(1 < p < \infty)$. Except when $p = 2$, there is no known explicit description of the idempotent elements σ of $\mathfrak{m}(\mathbf{E}, \mathbf{E}) = \mathfrak{m}(p, p)$ analogous to (16.8.14) and therefore no complete classification of all possible decompositions of $\mathbf{E}$ of the type described in 16.8.1. (When $p = 2$, all idempotent pseudomeasures are acceptable and all possible bracketings lead to decompositions.)

Theorem 12.10.1 guarantees the decomposition in which $I = \{0, 1, 2, \cdots\}$, $I_r = \{0, 1, \cdots, r - 1\}$ for $r = 1, 2, \cdots$, and $S_i = \{\pm i\}$ for $i \in I$. Other decompositions may be derived from this one by arbitrary *finite* bracketing.

Yet other decompositions result from the use of the idempotent measures of the type (16.8.14).

(4) For further study and results, see [Ro]; Rosenthal [1]; Rudin [8]; Price [3], Chapter 4. See also Exercise 16.30.

16.9 Absolute Multipliers

16.9.1. **Definition.** Given a set of distributions $\mathbf{F}$ (as in 16.1.1) and a complex-valued function φ on Z, φ is said to be an *absolute multiplier* of $\mathbf{F}$, if and only if $\varphi\hat{f} \in \ell^1(Z)$ for every $f \in \mathbf{F}$. This signifies that φ is a multiplier function of type $(\mathbf{F}, \mathbf{A})$, in the sense of 16.1.1; or, equivalently, that the operator U_φ maps $\mathbf{F}$ into $\mathbf{A}$.

16.9.2. **Examples.** (i) It follows from Exercise 3.14 that the absolute multipliers of $\mathbf{L}^1$ are precisely the elements of $\ell^1(Z)$.

(ii) By 14.3.5(2′), φ is an absolute multiplier of $\mathbf{C}$, if and only if $\varphi \in \ell^2(Z)$. Alternative proofs of this and various analogous results appear in Edwards, Hewitt and Ritter [1], Hewitt and Ritter [1] and Edwards and Helson [1]; cf. also Brown [1], [2].

(iii) Paley [1] proved in effect that φ is an absolute multiplier of

$$\mathbf{C}_{Z_+} = \{f \in \mathbf{C} \colon \operatorname{supp} \hat{f} \subseteq Z_+\},$$

where Z_+ denotes the set of nonnegative integers (cf. the spaces $\mathbf{H}^p$ defined in Exercise 3.9 and the space $\mathbf{A}^+$ mentioned in the Remark attached to Exercise 11.15), if and only if $\varphi \mid Z_+$ (the restriction of φ to Z_+) belongs to $\ell^2(Z_+)$.

A generalisation of this theorem to a general class of compact Abelian groups appears in [R], Section 8.7.8; see the next subsection. See also Fournier [3], and Helson [7].

Sometimes $\mathbf{C}_{Z_+}$ is denoted by $\mathbf{C}_A$ (the suffix "A" indicating "analytic type" or "power series type"); we shall here avoid this notation because of the obvious conflict with that employed in Chapter 15.

16.9.3. **Paley's theorem and its proof.** To reiterate, the theorem asserts that a complex-valued function φ on Z is an absolute multiplier of $\mathbf{C}_{Z_+}$, if and only if $\varphi \mid Z_+ \in \ell^2(Z_+)$.

The "if" part is close to trivial.

The proof of "only if" to follow is based upon that appearing in [R], Section 8.7.8. To lighten the notation we shall, throughout the proof to follow, write $\mathbf{E}$ in place of $\mathbf{C}_{Z_+}$.

To begin with, it follows from (12.9.9) (proved in 13.9.2) that, for every $p \in (0, 1)$, there exists a number $k_p \geqslant 0$, independent of h, such that

$$\left\| \sum_{n \in Z_+} \hat{h}(n) e_n \right\|_p \leqslant k_p \|h\|_1 \tag{16.9.1}$$

for all trigonometric polynomials h.

Assume that φ is an absolute multiplier of $\mathbf{E}$. An appeal to either the Boundedness Principle or the Closed Graph Theorem (Volume 1, Appendix B.2.1(1) or B.3.3) shows that there exists a number $k \geqslant 0$ such that, for all $f \in \mathbf{E}$,

$$\sum_{n \in Z_+} |\varphi(n) \hat{f}(n)| \leqslant k \|f\|_\infty . \tag{16.9.2}$$

Let F be a finite subset of Z_+, expressed as the range of an injective sequence $(n_k)_{1 \leqslant k \leqslant N}$. By Exercise 2.19, there exists a trigonometric polynomial P such that

$$\|P\|_1 \leqslant 2 \tag{16.9.3}$$

and

$$\hat{P}(n) = 1 \qquad \text{for all } n \in F. \tag{16.9.4}$$

Define, for all $\omega \in \mathscr{C}$ and all real x,

$$g_\omega(x) = \sum_{j=1}^{N} \rho_j(\omega) \varphi(n_j) \exp(i n_j x). \tag{16.9.5}$$

By (16.9.2) and the Hahn-Banach theorem (Volume 1, Appendix B.5.1), combined with 12.2.3, 12.2.9 and 12.3.8, for every $\omega \in \mathscr{C}$ there exists a measure μ_ω such that, for all $\omega \in \mathscr{C}$ and all $f \in \mathbf{E}$,

$$\|\mu_\omega\|_1 \leqslant k \tag{16.9.6}$$

and

$$\sum_{j=1}^{N} \rho_j(\omega) \varphi(n_j) \hat{f}(n_j) = \mu_\omega(\check{f}). \tag{16.9.7}$$

From (16.9.7) it follows in particular that, for all $\omega \in \mathscr{C}$,

$$\hat{\mu}_\omega(n_j) = \varphi(n_j) \rho_j(\omega) \qquad \text{for all } j \in \{1, \cdots, N\}$$

and

$$\hat{\mu}_\omega(n) = 0 \qquad \text{for all } n \in Z_+ \backslash \{n_1, \cdots, n_N\}.$$

Hence, for all $\omega \in \mathscr{C}$ and all $n \in Z_+$,

$$(P * \mu_\omega)^\wedge(n) = \hat{g}_\omega(n).$$

Therefore, by (16.9.1), (16.9.3) and (16.9.6),

$$\begin{aligned}\left(\int_0^{2\pi} |g_\omega(x)|^{1/2}\,dx\right)^2 &= \|g_\omega\|_{1/2} \leqslant k_{1/2}\|P * \mu_\omega\|_1 \\ &\leqslant k_{1/2}\|P\|_1\|\mu_\omega\|_1 \qquad \text{(see 12.7.3)} \\ &\leqslant 2k_{1/2}k = k' \qquad \text{for all } \omega \in \mathscr{C}. \qquad (16.9.8)\end{aligned}$$

Integrating with respect to ω over $\mathscr{C}$ and interchanging the order of integrations, (16.9.8) implies that

$$\int_0^{2\pi}\left(\int_{\mathscr{C}} |g_\omega(x)|^{1/2}\,d\lambda(\omega)\right)dx \leqslant k'^{1/2},$$

whence follows the existence of $x_0 \in R$ such that

$$\int_{\mathscr{C}} |g_\omega(x_0)|^{1/2}\,d\lambda(\omega) \leqslant k'^{1/2}. \qquad (16.9.9)$$

Defining, for all $\omega \in \mathscr{C}$,

$$h(\omega) = \sum_{j=1}^{N} \rho_j(\omega)\varphi(n_j)\exp(in_j x_0),$$

h is a Rademacher polynomial; and, by (16.9.9),

$$\int_{\mathscr{C}} |h(\omega)|^{1/2}\,d\lambda(\omega) \leqslant k'^{1/2}. \qquad (16.9.10)$$

On the other hand, by (14.2.1), there exists a number $k_2 \geqslant 0$ such that

$$\left(\int_{\mathscr{C}} |h(\omega)|^2\,d\lambda(\omega)\right)^{1/2} \leqslant k_2\left(\int_{\mathscr{C}} |h(\omega)|^{1/2}\,d\lambda(\omega)\right)^2. \qquad (16.9.11)$$

By (16.9.10) and (16.9.11),

$$\left(\int_{\mathscr{C}} |h(\omega)|^2\,d\lambda(\omega)\right)^{1/2} \leqslant k_2 k'.$$

This, together with (16.9.5) and 14.1.7 or 14.1.16, implies that

$$\sum_{n \in F} |\varphi(n)|^2 = \sum_{j=1}^{N} |\varphi(n_j)|^2 \leqslant k_2 k'^2. \qquad (16.9.12)$$

Since k_2 and k' are independent of F, it follows from (16.9.12) that

$$\sum_{n \in Z_+} |\varphi(n)|^2 < \infty,$$

which completes the proof.

16.9.4. **Remarks.** (1) It is clear that Paley's theorem holds, with $\mathbf{C}_{Z_+}$ replaced by

$$\mathbf{C}_S = \{f \in \mathbf{C}: \operatorname{supp} \hat{f} \subset S\},$$

whenever S is a subset of Z having the property that there exists $p > 0$ and $k \geqslant 0$ such that

$$\left\| \sum_{n \in S} \hat{h}(n) e_n \right\|_p \leqslant k \|h\|_1 \qquad \text{for all } h \in \mathbf{T}. \tag{16.9.1'}$$

No systematic study of the class of such sets S seems to have been published. A few properties of this class are discussed in Exercise 16.31.

(2) The appearance of $\mathbf{C}_{Z_+}$ in Paley's theorem naturally suggests the systematic consideration of multipliers of the Hardy spaces $\mathbf{H}^p$ defined in Exercise 3.19.

A good deal of work has been done on this topic, but space forbids an attempt at any account in this book. Any interested reader should consult, for example, Meyer [1]; Gaudry [5]; Hedland [1]; Yamaguchi [1]; and the references listed in these papers.

16.10 Multipliers of weak type (p, p)

A multiplier ϕ of type (p, p), as described in 16.1.1, defines and is defined by a linear operator U_ϕ from $\mathbf{L}^p$ into itself having the property that $(U_\phi f)^\wedge = \phi \cdot \hat{f}$ for every $f \in \mathbf{L}^p$. As is proved in 16.21, U_ϕ is necessarily continuous. In view of the terminology introduced in 13.2.1, such a multiplier ϕ might for emphasis be said to be of *strong type* (p, p).

Zafran [4] has studied multipliers ϕ which he describes as being of "weak type (p, p)", a description which is suggested by the terminology introduced in 13.7.5 above. If confusion is to be avoided, a little care is needed in approaching this new concept because, in spite of what might be read into the description "of weak type (p, p)", such a multiplier does not in general map $\mathbf{L}^p$ into itself, but rather into a bigger space $\mathbf{L}^{p, \infty}$ to be defined in 16.10.1. This in turn raises the possibility that ϕ may fail to be of strong type (p, p). It is, on the other hand, by no means obvious that this possibility is actually realised, a proof of which is Zafran's achievement.

Although Zafran's work applies to certain other familiar groups, we shall concentrate on the case of the circle group T.

Throughout this section, it will be assumed that $1 \leqslant p < \infty$.

16.10.1. **The space $\mathbf{L}^{p,\infty}$.** The space $\mathbf{L}^{p,\infty} \equiv \mathbf{L}^{p,\infty}(T)$ is defined to be the complex linear space of all complex-valued measurable functions f on T such that

$$\| f \|_{p,\infty}^{*} \equiv \sup \{t(D_f(t))^{1/p} : t \text{ real and } t > 0\} < \infty, \qquad (16.10.1)$$

D_f being defined as in 13.7.2. (As with the spaces $\mathbf{L}^p$, the elements of $\mathbf{L}^{p,\infty}$ are, in many contexts, to be thought of as equivalence classes, modulo equality almost everywhere, of functions f satisfying (16.10.1).) For further details about $\mathbf{L}^{p,\infty}$ (which is one amongst a class of so-called Lorentz spaces), see [StW], Chapter V, §3 where $\mathbf{L}^{p,\infty}$ is denoted by $\mathbf{L}(p, \infty)$.

It is evident that $\mathbf{L}^{p,\infty}$ is translation invariant. From Exercise 13.16 it follows that $\mathbf{L}^{p,\infty}$ is a subset of $\mathbf{L}^q$ for every q satisfying $0 < q < p$. It can also be proved that $\mathbf{L}^p$ is a proper subset of $\mathbf{L}^{p,\infty}$; see Exercise 16.32.

The function $f \to \| f \|_{p,\infty}^{*}$ defined by (16.10.1) is not subadditive and hence not a norm, but there is (see [StW], p. 204) a norm $\| \cdot \|_{p,\infty}$ on $\mathbf{L}^{p,\infty}$ which, if $1 < p < \infty$, makes $\mathbf{L}^{p,\infty}$ into a Banach space and which is such that

$$\| f \|_{p,\infty}^{*} \leqslant \| f \|_{p,\infty} \leqslant p(p-1)^{-1} \| f \|_{p,\infty}^{*}$$

for every $f \in \mathbf{L}^{p,\infty}$.

16.10.2 **Multipliers of weak type** (p, p). By definition, a complex-valued function ϕ on Z (and/or the associated operator U_ϕ) is said to be of *weak type* (p, p), if and only if it is a multiplier of type $(\mathbf{L}^p, \mathbf{L}^{p,\infty})$ in the sense described in 16.1.1. This is the case, if and only if there exists a number $K \geqslant 0$ such that, for every $f \in \mathbf{L}^p$, the function $g = U_\phi f$ satisfies

$$D_g(t) \leqslant K \cdot \| f \|_p^p \cdot t^{-p} \qquad \text{for every real } t > 0; \qquad (16.10.2)$$

cf. 13.7.5.

To make things a little more explicit, ϕ is of weak type (p, p), if and only if (16.10.2) holds for all $f \in \mathbf{T}$, $g = U_\phi f$ being in this case the trigonometric polynomial

$$\sum_{n \in Z} \phi(n) \hat{f}(n) e_n.$$

The extension of U_ϕ to $\mathbf{L}^p$ is then obtained as follows: given $f \in \mathbf{L}^p$, choose a $\mathbf{T}$-valued sequence (f_n) converging in $\mathbf{L}^p$ to f; then, by Exercise 13.19 and (16.10.2) applied for trigonometric polynomials, the sequence $(g_n) = (U_\phi f_n)$ converges in measure to a function g, the equivalence class of g being independent of the sequence (f_n); then $U_\phi f$ is, by definition, the equivalence class of g; and (16.10.2) continues to hold for every $f \in \mathbf{L}^p$; in particular, $U_\phi f \in \mathbf{L}^{p,\infty}$ for every $f \in \mathbf{L}^p$.

Evidently, every multiplier ϕ of (strong) type (p, p) is also of weak type (p, p). Regarding the converse, see the next subsection.

16.10.3 **Zafran's theorem.** The substance of 16.10.1 and 16.10.2 makes it evident that every multiplier of weak type (p, p) is a multiplier of strong type (p, q) for every $q < p$. On the other hand, the substance of 12.8.1, 12.8.4, 12.9.1 and 13.9.1, shows that the functions ϕ_0, such that $\phi_0(n)$ is 1 or 0 according as $n \in Z$ satisfies $n \geqslant 0$ or $n < 0$ respectively, is a multiplier of weak type (1, 1), is *not* of strong type (1, 1), and is of strong type (q, q) for all q such that $0 < q < 1$.

Zafran's achievement in [4] is to prove (among other things) that part of this phenomenon is reproduced for every p satisfying $1 < p < 2$. More precisely: for every p satisfying $1 < p < 2$, there exists a multiplier ϕ of weak type (p, p) which is not of strong type (p, p). The proof is quite elaborate and will not be presented here.

It is to be noted that, if ϕ is of weak type (2, 2), then (trivially) ϕ is bounded and hence (by 16.1.2(4)) of strong type (2, 2); thus, Zafran's result breaks down in the excluded case $p = 2$.

EXERCISES

16.1. Verify the statement made in 16.1.2(3). What is the representation of the corresponding operator U_ϕ in terms of convolution?

16.2. Prove that $(\mathbf{L}^2, \mathbf{L}^2) = \ell^\infty(Z)$.

16.3. Verify the inclusion relations stated in 16.1.2(5).

16.4. Suppose that $\phi \in (\mathbf{F}, \mathbf{D})$ and that $\lim_{k\to\infty} f_k = 0$ in $\mathbf{F}$ implies that $\lim_{k\to\infty} \|\hat{f}_k\| = 0$. Without using the closed graph theorem, prove that U_ϕ is continuous from $\mathbf{F}$ into $\mathbf{D}$ (compare the Remark following 16.2.1).

16.5. Verify that $\sigma_N f \to f$ weakly in $\mathbf{L}^\infty$ whenever $f \in \mathbf{L}^\infty$, and that $\sigma_N \mu \to \mu$ weakly in $\mathbf{M}$ whenever $\mu \in \mathbf{M}$. (See 16.2.5.)

16.6. Let U be a continuous linear operator from $\mathbf{F} = \mathbf{C}^k$ or $\mathbf{L}^p$ $(1 \leqslant p < \infty)$ into $\mathbf{G} = \mathbf{C}^h$, $\mathbf{L}^q$ $(1 \leqslant q < \infty)$, $\mathbf{M}$ or $\mathbf{D}$. Prove that U commutes with translations if and only if

$$U(t * f) = t * Uf$$

whenever t is a trigonometric polynomial and $f \in \mathbf{F}$; and that in this case

$$U(k * f) = k * Uf$$

whenever $k \in \mathbf{L}^1$ and $f \in \mathbf{F}$.

16.7. Using 16.3.2, show that every $U \in \mathfrak{m}(\mathbf{C}, \mathbf{L}^\infty)$ can be expressed in the form

$$Uf = \mu * f \qquad \text{for all } f \in \mathbf{C}$$

for some $\mu \in \mathbf{M}$, and that the analogous statement is true for every $U \in \mathfrak{m}(\mathbf{L}^\infty, \mathbf{L}^\infty)$.

Deduce that $(\mathbf{C}, \mathbf{L}^\infty) = (\mathbf{L}^\infty, \mathbf{L}^\infty) = \mathscr{F}\mathbf{M}$.

16.8. Prove that $(\mathbf{M}, \mathbf{M}) = \mathscr{F}\mathbf{M}$.

16.9. Suppose that $A \in \mathbf{D}$ has the property that $A * f \in \mathbf{C}$ for every $f \in \mathbf{L}^p$, where $1 \leqslant p \leqslant \infty$. Prove that $A \in \mathbf{L}^{p'}$.

16.10. Using the notation of 16.3.7, verify that $\mathbf{L}_b{}^\infty$ is a Banach space which satisfies the standing hypotheses listed in 16.1.1.

16.11. Construct detailed proofs of the results stated in 16.3.9. (You will need to call upon the result stated in Exercise 12.23.)

16.12. Write out a detailed proof of 16.4.2.

16.13. Show that $U \in \mathfrak{m}(\mathbf{A}, \mathbf{A})$ (where $\mathbf{A}$ is as in Section 10.6) if and only if U is expressible in the form

$$Uf = S * f, \qquad \text{for all } f \in \mathbf{A},$$

where S is a pseudomeasure.

16.14. Show how to use Stein's theorem cited in 16.2.8 and Lebesgue's theorem cited in 6.4.2 to prove that the Hardy-Littlewood maximal operators defined in Example 13.7.6(2) are of weak type (1, 1) on $\mathbf{L}^1$.

16.15. Verify the statement made in 16.5.1 concerning the nature of the elements of $\mathbf{V}_\infty$.

Hint: If $\Phi \in \mathbf{V}_\infty$, consider the functions $\Phi_N = F_N * \Phi$; these are trigonometric polynomials which are of uniformly bounded variation. Show that a subsequence of (Φ_N) may be selected which converges pointwise to a function of bounded variation.

16.16. Verify that $\hat{\mu}\phi \in (p, p)$ and

$$\|\hat{\mu}\phi\|_{p,p} \leqslant \|\mu\|_1 \|\phi\|_{p,p}$$

whenever $\mu \in \mathbf{M}$ and $\phi \in (p, p)$. (The notation is as in Section 16.6.)

16.17. Exhibit a sequence $(f_n)_{n=1}^\infty$ of nonnull trigonometric polynomials such that

$$\lim_{n \to \infty} \frac{\|f_n\|_{p,p}}{\|f_n\|_1} = 0$$

for every p satisfying $1 < p < \infty$, where

$$\|f\|_{p,p} \equiv \sup\{\|f * g\|_p : g \in \mathbf{L}^p,\ \|g\|_p \leqslant 1\}.$$

16.18. Suppose that $U \in \mathfrak{m}(p, q)$, where $1 \leqslant p < \infty$ and $1 \leqslant q \leqslant \infty$. Show that there exists a sequence $(h_i)_{i=1}^\infty$ of trigonometric polynomials such that

$$Uf = \lim_{i \to \infty} h_i * f \quad \text{in } \mathbf{L}^q,$$

uniformly with respect to $f \in \mathbf{Q}$ whenever $\mathbf{Q}$ is a relatively compact subset of $\mathbf{L}^p$, and

$$\|h_i * f\|_q \leqslant \|U\|_{p,q} \|f\|_q$$

for all i and all $f \in \mathbf{L}^p$.

16.19. Suppose that $U \in \mathfrak{m}(\mathbf{L}^p, \mathbf{L}^\infty)$, where $1 \leqslant p < \infty$. Show that, if one agrees to identify any two functions that agree almost everywhere, then $U \in \mathfrak{m}(\mathbf{L}^p, \mathbf{C})$.

Hint: See Exercise 3.5.

16.20. Show that, with the convention mentioned in Exercise 16.19, $\mathfrak{m}(\mathbf{C}, \mathbf{L}^\infty) = \mathfrak{m}(\mathbf{C}, \mathbf{C}) = \mathbf{M}$.

Hint: This is as for Exercise 16.19, together with a reference to 16.3.2.

16.21. Suppose that $1 \leqslant p < \infty$, $1 \leqslant q \leqslant \infty$, $\phi_k \in (p, q)$ $(k = 1, 2, \cdots)$, that

$$\phi = \lim_{k \to \infty} \phi_k \qquad \text{pointwise on } Z,$$

and that

$$\ell \equiv \lim_{k \to \infty} \|\phi_k\|_{p,q} < \infty.$$

Prove that $\phi \in (p, q)$, that

$$\|\phi\|_{p,q} \leqslant \ell,$$

and that $\lim_{k \to \infty} U_{\phi_k} f = U_\phi f$ in $\mathbf{L}^q$ for each $f \in \mathbf{L}^p$.

What can you prove of a similar nature in case $p = \infty$?

16.22. Suppose that $U \in \mathfrak{m}(1, p)$, where $1 < p \leqslant \infty$. Show that there exist positive integers n such that $U^n \in \mathfrak{m}(\mathbf{L}^1, \mathbf{C})$ and make an estimate of the smallest positive integer n with this property.

Is anything of a similar nature true for operators $U \in \mathfrak{m}(p, q)$, where $1 < p < q \leqslant \infty$?

16.23. Suppose that $A \in \mathbf{D}$ and that $\hat{A}^{-1}(0)$ and the range of $\hat{A}$ are finite. Show that $A \in \mathfrak{m}(p, q)$ is false whenever $1 \leqslant p < q \leqslant \infty$.

16.24. Suppose that $1 \leqslant p \leqslant 2$ and that F is a complex-valued function on Z. Prove that $F\omega \in (p, p)$ for every ± 1-valued function ω on Z if and only if $F \in (p, 2)$.

What is the appropriate version of this result when $2 < p \leqslant \infty$?

16.25. Let U be a compact multiplier of type (p, q), where $1 \leqslant p$, $q \leqslant \infty$, represented by convolution with a pseudomeasure σ. Show that $\hat{\sigma} \in \mathbf{c}_0(Z)$.

State and prove a converse for the case in which $p \geqslant 2 \geqslant q$.

Remark. Stronger results are known; see the references cited in 16.7.2.

16.26. Let **E** and **F** each denote one of the spaces $\mathbf{C}^k$ ($k \in Z$, $k \geqslant 0$), $\mathbf{L}^p$ ($1 \leqslant p \leqslant \infty$), or **M**. For $m \in Z$, $m \geqslant 0$, let $\mathbf{W}_{\mathbf{E}}^{(m)}$ denote the set of $f \in \mathbf{E}$ such that $D^r f \in E$ for $r \in Z$, $0 \leqslant r \leqslant m$. (Regarding this notation, see [So], p. 45 and [L_2], p. 50.)

Show that $(\mathbf{W}_{\mathbf{E}}^{(m)}, \mathbf{W}_{\mathbf{F}}^{(m)}) = (\mathbf{E}, \mathbf{F})$.

16.27. Let **AC** denote the space of absolutely continuous functions endowed with the norm

$$\|f\| = |f(0)| + \|Df\|_1.$$

Show that $\mathfrak{m}(\mathbf{AC}, \mathbf{L}^\infty) = \mathfrak{m}(\mathbf{AC}, \mathbf{C})$ consists precisely of those distributions of the form const $+ D\phi$, where $\phi \in \mathbf{L}^\infty$ [compare 12.8.5(3)].

16.28. Let each of **E** and **F** denote one of **C**, $\mathbf{L}^p$ ($1 \leqslant p \leqslant \infty$), or **M**, and let $A \in \mathfrak{m}(\mathbf{E}, \mathbf{F})$. Consider the equation

$$A * f = g. \tag{1}$$

Prove the following statements.

(a) If (1) is soluble for $f \in \mathbf{E}$ whenever $g \in \mathbf{F}$, then

$$\inf_{n \in Z} |\hat{A}(n)| > 0, \tag{2}$$

and therefore (1) is *uniquely* soluble for $f \in \mathbf{E}$ whenever $g \in \mathbf{F}$.

(b) If (1) is soluble for $f \in \mathbf{E}$ whenever $g \in \mathbf{F}$, then there exists $B \in \mathfrak{m}(\mathbf{F}, \mathbf{E})$ such that

$$A * B = \varepsilon. \tag{3}$$

(c) If A has the form $\delta + h$, where δ is a discrete measure (see Exercise 12.51) and $h \in \mathbf{L}^1$, and if (2) is satisfied, then (1) has a unique solution $f \in \mathbf{E}$ whenever $g \in \mathbf{E}$.

16.29. Discuss analogues, for the case in which T and Z interchange their roles, of the results in 16.3.2 to 16.3.6 and 16.4.1 to 16.4.6.

16.30. Suppose that E is a subset of Z and that χ_E denotes the characteristic function of E relative to Z. Discuss connections between assertions of the type $E \in \Lambda(p)$ and those of the type $\chi_E \in (q, r)$.

As an instance, prove that if $p > 2$ and $E \in \Lambda(p)$, then $\chi_E \in (p', p)$.

16.31. Refer to 16.9.4. Denote by $\mathfrak{P}$ *the set of all subsets* S of Z for each of which there exist $p > 0$ and $k > 0$ such that

$$\left\| \sum_{n \in S} \hat{h}(n) e_n \right\|_p \leqslant k \|h\|_1 \qquad \text{for all } h \in \mathbf{T}.$$

Prove the following (in which χ_S denotes the characteristic function of S relative to Z):

(a) If $S \in \mathfrak{P}$, then $Z \backslash S \in \mathfrak{P}$.

(b) If $\chi_S \in \mathscr{F}\mathbf{M}$, then $S \in \mathfrak{P}$.

(c) If $S \in \mathfrak{P}$ and S is $\Lambda(1)$, then $\chi_S \in \mathscr{F}\mathbf{M}$ and hence (see 12.7.4(3) and Exercise 15.9) S is finite.

(d) If $S \in \mathfrak{P}$ and S is q-Sidon for some $q < 2$ (see 15.8.1 above), then S is finite.

(e) If the operator

$$h \to \sum_{n \in S} \hat{h}(n) u_n$$

is of weak type (1, 1), then $S \in \mathfrak{P}$.

16.32. Assume that $1 \leqslant p < \infty$. Exhibit a function f such that $f \in \mathbf{L}^{p,\infty}$ and $f \notin \mathbf{L}^p$.

Bibliography

Background reading in general topology, functional analysis, and integration theory.

[AB] ASPLUND, E. AND BUNGART, L. *A First Course in Integration.* Holt, Rinehart and Winston, Inc., New York (1966).

[$DS_{1,2}$] DUNFORD, N. AND SCHWARTZ, J. T. *Linear Operators, Parts I, II.* Interscience Publishers, Inc., New York (1958, 1963).

[E] EDWARDS, R. E. *Functional Analysis: Theory and Applications.* Holt, Rinehart and Winston, Inc., New York (1965).

[GP] GOFFMAN, C. AND PEDRICK, G. *First Course in Functional Analysis.* Prentice-Hall, Inc., Englewood Cliffs, New Jersey (1965).

[HS] HEWITT, E. AND STROMBERG, K. *Real and Abstract Analysis.* Springer-Verlag, Berlin (1965).

[K] KELLEY, J. L. *General Topology.* D. Van Nostrand Company, Inc., Princeton, New Jersey (1955).

[W] WILLIAMSON, J. H. *Lebesgue Integration.* Holt, Rinehart and Winston, Inc., New York (1962).

References on Fourier Series and other specialized topics.

[A] ALEXITS, G. *Convergence Problems of Orthogonal Series.* Pergamon Press, Inc., New York (1961).

[B] BACHMAN, G. *Elements of Abstract Harmonic Analysis.* Academic Press, Inc., New York (1964).

[$Ba_{1,2}$] BARY, N. *A Treatise on Trigonometric Series, Vols. 1 and 2.* Pergamon Press, Inc., New York (1964).

[Be] BENEDETTO, J. *Harmonic Analysis on Totally Disconnected Sets.* Lecture Notes in Mathematics 202. Springer-Verlag, Berlin–Heidelberg–New York (1971).

[Bi] BIRTEL, F. T. (editor). *Function Algebras.* Scott, Foresman and Company, Glenview, Illinois (1966).

[Bo] BOURBAKI, N. *Intégration. Chapitres* **7**, **8**. Act. Sci. et Ind. 1306. Hermann & Cie, Paris (1963).

[Bo_1] BOURBAKI, N. *Théories Spectrales. Chapitres* **1**, **2**. Act. Sci. et Ind. 1332. Hermann & Cie, Paris (1967).

[Br] BREMERMANN, H. *Distributions, Complex Variable and Fourier Transforms.* Addison-Wesley Publishing Company, Inc. (1965).

[C] CARTAN, H. *Théorie spectrale des C-algèbres commutatives.* Séminaire Bourbaki, Exposé 125. Paris (1956).

[CH] COURANT, R. AND HILBERT, D. *Methods of Mathematical Physics, Vol. II.* Interscience Publishers, New York (1962).

[dBR] DE BRANGES, L. AND ROVNYAK, J. *Square Summable Power Series.* Holt, Rinehart and Winston, Inc., New York (1966).

[D] DONOGHUE, W. F., JR. *Distributions and Fourier Transforms.* Academic Press, Inc., New York and London (1969).

[DR] DUNKL, C. F. AND RAMIREZ, D. E. *Topics in Harmonic Analysis.* Appleton-Century-Crofts, New York (1971).

[DW] DORAN, R. S. AND WICHMANN, J. *Approximate identities and factorization in Banach Modules.* Springer Lecture Notes Vol. 768, Springer-Verlag, Berlin (1979).

[E_1] EDWARDS, R. E. *Integration and Harmonic Analysis on Compact Groups.* Cambridge University Press (1972).

[EG] EDWARDS, R. E. AND GAUDRY, G. I. *Littlewood–Paley and Multiplier Theory.* Ergebnisse der Math. und ihrer Grenzgebiete **90**. Springer-Verlag, Berlin–Heidelberg–New York (1977).

[Eh] EHRENPREIS, L. *Fourier Analysis in Several Complex Variables.* Wiley Interscience Publishers, New York (1970).

[Er] ERDÉLYI, A. *Operational Calculus and Generalized Functions.* Holt, Rinehart and Winston, Inc., New York (1962).

[G] GROSS, L. *Harmonic Analysis on Hilbert Space.* Mem. Amer. Math. Soc., No. 46, Amer. Math. Soc. (1963).

[Ga] GARSOUX, J. *Espaces Vectoriels Topologiques et Distributions.* Dunod, Paris (1963).

[Go] GOLDBERG, R. R. *Fourier Transforms.* Cambridge Tracts in Mathematics and Mathematical Physics, No. 52. Cambridge University Press, New York (1961).

[Go_1] GOLDBERG, R. R. *Recent results on Segal algebras.* Lecture Notes in Mathematics **399**, Springer-Verlag (1974).

[Gr] GRAHAM, C. C. *Harmonic analysis and locally compact Abelian groups.* M.A.A. Stud. Math., Vol. 13, pp. 161–197. Math. Assoc. of America, Washington, D.C. (1976).

[GRS] GELFAND, I. M., RAIKOV, D. AND SHILOV, D. *Commutative Normed Rings.* Chelsea Publishing Company, New York (1964).

[GS] GELFAND, I. M. ET CHILOV, G. E. *Les Distributions, Tomes 1–3.* Dunod, Paris (1962, 1964, 1965).

[GV] GELFAND, I. M. AND VILENKIN, N. YA. *Generalized Functions, Vol. 4. Applications of Harmonic Analysis.* Academic Press, Inc., New York (1964).

[Ha] HARDY, G. H. *Divergent Series.* Oxford University Press, New York (1949).

[Hal] HALPERIN, I. *Introduction to the Theory of Distributions.* University of Toronto Press (1962).

[HaR] HARDY, G. H. AND ROGOSINSKI, W. *Fourier Series.* Cambridge University Press, New York (1944).

[He] HEINS, M. *Topics in Complex Function Theory.* Holt, Rinehart and Winston, Inc., New York (1962).

[Hel] HELSON, H. *Lectures on Invariant Subspaces.* Academic Press, Inc., New York (1964).

[Hew] Hewitt, E. *A Survey of Abstract Harmonic Analysis. Surveys in Applied Mathematics, IV*. John Wiley & Sons, Inc., New York (1958), 107–168.

[Hi] Hirschman, I. I., Jr. (editor). *Studies in Real and Complex Analysis. Studies in Mathematics, Vol. 3.* Math. Association of America; Prentice-Hall, Inc., Englewood Cliffs, New Jersey (1965).

[HLP] Hardy, G. H., Littlewood, J. E. and Pólya, G. *Inequalities.* Cambridge University Press, New York (1934).

[Ho] Hoffman, K. *Banach Spaces of Analytic Functions.* Prentice-Hall, Inc., Englewood Cliffs, New Jersey (1962).

[HR] Hewitt, E. and Ross, K. A. *Abstract Harmonic Analysis, I, II.* Springer-Verlag, Berlin (1963, 1979).

[I] Izumi, S.-I. *Introduction to the Theory of Fourier series.* Institute of Mathematics, Academia Sinica, Taipei (1965).

[J] Jones, D. S. *Generalised functions.* McGraw-Hill Book Co., New York–Toronto, Ontario–London (1966).

[Kac] Kac, M. *Statistical Independence in Probability, Analysis and Number Theory.* The Math. Assoc. of America (Inc.); Carus Math. Monograph 12. John Wiley and Sons, Inc., New York (1959).

[Kah] Kahane, J.-P. *Algebras de Convolucion de Sucessiones, Funciones y Medidas Sumables.* Cursos y Seminarios de Matematica, Fasc. 6. Univ. de Buenos Aires (1961).

[Kah_1] Kahane, J.-P. *Lectures on Mean Periodic Functions.* Tata Institute of Fundamental Research Lectures on Mathematics and Physics, No. 15, Bombay (1959).

[Kah_2] Kahane, J.-P. *Séries de Fourier absolument convergentes.* Ergebnisse der Mathematik und ihrer Grenzgebiete, Bd. 50. Springer-Verlag, Berlin–Heidelberg–New York (1970).

[Kah_3] Kahane, J.-P. *Some random series of functions.* D. C. Heath and Company, Lexington, Mass. (1968).

[KS] Kahane, J.-P. and Salem, R. *Ensembles Parfaits et Séries Trigonométriques*, Hermann & Cie, Paris (1963).

[KSt] Kaczmarz, S. and Steinhaus, H. *Theorie der Orthogonalreihen.* Chelsea Publishing Company, New York (1951).

[Kz] Katznelson, Y. *An Introduction to Harmonic Analysis.* John Wiley and Sons, Inc., New York (1968).

[La] Larsen, R. *An Introduction to the Theory of Multipliers.* Die Grund. der math. Wiss. in Einzeldarstellungen, Bd. 175. Springer-Verlag, Berlin–Heidelberg–New York (1971).

[Li] Littlewood, J. E. *Some Problems in Real and Complex Analysis.* Lecture notes, Univ. of Wisconsin, Madison (1962).

[Lig] Lighthill, M. J. *An Introduction to Fourier Analysis and Generalized Functions.* Cambridge University Press, New York (1958).

[Lo] Loomis, L. H. *An Introduction to Abstract Harmonic Analysis.* D. Van Nostrand Company, Inc., Princeton, New Jersey (1953).

[LP] Lindahl, L.-A. and Poulsen, F. *Thin Sets in Harmonic Analysis.* Marcel Dekker, Inc., New York (1971).

[LR] Lopez, J. M. and Ross, K. A. *Sidon Sets.* Lecture Notes in Pure and Applied Mathematics. Marcel Dekker, Inc., New York (1975).

[M] Mandelbrojt, M. *Séries de Fourier et classes quasianalytiques de fonctions.* Gauthier-Villars, Paris (1935).

[Ma] Malgrange, B. *Ideals of Differentiable Functions.* Tata Institute of Fundamental Research Studies in Mathematics. Oxford University Press, New York (1967).

[MG] McGehee, O. C. and Graham, C. Essays in Commutative Harmonic Analysis. Springer-Verlag, New York (1979).

[Mi] Michael, E. *Locally multiplicatively-convex topological algebras. Memoirs Amer. Math. Soc., No.* **7**. American Mathematical Society, Providence, Rhode Island (1952).

[Mo] Mosak, R. D. *Banach algebras.* Chicago Lectures in Mathematics, University of Chicago Press, Chicago and London (1976).

[Moz] Mozzochi, C. P. *On the pointwise convergence of Fourier series.* Lecture Notes in Mathematics, **199**. Springer-Verlag, Berlin–Heidelberg–New York (1971).

[MT] Martineau, A. et Trèves, F. *Éléments de la Théorie des Espaces Vectoriels Topologiques et des Distributions, Fascicules I, II.* Centre de la Documentation Universitaire et S.E.D.E.S. Reunis, Paris (1962, 1964).

[N] Naimark, M. A. *Normed Rings.* P. Noordhoff, N. V., Groningen, Netherlands (1959).

[P] Pitt, H. R. *Tauberian Theorems.* Oxford University Press, New York (1958).

[R] Rudin, W. *Fourier Analysis on Groups.* Interscience Publishers, New York (1962).

[R_1] Rudin, W. *Real and Complex Analysis.* McGraw-Hill Book Company, New York (1966).

[Re] Reiter, H. *Classical Harmonic Analysis on Locally Compact Groups.* Oxford University Press (1968).

[Re_1] Reiter, H. *L^1-algebra and Segal algebra.* Lecture Notes in Math. **231**, Springer-Verlag (1971).

[Ri] Rickart, C. E. *Banach Algebras.* D. Van Nostrand Company, Inc., Princeton, New Jersey (1960).

[Ro] Rosenthall, H. P. *Projections onto translation-invariant subspaces of $L^p(G)$.* Mem. Amer. Math. Soc. No. 63 (1966).

[$S_{1,2}$] Schwartz, L. *Théorie des Distributions, Tomes I et II.* Hermann & Cie, Paris (1950 and 1951).

[SHA] *Studies in Harmonic Analysis.* Vol. 13 (edited by J. M. Ash). The Math. Assoc. of America (Inc.) (1976).

[SMA] *Studies in Modern Analysis, Vol. I.* (Edited by R. C. Buck.) Math. Assoc. of America (1962).

[So] Sobolev, S. L. *Applications of Functional Analysis in Mathematical Physics.* Amer. Math. Soc. Translations of Math. Monographs, Vol. VII (1963).

[St] Stein, E. M. *Singular Integrals and Differentiability Properties of Functions.* Princeton University Press (1970).

[StW] Stein, E. M. and Weiss, G. *Introduction to Fourier Analysis on Euclidean Spaces.* Princeton University Press (1971).

[T] Tonelli, L. *Serie Trigonometriche.* Bologna (1928).

[Ta] Taylor, J. L. *Measure algebras.* Amer. Math. Soc., Providence, R.I. (1973).

[Tm] Titchmarsh, E. C. *An Introduction to the Theory of Fourier Integrals.* Oxford University Press, New York (1948).

[Tr] Trèves, F. *Topological Vector Spaces, Distributions and Kernels*. Academic Press, New York (1967).

[TV] Tolstov, G. P. *Fourier Series*. Prentice-Hall, Inc., Englewood Cliffs, New Jersey (1962).

[VW] van der Warden, B. L. *Moderne Algebra*. Springer-Verlag, Berlin (1937).

[W] Wermer, J. *Seminar über Funktionen-Algebren*. Springer-Verlag, Berlin (1964).

[Wa] Wang, H. C. *Homogeneous Banach algebras*. Lecture Notes in Pure and Applied Mathematics. Marcel Dekker Inc., New York (1977).

[War] Ward, J. A. *Banach spaces of pseudomeasures on compact groups with emphasis on homogeneous spaces*. Ph.D. Thesis, Australian National University (1979).

[We] Weil, A. *L'Intégration dans les Groupes Topologiques et ses Applications*. Hermann & Cie, Paris (1951).

[Wi] Wiener, N. *The Fourier Integral and Certain of Its Applications*. Cambridge University Press, New York (1933).

[$Z_{1,2}$] Zygmund, A. *Trigonometrical Series, Vols. I and II*. Cambridge University Press, New York (1959).

Research Publications

The following list makes no attempt at completeness; it is rather a supplement to the bibliographies of such standard references as [Z], [Ba], *and* [R]. *Where it is necessary, references are numbered so as to follow on from the bibliography in Volume I.*

AHERN, P. R.

[1] On the generalized F. and M. Riesz theorem. *Pacific J. Math.* **15** (1965), 373–376.

AKEMAN, C. A.

[1] Some mapping properties of the group algebras of a compact group. *Pacific J. Math.* **22** (1967), 1–8.

ALLEN, G. R.

[1] A spectral theory for locally convex algebras. *Proc. London Math. Soc.* **15** (1965), 399–421.

[2] On a class of locally convex algebras. *Proc. London Math. Soc.* **17** (1967), 91–114.

ASKEY, R. AND WAINGER, S.

[1] Integrability theorem for Fourier series. *Duke Math. J.* **33** (1966), 223–228.

BACHELIS, G. F. AND EBENSTEIN, S. E.

[1] On $\Lambda(p)$ sets. *Pacific J. Math.* **54** (1974), 35–38.

BELLMAN, R.

[1] A note on a theorem of Hardy on Fourier constants. *Bull. Amer. Math. Soc.* **50** (1944), 741–744.

BENEDETTO, J. J.

[1] Tauberian translation algebras. *Ann. Mat. Pura Appl., IV Ser.* **74** (1966), 255–282.

BEURLING, A.

[1] Sur les intégrales de Fourier absolument convergentes et leur application à une transformation fonctionelle. Neuvième Congress Math. Scand. (Helsinki, 1938), 345–366.

[2] Construction and analysis of some convolution algebras. *Ann. Inst. Fourier* **14** (1964), Nr 2 (1965), 1–32.

[3] Un théorème sur less fonctions bornées et uniformément continues sur l'axe réel. *Acta Math.* **77** (1945), 127–136.

BILLARD, P.

[1] Séries de Fourier aléatoirement bornés, continues, uniformément convergentes. *Ann. Sci. École Norm. Sup.* (**3**) **82** (1965), 131–179.

BIRTEL, F. T.

[1] Banach algebras of multipliers. *Duke Math. J.* **28** (1961), 203–211.

[2] Isomorphisms and isometric multipliers. *Proc. Amer. Math. Soc.* **13** (1962), 204–210.

[3] On a commutative extension of a Banach algebra. *Proc. Amer. Math. Soc.* **13** (1962), 815–822.

BRAINERD, B. AND EDWARDS, R. E.

[1] Linear operators which commute with translations, I and II. *J. Austr. Math. Soc.* **VI** (1966), 289–350.

BROWN, G.

[1] Fourier transforms on the line and Gronwall's theorem. *J. London Math. Soc.* (2) **16** (1977), 475–482.

[2] Integrability of Fourier transforms under an ergodic hypothesis. *J. Austral. Math. Soc.* **26** (2) (1978), 129–153.

[3] Construction of Fourier multipliers. *Bull. Austral. Math. Soc.* **16** (1977), 463–472.

[4] Riesz products and generalized characters. *Proc. London Math. Soc.* (3) **30** (1975), 209–238.

BROWN, G. AND HEWITT, E.

[1] Some new singular Fourier–Stieltjes series. *Proc. Nat. Acad. Sci. U.S.A.* **75** (11) (1978), 5268–5269.

[2] Continuous singular measures with small Fourier–Stieltjes transforms. *Advances in Mathematics* **37** (1) (1980), 27–60.

BRYANT, J.

[1] On convolutions and Fourier series. *Bull. Amer. Math. Soc.* **73** (1967), 149–150.

BURNHAM, J. T.

[1] Closed ideals in subalgebras of Banach algebras, I. *Proc. Amer. Math. Soc.* **32** (1972), 551–555.

[2] Closed ideals in subalgebras of Banach algebras, II. *Monatsh. Math.* **78** (1974), 1–3.

[3] Segal algebras and dense ideals in Banach algebras. *Lecture Notes in Math.* **399**, Springer-Verlag (1974), 33–58.

BURNHAM, J. T. AND GOLDBERG, R. R.

[1] Basic properties of Segal algebras. *J. Math. Anal. App.* **42** (1973), 323–329.

CALDERÓN, A. P.

[1] Intermediate spaces and interpolation, the complex method. *Studia Math.* **XXIV** (1964), 113–190.

[2] Singular integrals. *Bull. Amer. Math. Soc.* **72** (1966), 427–465.

[3] Lebesgue spaces of differentiable functions and distributions. *Proc. Sympos. Pure Math., Vol.* **IV**, 33–49. Amer. Math. Soc., Providence R.I. (1961).

CALDERÓN, A. P. AND ZYGMUND, A.

[1] On the existence of certain singular integrals. *Acta Math.* **88** (1952), 85–139.

[2] Singular integrals and periodic functions. *Studia Math.* **XIV** (1956), 259–271.

CAMPANATO, S. AND MURTHY, M. K. V.

[1] Una generalizzazione del teorema di Riesz–Thorin. *Ann. Scuola Norm. Sup. Pisa* (3) **19** (1965), 87–100.

CARLESON, L.

[2] Sets of uniqueness for functions regular in the unit circle. *Acta Math.* **87** (1952), 325–345.

CARTWRIGHT, D., HOWLETT, R. B. AND MCMULLEN, J. R.

[1] Extreme values for the Sidon constant. To appear.

COHEN, P. J.

[2] On homomorphisms of group algebras. *Amer. J. Math.* **82** (1960), 213–226.

[3] On a conjecture of Littlewood and idempotent measures. *Amer. J. Math.* **82** (1960), 191–212.

[4] A note on constructive methods in Banach algebras. *Proc. Amer. Math. Soc.* **12** (1961), 159–163.

COIFMAN, R. R.

[1] Remarks on weak type inequalities for operators commuting with translations. *Bull. Amer. Math. Soc.* **74** (1968), 710–714.

COIFMAN, R. R., CWIKEL, M., ROCHBERG, R., SAGHER, Y. AND WEISS, G.

[1] Complex interpolation for families of Banach spaces. *Proc. Symp. Pure Math.* **35** (1979), 269–282. (Amer. Math. Soc. Publication)

COIFMAN, R., ROCHBERG, R. AND WEISS, G.

[1] Applications of Transference: The L^p version of von Neumann's inequality and the Littlewood–Paley–Stein theory. To appear.

COIFMAN, R. R. AND WEISS, G.

[1] Transference methods in analysis. *CBMS Regional Conference Series in Math.*, No. **31**, Amer. Math. Soc. (1977).

CORDES, H. O.

[1] The algebra of singular integral operators on R^n. *J. Math. Mech.* **14** (1965), 1007–1032.

COURY, J. E.

[1] Walsh series with coefficients tending monotonically to zero. *Pacific J. Math.* **54** (2) (1974), 1–16.

COWLING, M.

[1] The Kunze–Stein phenomenon. *Rendi. del Sem. Mat. e Fis. di Milano* **46** (1976), 35–41.

[2] The Kunze–Stein phenomenon. *Annals of Math.* **107** (1978), 209–234.

[3] La synthèse des convoluteurs de L^p de certains groupes pas moyennables. *Boll. U. M. I.* **14**-A (1977), 551–555.

[4] The Plancherel formula for a Group not of Type I. *Boll. U. M. I.* **15**-A (1978), 616–623.

[5] An application of Littlewood–Paley theory in harmonic analysis. *Math. Ann.* **241** (1979), 83–96.

CURTIS, P. C., JR. AND FIGÀ-TALAMANCA, A.

[1] Factorization theorems for Banach algebras. *Function Algebras*, pp. 169–185. Scott, Foresman and Company, Glenview, Ill. (1966).

DE LEEUW, K.

[1] On L_p multipliers. *Ann. of Math.* **81** (1965), 364–379.

[2] Homogeneous algebras on compact abelian groups. *Trans. Amer. Math. Soc.* **87** (1958), 372–386.

DE LEEUW, K. AND GLICKSBERG, I.

[1] Quasi-invariance and analyticity of measures on compact groups. *Acta Math.* **109** (1963), 179–205.

DE LEEUW, K. AND HERZ, C.

[1] An invariance property of spectral synthesis. *Illinois J. Math.* **9** (1965), 220–229.

DOMAR, Y.

[1] Harmonic analysis based on certain commutative Banach algebras. *Acta Math.* **96** (1956), 1–66.

[2] On spectral analysis in the narrow topology. *Math. Scand.* **4** (1956), 328–332.

DOSS, R.

[1] Some inclusions in multipliers. *Pacific J. Math.* **32** (1970), 643–646.

DRURY, S. W.

[1] Sur la synthèse harmonique. *C. R. Acad. Sci. Paris* Sér. A-B **271** (1970), A42–A44.

[2] Sur les ensembles parfait et les séries trigonométriques. *C. R. Acad. Sci. Paris* Sér. A-B **271** (1970), A94–A95.

[3] Sur les ensembles de Sidon. *C. R. Acad. Sci. Paris* Sér. A-B **271** (1970), A162–A163.

DUDLEY, E. AND HALL, P.

[1] The Gaussian law and lacunary sets of characters. *J. Austral. Math. Soc.* A **27** (1) (1979), 91–107.

DUNKL, C. F. AND RAMIREZ, D. E.

[1] Sections induced from weakly sequentially complete spaces. *Studia Math.* **49** (1973), 95–97.

EDWARDS, R. E.

[2] Translates of L^∞ functions and of bounded measures. *J. Austr. Math. Soc.* **IV** (1964), 403–409.

[3] Boundedness principles and Fourier theory. *Pacific J. Math.* **21** (1967), 255–263.

[4] Spans of translates in $L^p(G)$. *J. Austr. Math. Soc.* **V** (1965), 216–233.

[5] Uniform approximation on noncompact spaces. *Trans. Amer. Math. Soc.* **122** (1966), 249–276.

[6] Approximation by convolutions. *Pacific J. Math.* **15** (1965), 85–95.

[7] Changing signs of Fourier coefficients. *Pacific J. Math.* **15** (1965), 463–475.

[8] Bounded functions and Fourier transforms. *Proc. Amer. Math. Soc.* **9** (1958), 440–446.

[9] Supports and singular supports of pseudomeasures. *J. Austr. Math. Soc.* **VI** (1966), 65–75.

[10] Bipositive and isometric isomorphisms of some convolution algebras. *Canad. J. Math.* **17** (1965), 839–846.

[11] Endomorphisms of function-spaces which leave stable all translation-invariant manifolds. *Pacific J. Math.* **14** (1964), 31–48.

[12] On functions which are Fourier transforms. *Proc. Amer. Math. Soc.* **5** (1954), 71–78.

[13] A class of multipliers. *J. Austr. Math. Soc.* **VII** (1968), 584–590.

[14] Inequalities related to those of Hausdorff–Young. *Bull. Austr. Math. Soc.* **6** (1972), 185–210.

[15] Criterion for Fourier transforms. *J. Austral. Math. Soc.* **7** (1967), 239–246.

EDWARDS, R. E. AND HELSON, H.

[1] Absolute Fourier multipliers. To appear *Res. der. Math.* (**1**).

EDWARDS, R. E. AND HEWITT, E.

[1] Pointwise limits for sequences of convolution operators. *Acta Math.* **113** (1965), 181–218.

EDWARDS, R. E., HEWITT, E. AND RITTER, G.

[1] Fourier multipliers for certain spaces of functions with compact supports. *Inventiones Math.* **40** (1977), 37–57.

EDWARDS, R. E., HEWITT, E. AND ROSS, K. A.

[1] Lacunarity for compact groups, II. *Pacific J. Math.* **41** (1972), 99–109.

EDWARDS, R. E. AND PRICE, J. F.

[1] A naively constructive approach to boundedness principles, with applications to harmonic analysis. *L'Ens. Math.* **XVI**, *fasc.* 3–4 (1970), 255–296.

EDWARDS, R. E. AND ROSS, K. A.

[1] Helgason's number and lacunarity constants. *Bull. Austral. Math. Soc.* **9** (1973), 187–218.

[2] p-Sidon sets. *J. Functional Anal.* **15** (1974), 404–427.

EHRENPREIS, L.

[1] Mean periodic functions. *Amer. J. Math.* **78** (1955), 292–328.

ELLIOTT, R. J.

[1] Some results in spectral synthesis. *Proc. Cambridge Phil. Soc.* **61** (1965), 395–424.

EMEL'JANOV, V. F.

[1] On convergence of lacunary trigonometric series on sets. *Soviet Math. Dokl.* **6** (1965), 1437–1438.

FEKETE, M.

[1] Über die Faktorenfolgen welche die "Klasse" einer Fourierschen Reihe unverändert lassen. *Acta Sci. Math.* (Szeged) **1** (1923), 148–166.

FIEDLER, H., JURKAT, W. AND KÖRNER, O.

[1] On Salem's problem for Fourier and Dirichlet series. *Period. Math. Hungar.* **8** (1977), No. 3–4, 229–242.

Figà-Talamanca, A.

[1] On the subspaces of L^p invariant under multiplication of transforms by bounded continuous functions. *Rend. Sem. Mat. d. Univ. di Padova* (1965), 176–189.

[2] Translation invariant operators in L^p. *Duke Math. J.* **32** (1965), 495–502.

[3] Multipliers of p-integrable functions. *Bull. Amer. Math. Soc.* **70** (1964), 666–669.

Figà-Talamanca, A. and Gaudry, G. I.

[1] Density and representation theorems for multipliers of the type (p, q). *J. Austr. Math. Soc.* **VII** (1967), 1–6.

[2] Multipliers of L^p which vanish at ∞. *J. Functional Anal.* **7** (3) (1971), 475–486.

[3] Multipliers and sets of uniqueness of L^p. *Michigan Math. J.* **17** (1970), 179–191.

[4] Extension of multipliers. (Italian summary). *Boll. Un. Mat. Ital.* (4) **3** (1970), 1003–1014.

Figà-Talamanca, A. and Price, J. F.

[1] Applications of random Fourier series over compact groups to Fourier multipliers. *Pacific J. Math.* **43** (1972), 531–541.

Figà-Talamanca, A. and Rider, D.

[1] A theorem of Littlewood and lacunary series for compact groups. *Pacific J. Math.* **16** (1966), 505–514.

[2] A theorem on random Fourier series on compact groups. *Pacific J. Math.* **21** (1967), 487–492.

Fine, N. J.

[1] On the Walsh functions. *Trans. Amer. Math. Soc.* **65** (1949), 372–414.

Forelli, F.

[1] Homomorphisms of ideals in group algebras. *Illinois J. Math.* **9** (1965), 410–417.

Fournier, J. J. F.

[1] Local complements to the Hausdorff–Young theorem. *Michigan Math. J.* **20** (1973), 263–276.

[2] On a theorem of Paley and the Littlewood conjecture. *Amer. J. Math.* **82** (1960), 191–212.

[3] Extensions of a Fourier multiplier theorem of Paley. *Pacific J. Math.* **30** (1969), 415–431.

Freeman, M.

[1] Some conditions for uniform approximation on a manifold. *Function Algebras*, pp. 42–60. Scott, Foresman and Company, Glenview, Ill. (1966).

Gaudry, G. I.

[1] Quasimeasures and operators commuting with convolutions. *Pacific J. Math.* **18** (1966), 461–476.

[2] Multipliers of type (p, q). *Pacific J. Math.* **18** (1966), 477–488.

[3] *Quasimeasures and multiplier problems.* Doctoral thesis, Australian National University (1966).

[4] Isomorphisms of multiplier algebras. *Canad. J. Math.* **20** (1968), 1165–1172.

[5] H^p multipliers and inequalities of Hardy and Littlewood. *J. Austral. Math. Soc.* **10** (1969), 23–32.

[6] Bad behaviour and inclusion results for multipliers of type (p, q). *Pacific J. Math.* **35** (1) (1970), 83–94.

GELFAND, I. M.

[1] Normierte Ringe. *Mat. Sbornik* (N.S.) **9** (1941), 3–24.

GILBERT, J. E.

[1] Spectral synthesis problems for invariant subspaces on groups. II. *Function Algebras*, pp. 257–264. Scott, Foresman and Company, Glenview, Ill. (1966).

[2] Convolution operators on $L^p(G)$ and properties of locally compact groups. *Pacific J. Math.* **24** (1968), 257–268.

[3] Nikišin–Stein theory. Informal lecture notes.

GLAESER, G.

[1] Synthèse spectrale des idéaux de fonctions lipschitziennes. *C. R. Acad. Sci. Paris* **260** (1965), 1539–1542.

GLICKSBERG, I.

[1] Homomorphisms of certain measure algebras. *Pacific J. Math.* **10** (1960), 167–191.

[2] Fourier–Stieltjes transforms with small supports. *Illinois J. Math.* **9** (1965), 418–427.

[3] The abstract F. and M. Riesz theorem. *J. Functional Analysis* **1** (1967), 109–122.

GLICKSBERG, I. AND WIK, I.

[1] Multipliers of quotients of L_1. *Pacific J. Math.* **38** (1971), 619–624.

GOES, G.

[1] Komplementäre Fourierkoeffizientenräume und Multiplikatoren. *Math. Ann.* **137** (1959), 371–384.

[2] On Fourier–Stieltjes series with finitely many distinct coefficients and on almost periodic sequences. *J. Math. An. Applied.* **19** (1967), 26–34. Addendum *ibid.* **21** (1968), 618.

[3] Fourier–Stieltjes Transforms of Discrete Measures; Periodic and Semiperiodic functions. *Math. Ann.* **174** (1967), 148–156.

[4] Über einige Multiplikatorenklassen. *Math. Z.* **80** (1963), 324–327.

GOLDBERG, R. R. AND SIMON, A. B.

[1] The Riemann–Lebesgue theorem on groups. *Acta Sci. Math.* (*Szeged*) **27** (1966), 35–39.

GRAHAM, C. C.

[1] Helson sets and simultaneous extensions to Fourier transforms. *Studia Math.* **43** (1972), 57–60.

GREENLEAF, F. P.

[1] Norm decreasing homomorphisms of group algebras. *Bull. Amer. Math. Soc.* **70** (1964), 536–539.

[2] Closed subalgebras of group algebras which are group algebras. *Function Algebras*, pp. 276–281. Scott, Foresman and Company, Chicago (1966).

[3] Norm decreasing homomorphisms of group algebras. *Pacific J. Math.* **15** (1965), 1187–1219.

GROSS, K. I.

[1] On the evolution of noncommutative harmonic analysis. *Amer. Math. Monthly* **85** (1978), 525–548.

HAHN, L.-S.

[1] On multipliers of p-integrable functions. *Trans. Amer. Math. Soc.* **128** (1967), 321–335.

HARASYMIV, S. R.

[1] A note on dilations in L^p. *Pacific J. Math.* **21** (1967), 493–501.

HEDLUND, J. H.

[1] Multipliers of H^p spaces. *J. Math. Mech.* **18** (1969), 1067–1074.

HELGASON, S.

[1] Multipliers of Banach algebras. *Ann. of Math.* **64** (1956), 240–254.

[2] Topologies of group algebras and a theorem of Littlewood. *Trans. Amer. Math. Soc.* **86** (1957), 269–283.

[3] Lacunary Fourier series on noncommutative groups. *Proc. Amer. Math. Soc.* **9** (1958), 782–790.

HELSON, H.

[1] Proof of a conjecture of Steinhaus. *Proc. Nat. Acad. Sci. U.S.A.* **40** (1954), 205–206.

[2] Note on harmonic functions. *Proc. Amer. Math. Soc.* **4** (1953), 686–691.

[3] On a theorem of Szëgo. *Proc. Amer. Math. Soc.* **6** (1955), 235–242.

[4] Fourier transforms on perfect sets. *Studia Math.* **XIV** (1954), 209–213.

[5] Compact groups with ordered duals. *Proc. London Math. Soc.* **XIVA** (1965), 144–156.

[6] Foundation of the theory of Dirichlet series. *Acta Math.* **118** (1967), 61–77.

[7] Conjugate series and a theorem of Paley. *Pacific J. Math.* **8** (1958), 437–446.

HELSON, H. AND KAHANE, J.-P.

[1] A Fourier method in Diophantine problems. *J. Analyse Math.* **XV** (1965), 245–262.

HERZ, C. S.

[2] The spectral theory of bounded functions. *Trans. Amer. Math. Soc.* **94** (1960), 181–232.

[3] Remarques sur la note précédente de Varopoulos. *C. R. Acad. Sci. Paris* **260** (1965), 6001–6004.

[4] Drury's lemma and Helson sets. *Studia Math.* **42** (1972), 205–219.

HEWITT, E.

[1] The ranges of certain convolution operators. *Math. Scand.* **15** (1965), 147–155.

[2] The asymmetry of certain algebras of Fourier–Stieltjes transforms. *Michigan Math. J.* **5** (1958), 149–158.

HEWITT, E. AND HIRSCHMAN, I. I. JR.

[1] A maximum problem in harmonic analysis. *Amer. J. Math.* **76** (1954), 839–852.

HEWITT, E. AND KAKUTANI, S.

[1] A class of multiplicative linear functionals on the measure algebra of a locally compact Abelian group. *Illinois J. Math.* **4** (1960), 553–574.

[2] Some multiplicative linear functionals on $M(G)$. *Ann. Math.* (2) **79** (1964), 489–505.

HEWITT, E. AND RITTER, G.

[1] Über die Integrierbarkheit von Fourier-Transformierten auf Gruppen; Teil I: Stetige Funcktionen mit Kompakten Träger und ein Bermerkung über hyperbolische Differentialoperatoren. *Math. Ann.* **224** (1976), 77–96.

[2] On the integrability of Fourier transforms on Groups; Part II: Fourier–Stieltjes transforms of singular measures. *Proc. Roy. Irish Academy* **76** (1976), No. 25, 265–287.

[3] The Orlicz–Paley–Sidon phenomenon for singular measures. *Symposia Mathematica, Vol.* **XXII** *(Convegno sull'Analisi Armonica e Spazi di Funzioni su Gruppi Localmente Compatte, Indam, Rome, 1976), pp. 21–31. Academic Press, London* 1977.

HEWITT, E. AND ZUCKERMAN, H. S.

[1] Some theorems on lacunary Fourier series, with extensions to compact groups. *Trans. Amer. Math. Soc.* **93** (1959), 1–19.

[2] On a theorem of P. J. Cohen and H. Davenport. *Proc. Amer. Math. Soc.* **14** (1963), 847–855.

[3] Singular measures with absolutely continuous convolution squares. *Proc. Camb. Phil. Soc.* **62** (1966), 399–420. Corrigendum. *Ibid.* **63** (1967), 367–368.

[4] Some singular Fourier–Stieltjes series. *Proc. London Math. Soc.* **19** (1969), 310–326.

HIRSCHMAN, I. I. JR.

[1] On multiplier transformations. *Duke Math. J.* **26** (1959), 221–242.

[2] Szegö functions on a locally compact Abelian group with ordered dual. *Trans. Amer. Math. Soc.* **121** (1966), 133–159; **123** (1966), 548.

HÖRMANDER, L.

[1] Estimates for translation invariant operators in L^p spaces. *Acta Math.* **104** (1960), 93–140.

HUNT, R. A.

[1] *Operators acting on Lorentz spaces.* Ph.D. thesis, University of Washington (1965).

HUNT, R. A. AND WEISS, G.

[1] The Marcinkiewicz interpolation theorem. *Proc. Amer. Math. Soc.* **15** (6) (1964), 996–998.

IGARI, S.

[1] Sur les fonctions qui opèrent sur l'espace $\hat{A}^2$. *Ann. Inst. Fourier*, Grenoble, **15** (1965), 525–536.

[2] L^p-multipliers. *Tôhoku Math. J.* (2) **21** (1969), 304–320.

[3] L^p-multipliers. *Tôhoku Math. J.* (2) **26** (1974), 555–561.

IZUMI, M. AND S.-I.

[1] On lacunary Fourier series. *Proc. Japan Academy* **41** (1965), 648–651.

[2] On the Leindler's theorem. *Proc. Japan Academy* **42** (1966), 533–534.

IZUMI, M. AND S.-I. AND KAHANE, J.-P.

[1] Théorèmes élémentaires sur les séries de Fourier lacunaires. *J. Analyse Math.* **14** (1965), 235–246.

JOHNSON, B. E.

[1] Isometric isomorphisms of measure algebras. *Proc. Amer. Math. Soc.* **15** (1964), 186–188.

[2] An introduction to the theory of centralisers. *Proc. London Math. Soc.* **14** (1964), 299–320.

[3] Centralisers of certain topological algebras. *J. London Math. Soc.* **39** (1964), 603–614.

[4] Continuity of transformations which leave invariant certain translation-invariant subspaces. *Pacific J. Math.* **20** (1967), 223–230.

[5] Symmetric maximal ideals in $M(G)$. *Proc. Amer. Math. Soc.* **18** (1967), 1040–1044.

[6] The Šilov boundary of $M(G)$. *Trans. Amer. Math. Soc.* **134** (1969), 289–296.

KAC, M.

[1] A remark on Wiener's Tauberian theorem. *Proc. Amer. Math. Soc.* **16** (1965), 1155–1157.

KACZMARZ, S.

[1] On some classes of Fourier series. *J. London Math. Soc.* **8** (1933), 39–46.

KAHANE, J.-P.

[1] Idempotents and closed subalgebras of $A(Z)$. *Function Algebras*, pp. 198–207. Scott, Foresman and Company, Glenview, Ill. (1966).

[2] Sur certaines classes de séries de Fourier absolument convergentes. *J. Math. Pures et Appl.* **35** (1956), 249–258.

[3] Sur un problème de Littlewood. *Nederl. Akac. Wetensch. Proc. Ser. A.60 = Indag. Math.* **19** (1957), 268–271.

[4] Lacunary Taylor and Fourier series. *Bull. Amer. Math. Soc.* **70** (1964), 199–213.

[5] On the construction of certain bounded continuous functions. *Pacific J. Math.* **16** (1966), 129–132.

[6] Ensembles de Ryll–Nardzewski et ensembles de Helson. *Colloq. Math.* **15** (1966), 87–92.

KAHANE, J.-P. AND KATZNELSON, Y.

[2] Contribution à deux problèmes concernant les fonctions de la classe A. *Israel J. Math.* **1** (1963), 110–131.

[3] Sur les ensembles de divergence des séries trigonométriques. *Studia Math.* **26** (1966), 305–306.

KAHANE, J.-P. AND MANDELBROT, B.

[1] Ensembles de multiplicite aléatoires. *C. R. Acad. Sci. Paris* **261** (1965), 3931–3933.

KAMAZOLOV, A. I.

[1] Multiplicative transformations of Fourier integrals in L^p spaces with weight. (Russian) *Izv. Vysǒ Učebn. Zaved. Matematika* No. 7 (**86**) (1969), 54–58.

KARAMATA, J.

[1] Suites de fonctionelles linéaires et facteurs de convergence de séries de Fourier. *J. Math. Pures et Appl.* **35** (1956), 87–95.

Katznelson, Y.

[1] Trigonometric series with positive partial sums. *Bull. Amer. Math. Soc.* **71** (1965), 718–719.

[2] Sur les ensembles de divergence des séries trigonométriques. *Studia Math.* **26** (1966), 301–304.

Kaufman, R.

[1] Examples in Helson sets. *Bull. Amer. Math. Soc.* **72** (1966), 139–140.

Kawada, Y.

[1] On the group ring of a topological group. *Math. Japon.* **1** (1948), 1–5.

Keogh, F. R.

[1] On strong and weak convergence of trigonometric series. *Proc. Roy. Irish Acad. Sect.* **A63** (1964), 75–85.

[2] Riesz products. *Proc. London Math. Soc.* **(3) 14a** (1965), 174–182.

Koizumi, S.

[1] On the Hilbert transform I, II. *J. Fac. Sci. Hokkaido Univ. Ser.* I **14** (1959), 153–224; **15** (1960), 93–130.

[2] Local estimations of conjugate functions, I. *Proc. Japan Acad.* **42** (1966), 891–895.

Konheim, A. G. and Weiss, B.

[1] Functions which operate on characteristic functions. *Pacific J. Math.* **15** (1965), 1279–1293.

Konyushkov, A. A.

[1] On the Lipschitz classes. (In Russian.) *Izv. Akad. Nauk SSSR Ser. Mat.* **21** (1957), 423–448.

Koosis, P.

[1] On the spectral analysis of bounded functions. *Pacific J. Math.* **16** (1966), 121–128.

Korevaar, J.

[1] Distribution proof of Wiener's Tauberian theorem. *Proc. Amer. Math. Soc.* **16** (1965), 353–355.

Körner, T. W.

[1] Some results on Kronecker, Dirichlet and Helson Sets. *Ann. Inst. Fourier (Grenoble)* **20** (1970), fasc. 2, 219–324 (1971).

Krée, P.

[1] Sur les multiplicateurs dans $\mathscr{F}L^p$. *C. R. Acad. Sci. Paris* **260** (1965), 4400–4403.

[2] Sur les multiplicateurs dans $\mathscr{F}L^p$ avec poids. *Ann. Inst. Fourier (Grenoble)* **16** (1966), fasc. 2, 91–121.

Krogstad, H. E.

[1] Multipliers on homogeneous Banach spaces on compact groups. *Ark. Mat.* **12** (1974), 203–212.

Kunze, R. A. and Stein, E. M.

[1] Uniformly bounded representations and harmonic analysis of the 2×2 real unimodular group. *Amer. J. Math.* **LXXXII** (1960), 1–62.

LANCONELLI, E.

[1] Su una classe di moltiplicatori di $\mathscr{F}L_p$ ed applicazioni (English summary). *Atti Accad. Naj. Lincei Rend. Cl. Sci. Fis. Mat. Natur.* 8 (**51**) (1971), 133–139.

LITTMAN, W.

[1] Multipliers in L^p and interpolation. *Bull. Amer. Math. Soc.* **71** (1965), 764–766.

LITTMAN, W., MCCARTHY, C. AND RIVIÈRE, N.

[1] L^p-multiplier theorems. *Studia Math.* **30** (1968), 193–217.

[2] The non-existence of L^p estimates for certain translation-invariant operators. *Studia Math.* **30** (1968), 219–229.

LIU, T. S.

[1] On vanishing algebras. *Proc. Amer. Math. Soc.* **14** (1963), 162–166.

LIZORKIN, P. I.

[1] Multipliers of Fourier integrals in the spaces $L_{p,\theta}$. (Russian). *Trudy Mat. Inst. Steklov.* **89** (1967), 231–248.

ŁOJASIEWICZ, S.

[1] Sur la valeur d'une distribution dans un point. *Bull. Acad. Polon. Sci. Cl. III.* **4** (1956), 239–242.

[2] Sur la valeur et la limite d'une distribution en un point. *Studia Math.* **16** (1957), 1–36.

LUMER, G.

[1] Analytic functions and the Dirichlet problem. *Bull. Amer. Math. Soc.* **70** (1964), 98–104.

MAHMUDOV, A. S.

[1] On the Fourier and Taylor coefficients of continuous functions. (In Russian.) *Izv. Akad. Nauk Azerbaĭdžan. SSSR Ser. Fiz.-Tehn. Mat. Nauk* **2** (1964), 23–29; **4** (1964), 35–44.

MALLIAVIN, P.

[1] Impossibilité de la synthèse spectrale sur les groupes Abéliens non compacts. *Publ. Math. Inst. Hautes ÉtudesSci. Paris* (1959), 61–68.

[2] Ensembles de résolution spectrale. *Proc. Internat. Congr. Math.* (Stockholm, 1962), 367–378. Inst. Mittag-Leffler, Djursholm (1963).

MALLIAVIN-BRAMARAT, M.-P. AND MALLIAVIN, P.

[1] Caractérisation arithmétique d'une classe d'ensembles de Helson. *C. R. Acad. Sci. Paris*, **264 A** (1967), 192–193.

MARCINKIEWICZ, J. AND ZYGMUND, A.

[1] Some theorems on orthogonal systems. *Fund. Math.* **28** (1937), 309–335.

MARCUS, M. B. AND PISIER, G.

[1] Random Fourier series with applications to harmonic analysis. *Centre for Statistics and Probability, Northwestern University, Evanston, Illinois* (1980).

Máté, L.

[1] Multiplier operators and quotient algebra. *Bull. Acad. Polon. Ser. Sci. Math. Astronom. Phys.* **13** (8) (1955), 523–526.

[2] Embedding multiplier operators of a Banach algebra B into its second conjugate space B^{**}. *Bull. Acad. Polon. Ser. Sci. Math. Astronom. Phys.* **13** (11–12) (1965), 809–812.

[3] Some abstract results concerning multiplier algebras. *Rev. Roumaine Math. Pures Appl.* **10** (1965), 261–266.

McGehee, O. C.

[2] Two remarks about Fourier analysis on thin sets. *Notices Amer. Math. Soc.* **14** (1967), 76.

McGehee, O. C., Pigno, L. and Smith, B.

[1] Hardy's inequality and the Littlewood conjecture. To appear. *Bull. Amer. Math. Soc.*

[2] Hardy's inequality and the L^1 norm of exponential sums. To appear. *Annals of Math.*

Meyer, Y.

[1] Endomorphismes des idéaux fermés de $L^1(G)$, classes de Hardy, et séries de Fourier lacunaires. *Ann. Scient. École Norm. Sup.* 4e série **1** (1968), 499–580.

[2] Spectres des mesures et mesures absolument continus. *Studia Math.* **30** (1968), 87–89.

Mirkil, H.

[1] The work of Šilov on commutative semisimple Banach algebras. *Notas de Matemática, No. 20. Fasciculo Publicado pelo Instituto de Matemática Pura e Aplicada do Conselho de Pesquisas.* Rio de Janeiro (1959).

Moeller, J. W.

[1] On the extrapolation of lacunary Fourier series. *J. Reine Angew. Math.* **222** (1966), 136–141.

Moeller, J. W. and Frederickson, P. O.

[1] A density theorem for lacunary Fourier series. *Bull. Amer. Math. Soc.* **72** (1966), 82–86.

Neubauer, D.

[1] The non-existence of projections from L^1 to H^1. *Proc. Amer. Math. Soc.* **12** (1961),

[2] Zur Spektraltheorie in lokalkonvexen Algebren, II. *Math. Ann.* **143** (1961), 251–263.

Newman, D. J.

[1] The non-existence of projections form L^1 to H^1. *Proc. Amer. Math. Soc.* **12** (1961), 98–99.

[2] An L^1 extremal problem for polynomials. *Proc. Amer. Math. Soc.* **16** (1965), 1287–1290.

[3] The closure of translates in l^p. *Amer. J. Math.* **86** (1964), 651–667.

Newman, D. J., Schwartz, J. T. and Shapiro, H. S.

[1] On generators of the Banach algebras l_1 and $L_1(0, \infty)$. *Trans. Amer. Math. Soc.* **107** (1963), 466–484.

OHTSUKA, M.
[1] On potentials on locally compact spaces. *J. Sci. Hiroshima Univ. Ser. A-I Math.* **25** (1961), 135–352.

OKIKIOLU, G. O.
[1] On Fourier transform multipliers of L^p. *J. Austral. Math. Soc.* **13** (1972), 219–223.

OKLANDER, E. T.
[1] L_{pq} interpolators and the theorem of Marcinkiewicz. *Bull. Amer. Math. Soc.* **72** (1966), 49–53.

O'NEIL, R.
[1] Convolution operators and $L(p, q)$ spaces. *Duke Math. J.* **30** (1963), 129–142.
[2] Two elementary theorems on the interpolation of linear operators. *Proc. Amer. Math. Soc.* **17** (1966), 76–82.

O'NEIL, R. AND WEISS, G.
[1] The Hilbert transform and rearrangement of functions. *Studia Math.* **XXIII** (1963), 189–198.
[2] The Marcinkiewicz interpolation theorem. *Proc. Amer. Math. Soc.* **15** (1964), 996–998.

PALEY, R. E. A. C.
[1] A note on power series. *Proc. London Math. Soc.* **7** (1932), 122–130.

PARROTT, S. K.
[1] Isometric multipliers. *Pacific J. Math.* **25** (1968), 159–166.

PEETRE, J.
[1] Espaces d'interpolation et théorème de Soboleff. *Ann. Inst. Fourier, Grenoble,* **16** (1966), 279–317.
[2] Applications de la théorie des espaces d'interpolation dans l'analyse harmonique. *Ricerche Mat.* **XV** (1966), 1–34.
[3] On convolution operators leaving $L^{p,\lambda}$ spaces invariant. *Ann. Mat. Pura Appl. Ser. IV,* **LXXII** (1966), 295–304.

PICHORIDES, S. K.
[1] Norms of exponential sums. *Publ. Math. Orsay* (1977), #73.
[2] On a conjecture of Littlewood. To appear. *Bull. Greek Math. Soc.*
[3] On the L^1 norm of exponential sums. *Annals Inst. Fourier, (Grenoble)* **30** (2) 1980, 79–89.

PIGNO, L.
[1] A multiplier theorem. *Pacific J. Math.* **34** (1970), 755–757.
[2] Restriction of L^p transforms. *Proc. Amer. Math. Soc.* **29** (1971), 511–515.

PISIER, G.
[1] Ensembles de Sidon et processus gaussiens. *C. R. Acad. Sci. Paris* **286** A (1978), 671–674.
[2] De nouvelles caractérisations des ensembles de Sidon. To appear.

PRICE, J. F.

[1] Multipliers between some spaces of distributions. *J. Austral. Math. Soc.* **9** (1969), 415–423.

[2] Some strict inclusions between spaces of L^p-multipliers. *Trans. Amer. Math. Soc.* **152** (1970), 321–330.

[3] (L^p, L^q)-*Multiplier Problems*. Doctoral thesis, Australian National University (1970).

[4] Littlewood's conjecture for Dirichlet kernels. *Australian Math. Soc. Gazette* **8** (2) (1981), 37–40.

PROHORENKO, V. I.

[1] Certain properties of Fourier coefficients. (In Russian; English summary.) *Vestnik Moskov. Univ. Ser. I Math. Meh.* (1964), No. 6, 51–60.

RAMIREZ, D. E.

[1] Uniform approximation by Fourier–Stieltjes transforms. *Proc. Camb. Phil. Soc.* **64** (1968), 323–333.

[2] Uniform approximation by Fourier–Stieltjes transforms. *Proc. Camb. Phil. Soc.* **64** (1968), 615–623.

[3] Weakly almost periodic functions and Fourier–Stieltjes transforms. *Proc. Amer. Math. Soc.* **19** (1968), 1087–1088.

RAUCH, H. E.

[1] Harmonic and analytic functions of several complex variables and the maximal theorem of Hardy and Littlewood. *Canad. J. Math.* **8** (1965), 171–183.

REID, G. A.

[1] Concepts of differentiability and analyticity on certain classes of topological groups. *Proc. Camb. Phil. Soc.* **61** (1965), 347–379.

REITER, H.

[1] Contributions to harmonic analysis, IV. *Math. Ann.* **135** (1958), 467–476.

[2] Subalgebras of $L^1(G)$. *Nederl. Akad. Wetensch. Proc. Ser.* **A68** = *Indag. Math.* **27** (1965), 691–696.

RICHARDS, I.

[1] On Malliavin's counterexample to spectral synthesis. *Bull. Amer. Math. Soc.* **72** (1966), 698–700.

RIDER, D.

[1] Central idempotents in group algebras. *Bull. Amer. Math. Soc.* **72** (1966), 1000–1002.

[2] Transformations of Fourier coefficients. *Pacific J. Math.* **19** (1966), 347–356.

[3] A relation between a theorem of Bohr and Sidon sets. *Bull. Amer. Math. Soc.* **72** (1966), 558–561.

[4] Gap series on circles and spheres. *Canad. J. Math.* **18** (1966), 389–398.

[5] Closed subalgebras of $L^1(T)$. *Yale University*, Department of Mathematics (1967).

RIEFFEL, M.

[1] Multipliers and tensor products of L^p spaces of locally compact groups. *Studia Math.* **33** (1969), 71–82.

RIGELHOF, R.
[1] Norm decreasing homomorphisms of measure algebras. *Trans. Amer. Math. Soc.* **136** (1969), 361–371.

RISS, J.
[1] Éléments de calcul différentiel et théorie des distributions sur les groupes Abéliens localement compacts. *Acta Math.* **89** (1953), 45–105.

RIVIÈRE, N. M. AND SAGHER, Y.
[1] Multipliers of trigonometric series and pointwise convergence. *Trans. Amer. Math. Soc.* **140** (1969), 301–308.

ROSENTHAL, H. P.
[1] Projections onto translation-invariant subspaces of $L^p(G)$, G noncompact. *Function Algebras*, pp. 265–275. Scott, Foresman and Company, Glenview, Ill. (1966).
[2] Caractérisation d'ensembles de Helson, par l'existence de certains projecteurs. *C. R, Acad. Sci. Paris Sér.* A-B **262** (1966), A286–A288.
[3] On trigonometric series associated with weak * closed subspaces of continuous functions. *J. Math. Mech.* **17** (1967), 485–490.

RUDIN, W.
[6] Trigonometric series with gaps. *J. Math. Mech.* **9** (1960), 203–228.
[7] Ideals with small automorphisms. *Bull. Amer. Math. Soc.* **72** (1966), 339–341.
[8] Projections on invariant subspaces. *Proc. Amer. Math. Soc.* **13** (1962), 429–432.

SAEKI, S.
[1] Translation invariant operators on groups. *Tôhoku Math. J.* **22** (1970), 409–419.
[2] On the union of two Helson sets. *J. Math. Soc. Japan* **23** (1971), 636–648.

SALEM, R.
[1] On a problem of Littlewood. *Amer. J. Math.* **77** (1955), 535–540.

SALEM, R. AND ZYGMUND, A.
[1] Some properties of trigonometric series whose terms have random signs. *Acta Math.* **91** (1954), 245–301.

SANDERS, J. W.
[1] Weighted Sidon sets. *Pacific J. Math.* **63** (1976), 255–279.

SAWYER, S.
[1] Maximal inequalities of weak type. *Ann. Math.* **84** (1966), 157–174.

SCHECHTER, M.
[1] Interpolation spaces by complex methods. *Bull. Amer. Math. Soc.* **72** (1966), 526–533.

SCHWARTZ, L.
[1] Sur une propriété de synthèse spectrale dans les groupes noncompacts. *C. R. Acad. Sci. Paris* **227** (1948), 424–426.
[2] Théorie générale des fonctions moyennes-périodiques. *Ann. Math.* **48** (1947), 857–929.
[3] Sur les multiplicateurs de FL^p. *Kungl. Fysiogr. Sällsk. Lund Forh.* **22** (1953), 124–128.

SHIMOGAKI, T.
[1] Hardy–Littlewood majorants in function spaces. *J. Math. Soc. Japan* **17** (1965), 365–373.

SIMON, A. B.
[1] Cesàro summability on groups: Characterization and inversion of Fourier transforms. *Function Algebras*, pp. 208–215. Scott, Foresman and Company, Chicago (1966).

SIMONENKO, I. B.
[1] Multidimensional discrete convolutions. (Russian). *Mat. Issled.* **3** (1968). ryp. 1 (7), 108–122.

SKVORCOVA, M. G.
[1] Some new theorems on class of multipliers transforming Fourier series. II. (Russian). *Kabardino-Balkarsk. Gos. Univ. Učen. Zap.* No. **30** (1966), 220–228.

SMITH, K. T.
[1] A generalization of an inequality of Hardy and Littlewood. *Canad. J. Math.* **8** (1956), 157–170.

SRINIVASAN, T. P. AND WANG, J.-K.
[2] On closed ideals of analytic functions. *Proc. Amer. Math. Soc.* **16** (1965), 49–52.

STAMPACCHIA, G.
[1] $\mathscr{L}^{(p,\lambda)}$-spaces and interpolation. *Comm. Pure Appl. Math.* **17** (1964), 293–306.

STEIN, E. M.
[1] On limits of sequences of operators. *Ann. Math.* **74** (1961), 140–170.
[2] Interpolation of linear operators. *Trans. Amer. Math. Soc.* **83** (1956), 482–492.

STEIN, E. M. AND WEISS, G.
[1] An extension of a theorem of Marcinkiewicz and some of its applications. *J. Math. Mech.* **8** (1959), 263–284.
[2] Interpolation of operators with change of measures. *Trans. Amer. Math. Soc.* **87** (1965), 159–162.

STEIN, E. M. AND ZYGMUND, A.
[1] Boundedness of translation invariant operators on Hölder spaces and L^p-spaces. *Ann. Math.* (2) **85** (1967), 337–449.

STRICKARTZ, R.
[1] Isomorphisms of group algebras. *Proc. Amer. Math. Soc.* **17** (1966), 858–862.

TAIBLESON, M.
[1] Fourier coefficients of functions of bounded variation. *Proc. Amer. Math. Soc.* **18** (1967), 766.
[2] On the theory of Lipschitz spaces of distributions on Euclidean n-space. II. Translation-invariant operators, duality and interpolation. *J. Math. Mech.* **14** (1965), 821–839.

TAYLOR, J. L.

[1] The Shilov boundary of the algebra of measures on a group. *Proc. Amer. Math. Soc.* **16** (1965), 941–945.

[2] The structure of convolution measure algebras. *Trans. Amer. Math. Soc.* **119** (1965), 150–166.

[3] Convolution measure algebras with group maximal ideal spaces. *Trans. Amer. Math. Soc.* **128** (1967), 257–263.

[4] L-subalgebras of $M(G)$. *Trans. Amer. Math. Soc.* **135** (1969), 105–113.

[5] Ideal theory and Laplace transforms for a class of measure algebras on a group. *Acta Math.* **121** (1968), 251–292.

[6] Non-commutative convolution measure algebras. *Pacific J. Math.* **31** (1969), 809–826.

[7] Measures which are convolution exponentials. *Bull. Amer. Math. Soc.* **76** (1970), 415–418.

[8] The cohomology of the spectrum of a measure algebra. *Acta Math.* **126** (1971), 195–225.

[9] Inverses, logarithms and idempotents in $M(G)$. *Rocky Mountain J. Math.* **2** (2) (1972), 183–206.

[10] Homology and cohomology for topological algebras. *Advances in Math.* **9** (1972), 137–182.

[11] On the spectrum of a measure. *Advances in Math.* **12** (1974), 451–463.

THORIN, G. O.

[1] An extension of a convexity theorem due to M. Riesz. *Kungl. Fysiografiska Saellskapet i Lund Forhaendlinger* **8** (1939), No. 14.

[2] Convexity theorems. Dissertation, Lund (1948), 1–57.

UCHIYAMA, S.

[1] À propos d'un problème de M. J. E. Littlewood. *C. R. Acad. Sci. Paris* **260** (1965), 2675–2678.

[2] On the mean modulus of trigonometric polynomials whose coefficients have random signs. *Proc. Amer. Math. Soc.* **16** (1965), 1185–1190.

VAROPOULOS, N. TH.

[1] Sur les ensembles parfaits et les séries trigonométriques. *C. R. Acad. Sci. Paris* **260** (1965), 4668–4670; 5165–5168; 5997–6000.

[2] Sur les ensembles parfaits et les séries trigonométriques. *C. R. Acad. Sci. Paris* **260** (1965), 3831–3834.

[3] Spectral synthesis on spheres. *Proc. Cambridge Phil. Soc.* **62** (1966), 379–387.

[4] The functions that operate on $B_0(\Gamma)$ of a discrete group Γ. *Bull. Soc. Math. France* **93** (1965), 301–321.

[5] Sur la réunion de deux ensembles de Helson. *C. R. Acad. Sci. Paris* Sér. A-B **271** (1970), A251–253; *Acta Math.* **125** (1970), 109–154.

WADA, J.

[1] A note on multipliers of ideals in function algebras. *Proc. Japan. Academy* **42** (10) (1966), 1134–1138.

WAELBROECK, L.

[1] Le calcul symbolique dans les algèbres commutatives. *J. de Math. Pure et Appl.* **33** (1954), 147–186.

[2] On the analytic spectrum of Arens. *Pacific J. Math.* **13** (1963), 317–319.

WALTER, G.

[1] Pointwise convergence of distribution expansions. *Studia Math.* **26** (1966), 143–154.

WANG, J.-K.

[1] Multipliers of commutative Banach algebras. Dissertation, Stanford Univ. (1959).

WARNER, C. R.

[1] Closed ideals in the group algebra $L^1(G) \cap L^2(G)$. *Trans. Amer. Math. Soc.* **121** (1966), 408–423.

WEISS, G.

[1] Harmonic Analysis. *Studies in Mathematics, Vol.* 3, pp. 124–178. The Mathematical Association of America; Prentice-Hall, Inc., Englewood Cliffs, N.J. (1965).

WEISS, M.

[1] On a problem of J. E. Littlewood. *J. London Math. Soc.* **34** (1959), 217–221.

WELLS, J.

[1] Restriction for Fourier–Stieltjes transforms. *Proc. Amer. Math. Soc.* **15** (1964), 243–246.

[2] Multipliers of ideals in function algebras. *Duke Math. J.* **31** (1964), 703–709.

WENDEL, J. G.

[1] On isometric isomorphisms of group algebras. *Pacific J. Math.* **1** (1951), 305–311.

[2] Left centralizers and isomorphisms of group algebras. *Pacific J. Math.* **2** (1952), 251–261.

WERMER, J.

[1] Banach Algebras and Analytic Functions. *Advances in Mathematics,* **1**. *Fasc, 1.* Academic Press, Inc., New York (1961), 51–102.

WIDOM, H.

[1] Toeplitz Matrices. *Studies in Real and Complex Analysis, Vol. 3,* pp. 179–209. The Mathematical Association of America; Prentice Hall, Inc., Englewood Cliffs, N.J. (1965).

[2] Toeplitz operators on H_p. *Pacific J. Math.* **19** (1966), 573–582.

WIK, I.

[1] On linear dependence in closed sets. *Arch. Mat.* **4** (1960), 209–218.

[2] Some examples of sets with linear independence. *Ark. Mat.* **5** (1964), 207–214.

WILLIAMS, L. R.

[1] Generalized Hausdorff–Young inequalities and mixed-norm spaces. *Pacific J. Math.* **38** (1971), 823–833.

WONG, J. S. W.

[1] On a characterisation of Fourier transforms. *Monatsch. Math.* **70** (1966), 74–80.

YAMAGUCHI, H.

[1] Some multipliers on $H_p^1(G)$. *J. Austral. Math. Soc.* **29** (1) (1980), 52–60.

YANO, S.

[1] On Walsh–Fourier series. *Tôhoku Math. J.* **3** (1971), 223–242.

YAP, L. P. H.

[1] Some remarks on convolution operators and $L(p, q)$ spaces. *Duke Math. J.* **36** (1969), 647–658.

YOUNG, F.

[1] Transformations of Fourier coefficients. *Proc. Amer. Math. Soc.* **3** (1952), 783–791.

ZAFRAN, M.

[1] The spectra of multiplier transformations on the L_p spaces. *Ann. Math.* **103** (2) (1976), 355–374.

[2] The functions operating on certain algebras of multipliers. *Bull. Amer. Math. Soc.* **82** (1976), No. 6, 939–940.

[3] The functions operating on multiplier algebras. *J. Functional Analysis* **26** (1977), No. 3, 289–314.

[4] Multiplier transformations of weak type. *Ann. Math.* (2) **101** (1975), 34–44.

ZYGMUND, A.

[1] On the preservation of classes of functions. *J. Math. Mech.* **8** (1959), 889–895.

CORRIGENDA TO 2nd (REVISED) EDITION OF VOLUME 1

Page	Line	For	Read
30	22	$\hat{\hat{f}}$	$\hat{f}$
45	10^-	$R/2\pi Z$	T
45	9^-	$R/2\pi Z$	T
75	13	$R/2\pi Z$	T
75	4^-	$R/2\pi Z$	T
77	10^-	$R/2\pi Z$	T
111	7	$\geqslant 0 \geqslant 0;$	$\geqslant 0;$
173	4	$\mathbf{A}(R/2\pi Z)$	$\mathbf{A}(T)$
194	15	$p_k(\lambda y)$	$p_k(\lambda x)$
219 LHC	1	$\mathbf{A}(R/2\pi Z)$	$\mathbf{A}(T)$
219 RHC	6	78	79
219 RHC	17	35, 128	36, 135
221 LHC	9^-	101	102
221 LHC	5^-	34	111
221 RHC	9	191	179
221 RHC	1^-	ADD 119, 123	
214	12^-	To appear *Pacific J. Math.*	*Pacific J. Math.* **21** (1967), 255–263.
215	17, 18	To appear *The Mathematical Intelligencer.*	*Arch. History Exact Sci.* **21** (1979), 129–159.

Symbols

Numerals in italic type refer to Volume 1. Numerals in boldface type refer to the exercises, the exercise labeled **x.y** appearing in Volume 1 if and only if $1 \leqslant \mathbf{x} \leqslant 10$.

Index

Numerals in italic type refer to Volume 1. Numerals in boldface type refer to the exercises, the exercise labeled **x.y** appearing in Volume 1 if and only if $1 \leqslant \mathbf{x} \leqslant 10$.

Graduate Texts in Mathematics

1 TAKEUTI/ZARING. Introduction to Axiomatic Set Theory. 2nd ed.
2 OXTOBY. Measure and Category. 2nd ed.
3 SCHAEFFER. Topological Vector Spaces.
4 HILTON/STAMMBACH. A Course in Homological Algebra. 3rd printing.
5 MACLANE. Categories for the Working Mathematician.
6 HUGHES/PIPER. Projective Planes.
7 SERRE. A Course in Arithmetic.
8 TAKEUTI/ZARING. Axiometic Set Theory.
9 HUMPHREYS. Introduction to Lie Algebras and Representation Theory. 3rd printing, revised.
10 COHEN. A Course in Simple Homotopy Theory.
11 CONWAY. Functions of One Complex Variable. 2nd ed.
12 BEALS. Advanced Mathematical Analysis.
13 ANDERSON/FULLER. Rings and Categories of Modules.
14 GOLUBITSKY/GUILLEMIN. Stable Mappings and Their Singularities.
15 BERBERIAN. Lectures in Functional Analysis and Operator Theory.
16 WINTER. The Structure of Fields.
17 ROSENBLATT. Random Processes. 2nd ed.
18 HALMOS. Measure Theory.
19 HALMOS. A Hilbert Space Problem Book. 2nd ed., revised.
20 HUSEMOLLER. Fibre Bundles. 2nd ed.
21 HUMPHREYS. Linear Algebraic Groups.
22 BARNES/MACK. An Algebraic Introduction to Mathematical Logic.
23 GREUB. Linear Algebra. 4th ed.
24 HOLMES. Geometric Functional Analysis and its Applications.
25 HEWITT/STROMBERG. Real and Abstract Analysis. 5th printing.
26 MANES. Algebraic Theories.
27 KELLEY. General Topology.
28 ZARISKI/SAMUEL. Commutative Algebra. Vol. I.
29 ZARISKI/SAMUEL. Commutative Algebra. Vol. II.
30 JACOBSON. Lectures in Abstract Algebra I: Basic Concepts.
31 JACOBSON. Lectures in Abstract Algebra II: Linear Algebra.
32 JACOBSON. Lectures in Abstract Algebra III: Theory of Fields and Galois Theory.
33 HIRSCH. Differential Topology.
34 SPITZER. Principles of Random Walk. 2nd ed.
35 WERMER. Banach Algebras and Several Complex Variables. 2nd ed.
36 KELLEY/NAMIOKA et al. Linear Topological Spaces.
37 MONK. Mathematical Logic.
38 GRAUERT/FRITZSCHE. Several Complex Variables.
39 ARVESON. An Invitation to C^*-Algebras.
40 KEMENY/SNELL/KNAPP. Denumerable Markov Chains. 2nd ed.
41 APOSTOL. Modular Functions and Dirichlet Series in Number Theory.
42 SERRE. Linear Representations of Finite Groups.
43 GILLMAN/JERISON. Rings of Continuous Functions.
44 KENDIG. Elementary Algebraic Geometry.
45 LOÈVE. Probability Theory I. 4th ed.
46 LOÈVE. Probability Theory II. 4th ed.

47 MOISE. Geometric Topology in Dimensions 2 and 3.
48 SACHS/WU. General Relativity for Mathematicians.
49 GRUENBERG/WEIR. Linear Geometry. 2nd ed.
50 EDWARDS. Fermat's Last Theorem. 2nd printing.
51 KLINGENBERG. A Course in Differential Geometry.
52 HARTSHORNE. Algebraic Geometry.
53 MANIN. A Course in Mathematical Logic.
54 GRAVER/WATKINS. Combinatorics with Emphasis on the Theory of Graphs.
55 BROWN/PEARCY. Introduction to Operator Theory I: Elements of Functional Analysis.
56 MASSEY. Algebraic Topology: An Introduction.
57 CROWELL/FOX. Introduction to Knot Theory.
58 KOBLITZ. p-adic Numbers, p-adic Analysis, and Zeta-Functions.
59 LANG. Cyclotomic Fields.
60 ARNOLD. Mathematical Methods in Classical Mechanics. 3rd printing.
61 WHITEHEAD. Elements of Homotopy Theory.
62 KARGAPOLOV/MERZLJAKOV. Fundamentals of the Theory of Groups.
63 BOLLOBÁS. Graph Theory.
64 EDWARDS. Fourier Series. Vol. I. 2nd ed.
65 WELLS. Differential Analysis on Complex Manifolds. 2nd ed.
66 WATERHOUSE. Introduction to Affine Group Schemes.
67 SERRE. Local Fields.
68 WEIDMANN. Linear Operators in Hilbert Spaces.
69 LANG. Cyclotomic Fields II.
70 MASSEY. Singular Homology Theory.
71 FARKAS/KRA. Riemann Surfaces.
72 STILLWELL. Classical Topology and Combinatorial Group Theory.
73 HUNGERFORD. Algebra.
74 DAVENPORT. Multiplicative Number Theory. 2nd ed.
75 HOCHSCHILD. Basic Theory of Algebraic Groups and Lie Algebras.
76 IITAKA. Algebraic Geometry.
77 HECKE. Lectures on the Theory of Algebraic Numbers.
78 BURRIS/SANKAPPANAVAR. A Course in Universal Algebra.
79 WALTERS. An Introduction to Ergodic Theory.
80 ROBINSON. A Course in the Theory of Groups.
81 FORSTER. Lectures on Riemann Surfaces.
82 BOTT/TU. Differential Forms in Algebraic Topology.
83 WASHINGTON. Introduction to Cyclotomic Fields.
84 IRELAND/ROSEN. A Classical Introduction Modern Number Theory.
85 EDWARDS. Fourier Series. Vol. II. 2nd ed.
86 VAN LINT. Introduction to Coding Theory.
87 BROWN. Cohomology of Groups.
88 PIERCE. Associative Algebras.